THE FIELD GUIDE TO MIXING SOCIAL AND BIOPHYSICAL

METHODS IN ENVIRONMENTAL RESEARCH

The Field Guide to Mixing Social and Biophysical Methods in Environmental Research

Edited by Rebecca Lave and Stuart N. Lane

OpenBook
Publishers

Critical Physical Geography: Interdisciplinary Approaches to Nature, Power and Politics Vol. 1
ISSN (print): 3049-7469
ISSN (digital): 3049-7477

ISBN Paperback: 978-1-80511-366-9
ISBN Hardback: 978-1-80511-367-6
ISBN Digital (PDF): 978-1-80511-368-3
ISBN Digital eBook (EPUB): 978-1-80511-369-0
ISBN HTML: 978-1-80511-370-6

DOI: 10.11647/OBP.0418

Cover image: Stuart N. Lane, When the "environment" tries to tell you that you are measuring the wrong thing (2023), all rights reserved
Cover design: Jeevanjot Kaur Nagpal

Contents

Acknowledgements

This *Field Guide* was a solid six years in the making. Some of that was due to our administrative workloads, which ballooned during this time period. The COVID-19 pandemic didn't help. We are very grateful to the contributing authors, who stuck with us during years of delays and sent us brilliant work. We also owe an immense debt to Peter Ashmore and Adam Romero, who provided speedy and insightful reviews of the whole manuscript in Spring 2024, helping us to frame both the three main sections and the *Field Guide* as a whole far more clearly.

We are grateful to Open Book Publishers for enabling us to publish the *Field Guide* open access (OA). There were far fewer OA options back in 2018 than there are even now in 2024. We were delighted to find Open Book Publishers and are very grateful for Alessandra Tosi's patience and support, Lucy Barnes and Sophia Bursey's guidance at the copy-editing stage, and Jeevanjot Nagpal's excellent cover design. We also appreciate all of Joshua Pontillo's help in getting the manuscript styled and ready to go and Toby Lane for helping with some of the final editing.

The fact that we stuck with this project over six years of delays speaks to our deep commitment to conducting integrated social and biophysical environmental research and to helping others do so, too. It also speaks to the strength of a collaboration that began more than a decade ago when Rebecca took the risk of reaching out to a senior professor she had never met, and Stuart decided to take a chance on attending a workshop organised by a very junior assistant professor. Unlike some of our respective intellectual gambles, this one worked!

List of Illustrations

1. Introduction to the *Field Guide*

Rebecca Lave and Stuart N. Lane

Despite disagreement over its use and start date, the core premise of the Anthropocene within which we live is widely accepted. The environment is now fundamentally co-produced by biophysical and social forces, whether at the scale of the globe in terms of human-induced climate change, or more locally in how we manage the land. Landscapes and hydroscapes are shaped by climate dynamics and legacies of colonialism, gender norms and hydrological processes, soil chemistry and property rights, environmental injustice and pollution. We cannot fully understand our world without analysing these interactions, and that analysis requires a mixture of methods: qualitative and quantitative, biophysical and social.

Yet even with this clear intellectual mandate for mixing methods in environmental research, it remains relatively rare. Indeed, we might hypothesise that there are more examples of how we should think about the Anthropocene than there are treatises on the kinds of methods that need to be mixed, and how, in order to study it. Both of us, the editors of this volume, have long histories of mixing methods in environmental research. We have experienced the challenges and frustrations of doing so, the intellectual battles within our own disciplines and with other disciplines when we have tried to do something that didn't quite fit with established disciplinary norms. But we have also experienced the immense pleasure of being able to shape what we do to the questions being asked so that we could answer them in original and creative ways. This way of working needs additional time, effort, understanding, patience and above all humility but it is truly worth it. By mixing methods in thoughtful ways, we believe that both our research practices and research findings have become more intellectually robust and potentially transformative than if we had remained within a single

 https://doi.org/10.11647/OBP.0418.01

methodological framework. Thus, we designed the *Field Guide* to inspire and enable you to conduct environmental research by mixing methods and to do so in a thoughtful and informed way.

The audience we envisioned as we assembled the *Field Guide* is primarily early-career researchers working at the interface of the biophysical and social sciences: graduate students, postdoctoral scholars, and early career faculty who are interested in doing environmental research differently and need some help to set out on that path. Parts of this book, particularly Section 2, will also be of interest to undergraduates, as the authors have provided compelling case studies of environmental research and the way the questions you ask and answer change when you are willing to think between disciplines and to combine different methods. We also hope the *Field Guide* will be of interest to more established researchers who have slowly (or even suddenly!) come to question the boundaries of the fields in which they were trained.

Listening when your field site "speaks back" to you

Many words are used to describe research that mixes, or attempts to mix, biophysical and social methods, including cross-disciplinary, multidisciplinary, interdisciplinary, and transdisciplinary. We address the differences among these terms in Chapter 3, but for now we use the phrase "mixed methods" to describe research that brings approaches from across the biophysical and social sciences into conversation in order to speak thoughtfully and robustly about a particular site, question, or topic.

Between the two of us, over the decades of both successful and unsuccessful mixed-methods environmental research, we have learned the hard way that what to mix and how cannot be prescribed in advance. Rather, the skill in mixing methods is to be sufficiently engaged in and reflexive with respect to the place you are studying, your field site. All too often in conventional research we are asked to distance ourselves, as the researchers, from what it is we are researching, in the hope of making our research more objective and our results more general. We do not challenge the importance of being able to step back from what we are researching in thinking through, for example, what it is we have found. But we do argue that crucial to mixing methods, and even modifying them, is a careful response to your field site and what you want to learn about it.

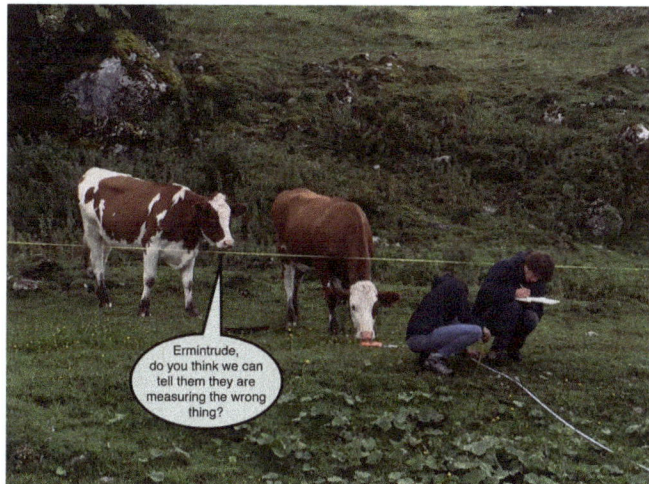

Fig. 1.1 Stuart N. Lane, 2023. When the "environment" tries to tell you that you are measuring the wrong thing.

Following Isabelle Stengers (2005), we argue that research makes most progress when we put ourselves in a situation where we slow down enough to be able to look up from our research papers or books and engage with what is around us, which enables us to think about what we are studying in different ways. Metaphorically, we believe this is about "listening" when your field site tries to "speak" to you and to tell you how it wants to be studied. Social scientists are used to the idea of their research subjects talking back, but this may be more of a stretch for biophysical scientists studying organic carbon, salmonids, or landslides. Yet it is still the case that non-human processes and objects can tell you what is needed to understand them if you put yourself in situations where they can do so. For us, this means spending time in the field, using our sensory perceptions and intuitions to understand not only what we need to do to study something but even what the very questions are that we need to ask. This is why we called this volume a field guide rather than a handbook: the field is key. Grounding mixed methods in "working with the field" is common in many of the disciplines that feed into environmental research (think of anthropology, archaeology, geography, and geology, for example); what we are advocating is not as radical as it may initially appear.

We have found that listening to our field sites requires learning to step back from the constraints of the disciplinary frames (how an academic field defines a good topic, a good question, and an appropriate set of methods, among other things; see Lane and Lave, Chapter 3) in which we were trained. Deep training in one or more disciplines is invaluable, but it is not sufficient. We work in an academy that is structured around disciplines, and for those wanting to remain in the academy it may be hard to progress without at least some of the belongings that come with a discipline. However, disciplines do constrain what we are able to do and may cause us to be more attached to disciplinary norms than to the things we are studying. When this happens, our disciplines begin to define what constitutes the world around us, rather than the world itself. As Isabelle Stengers (2013) points out, this lack of sensitivity to the empirical reality of what we are studying makes us less and not more scientific, however committed we are to what our disciplines tell us is "best practice". This is why in using this *Field Guide*, we want you to be undisciplined with respect to academic disciplines, but disciplined in your commitment to what it is you are studying. In this *Field Guide*, we encourage you to do the expansive thinking and undisciplined methodological selection necessary to answer the questions your field site poses. That is not to say that we expect you to develop brand new methods, but that we encourage you to be creative in bringing existing methods into conversation.

Field Guide structure

We conceptualised the *Field Guide* as a cookery book. It is just possible that you have come across a classic example of these: *Mrs Beeton's Dictionary of Everyday Cookery*. Isabella Beeton was a 19th-century writer and this dictionary was published posthumously in 1865. She is thought to be a pioneer in the history of cookery and her volume is still in publication. It was abridged, however, from a bigger undertaking, Mrs Beeton's Dictionary of Household Management, published in 1861. Still in print, the latter not only contains recipes, but extensive information on what you need to run a kitchen, to acquire ingredients, and to bring them together in ways that produce edible food. She makes the point that recipes and ingredients are not enough; kitchens also determine what it is you can cook. Thus, in the first section of the Field Guide we reflect upon research kitchens, the challenges and constraints to building them, and a range of issues from the more general to the more specific about what should go in them. In the second section, we provide example research recipes that not only illustrate how methods can be mixed but reflect upon the challenges of doing so. We wanted this section to illustrate what mixing methods can achieve and to concretise the issues discussed in Section 1. In the third section, inspired by the ingredients that our recipe authors used, we identify a suite of methods that might be used in environmental research when mixing methods.

How to use the *Field Guide*

As with most cookery books you are likely to get the maximum benefit from reading selectively and following your interests at any given moment rather than reading it cover-to-cover. Perhaps Section 1 is the exception as the chapters therein raise general issues, notably Chapters 3, 4 and 5, that you should be aware of when mixing methods. There is a strong emphasis on explaining how methods are mixed in different kinds of disciplinarity, and the structural challenges (related to disciplines) when you try to do this (Lane and Lave, Chapter 3). Biermann and Gibbes (Chapter 4) illustrate the critical point that only when findings from different methods are juxtaposed and triangulated can we start to get some confidence in what we are finding. Mixing methods can be an uphill struggle, not simply because it is methodologically challenging, but also because not everyone will support your attempts to do so (Lane and Lave, Chapter 3). There is an equally important reflection on ethical issues (Biermann and Gibbes, Chapter 4; Meadow et al., Chapter 5), all too often overlooked in many research projects, and the responsibilities that we acquire as researchers when we start to mix methods. Frameworks are provided for thinking about ethical responsibilities in both theory and practice. Section 1 also reflects on pedagogy and training, notably in practice, and these are likely to be most relevant as you plan a specific research project. We also include some short chapters on generic issues (environmental impact assessment, inclusivity, health and safety) that likely cut across all projects. These should be a part of research planning but, along with the chapters that address the tensions associated with research collaborations and working with people, also raise issues that you need to be continually aware of.

Section 2 presents sample interdisciplinary research "recipes" that we hope will inspire you: powerful examples of mixed-methods research that explore eco-social landscapes and hydroscapes. In each chapter, the author(s) explains how they came to realise that they needed interdisciplinary approaches, and why the specific combinations of methodological ingredients they used allowed them to answer their research questions in ways that more traditional disciplinary approaches did not. Importantly, the authors reflect upon some of the challenges they have experienced in mixing methods: where their research recipes have gone wrong, and how they fixed them by creatively combining their existing methodological "ingredients" or by adding new ones.

You may wish to read Section 2 straight through, as each chapter contributes a different way of doing interdisciplinary environmental research. Alternatively, you may wish to pick and choose based on the topics or geographical regions that interest you, or the methods you think you might use in your own research. However, we want to add a word of caution. As the chapters in Section 1 argue, the essence of mixing methods effectively in environmental research is allowing what is mixed and how (the "recipe") to be shaped by what it is you are studying, letting the environment "speak" to you. Thus, the recipes need to be seen as illustrations of mixing methods and not as skeletons or templates that you can simply transfer to your own research site. The skill in becoming a mixed-methods researcher comes not from being the kind of chef who can faithfully follow a given recipe, but from being the kind of chef who has the creativity, innovation, and willingness to experiment (see Salmond and Brierley, Chapter 6) in making new recipes. Doing so is not easy. It takes time and will, inevitably, mean being experimental in the sense of being willing to try things out and to learn from mistakes. As a community of researchers, we are not very good at reporting what does not work, including negative results. We probably should work harder to report our "brilliant failures".[1] Until that time, the best solution may be to exchange ideas and experiences with those who have done this kind of work already.

Inspired by Section 2 to cook up a mixed-methods research project for yourself, but not sure which approaches to combine? Section 3 of the *Field Guide* provides an "ingredient list": succinct overviews of individual biophysical and social methods used in the chapters in Section 2. These short chapters address each method's strengths and weaknesses, compatibility with other methods, and ethical considerations to keep in mind when employing them. We recommend you read selectively in Section 3 on an "as needed" basis. Although the more than two dozen methods covered are but a small portion of the vast universe of existing research approaches, there are still a lot of them! You may wish to start with descriptions of methods you have already heard of or know a bit about and then expand from there. These descriptions are not designed to be either complete or sufficient. Rather, they are designed to be entry points, to help you to decide if a particular method might be a good fit for your recipe and, if so, what issues you need to think about before you begin to research and to use a method further.

We trust you will take our advice with a grain of salt and use the *Field Guide* to develop your own research recipe, one best suited to you and your field site.

References cited

Biermann, C. and Gibbes, C., Chapter 4, this volume. 'Mixed methods in tension: lessons for and from the research process'.

Lane, S.N. and Lave, R., Chapter 3, this volume. 'Frames, disciplines and mixing methods in environmental research'.

Meadow, A., Wilmer, H. and Ferguson, D., Chapter 5, this volume. 'Expanding research ethics for inclusive and transdisciplinary research'.

---

1   This term is inspired by the Research for Development (r4d) programme between the Swiss National Science Foundation and the Swiss Development Corporation that supported North-South research partnerships. See https://k4d. ch/a-new-culture-of-failure-in-north-south-research-partnerships/

Salmond J. and Brierley, G., Chapter 6, this volume. 'Embracing and enacting critical and constructive approaches to teaching Critical Physical Geography'.

Stengers, I. 2005. 'The cosmopolitical proposal', in *Making Things Public*, ed. by B. Latour and P. Weibel (MIT Press), pp. 994–1003.

Stengers, I. 2013. *Une autre science est possible* (Editions La découverte).

SECTION 1: BUILDING YOUR RESEARCH KITCHEN

2. Introduction to building the research 'kitchen'

Stuart N. Lane and Rebecca Lave

Research is not developed in a vacuum but is rather shaped and influenced by the discipline in which the research is to be done. Here, we describe this by analogy as the kitchen (research environment) in which different ingredients (methods, see Section 3) will be brought together according to recipes (examples, see Section 2), ultimately defined by the cook (the researcher). Like kitchens in homes over many centuries, the space within which research is done, the research environment, does not exist in isolation. It is shaped by what research has been done before and the wider social, cultural and education system that houses it; as research is practised, so the research environment will change. In environmental research, the kitchen, or the research environment, often has to be built and re-built by the researcher.

The aim of Section 1 is to introduce a range of topics to help you to build your research kitchen and so create a research environment appropriate for answering challenging environmental questions. By analogy with a research kitchen, Lane and Lave (Chapter 3) begin by reflecting on whether or not you need to build a new one. The chapter reviews theoretically the notions of disciplines, multidisciplinarity, interdisciplinarity, and transdisciplinarity. Most disciplines already have well-established methodological approaches and methods (existing kitchens). In cross-disciplinary research, you might use the methods of two or more disciplines, each providing their own ways of working and results, so as to answer a question. In multidisciplinary research, there may be more emphasis on combining results from different disciplines to go beyond the answers that each individually provide. In both cross- and multidisciplinary research, the disciplinary kitchens are left intact. In interdisciplinary research, new kitchens are built by combining and developing methodological approaches and methods better suited

      https://doi.org/10.11647/OBP.0418.02

to answering the questions of interest. In transdisciplinary research, attention is given to who is allowed to decide how the new kitchen is built (i.e., how it is arranged and what is easy to make), opening up the possibility that what is being researched, and by whom, can influence how that research is done. Lane and Lave (Chapter 3) also consider the difficulties of doing research outside the bounds of individual disciplines; there are costs associated with building new kitchens!

Biermann and Gibbes (Chapter 4) then reflect upon why mixing methods in new kinds of research kitchens is important. They argue that mixing methods is not just about seeing whether different methodological approaches give you the same or similar answers. Instead, combining methods may reveal unexpected or unintended findings enabling a richer understanding of what is being studied, how disciplines constrain our research, and how wider social, economic, and political relations influence what we do and how we do it. Especially in a transdisciplinary framework, mixing methods may make us more attentive to those who have to live with the consequences of our research. Biermann and Gibbes advocate working with the points of disagreement and tensions that different methodological approaches reveal. This may benefit from being done collaboratively as it may be important to be sufficiently expert in the methods need to answer a research question. Such expertise may not only be needed for deployment but also to modify them to make them fit for purpose. Interdisciplinary and multidisciplinary research kitchens sometimes function best when they involve many chefs but such kitchens may take more time to build and may need to evolve as they are used. We should also not forget that "too many cooks may spoil the broth", and projects with too many collaborators may prevent effective mixing of methods.

Complex environmental questions will commonly require methods that work with people, many of whom live on a day-to-day basis with the very questions and challenges that a research project is seeking to address. Even if the methods being mixed don't directly involve people, they may lead to research outcomes that do. This is why the research kitchen needs to be sensitive to research ethics and this is addressed by Meadow et al. (Chapter 5), and in all the chapters on individual methods in Section 3. Some research areas already require consent from participants (e.g., in clinical research or in social science research involving interviews). However, as Meadow et al.

outline, research ethics is about more than just consent. They present the CARE approach (Collective Benefit, Authority to Control, Responsibility, and Ethics) to providing a broader framework for directing research that during its practice or afterwards could have consequences for humans (and potentially non-humans). Mirroring the observations of Lane and Lave (Chapter 3) and Biermann and Gibbes (Chapter 4), the authors argue that the research kitchen itself must be constructed in different ways if it is to properly engage with ethical questions.

Although this volume largely focuses on how to conduct mixed-methods environmental research, it requires training to construct a new kind of research kitchen and to be involved in mixed-methods research. Further, many of those who become experienced in mixing methods in environmental research will go on to teach them. This is why it is important to think about how we can improve teaching of mixed-methods research. Salmond and Brierley (Chapter 6) provide ideas for teaching mixed methods in environmental research as part of an initiative in Critical Physical Geography (CPG). CPG specifically advocates research at the interface of the social sciences and the biophysical sciences to address critical environmental questions (Lave et al., 2014) and this commonly implies drawing methods from very different kinds of disciplines. Salmond and Brierley propose adopting a more open-ended learning experience that directs students to look outwards rather than inwards, encouraging creativity, experimentation, and innovation. They share some of the challenges of this pedagogical approach, but also the rewards that can come from it. Salmond and Brierley's argument is extended in the chapter by Johnson et al. (Chapter 7), who describe the process of developing an interdisciplinary multi-site research project and show how the research kitchen needs to have built-in time, space, and motivation for training to move beyond conventional practices. Johnson et al. illustrate some of the ethical challenges this entails, as when moving across cultures requires participants to work outside of their normal "comfort zones". Questions of power, experience and language are answered through ways of working that are inclusionary, open-ended, and above all sensitive to not only where the research is being done but also those who are doing it. It is a good reminder that we may be building research kitchens in very different cultural and political contexts to those where we have been trained, which requires a particular commitment to reflexivity during the research process.

In the final part of Section 1, the chapters address three more practical and focused topics relevant to all mixed-methods research projects. First, for too long we have overlooked the considerable environmental costs of the research that we do. That research does involve choices and so Lane (Chapter 8) presents a framework for thinking through and then minimising the potential environmental impacts of field research. Second, environmental research is often place-based and involves fieldwork, yet we often overlook the challenges associated with doing fieldwork, especially in relation to diversity and inclusion. Thus, Miesen and Gevers (Chapter 9) bring to the research kitchen the importance of thinking through how we can make fieldwork in particular, and research in general, more inclusive. Finally, undertaking research requires us to think through the risks of doing so. Thus, Miesen (Chapter 10) presents a basic framework for identifying risks and mitigating them when planning and undertaking research.

References cited

Biermann, C. and Gibbes, C., Chapter 4, this volume. 'Mixed methods in tension: lessons for and from the research process'.

Johnson, A., Oven, K., Rosser, G., Basnet, D., Bhotia, N.D., Dong, T.B., Puri, A., Tamang, S., and Kincey. M. this volume. 'Reflections on pedagogy and practice for 'deep interdisciplinary' engagement within the Sajag-Nepal project'.

Lane, S.N., Chapter 8, this volume. 'The environmental impacts of fieldwork: making an environmental impact statement'

Lane, S.N. and Lave, R., Chapter 3, this volume. 'Frames, disciplines and mixing methods in environmental research'.

Lave, R., M. Wilson, E. Barron, C. Biermann, M. Carey, C. Duvall, L. Johnson et al. 2014. 'Intervention: Critical Physical Geography', *The Canadian Geographer/Le Géographe canadien*, 58.1, pp. 1–10. https://doi.org/10.1111/cag.12061

Meadow, A., Wilmer, H., and Ferguson, D., Chapter 5, this volume. 'Expanding research ethics for inclusive and transdisciplinary Research'.

Miesen, F., Chapter 10, this volume. 'Fieldwork safety planning and risk management'.

Miesen, F. and Gevers, M., Chapter 9, this volume. 'Inclusive practices in fieldwork'.

Salmond J. and Brierley, G., Chapter 6, this volume. 'Embracing and enacting critical and constructive approaches to teaching Critical Physical Geography'.

3. Frames, disciplines and mixing methods in environmental research

Stuart N. Lane and Rebecca Lave

Introduction

The aim of this chapter is to set out the relationship between how we frame research questions and research projects, how frames can change, and the power of disciplines in constraining the emergence of new frames. Our premise is that mixing methods in environmental research is a means of escaping the confines of particular frames and disciplines.

A frame can be defined as an idea that organises concepts, beliefs, observations etc. in order to give them sense and to draw out a key message (Bateson 1955). According to Goffman (1974), we render events meaningful by drawing upon one or more primary frames, which need not be clearly defined or even consciously employed. Frames, then, are organisational devices (Snow et al., 1986) that we use to make sense of the world around us. They are similar to paradigms, perhaps only differing in the senses that the social sciences tend to talk of frames and the biophysical sciences tend to talk of paradigms; and that paradigms may be more than just organisational, describing a particular ontological or epistemological view of the world.

We begin by defining a three-part classification of frames which allows us to illustrate some basic characteristics of how frames function and how they change. Such mutation happens within networks of researchers, laboratories, field sites, etc. We briefly describe work, often highly empirical, that has shown the importance of frames in research practices and revealed their substantial power in determining what are acceptable methods and how they may be mixed. In turn this ties

 https://doi.org/10.11647/OBP.0418.03

questions of frames to disciplines and the political economy of the academy. Escaping disciplinary confines is, we believe, necessary for mixing methods but also central to challenging disciplinary powers. Thus, we consider the different ways methods can be mixed "beyond disciplines" in cross-disciplinary, multi-disciplinary, interdisciplinary, and transdisciplinary research.

Kinds of frames

There are three broad kinds of frames that relate to environmental research: theoretical, empirical, and positional (which focuses on the relationship between a researcher and their research). A theoretical frame is clearly dissociated from the empirical specifics—the where or when—of a research question. Rusca and Mazzoleni (Chapter 14) contrast two different theoretical frames around water and governance. One comes from hydrology and focuses on explaining water resource issues as a function of a largely biophysical description of the water cycle and the human interventions, such as climate change, that modify it. A second comes from political ecology and focuses on explaining differential access to clean and safe water as a function of political and economic systems, such as those related to governance.

Both theoretical frames could apply in almost any place or time, but they would shape research design in quite distinct ways. The hydrological frame leads to a focus on numerical models of the flux of water between different compartments of the hydrological cycle and a whole suite of established practices and communities of practice concerned with hydrological modelling. The political ecology frame focuses on who has access to clean and safe water, how this access is differentially distributed, and the structural factors and histories of inequality that create unequal access. Rusca and Mazzoleni's chapter also demonstrates that the role of frames is not limited to how we research it but also the interventions that might follow. A hydrological frame, for instance, might focus on the need to restore the water cycle to a more natural state and how to mitigate for and adapt to future climate change; a political ecology frame might focus on the need to address inequities in access to that water.

Crucially, very different theoretical frames of the same problem can co-exist in ways that are either (1) mutually independent but neutral (both the hydrological frame and the political ecology frame can co-exist, one without the other); (2) mutually independent but antagonistic (is the lack of clean and safe water explained better by climate change or by distributional problems?); or (3) in ways that are mutually intertwined. As Rusca and Mazzoleni show, there is a political ecology of hydrological modelling; how hydrological models are applied and their results interpreted are bound with decisions over how water is managed.

Secondly, a frame can be more empirical, or even personal, constructed by one's experience of a particular place or places at a time or through time. We all develop such frames through the day-to-day rolling on of our normal lives. But frames based upon observation are also an important element of environmental research. For example, we can observe the consequences of human activities directly (e.g., habitat loss; floodplain management) or indirectly (e.g., the consequences of pollution due to agricultural runoff as seen in shifts in lake ecology). For the first author of this chapter, living in a village that was repeatedly cut off by flooding framed his realisation of the role of complex floodplain structures (housing, walls, ditches) in determining flood inundation. Existing treatment of such structures as things that simply slowed the flow (as "roughness elements") rather than blocking it (dry stone walls) was brought to reality on a dark night when the road across the floodplain had flooded after water caused a brick wall to fail, and he fell in the water because his bicycle wheel hit hidden bricks. It motivated him to develop a new approach to modelling flood inundation based upon explicit representation of blockage rather than upscaling of roughness elements to represent the effects of blockage (Yu and Lane 2006).

However, empirical frames may also develop through the process of doing research. Malone (Chapter 16) describes her growing interest in, and realisation of the need to understand, contamination in urban community gardens. This frame came not only from biophysical observations of urban soil contamination but also from her observations of poor community awareness of contamination and very different community definitions of what it constitutes. She then developed new empirical frames for her research question that challenged a more conventional biophysical frame based on contamination concentrations in soil. This discussion of empirical frames emphasises that all frames can and do change and that a marker of a good practice in research is letting what you are studying metaphorically "speak back" and reshaping your frame(s) as you experience what you are studying.

Thirdly, frames also differ along an important axis that goes from the positive to the normative. Positive frames seek to describe, state, or explain what is being studied. In a positive frame, the researcher seeks to elicit supposedly factual statements that exist independently of the researcher and their intentions. This approach implies separation or distance between the researcher and their subject, as is often a stated goal of conventional scientific method. In a normative frame, there is a shift from research being undertaken to describe "what is" to research being designed to determine "what should be", where there is a more-or-less moral judgement being made as part of the research process. Antithetical to positivist frames of research, Science and Technology Studies have questioned the extent to which any of the research we do is conducted without at least some normative frame (Latour and Woolgar 1979).

Some normative frames are epistemological, illustrated by "this is how we should study the world in order to know it", and commonly traced into particular schools of thought, sub-disciplines, or even disciplines. Rusca and Mazzoleni (Chapter 14) introduce the field of socio-hydrology, the proponents of which (e.g., Sivapalan et al. 2012) provide a good example of an epistemological normative frame. They wrote;

> We argue in this paper for a new science of socio-hydrology that treats people as an endogenous part of the water cycle, interacting with the system in multiple ways, including through water consumption for food, energy and drinking water supply, through pollution of freshwater resources, and through policies, markets, and technology ... We insist, however, that socio-hydrology must strive to be a quantitative science. While broad narratives may be important for context, quantitative descriptions are needed for testing hypotheses, for modelling the system and for predicting possible future trajectories of system states (Sivapalan et al. 2012, 1275).

The proponents state both what is needed in research (normative) and how it should be conducted (epistemological). One of the common mistakes in positivist research is to assume objectivity when there may be normative epistemological frames that are implicit or hidden. Such frames may define what are acceptable ways of knowing the world. When they are used normatively, they prevent you from allowing what you are studying having the capacity to frame how it is studied. Epistemological frames can have considerable power, enabling certain accounts of the world while constraining others. We return to the relationship of this challenge to mixed methods below.

The second kind of normative frame occurs when research is motivated by potentially transformative change. Such frames can be traced back to action research, originating with challenges surrounding improving inter-group relations within society (Lewin 1946). Lewin observes that the entry point for action research is not the desired change itself, followed by doing the research to support it. Rather, the entry point is where transformative change is accepted as needed but what that change could or should be is not known. Lewin wrote

> Two basic facts emerged from these contacts: there exists a great amount of good-will, of readiness to face the problem squarely and really to do something about it. If this amount of serious goodwill could be transformed into organized, efficient action, there would be no danger for intergroup relations in the United States. But exactly here lies the difficulty. These eager people feel to be in the fog. They feel in the fog on three counts: 1. What is the present situation? 2. What are the dangers? 3. And most important of all, what shall we do? (Lewin 1946: 34)

The quote illustrates that a normative frame originates with a problem that then motivates research that may lead to transformative change, rather than with the desired change itself and the research needed to justify it.

Malone (Chapter 16) provides an excellent example of the relationship between a normative research frame and transformative change when she talks about communities living with urban soil contamination. There is a gap between what they are told the problem is in conventional risk assessment terms (e.g., Environmental Protection Agency contamination guidelines) and how they physically experience contamination in their day-to-day lives through growing the food they need on urban soils. This is a problem that motivates a need for new kinds of frames that might support transformative change. The research that is then needed to identify what that change should be is very different to traditional risk assessment of contaminated soil.

Lewin (1946) also shows how the normative frames associated with action research often need certain kinds of research epistemologies. In noting the superficial nature of traditional social surveys, he describes the importance of more in-depth investigation of the "motivations behind the sentiments expressed" (Lewin 1946: 37). He relates an experiment in which delegates to a meeting in Connecticut (USA) concerned with race relations came from initially opposing backgrounds and interests but, through working together to understand what the challenges were, ended up identifying a series of actions that were broadly accepted. This is an example of what anthropologists call *emic* research, conducted "with people", in contrast to *etic* research where a culture is studied from the perspective of the outside looking in, that is "on people". Lewin (1946) argued that the broader acceptability of the research findings and actions that result may require more emic approaches. Although action research is quite often emic in nature (e.g., Malone, Chapter 16) it should be emphasised that it is not necessarily so. Dorling (2023), for example, provides an example of a more etic account of inequality in the United Kingdom which, at the same time, has a clear normative purpose in identifying what needs to be done to prevent the United Kingdom becoming "a failed state". That said, framing is not simply an academic exercise as the frame chosen may impact quite strongly on the social, political, and more general acceptability of the results.

Characteristics of frames

The different kinds of frames introduced above were chosen to draw out and illustrate some key properties of frames (Table 3.1); understanding these properties is important for understanding the rationale behind and challenges for mixing methods.

Table 3.1 Characteristics of frames in research

- Frames matter for how we conceptualise the world—they are needed and can't be avoided, but they also have enormous power, both constraining and enabling research.
- Frames determine how we study the world, the choice of methodological approach and of specific methods, and how we apply them.
- Frames set up the kinds of interventions that follow and so can have significant consequences for the impact of research.
- Frames may impact the acceptability of research results for those who are the subjects of the research and so sometimes frames need to change if they are to be transformative.
- Frames may be complementary, antagonistic or entwined. Sometimes the most interesting moments in research arrive when two or more frames are set in juxtaposition and cause the bases of those frames to be questioned.
- Frames can and do evolve, and this may lead to new kinds of frames.
- Mixing methods may contribute to both questioning frames and forcing their evolution, especially where they reveal contradictions in what it is we think we know.

Two specific questions need to be addressed following this discussion: how and when can frames evolve; and, given the power that frames have, how do they relate to the academy? For instance, certain kinds of disciplines or sub-disciplines may adopt some frames and exclude others. Thus, the question of how and when frames evolve is also related to disciplines and different kinds of disciplinarity and we show below how disciplines may influence the mixing of methods in environmental research.

Evolving frames

Snow et al. (1986) describe the four main ways in which frames may evolve. First, *frame bridging* involves the linkage of two or more frames that are compatible but unrelated to one another. This tends to leave the frames intact. Second, *frame amplification* involves the accumulation of information that supports and so reinforces a given frame. The frame also remains intact. Third, *frame extension* involves changes in the scope of a frame to include additional perspectives, a potentially emancipatory process wherein the nature of the frame begins to evolve. Finally, *frame transformation* is when existing frames can't be reconciled through bridging, amplification, or extension, catalysing a fundamental change in how we come to perceive or to understand the frame itself. In frame transformation, the evidence does not necessarily change, but how we interpret it does, and so we develop new kinds of research questions and projects.

Changing frames relates to a wider conversation about changing the overarching paradigms that guide research. Following Kuhn (1962) individual scientific activities tend to cluster themselves on what is taken as a current paradigm during periods of normal science. A paradigm is a frame, in the language of Goffman (1974), in that it allows us to organise and justify our ideas, to develop meaningful research questions and approaches and so to participate in the paradigm (and the research community) of which it is a part. Kuhn (1962) was clear that multiple paradigms (and hence frames) can exist, and this results in scientists embarking upon different kinds of research programmes. These programmes may co-exist without knowledge of each other; sometimes they may be brought together in ways that are compatible with each other (frame bridging).

During the process of normal science, research produces information and often this reinforces individual frames. However, it may also produce anomalies. These may lead to refinement of the frame (frame extension), but progressive accumulation of anomalies may lead to a crisis where it is realised that accumulated anomalies render the whole paradigm or frame no longer tenable with respect to what is now known. A paradigm shift occurs (frame transformation) with a fundamental change in how we understand the anomalies that have accumulated. The new paradigm ushers in a new phase of normal science, or a new kind of accepted frame.

Viewed from the outside, Kuhn's (1962) ideas lead to the view that most research frames exist through a long period of stasis (one could call it "boredom") ended by very short periods of very rapid change (one could call it "terror"). In this view, the conditions associated with terror are created by the accumulation of knowledge of the frame whilst in a period of apparent boredom or stasis. However, this account is descriptive and overlooks the fact that there is a political, cultural, and social economy with which science is bound and which in turn may impact this stasis.

Researching frames

The motivation of Snow et al.'s (1986) description of how frames evolve was to understand the conditions in which social movements might change rapidly. Empirical research, notably in Science and Technology Studies, has revealed the extent to which frames are a dominant characteristic of scientific research. Fleck (1935), in a study of the history of the Wasserman reactor as a test for the illness syphilis identified "thought collectives" or "thought styles" which can be labelled as frames. They can become active and powerful devices that can be used to control both what scientists are allowed to do and how the results of research are interpreted (Fleck 1935).

In an anthropological study of medical scientist Roger Guillemin's laboratory at the Salk Institute, California, Latour and Woolgar (1979) showed what produced such frames. Firstly, the material, technical, and human resources of the laboratory limited what research could be done (see Blond, Chapter 20 for another example of how available methods can constrain environmental reconstruction). Secondly, Latour and Woolgar (1979) signalled how research relies upon necessary hypotheses or assumptions to make it feasible, assumptions that may not actually hold for the real world being studied. Shackley et al.'s (1999) study of climate modelling is the classic example in relation to environmental research, where empirical adjustment of heat, water, and momentum fluxes in global climate models, which has no a priori physical basis, is required to reproduce the current climate (IPCC 2013). The acceptability of such assumptions was traced both to different epistemological traditions (Shackley et al. 1999; Sundberg 2009; Brysee et al. 2012) and to the desire that the models produced predictions useful for policymakers (Shackley et al. 1999). Shackley et al. framed this approach as "pragmatist" and showed that for pragmatists, abandoning such a frame becomes costly from the perspective of what a climate modeler can do (provide predictions of the future) or say (inform policy). In this volume, both Kelley (Chapter 19) and Walters et al. (Chapter 13) describe how, in forest ecology, a biophysical frame centred on spaceborne remote sensing assumes that what is seen from above represents what is below, as to do otherwise would constrain the relevance of spaceborne remote sensing; but this then leads to particular kinds of explanations that exclude others that can only be seen from within a forest canopy or experienced by forest dwellers. Reliance on particular frames produces particular kinds of knowledge.

Thirdly, and perhaps most importantly, Latour and Woolgar (1979) showed that facts emerge in networks (made up of scientists and the things that they study) and what constitutes a fact and how these facts are presented and diffused (see Chignell et al., Chapter 18) has a crucial impact upon knowledge. This is the essence of framing. Whilst a frame may evolve or even take on a new form, networks produce particular frames and those frames only survive as long as the networks can maintain them. Latour (1999) illustrated this when he participated in a scientific expedition to the Amazon to study the dynamics of tropical rain forest expansion and contraction. Latour describes how scientists from three different disciplines—a botanist, a soil scientist and a geomorphologist—each brought to the research process their own frames. Some of the frames were technical, notably those objects (e.g., equipment, measurement protocols) that each scientist felt were needed to make the field site measurable according to the wider approaches and presumptions associated with their discipline. Such standards and protocols permeate biophysical research and if left unchallenged can have a profound impact on research results (see Malone, Chapter 16). For Latour, each technological-methodological frame was necessary to make the results from the field site comparable to those obtained from different field sites by researchers from the same disciplines. However, framing is not simply a technological-methodological exercise but also the means by which observations are translated into knowledge that sits comfortably with each discipline (Latour 1999). What survives the translation is a combination of both the constraints of the real world (Gooding 1990; see also Blond, Chapter 20) and the way that disciplines determine what is and what is not acceptable translation.

To summarise this section, empirical studies of research in practice have demonstrated the critical role played by frames. These frames (1) provide methodological-technical constraints regarding what a researcher can do; (2) define the hypotheses and assumptions needed to allow the research to be done; and (3) constitute networks of researchers, concepts, theories, epistemologies, methods, and tools that shape what research is done, by whom, and how. In an academic context, there is one additional dimension: the relation between frames and disciplines.

Frames, disciplines and dinosaurs

The practice of academic research is highly organised around disciplines. Following Bolman and Deal's (1984) model of organisations, academic disciplines are founded on four different dimensions: structures, human resources, symbols, and politics. *Structures* refer to primarily top-down imposed organisational systems needed to render an organisation functional (e.g., a university department). They include roles, responsibilities, practices, routines, and incentives (Reinholz and Apkarian 2018). *Human resources* are commonly deployed to academic disciplines. Although there may be individual goals, agencies, needs, and identities (Reinholz and Apkarian 2018), disciplines normally contain people with shared or common ground. This often comes from a shared or similar background and training. *Symbols* are the language, knowledge, methodologies, and methods associated with a discipline. Finally, *politics* is always at play within disciplines (Reinholz and Apkarian 2018), important in maintaining disciplinary identity and reinforcing a discipline structurally. Disciplines, then, have significant power in determining what is acceptable knowledge and, crucially in terms of mixed methods, how it should be produced. Disciplines are often criticised for the ways in which they "inhibit communication, stifle innovation, thwart the search for integrated solutions to social problems, inhibit the economic contributions of universities and provide a fragmented education for undergraduates" (Jacobs 2014: 13).

The power of disciplines and the strategies they use to maintain themselves are well known (e.g., Jacobs 2014). A key outcome of such strategies is disciplinary practitioners, who accept shared knowledge and methods (Schoenberger 2001) and in so doing control the structures, human resources, symbols, and distribution of power that in turn reproduce those disciplines.

Campbell (1969) described disciplines as functioning like tribes that steer research towards those topics that the discipline currently deems to be central. This inwards steering is crucial because it constrains what a researcher has to think about (Bruce et al. 2004) and so stops them from realising that there may be other kinds of frames that could ultimately question the foundations on which a discipline is built. Swales (1997) developed *Tyrannosaurus Rex* as a metaphor for the ways in which English was becoming increasingly dominant as a world language, "gobbling up the other denizens of the academic linguistic grazing ground" (p. 374). It is hard not to see disciplines as dinosaurs for the ways in which, as part of their own reproduction, they consume those ideas, methods, and findings that come sufficiently close to warrant a threat.

Disciplines (and not just the frames themselves) play a role in maintaining certain kinds of frames (Polanyi 1966). This matters because disciplines frame environmental research in different ways. Armstrong et al. (2022), for instance, describe a disciplinary divide about how the environmental impacts of urbanisation are framed by authors from ecology backgrounds compared with urban planning backgrounds. For ecologists, urbanisation was framed as a negative through a focus on local land conversion and habitat loss. For urban planners, whilst recognising some ecological consequences, the focus was more on the possible benefits of denser human settlement patterns. Urban areas in Europe, for example, have lower carbon footprints that rural areas (Armstrong et al. 2022) and it is now recognised that even urbanised environments can have significant ecological value (Salomon Cavin 2013). Armstrong et al. (2022) attributed these differences to disciplinary training: ecologists are rarely (if ever) trained in urban planning and sustainable urbanism; for their part, urban planners are rarely (if ever) trained in ecology and the consequences of habitat loss. The lack of conversation between these two disciplinary frames is unfortunate in this example and is why it is common to bemoan the existence of "disciplinary silos" (Petts et al. 2008). Remaining in disciplinary silos serves the disciplines, but radically limits our ability to understand and manage the environmental issues that a discipline may seek to address.

Beyond disciplinary frames and mixing methods

There are four ways in which research may move beyond disciplinary frames and so start to mix methods in different ways: cross-disciplinary research; multidisciplinary research; interdisciplinary research; and transdisciplinary research. Each kind of "beyond disciplinarity" treats disciplinary frames in different ways, and so impacts upon which methods are used in a research project and how they are combined together.

Cross-disciplinary research: borrowing the methods developed by others

In cross-disciplinary research, the researcher stays within their home discipline and its frames, but borrows research topics or methods from other disciplines (Nicolini et al. 2012). The function served by being cross-disciplinary may be practical, notably in providing methods that allow a particular discipline to answer questions it cannot within its dominant frame. The discipline may develop good practice in the use of those methods but the methods themselves remain the "property" of the discipline that developed them.

A good example of cross-disciplinarity is in environmental research that uses remote sensing. Kasvi (Chapter 45) describes the use of Uncrewed Airborne Systems (UAS) in environmental research. The transformation of raw imagery into geometrically correct imagery (i.e., without distortions due to the sensor, or relief of the surface being mapped) requires processing techniques developed in the discipline of engineering surveying (notably photogrammetry, which is the basis of "Structure from Motion", SfM) and computer vision (for the automation of data extraction, which is the basis of Multi-View Stereo matching, MVS). Researchers then make use of methods that have and are being developed in different disciplines. As Kasvi notes, the correct use of SfM-MVS photogrammetry is not straightforward and in a truly cross-disciplinary approach, a researcher would develop care and good practice in their application of SfM-MVS photogrammetry. This may in turn require at least some training in disciplines other than their own. Many of the more technical methods chapters in this volume (e.g., hydraulic modelling, Lane, Chapter 30; hydrological modelling,

Melsen, Chapter 31; remote sensing, Braun, Chapter 39) have this cross-disciplinary characteristic of method transfer. Indeed, some have argued that such cross-disciplinarity is a necessity to maintain the capacity of disciplines to be innovative and disruptive (Park et al. 2023).

Multidisciplinary research: a set of methods each making their own contribution

Multidisciplinary research recognises that there are research questions and problems that merit being approached from different disciplinary perspectives and researchers from different disciplines need to work together to do so. Each discipline brings its own particular frame but those frames are left largely intact (Petts et al. 2008). There are then interactions between disciplines (Hunt and Shackley 1999) but this does not go beyond frame-bridging and normally does not lead to synthesis or synergy in developing research findings (Bruce et al. 2004). In a multidisciplinary project, methodological expertise is typically brought and retained by disciplines, each discipline providing what it can do best.

De Feo et al. (2018) provide a good example of a multidisciplinary research project for bringing sustainability criteria into urban regeneration. Defining sustainability in social, economic, and environmental terms, the problem was addressed using methods from sociology to inform an extensive questionnaire-based survey to identify socially sound regeneration alternatives; from economics to determine the costs of each alternative; and from the environmental sciences to determine their environmental impacts. In this case there was some synthesis into a single measure of sustainability, but the methods used came from and remained true to the disciplines that supplied them.

Interdisciplinary research: when the mix of methods is shaped by the problem being addressed

Interdisciplinarity is not an easy term to define as there are different understandings of what it is (e.g., Brewer 1999; Karlqvist 1999; Lattuca 2001; Tress et al. 2005). In its most general sense, interdisciplinarity involves comparing, contrasting and modifying the frames brought by disciplines in a search for coherence and synthesis around an "inter-discipline" (Petts et al. 2008). This does not necessarily require integrating the biophysical and social sciences as within those broad disciplinary groupings, interdisciplinarity can emerge. However, this general definition raises the question as to what the focus is of the coherence or synthesis. What is it that motivates the integration and what does the integration allow to be done? This is why interdisciplinarity has become more specifically associated with research that is motivated by a particular problem rather than a discipline's frame of what constitutes admissible research ideas, foci, methodology, or methods (Campbell 1969; Brewer 1999; Aligica 2004; Tress et al. 2005). This problem focus is viewed as necessary for environmental research to become more policy-relevant (Lane et al. 2006). The applied focus of such policy-relevant research was labelled as "Mode 2" by Gibbons et al. (1994) to set it apart from research motivated by traditional academic frames ("Mode 1"), centred on disciplines and the questions they deem important.

Interdisciplinarity implies a fundamental shift in research practice that inevitably requires a radical mixing of methods. In the classical model of scientific progress of Kuhn (1962) described above, frames change progressively by accumulating anomalies (i.e., where one or more parts of a frame is wrong and has to be modified) to the point at which the frame is no longer recognizable in its prior form. Such progressive evolution of a particular frame also implies progressive methodological development. When dealing with applied problems and where the motivation is finding solutions, the traditional scientific logic that progress is made through anomalies, through being wrong, is replaced by a new imperative to provide answers that are right however philosophically illogical that might be (Lane et al. 2006).

When the motivation is a research question or a research problem, not the method available, and the question or problem is used to define the methods needed, then there is likely a need to mix methods and to adapt them such that they can answer the question being asked. Blond (Chapter 20) illustrates this mixing of methods in an interdisciplinary approach to researching the reconstruction of historical landscapes as socio-ecological systems. Her starting point is a given landscape, such as cultivated terraces in Ethiopia. In order to understand why this landscape is the way it is, teasing out both historical and contemporary influences, it is necessary to mobilise a range of different disciplines and sub-disciplines (archaeology, sedimentology, geomorphology, history, social science), to select the relevant elements of each, and to apply them in ways that are feasible given the cases being investigated. This methodological pluralism is needed for her work, but it can be risky both personally or for a discipline if the particular method one individual or discipline brings is subsequently found to be inappropriate or unnecessary. It is not surprising that many disciplines find interdisciplinary research uncomfortable, a threat to their existence (Bruce et al. 2004).

Transdisciplinary research: mixing methods in a truly scientific fashion

The assumption in interdisciplinarity is that the disciplines themselves are left intact, though they may be found to be redundant as a research question becomes better defined, and the methods they bring may be challenged or even forced to evolve. But interdisciplinarity does not go as far as the hybridisation (Hunt and Shackley 1999) associated with creating new disciplines, when disciplinary boundaries are dissolved and new ('hybrid') disciplines are formed (Lane et al. 2006). This is commonly called transdisciplinarity. Whilst there may be some debate, or at least some ambivalence, over the extent to which interdisciplinarity must be problem-oriented, transdisciplinarity starts from the view that the frames and disciplines that exist to address a problem must fundamentally change in order to address that problem adequately (Petts et al. 2008). One of the reasons environmental problems are known as "wicked" is that how we frame a problem begets the way it is solved (Rittel and Webber 1973). Transdisciplinarity explicitly seeks to challenge conventional frames of problems. The question then is how, and what makes this different to interdisciplinarity.

A distinction between interdisciplinarity and transdisciplinarity is in how to make disciplines change. Whilst interdisciplinarity tends to leave disciplines intact, if sometimes redundant, or facing the limits of their own methodologies and methods, transdisciplinarity is often associated with methods that can challenge disciplinary dinosaurs, and especially the frames they bring. Most of us are constrained by, and have to remain true to, our disciplines. As Stengers (2005) argued, disciplines are then one of the basic reasons why we as scientists stop being scientific. Building upon a loose conceptualisation of the logic of falsification, she describes the challenge of putting ourselves in a position with respect to our research where we get those seeds of doubt that cause us to slow down, interrogate the frames upon which our research and our disciplines are built, and try to explore the world in a different way. It is right to discover that we are wrong, and being scientific means being true to the empirical nature of what we study and not the disciplines that tell us what to study and in what way. Blond (Chapter 20) also shows transdisciplinarity in her work. Not only does she mobilise a mix of methods in an interdisciplinary way but her work shows a methodological dynamism that comes from putting methods into tension with one another in response to what she is studying. It is this commitment to an informal empiricism in her method, one that slows her down and causes her to look at problems differently, that captures what Stengers (2005) argues it is to be scientific. It also captures the inherent uncertainties of environmental research, which is commonly poorly bounded, theoretically less developed, empirically challenging, and where flexibility in mixing methods is critical.

How can we do the kind of slowing down that Stengers (2005) advocates and which may be a necessary precursor of transdisciplinarity? There are two broad approaches, and both involve changing the positionality of the researcher with respect to the researched. The first recognises that one of the things that can slow us down is being placed in what Gieryn (2006) calls truth spots. These are "delimited geographical locations", places, that comprise "irreducibly: (1) the material stuff agglomerated there, both natural and human-built; and (2) cultural interpretations and narrations (more or less explicit) that give meaning to the spot" (Gieryn 2006, 29). This is the kind of frame more typical of case study research (Lane, Chapter 24), and where the frames brought by disciplines are challenged explicitly by the empirical experience of being at a place in time. A case study focus is one of the distinguishing characteristics of the mixed-methods examples in this volume (Rusca and Mazzoleni, Chapter 14; Booth and Druschke, Chapter 15; Malone, Chapter 16; Chignell et al., Chapter 18; Kelley, Chapter 19; Blond, Chapter 20). Central to this kind of work is metaphorically letting the environment or a place "speak back" to the researcher, guiding the researcher in terms of how it wants to be studied.

However, the notion of a researcher putting themselves geographically and historically in a position where the environment can speak back does not escape the tensions that come from being a member of a discipline or part of the academy. This is why transdisciplinarity is increasingly associated with a more radical change in positionality: letting those who are concerned by the research but who are not researchers themselves (research subjects) into the research process and giving them the power to influence how the research is conceptualised and undertaken, and what methods are used. This repositioning often leads to methods that are more participatory (see Sayre, Chapter 34; Landström, Chapter 35; Mokos, Chapter 36). Hence this kind of transdisciplinarity is strongly emic. Inviting the subjects of research into the research process rather than leaving them as subjects to be studied appears to be anathema to the basic tenets of the scientific method, which seems to suppose a separation between the researcher and the researched. However, given the power that disciplines have in the traditional conduct of research, it is less an anathema and more a transition of power away from those who do research and towards those to whom the research pertains. Stengers (2013) argues that this should be the basis of a more democratic science.

Landström et al. (2011) provide an example of a transdisciplinary research project centred on this second kind of repositioning of scientists with respect to the subject of their research. They followed two scientists through a participatory research project designed to reduce flood risk. They showed how the scientists' starting point was their disciplines and disciplinary experiences, the kinds of rural land management solutions to reduce flood risk that were circulating in the academy, the hydrological and hydraulic models that were accepted by their scientific communities, and the traditional data sources and data collection systems deemed necessary. Then in working with people to understand what was driving the flood risk and what could be used to mitigate it, the scientists turned away from their normal networks of practice, and the range of frames associated with them, towards a new network constructed around residents participating in the project and for whom flooding was a matter of concern (Landström et al., 2011). The methods that were chosen were a function of the participatory process and were forced to change radically as it was realised that existing methods locked in certain kinds of solutions. A new model framework and approach emerged from this re-orientation and in turn identified other kinds of flood risk management options for further testing that had previously been excluded.

The outcome was not only different options but also a new "public" who, through working together, were able to make a political intervention that shifted the whole flood risk management approach in the region and beyond. The scientists never wrote up their new modelling framework (until other scientists, much later on, started to use it; Dixon et al., 2016) because in the new network that they were a part of it just didn't seem necessary. The research was valorised in different ways (e.g., the work becoming a government demonstration project for others to use).

Conclusions

Mixing methods in environmental research does not happen in a vacuum. It happens in an academy with established norms and practices that shape what is and is not acceptable. If we develop the metaphor adopted in the introduction, we have sketched out how the construction of the "research kitchen" frames what can be "cooked" and how. Escaping the constraints imposed by frames, especially given the power of disciplines, is likely central in moving towards a more scientific approach to environmental research: one framed by what is being researched rather than how disciplines say it should be researched.

We outlined how this mixing can happen in four different ways. Cross-disciplinary research mixes methods by borrowing those of other disciplines. This allows new questions to be answered in what can be very mature research areas, but often these methods are developed by others and not always easy for the uninitiated to apply. It can lead to friction when those who feel they "own" methods see them being used in ways that were not intended.

In multi-disciplinary research, frames and their associated methods are left intact. Each frame brings its own methods and these can co-exist. There may be some synthesis of results but the methods are not really challenged and if the methods are "ingredients" they remain poorly mixed. Multi-disciplinary collaboration tends to be comfortable in disciplinary terms as the frames and methods are neither questioned nor changed.

Interdisciplinary research rests on the principle that there are gaps between disciplines where combining different frames and mixing methods together leads to new kinds of knowledge, even new kinds of frames, that challenge the status quo. Increasingly, interdisciplinary research has become seen as problem-oriented, where the problem itself defines what ingredients are needed and how they should be mixed.

In transdisciplinary research, the inherent uncertainties in undertaking environmental research, notably around what the question or problem is and so what methods are needed to study it, are dealt with by mixing and modifying methods through the research process. Whilst the lack of a priori research design may seem ascientific, it can actually be more scientific where it is grounded in letting the environmental speak back, telling you how it wants to be studied as you study it. This tends to require a change in position of the researcher with respect to what they are researching. This change may be geographical, through more intensive case-study-based research or, increasingly commonly, by letting those who are being researched have a direct say in how that should happen.

Behind these different ways of mixing methods are some important challenges. These include the power of disciplines in determining what are acceptable frames, what constitutes acceptable practice; debates over the extent to which research can adopt normative frames and still be labelled as research; the challenges that come from both more etic and more emic approaches to research and the importance of mixing them. Mixing methods is generative but also difficult and, in doing it, researchers need to be sensitive to wider debates regarding the nature and practice of research.

References cited

Aligica, P.D. 2004. 'The challenge of the future and the institutionalization of interdisciplinarity: notes on Herman Khan's legacy', *Futures*, 36, pp. 67–83. https://doi.org/10.1016/s0016-3287(03)00136-8

Armstrong, J.H., A.C. Nisi, and A. Millard-Ball. 2022. 'A disciplinary divide in the framing of urbanization's environmental impacts', *Conservation Science and Practice*, 4, pp. e624. https://doi.org/10.1111/csp2.624

Bateson, G. 1955. 'A theory of play and fantasy', *Psychiatric Research Reports*, 2, pp. 39–51.

Blond, N., Chapter 20, this volume. 'Mixing geoarchaeology, geohistory and ethnology to reconstruct landscape changes on the longue durée'.

Bolman, L.G. and T.E. Deal. 1984. *Modern Approaches to Understanding and Managing Organizations* (Jossey-Bass).

Booth, E. and Gottschalk Druschke, C., Chapter 15, this volume. '"A hydrologist and a rhetorician walk into a workshop," or how we learned to collaborate on a decade of mixed methods river research across the humanities and biophysical sciences'.

Braun, A., Chapter 39, this volume. '(Critical) Satellite remote sensing'.

Brewer, G.D. 1999. 'The challenges of inter-disciplinarity', *Policy Sciences*, 32, pp. 327–37. https://doi.org/10.1023/a:1004706019826

Bruce, A., C. Lyall, J. Tait, and R. Williams. 2004. 'Interdisciplinary integration in Europe: the case of the Fifth Framework Programme', *Futures*, 36, pp. 457–70. https://doi.org/10.1016/j.futures.2003.10.003

Brysse, K., N. Oreskes, J. O'Reilly, and M. Oppenheimer. 2012. 'Climate change prediction: Erring on the side of least drama?, *Global and Environmental Change: Human and Policy Dimensions*, 23, pp. 327–37. https://doi.org/10.1016/j.gloenvcha.2012.10.008

Campbell, D.T. 1969. 'Ethnocentrism of disciplines and the fish-scale model of omniscience', in *Interdisciplinary Relationships in the Social Sciences*, ed. by M. Sherif and C.W. Sherif (Aldine), pp. 328–48.

Chignell, S., Howkins, A. and Fountain, A., Chapter 18, this volume. 'Antarctic Mosaic: Mixing Methods and Metaphors in the McMurdo Dry Valleys'.

De Feo, G., S. de Gisi, S. de Vita, and M. Notarnicola. 2018. 'Sustainability assessment of alternative end-uses for disused areas based on multi-criteria decision-making method', *Science of The Total Environment*, 631.2, pp. 142–152. https://doi.org/10.1016/j.scitotenv.2018.03.016

Dixon, S.J., D.A. Sear, N.A. Odoni, T. Sykes, and S.N Lane. 2016. 'The effects of river restoration on catchment scale flood risk and flood hydrology', *Earth Surface Processes and Landforms*, 41, pp. 997–1008. https://doi.org/10.1002/esp.3919

Dorling, D. 2023. *Shattered Nation: Inequality and the Geography of a Failed State* (Verso).

Fleck, L. 1935. *Genesis and Development of a Scientific Fact* (University of Chicago Press).

Gibbons, M., C. Limoges, H. Nowotny, S. Schwartzman, P. Scott, and M. Trow. 1994. *The New Production of Knowledge: The Dynamics of Science and Research in Contemporary Societies* (Sage Publications).

Gieryn, T.F. 2006. 'City as truth-spot: Laboratories and field-sites in urban studies', *Social Studies of Science*, 36, pp. 5–38. https://doi. org/10.1177/0306312705054526

Goffman, E. 1974. *Frame Analysis: An Essay on the Organisation of Experience* (Harvard University Press).

Gooding, D. 1990. 'Mapping experiment as a learning process: how the first electromagnetic motor was invented', *Science, Technology, and Human Values*, 15, pp. 165–201. https://doi.org/10.1177/016224399001500202

Hunt, J. and S. Shackley. 1999. 'Re-conceiving science and policy: academic, fiducial and bureaucratic knowledge', *Minerva*, 37, pp. 141–64. https://doi. org/10.1023/a:1004696104081

Jacobs, J.A. 2014. *In Defense of Disciplines: Interdisciplinarity and Specialization in the Research University* (University of Chicago Press). https://doi. org/10.7208/chicago/9780226069463.001.0001

Karlqvist, A. 1999. 'Going beyond disciplines: the meanings of interdisciplinarity', *Policy Sciences*, 32, pp. 379–83. https://doi. org/10.1023/a:1004736204322

Kasvi, E., Chapter 45, this volume. 'Uncrewed Airborne Systems'.

Kelley, L., Chapter 19, this volume. 'Engaging remote sensing and ethnography to seed alternative landscape stories and scripts'.

Kuhn, T.S. 1962. *The Structure of Scientific Revolutions* (University of Chicago Press).

Lahsen, M. 2005. 'Seductive simulations? Uncertainty distribution around climate models,' *Social Studies of Science*, 35, pp. 895–922. https://doi. org/10.1177/0306312705053049

Landström, C., Chapter 35, this volume. 'Participatory modelling'.

Landstrom, C., S.J. Whatmore, S.N. Lane, N. Odoni, N. Ward, and S. Bradley. 2011. 'Coproducing flood risk knowledge: redistributing expertise in critical "participatory modelling"', *Environment and Planning A*, 43, pp. 1617–33. https://doi.org/10.1068/a43482

Lane, S. N. and Lave, R., Chapter 3, this volume. 'Frames, disciplines and mixing methods in environmental research'.

Lane, S.N., Chapter 30, this volume. 'Hydraulic modelling'.

Lane, S.N., C.J. Brookes, A.L. Heathwaite, and S.M. Reaney. 2006. 'Surveillant science: challenges for the management of rural environments emerging from the new generation diffuse pollution models', *Journal of Agricultural Economics*, 57, pp. 239–57. https://doi.org/10.1111/j.1477-9552.2006.00050.x

Lattuca, L.R. 2001. *Creating Interdisciplinarity: Interdisciplinary Research and Teaching among College and University Faculty* (Vanderbilt University Press). https://doi.org/10.2307/j.ctv167563f

Latour, B. and S. Woolgar. 1979. *Laboratory Life: The Social Construction of Scientific Facts* (Sage Publications).

Latour, B. 1999. *Pandora's Hope: Essays on the Reality of Science Studies* (Harvard University Press).

Lewin, K. 1946. 'Action research and minority problems', *Journal of Social Issues*, 2, pp. 34–4.

Malone, M., Chapter 16, this volume. 'Using mixed methods to confront disparities in public health interventions in urban community gardens'.

Melsen, L., Chapter 31, this volume. 'Hydrological modelling'.

Mokos, J., Chapter 36, this volume. 'Participatory modelling'.

Nicolini, D., J. Mengis, and J. Swan. 2012. 'Understanding the role of objects in cross-disciplinary collaboration', *Organization Science*, 23, pp. 612–629. https://doi.org/10.1287/orsc.1110.0664

Park, M., E. Leahey, and R.J. Funk. 2023. 'Papers and patents are becoming less disruptive over time', *Nature*, 613, pp. 138–44. https://doi.org/10.1038/s41586-022-05543-x

Petts, J., S. Owens, and H. Bulkeley. 2008. 'Crossing boundaries: interdisciplinarity in the context of urban environments', *Geoforum*, 39, pp. 593–601. https://doi.org/10.1016/j.geoforum.2006.02.008

Polanyi, M. 1966. *The Tacit Dimension* (University of Chicago Press).

Reinholz, D.L. and N. Apkarian. 2018. 'Four frames for systemic change in STEM departments', *International Journal of STEM Education*, 5.3. https://doi.org/10.1186/s40594-018-0103-x

Rittel, H.W.J. and M.M. Webber. 1973. 'Dilemmas in a general theory of planning', *Policy Sciences*, 4, pp. 155–169. https://doi.org/10.1007/bf01405730

Rusca, M. and Mazzoleni, M., Chapter 14, this volume. 'The interface between hydrological modelling and political ecology'.

Salomon C.J. 2013. 'Beyond prejudice: conservation in the city: a case study from Switzerland', *Biological Conservation*, 166C, pp. 84–9. https://doi.org/10.1016/j.biocon.2013.06.015

Sayre, N.F., Chapter 34, this volume. 'Participant observation and ethnography'.

Schoenberger, E. 2001. 'Interdisciplinarity and social power', *Progress in Human Geography*, 25, pp. 365–81. https://doi.org/10.1191/030913201680191727

Snow, D.A., D.B. Rocheford, S.K. Worden, and R.D. Benford. 1986. 'Frame alignment processes, micromobilization, and movement participation', *American Sociological Review*, 51, pp. 464–481. https://doi.org/10.2307/2095581

Shackley, S., J. Risbey, P. Stone, and B. Wynne. 1999. 'Adjusting to policy expectations in climate change modeling—an interdisciplinary study of flux adjustments in coupled atmosphere-ocean general circulation models', *Climatic Change*, 43, pp. 413–54. https://doi.org/10.1023/a:1005474102591

Sivapalan, M., H.H.G. Savenije, and G. Blöschl. 2012. 'Socio-hydrology: A new science of people and water: invited commentary', *Hydrological Processes*, 26, pp. 1270–6. https://doi.org/10.1002/hyp.8426

Stengers, I. 2005. 'The cosmopolitical proposal', in *Making Things Public*, ed. by B. Latour and P. Weibel (MIT Press), pp. 994–1003.

Stengers, I. 2013. *Une autre science est possible* (Editions La découverte).

Sundberg, M. 2009. 'The everyday world of simulation modeling: the development of parameterizations in meteorology', *Science Technology and Human Values*, 34, pp. 162–181. https://doi.org/10.1177/0162243907310215

Swales, J.M. 1997. 'English as Tyrannosaurus rex', *World Englishes*, 16, pp. 373–382. https://doi.org/10.1111/1467-971x.00071

Tress, B., G. Tress, and G. Fry. 2005. 'Integrative studies on rural landscape: policy expectations and research practice', *Landscape and Urban Planning*, 70, pp. 177–91. https://doi.org/10.1016/j.landurbplan.2003.10.013

Walters, G., Hymas, O., Touladjan, S., and Ndong, K., Chapter 13, this volume. 'Revealing the social histories of ancient savannas and intact forests using a historical ecology approach in Central Africa'.

Yu, D. and S.N. Lane. 2006. 'Urban fluvial flood modelling using a two-dimensional diffusion wave treatment: 2. Development of a sub-grid scale treatment', *Hydrological Processes*, 20, pp. 1567–83. https://doi.org/10.1002/hyp.5936

4. Mixed methods in tension: lessons for and from the research process

Christine Biermann and Cerian Gibbes

Introduction

Our world is replete with examples of interwoven biophysical and social processes. Phenomena that were once primarily framed as biophysical and investigated using biophysical scientific methods are increasingly conceptualised as shaped by social processes. Similarly, the material biophysical dimensions of phenomena generally recognised as social—for example, slavery (Bruno 2022), imperialism (Greer et al. 2023), and genocide (Colucci et al. 2021)—have begun to be studied as well. To address and understand the hybrid, often messy material worlds we inhabit requires research approaches that are themselves hybrid and often messy: "if we accept the Anthropocene's foundational premise... that the biophysical world is now profoundly social, surely our methods must be, too" (Biermann et al. 2021: 808).

But despite recognition of the value of integrative, interdisciplinary, and transdisciplinary environmental research, much research remains siloed in its own disciplinary domain. While it is relatively common for research to address the social implications of biophysical findings, or the biophysical context for a social scientific study, the pursuit of mixed biophysical-social methods remains notably less common (Biermann et al. 2021). One reason for this, we believe, is that tensions often arise as biophysical and social methods (and the results they yield) are put into conversation with one another. As researchers, it may be tempting to try to avoid or minimise such tensions because they challenge conventional ways of designing and carrying out research and may require a slowing

 https://doi.org/10.11647/OBP.0418.04

down of data collection, analysis, and publication-unfavourable outcomes in the current political economy of academic research (see Lane and Lave, Chapter 3; Lane 2017).

There is much to gain, however, in embracing the messiness inherent to mixed-methods approaches. We can come to see the world differently, ask new questions, disobey conventional wisdom, and reposition ourselves in relation to our research subjects (Lane 2017, drawing on Stengers 2005). In this chapter we explore some of the productive tensions that researchers engage with when pursuing mixed-methods environmental research. Following Elwood (2010), we consider mixed method approaches to be "those that rely upon multiple types of data, modes of analysis, or ways of knowing, but may use these elements in a variety of ways in relationship to one another, for multiple intellectual or analytical purposes" (95). Tensions may arise not only when methods from the social and biophysical sciences are mixed, but when any methods are mixed, and even when a seemingly singular method is used to produce or analyse different forms of data (e.g., using GIS to analyse quantitative and qualitative data (Preston and Wilson 2014)).

For example, researchers may struggle to accommodate differences in the resolution, scale, and areal or temporal extent of different datasets. More broadly, different methods are often associated with vastly different ways of knowing (epistemologies), histories, and values. Methods may also yield information about slightly different research objects. Critical physical geographer Lisa Kelley (Chapter 19) used both remote sensing and qualitative humanistic methods to investigate what seems to be a single topic: commodity crop expansion in the Sulawesi province of Indonesia. But these methods yielded information about different aspects of the situation. Remote sensing (Braun, Chapter 39) produced knowledge specifically about macro-scale patterns of cacao expansion and forest cover change that altered the spectral signature of pixels within an image. In contrast, interviews (Johnston and Longhurst, Chapter 32) and oral histories (Chakov et al., Chapter 33) revealed how people experienced land use and land cover changes and how social dynamics influenced their land use decisions. This mixing of methods produced some results that were consistent with or corroborated one another. Both remote sensing and humanistic methods revealed links between cacao and deforestation, for example. But the mixed-methods approach also yielded silences, gaps, and discrepancies among results, and it is these tensions that enabled Kelley to produce a richer understanding of the situation in Sulawesi, ultimately challenging the overly simplistic, mono-causal explanation for deforestation that has predominantly informed resource management policies in Indonesia.

In this way, mixing methods allows researchers to "fill gaps, add context, envision multiple truths, play different sources of data off each other, and provide a sense of both the general and the particular" (Elwood and Cope 2009: 5). The benefits of mixed-methods approaches include increased breadth, depth, flexibility, and a greater emphasis on the process of research (versus solely the outcomes or products). At the same time, however, mixed-methods approaches raise challenging questions for researchers. How do we engage in conversation with collaborators about not only which methods are used but also "how they are used to ask which kinds of questions and how the results are interpreted" (Nightingale 2003: 79)? How do we determine whether un-matching results indicate material differences or just different ways of knowing and seeing the world? In applied research, is it possible to recognise and address tensions between methods while still weighing and offering potential solutions to real-world problems? We contend that not only are these issues inevitable, but that researchers can and should recognise and explore them rather than minimise or attempt to eliminate them. Further, we argue that mixed-methods approaches are valuable not in spite of the tensions they engender, but indeed because of these tensions, as they require a more reflexive research practice and in turn allow for a richer understanding of the world.

In the following section, we provide a brief overview of different aims of mixed method approaches. We then discuss specific issues that may arise when mixing methods.

Mixed-methods approaches: triangulation and beyond

Triangulation is one common approach to mixed-methods research. Triangulation, however, has itself been defined and understood in multiple ways. As initially applied to social science research, triangulation was an extension of the land surveying and geodesy technique of using multiple known locations to identify an unknown point or position

(Freeman 2020). Land surveyors combine the location of a known control point and a measurement of the azimuth to calculate the position of a second control point. The two control points serve as known positions from which a third position can be calculated. The first three known positions form the baseline triangle, the sides of which enable the calculation of additional locations and distances. This creates a network of known positions from which unknown locations can be determined. The more known positions there are, the more precise the position of an unknown point becomes. When applied to research, triangulation uses multiple methods to cross-reference evidence, validate findings, and increase precision and credibility (Denzin 1989). This understanding of triangulation—referred to by Elwood (2010) as a validation-oriented approach—emphasises that consistency and corroboration of findings from multiple methods can increase the explanatory power of research. This approach is rooted in the assumption that there is a knowable and fixed reality that we can directly gather information about and represent through research (Nightingale 2003, 2009). However, mixed-methods research often proceeds much differently than land surveying, revealing a multiplicity of possible realities rather than a singular position.

The validation-oriented approach to triangulation stands in contrast to approaches that emphasise complementarity (rather than integration or validation) to generate new insights. Such approaches generally have a different epistemological starting point, recognising that all knowledge is situated and partial, and that different methods can help to reveal how knowledge is shaped by the context in which it is produced (Haraway 1988). Because the results of all methods are assumed "incomplete" (Nightingale 2003: 77), mixing methods can help to broaden the viewpoint of the researcher and produce a more detailed and multi-faceted account of the phenomena being studied. Here, what is important is not necessarily how data sources *match* or *correspond* with one another but rather the silences, gaps, and contradictions between them (Nightingale 2003; Elwood 2010). The relationship between the representations of reality depicted through different methods or data sources can be likened to a kaleidoscope (Gray 2002; Nightingale 2016). Each analytical starting point, and each method or data source, produces a slightly different (or at times vastly different) representation of reality, so that "when the kaleidoscope is turned, a new pattern can emerge—albeit one that is always partial and situated—and when different patterns are compared, new insights can emerge" (Nightingale 2016: 41). Just as in a kaleidoscope, no two patterns match exactly. This is not considered problematic but is instead viewed as a function of how different methods represent and produce reality in different ways. No one method is privileged or viewed as more or less accurate, and if results across multiple methods do match, it is not assumed to be evidence of a singular reality. Other terms that have been used to describe similar complementary approaches to mixing methods include epistemological pluralism (Nightingale 2016), crystallisation (Ellingson 2009), and collage (Freeman 2020).

With Elwood (2010), we also identify a third approach to mixed-methods research: hybrid approaches that pursue both validation and contradiction, consistencies and inconsistencies. Such approaches recognise that where we start from analytically and how we conceptualise a problem (as well as the broader social relations in which research is embedded) affect our research questions, methods, and findings (King and Tadaki 2018). Many Critical Physical Geography studies follow an approach such as this (e.g., Lave et al. 2018; Luthra et al. 2022; Malone and McClintock 2022). However, this hybrid approach can raise a challenging question: if there is no expectation that results will match or corroborate one another, how can we determine if discrepancies among results from different methods reflect material differences or simply different ways of seeing and making sense of the world (or some combination of the two)? For some, it may be that this question is moot—from their perspective, representations of reality may be all that there is. For others, this is a key tension that must be grappled with in the research process (see Lebek and Krueger, Chapter 12; Kelley, Chapter 19).

We now turn to specific concerns that arise when mixing methods, drawing on examples from our own research projects, as well as those of others, to illustrate how these issues may play out in the research process.

Tensions in resolution, scale, and areal or temporal extent of data

Scale, resolution, and spatial or temporal extent are defining characteristics of geographic research, and the methodological tensions they generate get a good deal of attention in geographic literature. This attention, however, does not indicate that these issues are easily settled. On the contrary, the complexity inherent in bringing together multiple scales, resolutions, and/or extents continues to present challenges as researchers debate not only how, but if, scale and concepts of hierarchy in scale are appropriate (Margulies et al. 2016). Despite ongoing debates, researchers commonly conduct multi- and cross-scale studies aimed at understanding the dynamics of social-environmental systems by integrating data of varying spatiotemporal scales, resolutions, and extents (e.g., Tian et al. 2015; Walker et al. 2018). This integration means that researchers are working with collections of observations that cover different units of analysis, geographic areas, and time periods.

Data with varying resolutions can result from a single approach or because of the use of different methods. Remote sensing (Braun, Chapter 39), for example, can yield measurements ranging from a single tree (high-resolution) to a landscape (moderate- or low-resolution). High spatial resolution data (such as drone imagery (Kasvi, Chapter 45)) are frequently associated with reduced spatial and temporal extents and decreased temporal frequency, while low-resolution imagery tends to have broader spatial coverage, longer temporal extents, and more frequent repeat observation (see Braun, Chapter 39 and Kasvi, Chapter 45). For example, the Advanced Very High Resolution Radiometer (AVHRR) offers multiple scans of the Earth per day with global coverage of Earth observations for more than 40 years, but the spatial resolution of the data is actually quite coarse. Because of these tradeoffs, remote sensing analyses regularly integrate moderate-resolution datasets. These offer landscape-scale observations over extended periods of time (~50 years) that are commonly interpreted in relation to dominant drivers of landscape change (e.g., climate patterns) or can be linked to social processes that may have lagged impacts (e.g., shifts in policy). To interpret and augment large-scale remote sensing analyses and draw causal inferences, researchers may rely on downscaling methods or turn to methods such as case studies, household surveys, archival analysis, interviews, and oral histories (see Cope, Chapter 22; Lane, Chapter 24; Johnston and Longhurst, Chapter 32; Chakov et al., Chapter 33; Winata and McLafferty, Chapter 43). Often (though not always) these methods produce higher-resolution observations across a relatively small areal extent (Magliocca et al. 2018). For some data, it may be challenging to attempt to integrate or reduce rich contextual knowledge into a specific state of time or into the Euclidean space of latitude and longitude coordinates (Preston and Wilson 2014; Margulies et al. 2016).

Our own research illuminates some of the challenges and tensions that arise from using data with different scales, resolutions, and extents. In our work, we have utilised dendrochronology (tree-ring dating and analysis) alongside remote sensing to study the response of species and landscapes to large-scale processes such as climate shifts. One benefit of this integrative approach is that dendrochronology offers a long time series of annually resolved data (i.e., long temporal extent and high frequency of observation) (Rozendaal et al. 2010), while remote sensing often yields more spatially extensive data. For example, in Southworth et al. (2013), colleagues and I (Cerian Gibbes) used tree-ring analysis and remotely sensed measures of net primary production (NPP) to examine how declining precipitation has affected vegetation in Botswana's Okavango Delta. A pattern of decline was identified in both annual tree growth and NPP in association with decreased precipitation across much of the study region (with exceptions found in the driest areas and in the parts of the study region characterised by wetland ecosystems). When guided by triangulation, remote sensing and dendrochronology cross-check one another, corroborating a third position: a relationship between precipitation and vegetation. However, when guided by a hybrid approach that pursues both validation and contradiction, consistencies and inconsistencies, we came to identify multiple, not necessarily parallel, stories of how vegetation is impacted across ecozones. We suggest that the inconsistencies (in this case in the patterns found across the study region) that become apparent when mixed methods are applied offer opportunities for reflection which can then direct future research questions.

As the mixing of methods becomes more ambitious and wider ranging, the probability of generating multiple parallel accounts increases. In a study of climate and landscape change within the savannas of southern Africa, colleagues and I (Cerian Gibbes) analysed historical climate data, remote sensing data, key informant interviews, and environmental histories (Gibbes et al. 2013, 2014). These methods resulted in datasets that spanned different time periods and geographic extents and differed also in their representations of the connectivity of the landscape. For example, an interpolated representation of rainfall from the 1950s to the present yielded a seemingly complete and continuous representation of climate across a significant portion of a continent (Gibbes et al. 2014). In contrast, environmental histories constructed with community members produced understandings of rainfall that focused on notable time periods and often varied in space and scale. Individuals' experiences and recollections of rainfall relied on multiple temporal scales marked in individuals' minds by important personal events and linked to their mobility within the study area. Thus, the ways in which rainfall was presented through different methods criss-crossed spatial and temporal extents. The question then became how to represent this knowledge and understand these complicated presentations of rainfall without reinforcing existing scalar hierarchies. A mixed-methods approach may identify a kaleidoscope of patterns and processes impossible to observe with the use of any one method, but how do we justly represent the range of patterns that emerge from mixed methods without segregating the results into individual stories?

Representing the different geographic extents and landscape connectivity present in these varied forms of data proved challenging and conflicted with our initially envisioned research objective to develop a *single* understanding of climate and landscape change. Instead, the mixed-methods approach revealed multiple rainfall realities associated with differing perspectives on rainfall, consequences of rainfall, and ways of measuring rainfall. To begin to address this challenge, the research process itself needs adequate space and time to adapt and to adjust the initial research objective—in other words, to allow the subject of the research to "speak back" to "engender that slightly different understanding of the world around us, one that makes the curious practice of science so creative and exciting" (see Lane and Lave, Chapter 3; Lane 2017: 99). This, however, can be especially challenging if research aims to be directly usable for policymaking and decision-making, as we will touch on in the next section.

Tensions between different values, epistemologies, and histories build into methods

When mixing methods, it is crucial to address the values, epistemologies, worldviews, and histories associated with or informing particular methods or data sources. In addition, it is also important to consider how a method or data source is perceived or valued by different entities or institutions. Understanding what is accepted as evidence and how this acceptance differs within and across institutions is necessary and may serve to mitigate the overuse of one methodological approach. Quantitative analytical methods provide standardised ways to represent complex phenomena and are frequently associated with objectivity. Other methods (e.g., qualitative methods such as ethnography; see Sayre, Chapter 34) are more commonly associated with subjective experiences and may be less privileged than those typically associated with 'truth' and objectivity. This perceived objectivity of quantitative approaches is often a false one, however. Quantification has become particularly attractive not because it is inherently more rigorous but rather, Porter (1995) argues, because it has allowed some actors (e.g., business and government) to make decisions while being—or appearing to be—shielded from external pressures or politics, thereby diffusing accountability for the outcomes of decisions.

Intentional use of mixed methods offers space to question the assumptions embedded within different methods, such as the idea that quantitative methods are more objective. Associations among methods, values, and worldviews are not fixed, and in fact are shaped by the ways in which methods are applied and how results are used. Methodological approaches can be re-envisioned in ways that interrupt previous associations, directly addressing and confronting the conceptual hegemony of certain methods and ways of knowing. For example, critical, humanistic, and feminist GIS scholars have examined how GIS, commonly critiqued as a positivist mode of knowledge production, can be re-imagined to create new GIS practices (Wilson 2017; Zhao 2022). In this case, the method itself is used to challenge its privileged position.

The design and conceptualisation of research generally determine the selection of methods, yet some methods are more readily implemented within certain conceptual frameworks (and frames; see Lane and Lave,

Chapter 3). Conceptual approaches that highlight nuanced multiple realities (including both the complementary and hybrid-stye approaches discussed above) are better supported by methods that permit the 'seeing' of multiple realities. In addition, research that assumes a shared experience might be more likely to utilise a method that foregrounds the common and not the individual (e.g., focus groups; see Longhurst and Johnston, Chapter 27). In contrast, when research participants are framed from the outset as empowered contributors to the research process (e.g., as in participatory action research or other community-based research frameworks), methods that highlight individual voices, insights, and experiences and empower participants to direct what is included in the "data" are more readily applied (e.g., as in photovoice) (Gibbes and Skop 2022).

Yet, the ease and apparent logic of connection between conceptualisation and method can result in the default use of a particular method. This overdependence on a single method or suite of methods may ultimately obscure relevant processes or details. For example, remote sensing and other quantitative approaches are commonly used to measure the extent of forest cover, and data from these methods are highly valued in environmental management and policy making. However, such data provide a partial perspective of complex, multi-scalar deforestation and reforestation processes. If forest extent as measured by remote sensing is the only method used to understand deforestation and reforestation processes, there is the possibility that complex processes will be overlooked (Blaikie 1985, Davis 2018). For example, Fairhead and Leach (2000) discuss how quantification of deforestation was used in international settings to determine conservation policy in West Africa. Such measurements neglected to consider locally specific historical descriptions of forest cover. As a result, the "baseline" reference for extent of forest cover was overstated, and thus the amount of deforestation was exaggerated. In a similar example from Nepal, Nightingale (2003) used oral histories (Chakov et al., Chapter 33), participant observation (Sayre, Chapter 34), interviews (Johnston and Longhurst, Chapter 32), aerial photo interpretation, and quantitative vegetation inventory to assess and understand forest change. Putting these methods directly in conversation—and in tension—with one another revealed "the importance of challenging 'dominant' representations of forest change—in this case aerial photo interpretation—not by rejecting them outright, but by demonstrating explicitly how they provide only one part of the story of forest change" (Nightingale 2003: 80).

We have argued that it is crucial to consider the contradictions, silences, and gaps among insights yielded by multiple methods and data sources. It is also important to consider broader differences between epistemologies. The concept of Two-Eyed Seeing (or *Etuaptmumk* in Mi'kmaw), and its use as a research framework, provides an example. Two-Eyed Seeing is a process of learning "to see from one eye with the strengths of Indigenous knowledges and ways of knowing, and from the other eye with the strengths of mainstream knowledges and ways of knowing, and to use both of these eyes together, for the benefit of all" (Bartlett et al. 2012: 335). This framework has been used in academic research on long-term change in delta ecosystems (Abu et al. 2019), fisheries management (Reid et al. 2021), health and medicine (Forbes et al. 2020), and wildlife conservation (Rayne et al. 2020). On the one hand, some consider Two-Eyed Seeing as a process of triangulation or validation, where Indigenous knowledge and Western scientific data are integrated with an eye toward corroboration (Kutz and Tomaselli 2019). On the other hand, Two-Eyed Seeing has also been interpreted as a process of bringing multiple perspectives together toward knowledge coexistence and complementarity, rather than integration (which risks assimilation) (Reid et al. 2021).

Even beyond bringing multiple perspectives together, the centring of different epistemologies offers a model for co-production of entirely new understandings of the world. Co-production refers to a collaborative process of knowledge production that is compositional in nature and present throughout the entirety of the research process (Klenk et al. 2017). For example, in a study of flood risk science, Lane et al. (2011) described the production of novel solutions when "certified" (academicians/scientists) and "non-certified" (local people impacted by floods) experts were brought together not only to address flooding but also to practice interdisciplinary public science, in which both academics and the public work together to co-produce knowledge. Regardless of the approach, however, there is tension between what Broadhead and Howard (2021: 111) refer to as "the desire to... generate a trans-cultural 'third space' of understanding... and the denial or suppression of major contradictions between predominantly holistic Indigenous and predominantly reductionist Eurocentric worldviews". It is these types of concerns that, we believe, cannot be avoided and thus should not be ignored when performing research that engages multiple methods or ways of knowing.

Tensions between methods that yield information about different research objects

Bringing methods into tension with each other also includes considering how different methods yield information about different research objects. This consideration does not necessarily facilitate triangulation but rather is central to some complementary or hybrid approaches. The goal at the outset is not for methods to corroborate one another but instead for each method to inform data collection and interpretation using another method. Our own research provides an example here.

I (Christine Biermann) initially set out to investigate how tree growth is affected by climate using dendrochronological methods. Specifically, I was interested in if and how relationships between climate and tree growth change over time. Do trees change what climatic factors they are responding to and the strength of their responses? And if so, does this hinder scientists' ability to reconstruct past climate using tree-rings as a proxy measure? In the process of developing and investigating these questions, I came to recognise that my formulation of these questions was shaped by social and political dynamics in the science of dendroclimatology and the broader political economy of climate science.

This led me to new questions that foregrounded the relationship between our collective knowledge of tree growth-climate relationships and the social dynamics of science. How are tree-ring scientists addressing and interpreting potential changes in growth responses over time, and what social factors are shaping their practices? To pursue these two distinct lines of inquiry—one focusing on tree growth-climate interactions and the other on science as a social practice—I relied on different methods. First, I performed dendrochronology and statistical analysis of trees' growth responses to climate. Second, I surveyed tree-ring scientists about their views, experiences, methodological approaches, and perceptions of stability or instability of trees' growth responses to climate over time.

As I tacked back and forth between methods and datasets, each came to shape how I approached the other. Right away this raised tensions and questions. Would my analysis of tree growth patterns be influenced by my survey findings? Would certain interpretations of the datasets rise to the surface over others? Working through these questions, I came to realise that the answer to both was a vehement yes. Survey data (both qualitative and quantitative) about scientific practices informed my interpretation of tree growth patterns, highlighted shortcomings of particular methods, and even brought forward unseen facets (e.g., politics) that influenced my own choice of dendrochronological methods and thereby my findings. Analysis of tree-growth patterns, on the other hand, allowed me to experience through my own scientific practice how different analytical starting points can lead to vastly different research findings and interpretations (Biermann 2018).

When these two distinct research objects were pursued in tandem, a third object emerged: partiality and the situatedness of knowledge. As I played each dataset off of the other, I came to see how each approach represented a particular way of seeing and understanding the world. I began to focus on the silences and partial truths of each method. What was ignored, concealed, or unaccounted for? On the one hand, having multiple research objects, and correspondingly using multiple research methods, provided a more holistic understanding of the relationships among trees, climate, and scientific methods. On the other hand, the outputs generated by this research were nuanced but untidy, deeply contextual but arguably lacking in clear conclusions.

We see similar issues arise in research on coupled human-environment systems, where distinct methods are often used to produce information about different parts of a system under study. Methods inform each other, and quantitative and qualitative approaches are commonly used alongside each other (Chiang et al. 2012). For example, household surveys can be used to inform modelling efforts. The concept of a coupled human-environment system emphasises connections and flows across human and environmental components of the system. Such systems are complex, so it is imperative that methods and the data they generate be in dialogue with each other.

This purposeful use of methods in tension with one another can make otherwise invisible relationships evident. As Elwood (2003: 96) states, "qualitative analysis of interviews may potentially illuminate meanings, relationships, and interactions not made visible through quantitative analysis of survey data. Alternatively, quantitative analysis of survey data might reveal patterns helpful in examining broader social relationships." However, Chiang et al. (2012) caution against the haphazard inclusion of too great a diversity of methods as this can lead to research participant fatigue and may introduce uncertainty. This caution highlights the need for mixed-methods approaches to be used in a deliberate manner, requiring researchers to consider how our understandings of research objects are informed by the method and how the methods will intermingle within the research process and setting (see Lane and Lave, Chapter 3; Lave and Lane, Chapter 21). A deliberate mixing of methods, however, should still be fluid. As we undertake a project and our knowledge of a situation develops and changes, our methods, research questions, and frameworks should evolve as well. As Lane and Lave (Chapter 3) discuss, this leads to a more reflexive research process in which our methods not only challenge our understandings of the world, but our research findings also challenge how we do research.

Conclusion

Mixed methods serve various aims. One of these aims is validation of results, as in a triangulation-focused approach. There are other purposes, however, that are less widely recognised. First, as stated at the outset, mixed methods help us to engage with the complexity present in the interconnected biophysical and social worlds. More specifically, various methods can allow us to observe and conceptualise complexity in different ways. Seeking out both overlaps and discrepancies across different data sources or methods allows us to work against the fragmented knowledge that our disciplinary silos create and uphold. Second, mixed methods can help us to consider the politics of science and knowledge production, illustrating how social, cultural, and political economic relations affect our research frameworks, questions, and findings. Third, we can use mixed methods to attend to the material impacts and political consequences of our research. For example, mixing methods can help to challenge the hegemony of particular methods or data sources and bring new forms of knowledge, as well as knowledge producers, into policy and decision-making processes.

Challenges will inevitably arise as we employ mixed methods toward these aims. While some mixed method approaches seek to measure the exact same phenomenon, others interrogate a shared topic through multiple research objects, epistemologies, or scales. Rather than attempting to minimise or ignore discrepancies among methods or data sources, we advocate for recognition and exploration of tension. Holding methods and their results in tension can build a richer understanding of the subject of study and allow us to reflect on how we design and conduct research. As Fuller (2008) indicates, the question of "how to do" [science] is as important as what we do in science. Mixing methods, and attending to the many issues that emerge in the process, can shift the focus of research away from a single outcome—a map, a model, an answer to a problem—and toward the process of research itself (Preston and Wilson 2014).

Recasting tension as opportunity does not suggest that mixing methods comes without risk or challenge. One challenge can be seen as researchers begin to intermingle biophysical and social framings of phenomena and thus need to venture into the realms of new (to them) methods. As Martin (2020: 13) cautions,

> when [biophysical science] researchers do not have adequate training, knowledge, and experience, their social scientific studies are often poorly designed, neglect vast bodies of social scientific knowledge, and are full of methodological flaws. Ultimately these problems may lead to misinterpretation of the results and unsubstantiated conclusions.

Similar issues may arise as social scientists wade into biophysical methods or data sources. We therefore urge researchers to develop wide-ranging, diverse interdisciplinary collaborations that challenge each member to expand their concept of data, knowledge, and method.

Accepting and even inviting frictions among different methods, worldviews, epistemologies, research objects, and scales can advance knowledge about the hybrid worlds we inhabit, even as it poses new risks and challenges. The questions society needs to address are increasingly difficult and the potential that mixed method approaches hold is too great to avoid on account of risk or difficulty. It is essential, however, that existing research systems—funders, reviewers, institutions—are fundamentally altered to reduce the risk to individual researchers and encourage genuine collaborative, reflexive, and iterative inquiry into the gaps, silences, and frictions generated by mixed-methods approaches. In other words, attention to tension should not be an afterthought, considered only when the results are being written up and reflected upon, but should be a central part of every stage of mixed method research.

References cited

Abu, R., M.G. Reed, and T.D. Jardine. 2019. 'Using two-eyed seeing to bridge Western science and Indigenous knowledge systems and understand long-term change in the Saskatchewan River Delta, Canada', *International Journal of Water Resources Development*, 36.2, pp. 1–20, https://doi.org/10.1080/0790 0627.2018.1558050

Bartlett, C., M. Marshall, and A. Marshall. 2012. 'Two-eyed seeing and other lessons learned within a co-learning journey of bringing together indigenous and mainstream knowledges and ways of knowing', *Journal of Environmental Studies and Sciences*, 2.4, pp. 331–340, https://doi.org/10.1007/s13412-012-0086-8

Biermann, C. 2018. 'Shifting climate sensitivities, shifting paradigms: Tree-ring science in a dynamic world', in *The Palgrave Handbook of Critical Physical Geography*, ed. by R. Lave, C. Biermann, and S.N. Lane (Palgrave Macmillan), pp. 201–225, https://doi.org/10.1007/978-3-319-71461-5_10

Biermann, C., L.C. Kelley, and R. Lave. 2020. 'Putting the Anthropocene into practice: Methodological implications', *Annals of the American Association of Geographers*, 111.3, pp. 808–818, https://doi.org/10.1080/24694452.2020.183 5456

Blaikie, P. 2016. *The Political Economy of Soil Erosion in Developing Countries* (Routledge).

Braun, A., Chapter 39, this volume. '(Critical) Satellite remote sensing'.

Broadhead, L.A. and S. Howard. 2021. 'Confronting the contradictions between Western and Indigenous science: a critical perspective on Two-Eyed Seeing', *AlterNative: An International Journal of Indigenous Peoples*, 17.1, pp. 111–119, https://doi.org/10.1177/1177180121996326

Bruno, T. 2022. 'Ecological memory in the biophysical afterlife of slavery', *Annals of the American Association of Geographers*, 113, pp. 1–11, https://doi.org/10.1080/24694452.2022.2107985

Chakov, A., Chang, T., Covey, H., Dickson, T., Goggins, S., Harris, N., Purna, S., Widell, S., and Druschke, C.G., Chapter 33, this volume. 'Oral history'.

Cheong, S.M., D.G. Brown, K. Kok, and D. Lopez-Carr. 2012. 'Mixed methods in land change research: towards integration', *Transactions of the Institute of British Geographers*, 37.1, pp. 8–12, https://doi.org/10.1111/j.1475-5661.2011.00482.x

Colucci, A.R., J.A. Tyner, M. Munro-Stasiuk, S. Rice, S. Kimsroy, C. Chhay, and C. Coakley. 2021. 'Critical physical geography and the study of genocide: lessons from cambodia', *Transactions of the Institute of British Geographers*, 46.3, pp. 780–793, https://doi.org/10.1111/tran.12451

Cope, M. and S. Elwood. 2009. *Qualitative GIS: A Mixed Methods Approach* (Sage Publications), https://doi.org/10.4135/9781412991476

Davis, D.K. 2018. 'Between sand and sea: constructing Mediterranean plant ecology', in *The Palgrave Handbook of Critical Physical Geography*, ed. by R. Lave, C. Biermann, and S.N. Lane (Palgrave Macmillan), pp. 129–151, https://doi.org/10.1007/978-3-319-71461-5_7

Denzin, N. 1989. *The Research Act*, 3rd ed. (Prentice Hall), https://doi.org/10.4324/9781315134543

Ellingson, L.L. 2009. *Engaging Crystallization in Qualitative Research: An Introduction* (Sage Publications), https://doi.org/10.4135/9781412991476

Elwood, S. (2010). 'Mixed methods: thinking, doing, and asking in multiple ways', in *The Sage Handbook of Qualitative Geography*, ed. by D. DeLyser, S. Herbert, S. Aitken, M. Crang, and L. McDowell (Sage Publications), pp. 94–113, https://doi.org/10.4135/9780857021090.n7

Forbes, A., S. Ritchie, J. Walker, and N. Young. 2020. 'Applications of Two-Eyed seeing in primary research focused on Indigenous health: a scoping review', *International Journal of Qualitative Methods*, 19, https://www.doi.org/10.1177/1609406920929110

Freeman, C. 2020. 'Multiple methods beyond triangulation: collage as a methodological framework in geography', *Geografiska Annaler: Series B, Human Geography*, 102.4, pp. 328–340, https://doi.org/10.1080/04353684.2020.1807383

Fuller, D. 2008. 'Public geographies: Taking stock', *Progress in Human Geography*, 32.6, pp. 834–44, https://doi.org/10.1177/0309132507086884

Gibbes, C., L. Cassidy, J. Hartter, and J. Southworth. 2013. 'The monitoring of land-cover change and management across gradient landscapes in Africa', in *Human-Environment Interactions*, ed. by E. Brondízio and E. Moran (Springer), https://doi.org/10.1007/978-94-007-4780-7_8

Gibbes, C., J. Southworth, P. Waylen, and B. Child. 2014. 'Climate variability as a dominant driver of post-disturbance savanna dynamics', *Applied Geography*, 53, pp. 389–401, https://doi.org/10.1016/j.apgeog.2014.06.024

Gibbes, C., and E. Skop. 2022. 'Disruption, discovery, and field courses: a case study of student engagement during a global pandemic', *The Professional Geographer*, 74.1, pp. 31–40, https://doi.org/10.1080/00330124.2021.1970593

Greer, K., A. Csank, K. Calvert, M. Maddison-MacFadyen, A. Smith, K. Monk, and S. Morrison. 2023. 'Understanding the historic legacies of empire from the timbers left behind: Towards critical dendroprovenancing in the British North Atlantic', *The Canadian Geographer/Le Géographe canadien*, 67, pp. 124–138, https://doi.org/10.1111/cag.12831

Haraway, D. 1988. 'Situated knowledges: the science question in feminism and the privilege of partial perspective', *Feminist Studies*, 14.3, pp. 575–599, https://doi.org/10.2307/3178066

Kasvi, E., Chapter 45, this volume. 'Uncrewed Airborne Systems'.

King, L. and M. Tadaki. 2018. 'A framework for understanding the politics of science (Core Tenet# 2)', in *The Palgrave Handbook of Critical Physical Geography*, ed. by R. Lave, C. Biermann, and S.N. Lane (Palgrave Macmillan), pp. 67–88, https://doi.org/10.1007/978-3-319-71461-5_4

Klenk, N., A. Fiume, K. Meehan, and C. Gibbes. 2017. 'Local knowledge in climate adaptation research: moving knowledge frameworks from extraction to co-production', *Wiley Interdisciplinary Reviews: Climate Change*, 8.5, pp. 475, https://doi.org/10.1002/wcc.475

Kutz, S. and M. Tomaselli. 2019. '"Two-eyed seeing" supports wildlife health', *Science*, 364.6446, pp. 1135–1137, https://doi.org/10.1126/science.aau6170

Lane, S.N. 2017. 'Slow science, the geographical expedition, and critical physical geography', *The Canadian Geographer/Le Géographe canadien*, 61.1, pp. 84–101, https://doi.org/10.1111/cag.12329

Lane, S.N. and Lave, R., Chapter 3, this volume. 'Frames, disciplines, and mixing methods in environmental research'.

Lave, R., M. Doyle, M. Robertson, and J. Singh. 2018. 'Commodifying streams: a critical physical geography approach to stream mitigation banking in the USA', in *The Palgrave Handbook of Critical Physical Geography*, ed. by R. Lave, C. Biermann, and S.N. Lane (Palgrave Macmillan), pp. 443–463, https://doi.org/10.1007/978-3-319-71461-5_21

Leach, M. and J. Fairhead. 2000. 'Challenging neo-Malthusian deforestation analyses in West Africa's dynamic forest landscapes', *Population and Development Review*, 26.1, pp. 17–43, https://doi.org/10.1111/j.1728-4457.2000.00017.x

Longhurst, R. and Johnston, L., Chapter 27, this volume. 'Focus groups'.

Johnston and Longhurst, Chapter 32, this volume. 'Interviews: Structured, semi-structured and open-ended'.

Luthra, A., K. Cunningham, A.M. Fraser, A. Pandey, S. Rana, and V. Singh. 2022. 'Ecological livelihoods of farmers and pollinators in the Himalayas:

Doing critical physical geography using citizen science', *The Canadian Geographer/Le Géographe canadien*, 67, pp. 35–51, https://doi.org/10.1111/cag.12799

Magliocca, N.R., E.C. Ellis, G.R Allington, A. de Bremond, J. Dell'Angelo, O. Mertz, and P.H. Verburg. 2018. 'Closing global knowledge gaps: producing generalized knowledge from case studies of social-ecological systems', *Global Environmental Change*, 50, pp. 1–14, https://doi.org/10.1016/j.gloenvcha.2018.03.003

Malone, M. and N. McClintock. 2022. 'A critical physical geography of no-till agriculture: Linking degraded environmental quality to conservation policies in an Oregon watershed', *The Canadian Geographer/Le Géographe canadien*, 67, pp. 74–91, https://doi.org/10.1111/cag.12789

Margulies, J.D., N.R. Magliocca, M.D. Schmill, and E.C. Ellis. 2016. 'Ambiguous geographies: connecting case study knowledge with global change science', *Annals of the American Association of Geographers*, 106.3, pp. 572–596, https://doi.org/10.1080/24694452.2016.1142857

Martin, V. 2020. 'Four common problems in environmental social research undertaken by natural scientists', *Bioscience*, 70.1, pp. 13–16, https://doi.org/10.1093/biosci/biz128

Nightingale, A. 2003. 'A feminist in the forest: situated knowledges and mixing methods in natural resource management', *ACME: An International Journal for Critical Geographies*, 2.1, pp. 77–90, https://doi.org/10.14288/acme.v2i1.709

Nightingale, A.J. 2009. 'Methods: Triangulation', in *International Encyclopedia of Human Geography*, ed. by R. Kitchen and N. Thrift (Elsevier), pp. 489–492, https://doi.org/10.1016/B978-0-08-102295-5.10437-8

Nightingale, A.J. 2016. 'Adaptive scholarship and situated knowledges? Hybrid methodologies and plural epistemologies in climate change adaptation research', *Area*, 48.1, pp. 41–47, https://doi.org/10.1111/area.12195

Porter, T.M. 1995. *Trust in Numbers* (Princeton University Press), https://doi.org/10.1515/9781400821617

Rayne, A., G. Byrnes, L. Collier-Robinson, J. Hollows, A. McIntosh, M. Ramsden, M. Rupene, P. Tamati-Elliffe, C. Thoms, and T.E. Steeves. 2020. 'Reimagining conservation translocations through two-eyed seeing', *People and Nature*, 2.3, pp. 512–526, https://doi.org/10.1002/pan3.10126

Reid, A.J., L.E. Eckert, J.F. Lane, N. Young, S.G. Hinch, C.T. Darimont, S.J. Cooke, N.C. Ban, and A. Marshall. 2021. '"Two-Eyed Seeing": An Indigenous framework to transform fisheries research and management', *Fish and Fisheries*, 22.2, pp. 243–261, https://doi.org/10.1111/faf.12516

Rozendaal, D.M.A. and P.A. Zuidema. 2010. 'Dendroecology in the tropics: A review', *Trees*, 25, pp. 3–16, https://doi.org/10.1007/s00468-010-0480-3

Sayre, N. F., Chapter 34, this volume. 'Participant observation and ethnography'.

Southworth, J., L. Rigg, C. Gibbes, P. Waylen, L. Zhu, S. McCarragher, and L. Cassidy. 2013. 'Integrating dendrochronology, climate and satellite remote sensing to better understand savanna landscape dynamics in the Okavango Delta, Botswana', *Land*, 2.4, pp. 637–655, https://doi.org/10.3390/land2040637

Tian, Q., D.G. Brown, L. Zheng, S. Qi, Y. Liu, and L. Jiang. 2015. 'The role of cross-scale social and environmental contexts in household-level land-use decisions, Poyang Lake Region, China', *Annals of the Association of American Geographers*, 105.6, pp. 1240–1259, https://doi.org/10.1080/00045608.2015.1060921

Walker, X.J., B.M. Rogers, J.L. Baltzer, S.G. Cumming, N.J. Day, S.J. Goetz, J.F. Johnstone, E. Schurr, M.R. Turetsky, and M.C. Mack. 2018. 'Cross-scale controls on carbon emissions from boreal forest megafires', *Global Change Biology*, 24.9, pp. 4251–4265, https://doi.org/10.1111/gcb.14287

Wilson, M.W. 2017. *New Lines: Critical GIS and the Trouble of the Map* (University of Minnesota Press), https://doi.org/10.5749/j.ctt1pwt6q4

Zhao, B. 2022. 'Humanistic GIS: Toward a research agenda', *Annals of the American Association of Geographers*, 112.6, pp. 1576–1592, https://doi.org/10.1080/24694452.2021.2004875

5. Expanding research ethics for inclusive and transdisciplinary research

Alison M. Meadow, Hailey Wilmer, and Daniel B. Ferguson

Introduction

The complexity of 21st-century environmental challenges has accelerated as a result of climate change, biodiversity loss, and the escalation of resource extraction and pollution as most of the world has industrialised. The wicked problems (Whyte and Thompson 2012) that arise as a result of these trends have driven many toward research approaches that reach beyond standard Western science to include multiple knowledges, multiple disciplines, and multiple methods of collecting and interpreting data. This mode of research is referred to by several different names, including collaborative research, community-based participatory research (Mokos, Chapter 36), and mixed-methods research. We use the term transdisciplinary (Td) research as an overarching term to capture these different research traditions. We have selected Td research as our model because this mode of research explicitly includes multiple research disciplines, practical experiences, cultures, and epistemologies in order to match the complexity of the problem with the diverse knowledge systems better suited to address them. Td is "based on a democratic scientific practice of engaging multiple knowledges, through equitable and mutually respectful partnership for action, in the design and implementation of research objectives, questions, methods, and desired outputs or outcomes" (Wilmer et al. 2021: 454).

The practice of engaged research approaches, like Td, helps to highlight the weaknesses within our standard research ethics frameworks, which were built on an assumption of a one-way flow of knowledge from science

    https://doi.org/10.11647/OBP.0418.05

to society. The intent of Td research approaches to integrate diverse knowledge systems pushes Western research into challenging ethical territory where there are few (if any) clear delineations between research subjects, participants, collaborators, and knowledge producers.

In the U.S., the guiding principles governing research that involves humans are the Belmont Principles, which were codified into law in the Federal Policy for the Protection of Human Subjects (aka The Common Rule) in 1991 (45 C.F.R. 46). The Common Rule requires researchers to gain informed consent from research participants, reveal any potential risks and benefits to them, protect their privacy and confidentiality, and ensure that participants are selected fairly and equitably. In the context of Td and other forms of engaged research, these traditional human subjects protocols are necessary for thinking about how we work with communities, but they are insufficient for work with non-academic, societal partners[1] as well as how we think about the ethics of working with non-human systems. We join researchers from several disciplines (Britton and Johnson 2023; Brydon-Miller 2008; Campbell and Goundwater-Smith 2007; Chief, Meadow, and Whyte 2016; McGregor 2013; Mikesell, Bromley, and Khodyakov 2013) in noting the ways in which our current research ethics frameworks fail to account for:

1. The inclusion of people outside the standard research enterprise as participants and collaborators, not just research subjects.

2. Research efforts that include people but do not meet federal definitions of "human subjects research" and are therefore not reviewed by ethics review boards.

3. Culturally diverse contexts in which knowledge and consent to participate in research are communal actions, not just individual decisions.

4. Research efforts that take an active role in providing immediate benefits to specific communities.

---

1   Throughout this chapter, we refer interchangeably to non-academic partners, community partners, and societal partners, which are all common terms for participants in collaborative, transdisciplinary research whose experience and expertise comes from outside of traditional academic (or Western science) contexts.

Given these limitations and building on the framework presented by Wilmer et al. (2021), we offer a set of reflective questions intended to guide Td research planning and team conversations toward a more expansive view of research ethics that accounts for the challenges and promises of engaged research. While our focus is explicitly on Td, much of what we describe below is relevant to interdisciplinary science or more broadly collaborative research that seeks to bring together multiple ways of knowing to address a complex problem (see for example Britton and Johnson 2023). We seek to cite and acknowledge innovative ideas that have emerged from diverse and sometimes very specific fields that can help researchers improve and decolonise research methods with communities. Our framework, and the guiding questions and actions below, are all premised on a collaborative research team's willingness to engage in thoughtful and iterative self-reflection.

This chapter is structured with subsections focused on four interrelated but distinct themes—representation, self-determination, deference, and reciprocity—and two cross-cutting themes, ethics beyond human dimensions and research skills as ethical practice. Under each subsection we offer a series of questions and actions research teams can take to apply the framework within their projects.

Representation

A central component of research is that we share our findings with others in order to share knowledge. The ways in which we portray the people, communities, and places that have been part of our research —the way we represent them to others who do not know them—can have a lasting effect on those people, communities, and places. One example of the impact of misrepresentation is the ways in which European and American historians exoticised or "othered" the peoples and history of the West Asian world (Said 1979). In contemporary literature, "damage-centred" research can lead to a community being defined by its worst experiences without recognition of its strengths and capacities (Britton and Johnson 2023; Tuck 2009). The portrayals of people and environments that emerge from research can color the ways in which they are viewed for many years. Therefore, we define the practice of ethical representation as "mindfulness of how we represent other people and communities as well as non-human elements of our research" (Wilmer et al. 2021: 454).

How do the people with whom you are working identify and represent themselves, their lands, and their environments?

Researchers can begin the process of ethical representation by using local names, terms, and descriptors throughout the research process (including in publications). We can also consider how we explain or describe current conditions and contexts. An example from climate change research has been the labelling of frontline communities as "vulnerable" or "at-risk," implying a static condition, while ignoring the policies, actions, and ideas that create dynamic conditions of vulnerability, precarity, and dependent relationships (Ranganathan and Bratman 2021; Simon and Dooling 2016). To remedy this form of misrepresentation, researchers can acknowledge the roots of contemporary conditions in their portrayals, such as the role that policies like "red-lining" have played in creating unhealthy conditions in low-income neighborhoods today (Bullard 2000; Hoffman, Shandas, and Pendleton 2020) and ensure that community strengths are recognised along with descriptions of adverse conditions (Ranganathan and Bratman 2021).

In biophysical contexts, misrepresentation might appear as reference to a place or landscape as "pristine" or "uninhabited", which might erase Indigenous communities from the landscape, with implications for land tenure policies, Indigenous sovereignty, and biodiversity outcomes (Norgaard 2014). Similarly, representing a landscape as "degraded" can affect how it is managed or whether it is protected in the future. For example, the common view that dryland rangelands or desert ecosystems are "wastelands" mischaracterises these social-ecological systems, which cover a vast area of the Earth's terrestrial surface and serve a vital role in Earth systems processes and livelihoods for diverse cultures (Hoover et al. 2020). Inaccurate or incomplete views of dryland systems that attribute conditions to, for example, overgrazing can prop up misunderstandings about local management and pastoral economies, and contribute to issues of sedentarisation, reduced flexibility and land access. Alternatively, science and local-knowledge informed understanding of land use-ecological feedbacks holds promise to more fully and accurately understand and manage these dynamics (Briske et al. 2020; Coppock 2016).

Researchers can ensure that societal partners have opportunities to validate descriptions of themselves, their community, and their landscapes throughout the research project and prior to publishing or sharing findings outside the community. This is part of ensuring that engagement and consent to be represented in a research project is an ongoing, rather than discrete, process.

Are all members of your research team prepared and committed to engagement with community or non-academic partners?

A simple, but often overlooked, aspect of ethical representation is the act of building relationships and trust between researchers and societal partners (Tachera 2021). Setting the expectation that all members of a research team will spend time with project collaborators is a way to help ensure that research will reflect the humanity of participants. Creating space and time within the research project for social interaction and informal sharing of knowledge and experiences is important to developing trust and increasing learning (Stern, Briske, and Meadow 2021). This could be as simple as building in time to socialise, particularly by sharing meals, throughout the research process and can extend to activities like supporting community events and gatherings as a way to demonstrate respect, build trust, and increase learning between groups. Embedding in different professional roles or organisations is another effective method for encouraging trust-building and social learning (Stern, Briske, and Meadow 2021). Making an effort to learn another person's job or responsibilities provides an invaluable perspective on their knowledge and expertise. It is important that research teams consciously plan and appropriately resource these activities so that all participants can meet the expectation of engaging with and learning from each other (Ferguson, Meadow, and Huntington 2022).

What are the risks to partners from participating in the research?

How a people or a landscape is represented in research can pose direct risks to communities and their environments. For example, while identifying sacred or culturally significant sites can be part of protecting them from development or other land use changes or may be part of other research activities, making those locations known through publications or data sharing opens the site up to damage and exploitation (Plaut 2009; UCLA School of Law Tribal Legal Development Clinic 2020). Similarly, sharing protected knowledge, such as sacred stories or other knowledge considered to be appropriate only for community members or particular contexts, harms communities by making public knowledge that is not intended to be. If there is a possibility that protected knowledge might emerge during the research process, co-developing a plan with partners for how such knowledge will be handled is one straightforward path to avoiding eventual conflict and harm (see Self-Determination, below). Where they exist, researchers should also follow the guidelines for federal agencies regarding protection of sensitive sites (Plaut 2009).

Research can also pose interpersonal risks to community participants. If research is conducted unethically or the outcomes are poor for the community, such as through misrepresentation of the community or the extraction of community knowledge for the researchers' gain, those community members who invited researchers into a community or worked directly with them during the research process may have their own reputations and standing within the community severely damaged. It is incumbent upon a researcher or a research team to recognise that collaboration may put some partners in challenging social or political situations and work closely with them to mitigate these risks.

Risk perception is highly subjective and, when it comes to representation and sharing of information, researchers' perceptions might be very different from those of their research partners (Campbell 2017). However, because "as researchers, we have the privilege of entering and exiting our partners' world at will" (Wilmer et al. 2021: 458), we should be prepared to defer to their perceptions of risk.

Self-determination

Effective Td research can enhance and support community self-determination. In the context of Td research ethics, we consider self-determination through the lens of community or collective choice. When conducting research in partnership with a community—in particular with sovereign Indigenous communities—they have the right to: "(1) decide whether and how research is conducted within their territories and involves their citizens; (2) require collective consent... instead of just individual consent to participate in research activities; and (3) control data about the community, community knowledge, and the community's ecological relationship with its territory" (Wilmer et al. 2021: 459). When working with Indigenous communities, the right to self-determination and sovereignty is the controlling principle and it is spelled out in the US context in federal policy and internationally in the UN Declaration on the Rights of Indigenous Peoples (Wiessner 2008). Researchers working with Indigenous communities and/or on Indigenous lands *must* seek consent from the authorities within the community who govern research activities (often a research review board) and follow the research protocols of the community. While, as with all communities, there are complex relationships between and among community members and governing authorities, as outside researchers, it is our responsibility to follow current guidelines and protocols articulated by these communities regarding data collection, management, and sharing. If you are working with a community that has not established research protocols, you can work with community partners to initiate conversations about community values related to self-determination. The bedrock principle of self-determination in this context is: a community has a fundamental right to collaborate on their own terms and for their own benefit.

What role might communal knowledge (e.g., traditional and local knowledge) play in your project?

If your work stands to benefit from engaging with the collective knowledge and wisdom of a community, the CARE (Collective Benefit, Authority to Control, Responsibility, and Ethics) principles (Carroll et al. 2020) provide a structure to guide your team's decision making (see Table 5.1). The CARE principles are specifically rooted in the movement for Indigenous Data Sovereignty (Walter et al. 2021) and are designed to "address historical inequities by creating value from Indigenous data in ways that are grounded in Indigenous worldviews and by realising opportunities for Indigenous Peoples within the knowledge economy" (Carroll et al. 2020: 8). Therefore, if you intend to work with Indigenous peoples, CARE provides a critical framework for ensuring that your project supports and empowers those communities in ways that are appropriate to them. If you intend to engage with collective knowledge that is not rooted in an Indigenous community, the CARE principles still provide a solid foundation on which to build your collaboration because they focus on respect for community knowledge and equitable collaboration.

Table 5.1 CARE Principles for Indigenous Data Governance
(Carroll et al. 2020)

Principle	*Supporting concept*
Collective Benefit	For inclusive development and innovation
	For improved governance and citizen engagement
	For equitable outcomes
Authority to control	Recognising rights and interests
	Data for governance
	Governance of data
Responsibility	For positive relationships
	For expanding capability and capacity
	For Indigenous languages and worldviews
Ethics	For minimising harm and maximising benefit
	For justice
	For future use

Who is consenting to the terms of your project?

Seeking appropriate consent for collaborative research that involves community knowledge is more complex than the standard informed consent required in human subjects research. If your work involves a sovereign Indigenous community, there will often be a research review protocol managed by the community government that you must follow in addition to the standard review required by universities or funders.[2] In most cases the terms of these protocols are intended to ensure that the community benefits from your research, so it is critical to work with your partners to fully understand what the community seeks, how the work supports their interests, and how you will be able to follow through with commitments made to the community.

When working with other communities, particularly minoritised or frontline communities, research teams can address community participation and benefits by working with one of the growing number of community-based research review boards (Britton and Johnson 2023). These non-profit or volunteer organisations exist to help protect the interests of frequently researched communities (see for example del Campo et al. 2013). While not every community has created such a research review process, research teams can develop the practice of seeking out community review boards or requesting review through specialised IRB processes designed for community-based research, when available.

How will data and information be managed?

Respecting the collective knowledge and contributions of community partners also requires intentional dialogue and careful planning for how data and information is to be used, managed, and shared (see Authority to Control in the CARE principles). The ethical treatment of community data includes actions such as facilitated conversations about: data management and sharing; how to handle Indigenous knowledge and other forms of collective knowledge; expectations about use of data and information in any outputs (e.g., manuscripts, reports, data sets); and funding requirements for data sharing, management and storage. For example, having trained community members lead interviews and subsequent data management processes when local or Indigenous knowledge is likely to arise can help protect and respect that knowledge (Murveit et al. 2023).

---

2   Several examples of Indigenous community research protocols can be found on the website of the Native Peoples Technical Assistance Office at the University of Arizona https://naair.arizona.edu/az-tribal-research-policies

Td projects can produce a substantial amount of both qualitative and quantitative data. It is critical that community contributors are provided ample and ongoing opportunities to assert their rights of self-determination to ensure that data and information are treated according to their wishes. At times, the responsibility to respect self-determination and protect community knowledge may bump up against the researcher's commitments to funders or their own academic inquiries. However, the desire to conduct research cannot overwhelm the rights of participants and communities to control information about their lives and experiences (Hudson et al. 2020). Researchers may need to agree not to publish certain information or to obtain consent before doing so. Seeking alternative ways to generate research outputs—such as through reports directly to the community rather than publications (see Reciprocity, below)—can be one way to prioritise the rights of research participants. Patton has summarised this principle well: "A researcher's scientific observation is some person's [or community's] real-life experience. Respect for the latter must precede respect for the former" (Patton 2015: 243).

Reciprocity

Too often, academic research has overemphasised the production of new knowledge as its primary goal, without regard to how that focus might affect other participants involved in the research. Many researchers have started to reflect on the ethics of using others' life experiences solely as data for research, even when the research is intended to broadly benefit society—or using others' life experiences to benefit their own careers through publications (Campbell 2017). To move away from extractive research models and toward problem solving, knowledge sharing, and positive societal and environmental impacts, it is important to integrate the concept of reciprocity into our research processes. Reciprocity is the principle that participants should receive direct benefit from the research to which they contribute. In Td research, reciprocity often takes the form of tangible benefits to participating communities or may even extend to conducting research in a way that alters the distribution of resources, power, and opportunities in the pursuit of greater social equity. Applying the ethical principle of reciprocity means committing to research outputs and outcomes that meet societal partners' needs—as defined and expressed by those partners (Wilmer et al. 2021).

What do your partners say they would like to happen because of this research? Is your project scoped to address those expectations?

Enacting reciprocity requires that research teams plan and budget for project outputs and outcomes that meet societal partners' expectations (Ferguson, Meadow, and Huntington 2022) such as data or reports aimed at a public audience or otherwise created in ways that meets community needs; publication of community language texts; or providing research assistance beyond the scope of the immediate project (Lomawaima 2000; Virapongse et al. 2022).

Ensuring reciprocity in a research effort requires additional planning so that the team can follow-through on their commitments to their partners. Based on partners' goals for project outputs and outcomes, evaluate your team's skills and consider ways to build training into your project to build capacity to meet your partners' goals and needs. For example, team leaders can support cultural competency training for team members prior to starting community-based work or support team members to develop the skills to present research findings in accessible formats over the course of the project. Additionally, teams can add members, as necessary, to ensure key competencies are included such as hiring new members with local knowledge or language skills or with the research communication skills required to ensure the findings are maximally useful to community partners (Djenontin and Meadow 2018).

What will you, as the researcher, get from this research?

It is important to be transparent with yourself and with your partners about what you stand to gain from completing a research project. For many academic researchers, the gain comes in the form of academic advancement, often through the vehicle of peer-reviewed publications. This can lead to the assumption that publications have the same value to everyone involved in a research project. While including societal partners as authors in research publications is a critical aspect of ethical representation, consider how to incorporate additional forms of reciprocity in a project as well because often the benefits of publications are not equal for academics (large benefit) and societal partners (possibly no direct benefit).

How are research funds being distributed in this project? Who is receiving those funds?

In addition to budgeting and planning to produce outputs for the benefit of societal partners, research teams can also examine their overall budgets for opportunities to budget-share with societal partner organisations and individual participants. Examples of financial reciprocity might include compensating participants for their time and expertise at meetings, workshops, or as interviewees. To budget for such compensation, consider consulting estimates of the value of volunteer hours in your communities.[3] Community partners can also be hired as part of research teams. It is important to explore any institutional hurdles to this process in the proposal stage, such as the logistics of making payments to people outside of the research institution or whether community partners are able to accept checks, so you have time to overcome them before launching the project.

Deference

Transdisciplinary research requires cooperation and collaboration among people who come from different knowledge traditions, each of which offers unique and valuable insights. As a result, we consider deference to these different epistemic communities to be a core principle for carrying out successful Td research. Deference in this context means "awareness [of] and respect for expertise, methods, and different epistemologies...among your academic colleagues and societal partners" (Wilmer et al. 2021: 460). Deference does not mean subsuming your own way of thinking or developing knowledge to others, but rather being cognizant of the practical reality that all knowledge communities have specific ways of developing and validating knowledge. Successful and

---

3   A current estimate from The Independent Sector, a national organisation of non-profit entities, estimates the value of volunteer hours in our community at $29.95 per hour. https://independentsector.org/resource/value-of-volunteer-time/

ethical Td research requires a measure of self-awareness and humility that extends beyond courtesy for other opinions to genuine recognition that there are ways to develop valid knowledge that go beyond your own specific training and experience.

In a recent essay, Krlev and Spicer (2022) develop an analogous idea—"epistemic respect"—which involves making a conscious effort to understand and appreciate [a] claim, even though one might disagree or have difficulties relating to it. "Epistemic respect entails (1) paying due attention, (2) valuing a knowledge claim and (3) behaving in a thoughtful way towards it" (Krlev and Spicer 2022: 4). In the research process, failing to recognise different ways of understanding or failing to value different experiences and expertise can mean missing key pieces of information or knowledge or closing off fruitful lines of inquiry or pathways to solutions. This closing off of lines of inquiry and practice is contrary to the principles of Td research, in which the goal is to learn from each other and co-develop knowledge together (Moon et al. 2021). The rationale for collective learning and respect for divergent ways of knowing is rooted in the reality that problems that require a Td approach are definitionally too complex to be addressed by strictly disciplinary or other more homogenous ways of thinking. Scholars have long argued that communities of practice and social learning systems (see for example, Pahl-Wostl et al. 2007; Wenger 2000) are fundamentally necessary to advance knowledge development in highly complex environments that require iterative approaches and constant innovation. An implicit characteristic of these approaches is the need for collaborators to be capable of situating their knowledge within a broader cognitive ecosystem so that the collective knowledge is greater than the sum of its parts.

What assumptions about knowledge development and validation do you and the other members of your team hold?

Much of the success of Western science has grown from focused disciplinary expertise and the depth of knowledge that results from it. Unfortunately, the focus on deep disciplinary training can result in a lack of awareness among academic researchers about the range of other epistemic traditions. The willingness to openly discuss the limits of our methods may be difficult, but it is critical if Td knowledge development is the goal. Facilitating conversations about how you and your academic and non-academic partners think about knowledge development and validation can be a useful step towards understanding potential pathways that allow for reconciling different ways of thinking. Engaging a team in practical discussions about data gathering approaches, analytical techniques, and assumptions can reveal hidden biases (e.g., an implicit privileging of quantitative over qualitative approaches) (Eigenbrode et al. 2007). For projects that intend to rely on local, traditional, and/or Indigenous knowledge it is especially important to provide a collaborative environment that allows everyone to explain how they know what they know and how that type of knowledge might contribute to the collective goals of the team.

How do you hope to bring different ways of knowing together in your work?

Integration of knowledge is often considered the most challenging aspect of Td research, but "successful team integration is not necessarily measured by the degree of fusion" of knowledge (Wilmer et al. 2021: 461; see also Biermann and Gibbes, Chapter 4). The parallel knowledge development that is the hallmark of multi-disciplinary research (Tress, Tress, and Fry 2005) can be easy to fall back on unless there is a focused effort by team leaders and collaborators to consistently articulate how the various pieces of the project puzzle fit together. Developing this synoptic view of a Td project relies on a level of deference to ways of thinking that may be foreign to members of the team, but open dialogue about how everyone thinks about the problem and the goals can present opportunities for finding connections (see Johnson et al., Chapter 7). Eigenbode et al. (2007) provide a "toolbox" for cross-philosophical dialogue in collaborative teams, an approach which can be employed or modified to improve team communication with one another and with non-academic partners. Collaboratively developing research questions, co-designing research methods, and co-developing outputs all provide specific opportunities for team members to discuss their knowledge traditions and learn from each other (see for example the process of creating boundary objects in

Star and Griesemer 1989). Using opportunities for co-learning and co-design throughout a project—from the design stage through the production of final outputs—can ensure that different knowledge traditions are ingrained in the project.

Is your team engaging with intellectual traditions across and beyond disciplines?

Meaningful engagement with the intellectual traditions and texts from different fields and knowledge communities will enrich research by introducing diverse perspectives and potentially hidden ways of engaging a problem. The inverse is also true: heavy reliance on a single or perhaps a few intellectual traditions may unnecessarily limit the scope and impact of a Td project. For example, a project that is led by biophysical scientists may rely on the framings and intellectual approaches from those fields even when human systems may be central to the work. More careful integration of social science traditions in this kind of work creates opportunities for using methods and theories that can ultimately lead to greater depth of understanding for the entire team.

Recent research suggests that, while academic science continues to suffer from a persistence of disciplinary and ethnic homophily, greater scientific impact is associated with increased diversity in collaborations (AlShebli, Rahwan, and Woon 2018). Recognising and then acting to overcome our tendencies to associate with others who think like us and come from similar backgrounds requires a measure of epistemic deference that can result in both better answers and more compelling collective learning by everyone involved. An aspect of this is consciously seeking out and lifting the voices of thinkers from diverse backgrounds who are commonly either under-cited or left entirely out of academic literature even when their ideas are salient (Helmer et al. 2017; Whyte 2017). Part of an ethical research practice can be seeking out and incorporating diverse voices and perspectives starting from the literature review stage of research projects (Kwon 2022) .[4]

---

4   The Just Environments Lab makes available a searchable database of Black, Indigenous, and/or Latinx scholars working in the fields of environmental and climate justice. https://www.just-environments.org/the-syllabus

Beyond human dimensions

Relationships among communities and the ecosystems in which we live are contextual, dynamic, and socially constructed. The majority of the expanded ethical framework discussed in this chapter deals with partnerships with people, however, Td research very often includes natural resource and ecological fields. The cross-cutting theme of beyond human dimensions prompts us to consider research ethics more broadly as part of our relations and responsibility to the biophysical environment and the living beings, beyond humans, within it. Ethical practice may include processes for complying with animal care, environmental regulations, or other relevant policy. In Td approaches, various members of the research team may have different levels of experience at the intersection of human and non-human work. Furthermore, Western epistemologies may not value beyond-human relations in a way that prioritises their well-being and reciprocity in daily life (Whyte 2018), our economies, or our research. Here we elucidate how the four principles discussed above might surface in the context of relationships and research with the non-human elements of our world.

What aspects of ethical representation apply to beyond-human actors in our research and the associated human communities?

As discussed in the section on Representation, the ways in which researchers portray landscapes and ecosystems can have implications for their conservation and management. Td approaches that focus on authentic and meaningful interaction among humans and the environment can provide opportunities to develop place-based knowledge of landscapes and knowledge of how communities (human and non-human) interact with, depend upon, and steward relationships and well-being. For example, on public rangeland landscapes in the US West, researchers, ranchers, conservation advocates, and public agency employees involved in the Collaborative Adaptive Rangeland Management (CARM) project spent time together walking, talking, observing, and experimenting on a specific piece of shortgrass prairie. The team learned to set management objectives for that land in a more place-based and scientifically informed way over time (Wilmer et al.

2019). Such a process of relationship building can help participants consider the representation of non-human actors in research, and to inspire new questions that enhance ethical engagement with biodiversity. In CARM, these questions included: "How can our project incorporate goals for multiple species and ecological outcomes?; What does it mean to quantify or to value conservation outcomes of a management approach or experimental treatment for individual birds or a species?; How do we balance the goal of learning about the ecosystem with the goal of managing the ecosystem sustainably in an experimental setting?" This experience ultimately improved the CARM team's knowledge and ability to manage for grassland bird conservation outcomes (Davis et al. 2021). The Td collaboration within CARM also helped participants see that speaking about the environment involves speaking about someone's environment and therefore about that someone, too, whether that someone is a grassland bird or the rancher who manages the plover's habitat.

Considering the concepts of self-determination and reciprocity, how does the research interact with non-human communities?

When planning a research project, it is important to assess whether the process of learning or experimentation will have a detrimental effect on the system and whether there are ways to enhance and conserve well-being through the study design (see Lane, Chapter 8). For example, will the impacts of an ecological experiment or monitoring project require restoration? How will experimental infrastructure impact wildlife? How can stress, pain and harm to individual animals be minimised? Who will be responsible for removing or maintaining infrastructure once the project has concluded? In the US, policies, guidelines, and trainings for ethical research involving animal subjects are well established and widely available to researchers (see: National Institutes of Health 2015; National Research Council 2011). The impact of research on the environment may be subject to additional regulation and policy depending on the location. Additionally, researchers can consider how sharing information about a particular place or landscape could lead to unsustainable stress on an ecosystem, excessive harvesting of wildlife species, or habitat alteration. Maintaining confidentiality in the interest of protecting the place in question is part of maximising benefits and acting with reciprocity in mind.

How can deference to and respect for local relationships with the natural world and knowledge be genuinely incorporated into the project?

Working beyond human dimensions creates challenges for Td researchers from many backgrounds. Researchers from social science and humanities backgrounds entering Td collaborations may be encountering the ethics of working with animals and environments for the first time and may need additional training. In contrast, Td research team members from ecology, conservation, and other biophysical fields that operate from more positivist, technocratic, or biodiversity-centred traditions may be encountering the complexities of working with human communities on wildlife or ecosystem issues for the first time. During Td research, we are likely to encounter situations where some partners have very different views of landscapes, wildlife, or land use goals. These diverse views can be challenging—and can also be an opportunity for learning for all participants. As discussed in the Representation section, creating time and space for all project participants to interact with and learn from each other can help the process of finding common ground and common interests. The point of this engagement is not to determine if local partners agree with researchers on ecosystem or management goals, or if they hold the same values, but to better understand the biophysical environment and its inhabitants. For example, when researchers presented findings about the positive effects of prescribed fire on shortgrass prairie forage quality to local ranchers in a co-produced grazing experiment, ranchers originally rejected the idea of applying more fire to the landscape, a goal of conservation groups who wanted to promote habitat for a specific grassland bird (Wilmer et al. 2018). But by listening to the ranchers' point of view, researchers began to understand that ranchers perceived more risk than benefit to the use of fire because of their experiences with high levels of forage variability. This moment of social learning paved the way for new strategies that accommodated ranchers' risk thresholds while promoting the use of fire in ways that were acceptable to conservation and agricultural partners in the project.

Skills as ethical practice

Another cross-cutting theme includes skill development among team members. This includes skills in research and team processes. As discussed in the Deference section, Td research requires respect for different ways of constructing knowledge—including how data is collected and analysed, (i.e., research skills). For researchers new to working with people, it is critically important to assess the team's ability to interact, collect data, and use that data with respect for all participants—human and non-human (see for example: Martin 2020). Building strong Td teams also requires management skills that support a long-term commitment to fostering pluralistic growth among the team members. Even when the research seems urgent and highly relevant, it can take years of preparation and stewardship to secure institutional support, develop community trust and collaborative engagement, and initiate the project with a well-rounded team. Once teams are put together, the work of building capacity and relationships continues as members learn new skills, methods, and ways of engaging with the problem.

Does your team include specialists in social science research and community engagement?

Just as an animal scientist completes animal care trainings and learns to handle the sheep she studies with low-stress techniques, so too do Td researchers need to develop skills for socially engaged research practice. High levels of emotional intelligence, research creativity, and cultural competency within a Td research team can help steward successful projects. In cases where we work directly with people and ask them to share their lived experiences and knowledge with us, perhaps through interviews or focus groups (see Longhurst and Johnston, Chapter 27; Johnston and Longhurst, Chapter 32), it is important that research teams have appropriate methodological skills to promote effective partnerships, manage and support social learning, and avoid emotional or psychological harm to interviewees. Harm can stem from asking overly personal or culturally inappropriate questions or framing inquiries in ways that imply disrespect (Shore 2006). When we ask people to share their personal experiences and knowledge, it is our ethical responsibility to collect the data well and interpret it accurately. Some common errors in qualitative data collection include using leading questions, passing personal judgment, or asking questions that are not relevant to the research. As Campbell (2017: 92) notes, "Methodological rigor ensures a high-quality product that is worthy of the time given by . . . individuals and groups." If a Td team does not include people with such skills, we put our societal partners at risk, as well as our reputations as ethical research partners.

What additional certifications and competencies are necessary for success within the project?

Research teams can ensure that they budget time and resources to ensure training in human subjects, animal care, or other backgrounds in research ethics. Examples of existing resources include the Collaborative Institutional Training Initiative (CITI Program) courses;[5] in particular the Native American Research and Community Engaged Research modules. Teams can also find and participate in professional development opportunities or graduate courses on engaged scholarship available through professional or scholarly organisations.[6] Furthermore, there may be specific training in methodological (e.g., non-positivist, qualitative research approaches), methods-focused (interview skills), or leadership (team conflict resolution) skills that can be refreshed or improved within the team.

Where can the team develop and improve interpersonal, leadership, and management skills?

Intentional skill building and application of reflexive, listening, emotional intelligence, and empathy skills complement more commonly addressed team leadership, conflict resolution, and time management skills (Ulibarri et al. 2019). Researchers can explicitly engage in training

---

5   https://about.citiprogram.org/
6   One example is a professional development webinar series on actionable science from the Northwest Climate Adaptation Science Center: https://nwcasc.uw.edu/resources/actionable-science-webinars/

as individuals or groups to enhance team cohesion and respect and can use ongoing team reflections to help check in on skill development and emergent challenges. Lab or team handbooks can be developed to help codify and transfer values, approaches, and context to team members over time (CLEAR 2021). The Center for Advancing Research Impacts in Society (ARIS)[7] curates free resources related to community engaged research, broader and societal impacts of research, science communication, and broadening participation in science.

How will your project be managed?

The leadership of transdisciplinary research teams may be distinctly less hierarchical and more complicated than in other organisations as various generations, institutions, roles, and emergent challenges come into play over time. Furthermore, this sort of work can sometimes put large amounts of work or risk on early career researchers and community members who may be the least experienced in dealing with research ethics and institutional structures. Likewise, project care work can disproportionately be shouldered by some in the group. Consider developing team-specific management plans and approaches that anticipate the complexity and ethical risks of transdisciplinary activities and allocate them equitably, with attention to career stage and intersectional burdens of care.

Conclusion

As our research modes change to better address the complex social and environmental challenges we currently face, we also need to adjust our research ethics. Td research—in which researchers and societal partners work collaboratively and equitably to develop knowledge and solve problems—represents a relationship between researchers and participants that is not reflected in our current ethical frameworks. In this chapter, we have proposed that researchers adopt an expanded set of ethical principles more attuned to our new modes of research. We suggest that researchers consider how our Representation of people and

---

7   https://researchinsociety.org/resources/

environments affects how those communities are seen by the outside world and that we use collaboration tools to both understand the risks to people and places subject to misrepresentation and mitigate those risks by ensuring that depictions reflect local understandings.

While standard ethical protocols require informed consent by participants, we note that this standard is insufficient when working with communal knowledge and denies communities the right to exercise Self-determination in the context of research projects. Researchers can incorporate the CARE principles into their work—always when working with Indigenous communities and as an excellent set of guidelines when working with any societal partners.

Td and other forms of collaborative research often share a principle that research should focus on tangible problem-solving. We take this principle a step farther and recommend incorporating the principle of Reciprocity into research efforts. This would ensure that participants (individuals or communities) receive some direct benefit from the research in exchange for sharing their knowledge, experiences, and expertise.

Td research requires ongoing collaboration between people from different knowledge traditions. Our proposed principle of Deference asks research teams to interrogate their own knowledge traditions as part of a process of learning from and respecting other traditions.

In addition to the four individual principles, we suggest that researchers consider two cross-cutting themes. First, these principles can and do apply even when working beyond human dimensions because landscapes, animals, and their relationships with human communities are all part of Td research. Second, we consider strengthening our research skills, particularly when incorporating unfamiliar research traditions into our work, part of ethical practice because we owe our partners the highest standard of care when collecting and analysing the data, information, and knowledge they bring to research projects.

The complex and multi-faceted problems facing our communities and environment today require modes of research that embrace multiple perspectives, knowledges, and forms of expertise. Using mixed methods, collaborative, and transdisciplinary research approaches to their fullest potential requires a recalibration of many of our research structures and infrastructures, including what should be the foundation of our research—our ethical principles. We have proposed a set of expanded ethical principles that address the complex relationships that are common in transdisciplinary research. By treating our research partners and places of research ethically and with respect we can contribute to robust and enduring benefits for our environments, communities, and planet.

References cited

The Protection of Human Subjects. *45 C.F.R. 46.*

AlShebli, B.K., T. Rahwan, and W.L. Woon. 2018. 'The preeminence of ethnic diversity in scientific collaboration', *Nature Communications*, 9.1, pp. 5163, https://www.doi.org/10.1038/s41467-018-07634-8

Briske, D.D., D.L. Coppock, A.W. Illius, and S.D. Fuhlendorf. 2020. 'Strategies for global rangeland stewardship: Assessment through the lens of the equilibrium–non-equilibrium debate', *Journal of Applied Ecology*, 57.6, pp. 1056–1067.https://doi.org/10.1111/1365-2664.13610

Britton, J. and H. Johnson. 2023. 'Community autonomy and place-based environmental research: recognizing and reducing risks', *Metropolitan Universities*, 34.2, pp. 118–137, https://www.doi.org/10.18060/26440

Brydon-Miller, M. 2008. 'Ethics and action research: deepening our commitment to principles of social justice and redefining systems of democratic practice', in *The Sage Handbook of Action Research*, ed. by P. Reason and H. Bradbury (Sage Publications), pp. 199–210.

Bullard, R. 2000. *Dumping in Dixie* (Routledge).

Campbell, A. and S. Goundwater-Smith. 2007. 'Introduction', in *An Ethical Approach to Practitioner Research: Dealing with Issues and Dilemmas in Action Research*, ed. by A. Campbell and S. Groundwater-Smith (Routledge), pp. 1–7.

Campbell, S.P. 2017. 'Ethics of research in conflict environments', *Journal of Global Security Studies*, 2.1, pp. 89–101. https://doi.org/10.1093/jogss/ogw024

Carroll, S., I. Garba, O. Figueroa-Rodríguez, J. Holbrook, R. Lovett, S. Materechera, M. Parsons, K. Raseroka, D. Rodriguez-Lonebear, and R. Rowe. 2020. 'The CARE principles for Indigenous data governance', *Data Science Journal*, 19, p. 43. https://doi.org/10.5334/dsj-2020-043

Chief, K., A. Meadow, and K. Whyte. 2016. 'Engaging Southwestern tribes in sustainable water resources topics and management', *Water*, 8.8, p. 350. https://doi.org/10.3390/w8080350

CLEAR. 2021. *CLEAR Lab Book: A Living Manual of Our Values, Guidelines, and Protocols*, https://civiclaboratory.nl/clear-lab-book

Coppock, D.L. 2016. 'Cast off the shackles of academia! Use participatory approaches to tackle real-world problems with underserved populations', *Rangelands*, 38.1, pp. 5–13. https://doi.org/10.1016/j.rala.2015.11.005

Davis, K.P., D.J. Augustine, A.P. Monroe, and C.L. Aldridge. 2021. 'Vegetation characteristics and precipitation jointly influence grassland bird abundance beyond the effects of grazing management', *The Condor*, 123.4, pp. 1–15. https://doi.org/10.1093/ornithapp/duab041

del Campo, F.M., J. Casado, P. Spencer, and H. Strelnick. 2013. 'The development of the Bronx Community Research Review Board: a pilot feasibility project for a model of community consultation', *Progress in Community Health Partnerships*, 7.3, pp. 341.

Djenontin, I.N.S. and A.M. Meadow. 2018. 'The art of co-production of knowledge in environmental sciences and management: Lessons from international practice', *Environmental Management*, 61.6, pp. 885–903.

Eigenbrode, S.D., M. O'rourke, J. Wulfhorst, D.M. Althoff, C.S. Goldberg, K. Merrill, W. Morse, M. Nielsen-Pincus, J. Stephens, and L. Winowiecki. 2007. 'Employing philosophical dialogue in collaborative science', *BioScience*, 57.1, pp. 55–64.

Ferguson, D.B., A.M. Meadow, and H.P. Huntington. 2022. 'Making a difference: planning for engaged participation in environmental research', *Environmental Management*, 69.2, pp. 227–243.

Helmer, M., M. Schottdorf, A. Neef, and D. Battaglia. 2017. 'Gender bias in scholarly peer review', *eLife*, 6.

Hoffman, J.S., V. Shandas, and N. Pendleton. 2020. 'The effects of historical housing policies on resident exposure to intra-urban heat: a study of 108 US urban areas', *Climate*, 8.1, p. 12.

Hoover, D.L., B. Bestelmeyer, N.B. Grimm, T.E. Huxman, S.C. Reed, O. Sala, T.R. Seastedt, H. Wilmer, and S. Ferrenberg. 2020. 'Traversing the wasteland: a framework for assessing ecological threats to drylands', *BioScience*, 70.1, pp. 35–47.

Hudson, M., N.A. Garrison, R. Sterling, N.R. Caron, K. Fox, J. Yracheta, J.Anderson, P. Wilcox, L. Arbour, and A. Brown. 2020. 'Rights, interests and expectations: Indigenous perspectives on unrestricted access to genomic data', *Nature Reviews Genetics*, 21.6, pp. 377–384.

Krlev, G. and A. Spicer. 2022. 'Reining in reviewer two: how to uphold epistemic respect in academia', *Journal of Management Studies*, 60.6, pp. 1624–1632.

Kwon, D. 2022. 'The rise of citational justice: how scholars are making references fairer', *Nature*, 603.7902, pp. 568–571.

Lane, S.N., Chapter 8, this volume. 'The environmental impacts of fieldwork: making an environmental impact statement'.

Lomawaima, K.T. 2000. 'Tribal sovereigns: Reframing research in American Indian education', *Harvard Educational Review*, 70.1, pp. 1–23.

Martin, V.Y. 2020. 'Four common problems in environmental social research undertaken by natural scientists', *BioScience*, 70.1, pp. 13–16.

McGregor, H.E. 2013. 'Situating Nunavut education with indigenous education in Canada', *Canadian Journal of Education*, 36.2, pp. 87–118.

Mikesell, L., E. Bromley, and D. Khodyakov. 2013. 'Ethical community-engaged research: a literature review', *American Journal of Public Health*, 103.12, pp. 7–14.

Mokos, J., Chapter 36, this volume. 'Participatory methods'.

Moon, K., C. Cvitanovic, D.A. Blackman, I.R. Scales, and N.K. Browne. 2021. 'Five questions to understand epistemology and its influence on integrative marine research', *Frontiers in Marine Science*, 8, pp. 1–9.

Murveit, A.M., S. Delphin, C. Domingues, S.D. Bourque, S.D. Faulstich, G. Garfin, N. Huntly, A.M. Meadow, and V. Preston. 2023. 'Stories as data: Indigenous research sovereignty and the "Intentional Fire" podcast', *Environment and Planning F*, 2.1–2, pp. 180–202.

National Institutes of Health. 2015. *PHS Policy on Humane Care and Use of Laboratory Animals* (U.S. Department of Health and Human Services).

National Research Council. 2011. *Guide for the Care and Use of Laboratory Animals: Eighth Edition* (The National Academies Press).

Norgaard, K.M. 2014. 'The politics of fire and the social impacts of fire exclusion on the Klamath', *Humboldt Journal of Social Relations*, 36, pp. 77–101.

Pahl-Wostl, C., M. Craps, A. Dewulf, E. Mostert, D. Tabara, and T. Taillieu. 2007. 'Social learning and water resources management', *Ecology and Society*, 12.2.

Patton, M.Q. 2015. *Qualitative Research and Evaluation Methods* (Sage Publications).

Plaut, E. 2009. 'Tribal-agency confidentiality: a Catch-22 for sacred site management?', *Ecology Law Quarterly*, 36.1, pp. 137–166.

Ranganathan, M., and E. Bratman. 2021. 'From urban resilience to abolitionist climate justice in Washington, DC', *Antipode*, 53.1, pp. 115–137.

Said, E.W. 1979. *Orientalism* (Vintage).

Shore, N. 2006. 'Re-conceptualizing the Belmont Report: A community-based participatory research perspective', *Journal of Community Practice*, 14.4, pp. 5–26.

Simon, G. and S. Dooling. 2016. *Cities, Nature and Development: The Politics and Production of Urban Vulnerabilities* (Routledge).

Star, S.L. and J.R. Griesemer. 1989. 'Institutional ecology, 'translations' and boundary objects: amateurs and professionals in Berkeley's Museum of Verebrate Zoology', *Social Studies of Science*, 19.3, pp. 387–420.

Stern, M. J., D.D. Briske, and A.M. Meadow. 2021. 'Opening learning spaces to create actionable knowledge for conservation', *Conservation Science and Practice*, 3.5, p. e378.

Tachera, D. 2021. 'Reframing funding strategies to build reciprocity', *Eos*, 102.

Tress, G., B. Tress, and G. Fry. 2005. 'Clarifying integrative research concepts in landscape ecology', *Landscape Ecology*, 20.4, pp. 479–493.

Tuck, E. 2009. 'Suspending damage: A letter to communities', *Harvard Educational Review*, 79.3, pp. 409–427.

UCLA School of Law Tribal Legal Development Clinic. (2020). *The Need for Confidentiality Within Tribal Cultural Resource Protection*, https://law.ucla. edu/sites/default/files/PDFs/Native_Nations/239747_UCLA_Law_ publications_Confidentiality_R2_042021.pdf

Ulibarri, N., A.E. Cravens, A.S. Nabergoj, and A. Royalty. 2019. *Creativity in Research* (Cambridge University Press).

Virapongse, A., R. Gupta, Z.J. Robbins, J. Blythe, R.E. Duerr, and C. Gregg. 2022. 'How can earth scientists contribute to community resilience? Challenges and recommendations [policy and practice reviews]', *Frontiers in Climate*, 4, pp. 1–18.

Walter, M., T. Kukutai, S.R. Carroll, and D. Rodriguez-Lonebear. 2021. *Indigenous Data Sovereignty and Policy* (Taylor and Francis).

Wenger, E. 2000. 'Communities of practice and social learning systems', *Organization*, 7.2, pp. 225–246.

Whyte, K.P. 2018. 'Critical investigations of resilience: A brief introduction to indigenous environmental studies and sciences', *Daedalus*, 147.2, pp. 136–147.

Whyte, K.P. 2017. 'Systematic discrimination in peer review: Some reflections', *Daily Nous*, May, 7.

Whyte, K.P. and P.B. Thompson. 2012. 'Ideas for how to take wicked problems seriously', *Journal of Agricultural and Environmental Ethics*, 25, pp. 441–445.

Wiessner, S. 2008. 'Indigenous sovereignty: A reassessment in light of the UN declaration on the rights of Indigenous People', *Vand. J. Transnat'l L.*, 41, pp. 1141.

Wilmer, H., J.D. Derner, M.E. Fernández-Giménez, D.D. Briske, D.J. Augustine, and L.M. Porensky. 2018. 'Collaborative adaptive rangeland management fosters management-science partnerships', *Rangeland Ecology and Management*, 71.5, pp. 646–657.

Wilmer, H., A.M. Meadow, A.B. Brymer, S.R. Carroll, D.B. Ferguson, I. Garba, C. Greene, G. Owen, and D.E. Peck. 2021. 'Expanded ethical principles for research partnership and transdisciplinary natural resource management science', *Environmental Management*, 68.4, pp. 453–467.

Wilmer, H., L.M. Porensky, M.E. Fernández-Giménez, J.D. Derner, D.J. Augustine, J.P. Ritten, and D.P. Peck2019. 'Community-engaged research builds a nature-culture of hope on North American Great Plains rangelands', *Social Sciences*, 8.1, pp. 1–26.

6. Embracing and enacting critical and constructive approaches to teaching Critical Physical Geography

Jennifer Salmond and Gary Brierley

Introduction

As we become more embedded in the Anthropocene it makes increasingly less sense to reinforce traditional nature-culture divides by teaching students how to measure, observe, and understand the eco-social worlds in which they live using techniques that are grounded solely in either the social or biophysical sciences. Such an approach is ill-suited to support efforts to describe, to explain and to predict drivers of environmental change. It fails to recognise contested values in untangling 'natural' (more-than-human) processes from anthropogenically generated or altered processes in a co-created world. Beyond this, such framings and assertions are non-sensical to many Indigenous (including Māori) students who conceptualise humans as part of nature, not separate from it, creating an ontological disconnect between humans and their 'more-than-human' surroundings (Salmond 2014; 2017).

This presents a challenge for educators who, given the current structural organisation of knowledge in universities, are required to teach either specialist social-cultural research skills or biophysical science skills. Familiarity with a plethora of specialist skills is required to interrogate and to interpret the deluge of data now available. However, siloed instruction provides little guidance for students as to how to combine methods effectively or how to compare and value the disparate forms of evidence and knowledge generated. This leaves

 https://doi.org/10.11647/OBP.0418.06

students ill-prepared for careers in environmental decision-making that require an ability to assess, to compare and to value information that is often limited and disparate, and which derives from a range of different sources, in order to make and justify choices.

Quite simply, graduates who have sole knowledge of either only social or only biophysical sciences are ill-equipped to 'do science' in the real world. New approaches to teaching environmental issues are required to develop skills to address wicked problems (Sharp et al. 2021) in an ethical, appropriately contextualised manner (Malone 2021). For example, Critical Physical Geography (CPG) has advocated bringing together work in the social and biophysical sciences with the aim of supporting eco-social transformation (Lave et al. 2018). Indeed, it provides a unique framework to address this educational gap (Gillett et al. 2018; Malone 2021).

This chapter draws on the theory of CPG, and the mixing of methods it encourages, to demonstrate how a third (final) year mixed-methods undergraduate Physical Geography course can be co-created with students to provide a pedagogically sound, student-focused, enriching educational experience that is rewarding for both students and educators. Three CPG tenets adapted and re-worded from Lave et al. (2018) underpin the educational framework, knowledge content, teaching methods, and learning outcomes used in our course:

- Environments are simultaneously physically and socially created, imagined and understood

- Scientific research is not neutral, objective or universal and requires a reflexive engagement with method and outcome to provide insight into the role of power and politics in determining:

  o how and why theories are developed; which last and which dominate,

  o why some methods and data are prioritised and privileged over others, and

  o the consequences of a) and b) for our understanding, prediction and management of the worlds of which we are a part.

     o   Recognition that how we measure and interpret our worlds
has profound impacts on how values and understandings
are constructed and the kinds of policies and interventions
that follow.

In this chapter we outline the theoretical and pedagogical benefits of
using a CPG framework and theory to teach mixed-methods approaches
to research. We show how our course relates learnings from situated
case studies to inform a general approach to teaching in this context
(see Couper 2023). We argue that whilst challenges must be addressed
in development and use of integrative teaching approaches, CPG
educational frameworks move students towards more inclusive and
abductive approaches to science (Brierley et al. 2021). They enable
students to embrace notions of knowledge beyond concepts of 'truth' and
'universality' and provide versatility for students to work in areas where
information and understanding may yet be limited and incomplete.

Situating eco-social teaching in the academy

A CPG perspective inherently questions traditional ways of organising
knowledge and teaching in universities around nature-culture or science-
society divides (Gibbons 2013; Malone et al. 2021). Differentiating
pedagogic practices and outcomes in Faculties of Science and Humanities
or the Arts is no longer fit for the purpose of educating students in an eco-
social world. Geography, and Physical Geography more specifically, has
long attempted to resolve the tension of dividing the organisation and
production of knowledge into cultural or biophysical categories (albeit
in limited ways) (Tadaki et al. 2012) via the study of biophysical (more-
than-human) worlds within cultural contexts. Geography therefore
seems an obvious home for mixed method teaching and research.

Geography departments are well used to negotiating the challenges
of residing administratively either in an Arts or Science Faculty while
teaching successfully across this notional boundary. Teaching social and
biophysical science skill sets under single umbrella courses provides both
necessary specialist skills and contextual knowledge that emphasises
the importance of accessibility of language and communication.
As an inherently relational discipline that emphasises concerns for
appropriately contextualised knowledges-in-place, geography has

a strong history of moving knowledge production towards more integrative spaces (Lane et al. 2018).

Embracing approaches to teaching Critical Physical Geography

Pedagogy is shaped by circumstance, and it is no coincidence that novel approaches to CPG in teaching and research were developed within the context of the University of Auckland – Waipapa Taumata Rau. In Aotearoa New Zealand Treaty obligations and commitments to Te Mana o te Wai fashion a distinctive (plural) approach to knowledge generation, science, teaching and management (Hikuroa et al. 2021). These are closely aligned with the tenets of CPG with its concerns for positionality and situatedness, which explicitly recognise and respect relations between Western science and mātauranga Māori (Wilkinson et al. 2020). This entails creating safe and respectful spaces to share and interweave multiple knowledges (e.g., Koppes 2022). Such framings are widely supported by our current institutional directions at the University of Auckland (UoA) where recent moves to 'indigenise the campus' have established a graduate profile that explicitly recognises the importance of transdisciplinary and builds directly upon a Māori lens (Hoskins and Jones 2022; see Table 6.1).

Table 6.1. Recently revised Graduate Profile at the University of Auckland (Waipapa Taumata Rau). Note the alignment between 'capabilities' and core principles of Geography.

Themes	Capabilities
Waipapa Herenga Waka: The Mooring Post Connecting to place for thriving and equitable communities	People and place
Waipapa ki Uta: The Landing Place Connecting to place for sustainable and enduring partnerships	Sustainability

Waipapa Ngā Maunga Whakahī: Land of Proud Mountains	Knowledge and practice
Knowledge engagement for excellent practice	Critical thinking
Robust inquiry for innovative responses	Solution seeking
Waipapa Tātai Hono: Ancestral Ties	Communication
Communicating and engaging in the service of relationships	Collaboration
Waipapa Tāngata Rau: The Place of Great People	Ethics and professionalism
Flourishing relationships through principled action	

Although the Geography Department at the University of Auckland, housed in the School of Environment, has a rich history of interweaving Physical and Human Geography approaches and thinking critically about environmental education, the emergence of a strong CPG programme enacted in both teaching and research was neither specifically planned nor intended. Rather it was accidental, the product of the interests and aspirations of a particular group of staff and students co-located in a particular place at a particular time (Blue et al. 2012; Tadaki et al. 2012; 2015). Unified by a common motivation to propose constructive and creative (innovative) solutions for environmental problems that embraced and were rooted in critical enquiry, the early work of the pioneers of CPG (Lave et al. 2014) quickly took root in our work. A strong collective commitment to engaging with hands-on research and teaching through purposeful engagements with 'things that matter' facilitated deliberation and negotiation between idealism and pragmatism (Blue 2018), engendering hopeful and caring responses with an aspirational and proactive focus on solutions (Brierley and Fryirs 2022). The aim was to move staff and students beyond preconceived ideas towards non-prescriptive, non-deterministic approaches to 'doing things better'.

Pedagogical benefits of mixing methods in a CPG approach

In this international, national and institutional context the ideas and philosophy behind CPG present fertile soil to grow and develop purposeful and aspirational interventions. In our teaching at UoA we had the opportunity to design a new third year course in Advanced Physical Geography. We organised our practices around four integrative eco-social themes of CPG which all require mixed method approaches to be effectively taught, understood, and utilised by the students. The four themes put an emphasis on i) place (knowledges-in-place and meanings of place) which provide a platform to define and declare positionality in ways that situate knowledge within specific contexts, ii) knowledge pluralism which requires an open-minded approach to learning, iii) doing research which requires students to investigate independently rather than following instructions, and iv) the importance of values as an integral part of all research and knowledge development. The context and rationale for choosing these themes are provided below, along with their implications for use of mixed method approaches. Examples of how this is operationalised are provided in the following sections.

1. The importance of place and situated knowledge

Concerns for place present a critical starting point for holistic eco-social framings of environmental issues. Relevant, co-produced knowledges build upon locally grounded ventures that incorporate field experiences and understandings (Brierley and Fryirs 2014). Such insights guide processes of knowledge transfer, striving to ensure that management applications are "fit-for-purpose" (Fuller et al. 2023). Such a place-based perspective also re-frames time as local to and a product of space, enabling notions of "stability" and "resilience" and conservation/preservation of "pristine (more-than-human) nature" to be challenged (Brierley and Fryirs 2024). It further asks students to challenge the validity of generalised, universal, time free solutions.

2. Importance and value of knowledge pluralism and acknowledging the context in which knowledge is constructed

Working at the interface of multiple strands of information and insight prompts effective use of abductive reasoning (which seeks to identify the most likely explanation from incomplete and uncertain information) to make best use of all the available knowledge for a given problem (Brierley et al. 2021). In educational terms this requires students to recognise and value types of knowledge and data which might look unfamiliar, be reported in 'non-scientific' forms and contexts, and may be difficult to convert into numerical classes or categories. For example, in an Aotearoa New Zealand context where Māori (the indigenous peoples of Aotearoa New Zealand) are recognised as rightsholders under the Treaty of Waitangi (Te Tiriti o Waitangi 1840), contemporary approaches to co-generation and use of knowledge aim to include both Western science and mātauranga Māori. Mātauranga Māori is a lens that conceives humans as part of nature, not separate from it, incorporating deep respect and ancestral connections to the environment (Ruru 2018; Stewart-Harawira 2020; Te Aho 2019). Due regard for ethical considerations in teaching environmental issues explicitly requires us to reflect upon the value and purpose of decolonising knowledges (see Chapter 5; Parsons et al. 2021; Tuhiwai-Smith 1999). This requires us to incorporate appreciation and understandings of the principles of kaupapa Māori (a framework for Māori-led research and evaluation that respects and values the cultural perspectives and relationships of the participants; see Bishop and Glynn 2000), explicitly emphasising concerns for a range of perspectives on knowledge and knowledge generation processes (Fryirs et al. 2019b; Mitchell et al. 2023; Sharp et al. 2021).

Secondly, CPG frameworks explicitly call for education and knowledge development practices which emphasise an awareness of individual and collective situatedness and positionality and its influence on knowledge production. Generation and use of appropriately contextualised understandings require familiarity with new pedagogic approaches that work at the interface of multiple (plural) knowledges, seeking ways to work across worlds (Koppes 2022). This entails respectful regard for the value and purpose of decolonising knowledges (Tuhiwai-Smith 1999), broadening perspectives on knowledge and knowledge generation processes.

Finally, an unintended co-benefit of CPG is that by embracing a kaupapa Māori approach to pedagogy, as informed by CPG theory, students engage with interactive, hands-on, bottom-up approaches to sharing and listening at the core of teaching practices (Bishop and Glynn 2000). Student-led group work teaches key skills in communication through "learning to participate" initiatives (Brierley et al. 2002). Ensuring that multiple voices are heard and valued within the context of knowledge pluralism also enables different student perspectives to create generative educational spaces and active learning environments (Couper 2023).

3. Doing not just thinking

Our previous work has shown that when teaching about environmental issues in Aotearoa New Zealand, how the concepts are taught is more important than the nature of the environmental concern in enabling students to engage with and learn from the material (Sharp et al. 2021). A course centred around a local issue enables students to engage effectively, and to co-develop their own embodied and situated knowledge in the field. Such framings allow students to learn at their own pace and to conceptualise both the problem and solution from a place-based perspective (Sharp et al. 2021; Gibson-Graham 2011). Teaching in an iterative fashion, with students providing feedback to the instructors (similar to the methods outlined by Gillett et al., 2018), further enables co-production of knowledge as students engage with the material in developing innovative solutions.

Providing opportunities for students to visit the place that they are studying, either in person or virtually, along with their teachers and fellow knowledge co-creators is an important part of a CPG approach to teaching and demonstrating mixed-methods research in practice. It gives students the ability and confidence to question the utility of traditional scientific concepts, theories and methodologies, to assess and to re-assess information and data, and to interpret and re-interpret results with confidence in group settings.

In a field-based context, students can also be asked to develop reflexive approaches to knowledge-making. By challenging students to collect data using different approaches they can reflect on what is gained by integrating social science with biophysical data, the limitations of technocratic, engineering and technological lenses when considering environmental management as well as the value of storytelling (Blue and Brierley 2016; Fuller et al. 2023). Such a participatory approach to learning helps to create active and independent learners—critical thinkers with strong problem-solving capabilities.

A focus on enactive teaching methods enables students to see how their decisions can result in generative change, laying foundations for approaches which de-centre both human life and traditional biophysical scientific approaches (Sharp et al. 2022). This mixed-methods approach enables a more inclusive, less colonialised and more values-oriented form of environmental management.

4. Benefits of re-inserting values, beliefs and context back into research (and teaching)

A CPG perspective requires acknowledgement of values and mindsets (Chan et al. 2016; Tadaki and Sinner 2014). It legitimises bringing feelings, values and experiences into the classroom, prioritising teamwork over individual agency. Students with different but complementary skill sets invariably have different perspectives and have to communicate sensitively with each other to complete tasks. Students and teachers can benefit from the position that there is no singular truth, set of knowledge, or theoretical framework which frames any given environmental issue. Thus, the range of perspectives developed through a co-learning experience in which everyone is encouraged to bring their own perspectives and acknowledge their limitations provides practice for developing productive dialogues with others, enables validation of others' ideas and contributions and gives agency for individuals and their group learning. Such a CPG (mixed-methods) approach makes space for multiple ways for students to learn and act. It acknowledges that there isn't a single, simple prescriptive/deterministic answer but multiple opportunities to make a difference. This enhances the realisation of agency, empowering students to imagine more hopeful futures (Sharp et al. 2021).

A CPG approach also enables our teaching frameworks to explore the relationships with more-than human aspects of our worlds within ethical frameworks. Emphasis on the social and political underpinnings of research and decision making provides a lens to explore outcomes which are less exploitative of social and ecological relations (Meadow et al., Chapter 5; Sharp et al. 2022). For example, the recognition that environmental decisions are based on information and knowledge collected using particular methods and instruments whose uses are governed by rules and underpinned by theoretical concepts and disciplinary framings enables students to question whether genuinely equitable solutions are being sought, who or what is gained and or lost for particular outcomes (Blue and Brierley 2016).

Such perspectives take deliberations in environmental science and management to a different, more-than-human space that envisages and enacts no ontological separation between people and nature (in a New Zealand context this is tangata whenua—people of the land). Alongside this, institutional framings and political policies in Aotearoa New Zealand have long emphasised a commitment to sustainability principles, initially enacted through the Resource Management Act (1991; Jackson and Dixon 1997; Knight 2016). Recent deliberations on the revised framework for managing biophysical resources in New Zealand contemplate visionary commitments to 'more-than-human' relations (e.g., Thomas 2015) and the Rights of Nature (e.g., river rights; Ruru 2018; Salmond 2014). To date, such framings are largely aspirational, and they are yet to be enacted (e.g., Samuelson et al. 2023).

Increasingly, geoethical deliberations that apply an ecocentric lens extend beyond concerns for anthropocentricism (Sharp et al. 2022). Engaging with aspirational prospects for a future of authentic and conscious unity with nature implicitly envisages 'living with the world of which we are a part' rather than seeking to assert human authority over the world through efforts to manage it (Brierley 2020).

Operationalising integrative practices in teaching

While the scientific and political desirability and utility of integrating Critical Human and Physical Geography is well demonstrated (Malone et al. 2021), in practice operationalising integrative approaches and methods can be challenging. This is especially true in an educational framework which has traditionally evaluated student performance based on knowledge accumulation, assimilation and recall, rather than generative practices. To date the literature on teaching CPG or using CPG in educational frameworks remains sparse (Gillet et al. 2018), especially in the undergraduate space where the challenge of integrated teaching within either Arts or Science degrees remains a significant barrier (op. cit.). Mixing methods in general, and approaches like CPG in particular, also call for radical changes in the way we teach the basics such as the types of questions we ask of students and that we ask them to create as well as in terms of using and mixing multiple methods and knowledge systems. It also requires students to take a geoethical approach while allowing space for hopeful, aspirational knowledge generation which emphasises constructive, positive, generative agenda-setting practices and solutions.

In this context we started from the position that our course in CPG was framed around the explicit aim of examining the challenges of 'doing science' in the real world. This had the benefit of making the material interesting, relevant and accessible to students from a range of different backgrounds. Our teaching methodologies are centred around practical, field-based examples of local issues which enable students to explore the development and mixing of methods themselves, giving due regard for evidence bases that support decision-making in environmental problem solving. We emphasise the need for students to think about how we construct our environments as an object of study and ask them to combine their specialist biophysical knowledge and technical skills (the bases for which have already been acquired in first and second year classes) with an understanding of the social forces (human relationships and notions of power and privilege) that created them (Malone et al. 2021; Tadaki et al. 2012; 2015). The idea is to put emphasis on generating and collecting place-based knowledge while exploring how to develop coherent arguments using evidence-based approaches in environmental management decision making. Students are required to recognise and to acknowledge their own positionality alongside that of knowledge brokers and decision makers. They learn, through their own co-creation of knowledge about a local environmental issue, how the evidence they collect is biased by particular methodologies, concepts, instruments and disciplinary framings which also affect interpretations, applications and outcomes. This provides graduating students with a more nuanced understanding of the challenges of living in the non-stationary conditions of the Anthropocene where competing social, cultural and economic priorities frame and re-frame the local area as a notionally neutral canvas for activities, resource or hazard.

Case study: Third- (final-) year course in Advanced Physical Geography

In this section we provide a case study example of how we operationalise our approach to teaching CPG. Like other examples of CPG in tertiary education, we centred our approach around a single case study of an applied 'real science' problem that researchers were already engaged with (Gillett et al. 2018) and emphasised the tenets of CPG from the outset. Unlike Gillett et al. (2018), we were unable to provide instructors from both Human and Physical Geography due to local constraints on workload. However, both instructors were applied scientists and routinely work with a wide range of inter- and trans-disciplinary researchers and end-users, and had been heavily engaged with the relevant literatures for several years. We were therefore able to sustain a high level of integrated education as required by CPG approaches (Lave et al., 2014). We also had students in the class who had a primarily Human Geography education as well as those with a primarily Physical Geography education. Our personal underlying ontological frameworks view the world as complex, multi-faceted, and non-binary, allowing us to draw on epistemological frameworks which embrace and legitimise multiple forms of knowledge. We had also worked together on similar research-based ideas while supporting postgraduate students (Tadaki et al. 2011; 2012).

The course was designed to be open to all students who had taken at least one course in Physical or Human Geography at second year, and the typical class size is small (25 to 50 students). The educational setting for the course was a 12-week block comprised of 24 hours of lectures, a 6-to-8-hour field experience and 10 tutorials each of 2 hours duration. The remaining hours were allocated to approximately 36 hours of independent reading and thinking, 32 hours of work on laboratory assignments, and 38 hours of exam preparation. The objectives and outcomes of the course summarised in Table 6.2 align with the UoA graduate profile shown in Table 6.1.

Table 6.2 Learning objectives and outcomes for the third year CPG course at UoA

Learning Objectives	Key questions addressed	Learning Outcomes: Students can...	Key principles and practices students able to demonstrate by end of the course
Recognise limitations in understandings of environmental processes and how human activities modify them	What types of data, knowledge and information are needed to support decision making? How do human activities effect, and how are they affected by, their environmental settings? How can we improve our understanding, and prediction, of the world around us?	... critically discuss the key theoretical frameworks used to describe and predict environmental systems	Recognise the absence of a single, prescriptive, deterministic approach Develop appropriately contextualised knowledges, reflecting upon the hidden assumptions, recognising and respecting 'voices unheard' Recognise that methods are value-laden—they embed certain kinds of normative social-environmental relationships Reflect upon ethical considerations—how choices impact on outcomes, interpretations and conclusions
Relate theoretical frameworks to local contexts		... identify and evaluate the impact of human activities on environmental processes	
Identify data needs and limitations (uncertainties) to support evidence-based decision making		... outline the issues surrounding production and use of data for environmental decision making	
Incorporate multiple (plural) knowledges in decision making		... demonstrate an understanding of the importance of mātauranga Māori and the Treaty of Waitangi	

Course design

The course was built around two hands-on locally grounded group-based exercises which incorporate field work and experiences at a site close to the university (<10 km away). Given the professional interests of the lecturers, the course was centred around river management and air pollution management. However, other specialist nodes could readily be accommodated depending on staff interests. The site was chosen extremely carefully, as it needed to be a coherent, self-contained but multifaceted local site at an appropriate scale to consider eco-social issues, relations and connections across the catchment as a whole (in this instance, 12 km², see Fig. 6.1). Details of the study catchment are provided in Box 6.1.

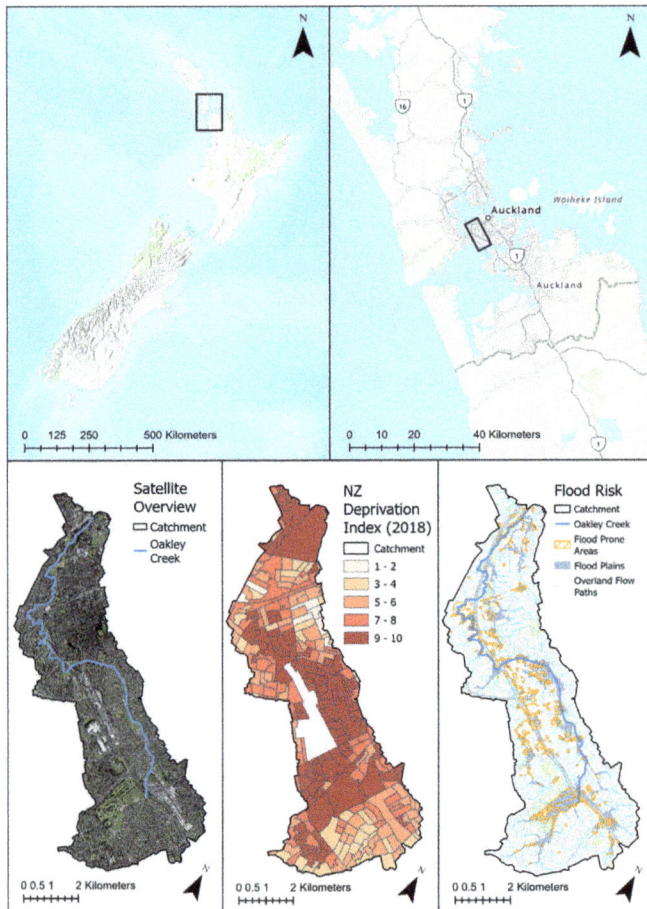

Fig. 6.1 Gary Brierley, 2024. The location of Oakley Creek – Te Auaunga GIS including maps of a) National location, b) regional location, c) Creek catchment, d) Deprivation index (2018), e) Flood risk (2018)

Central to the approach was mixing methods. The students drew on a combination of traditional biophysical methodologies to assess the quality of the river catchment (river health and air quality). They were tasked to reflect on the types of knowledge they have constructed and the limitations of approaches that have been utilised to date. They were then required to attempt to derive a more integrated eco-social-political analysis of the region. For the rivers section of the course, the evaluation focused on a combination of group work and individual written work to develop a coherent catchment management plan based on CPG approaches using mixed methods. This was followed by an individual reflection and critique on the academic framing of their plan (see Box 6.2). For the air pollution section, students undertook group work to collect air quality data along the creek using traditional scientific approaches. They were then required to individually analyse the data in different ways to show how multiple interpretations and conclusions could be drawn from the same data set (see Box 6.3; Breznau et al. 2023). This exercise highlighted different approaches to, and implications of, sample design and data resolution in assessing urban air quality (what to measure where, when, how, and why) and the impact of decisions made in processing, analysing and reporting upon data interpretation (cf., Braun 2021).

Cultural context history

As experienced everywhere in the long, thin country, you're never far from the coast in New Zealand and associated cultural connections are part of the national psyche. Such relations are especially prominent for Māori, as their arrival in waka (ocean craft) to what came to be known as New Zealand shores prompted reflections upon terrestrial relations that look upstream 'From the Sea to the Mountains' (Fig. 6.2A, B).

Ngāti Whātua o Ōrākei are acknowledged as Mana Whenua (Indigenous iwi, hapu) and kaitiaki (guardians) of Oakley Creek. In early days, the creek was used as a corridor or transit (portage) route to access the upper reaches of adjacent catchments, providing access to the west coast. The estuary and the lower course of the river have long-supported navigability and trade links to other parts of the region (Fig. 6.2C). In early days of colonial settlement this area supported the distribution of agricultural products from the catchment. Fertile soils supported Māori gardens in the lower catchment. Waitematā Harbour also provided abundant food resources (kaimoana)—flounder, sprats, eels, birds, etc. Significant industrial developments such as a water mill, flour milling, quarrying, brickmaking and a tannery subsequently accompanied land use changes adjacent to the river mouth.

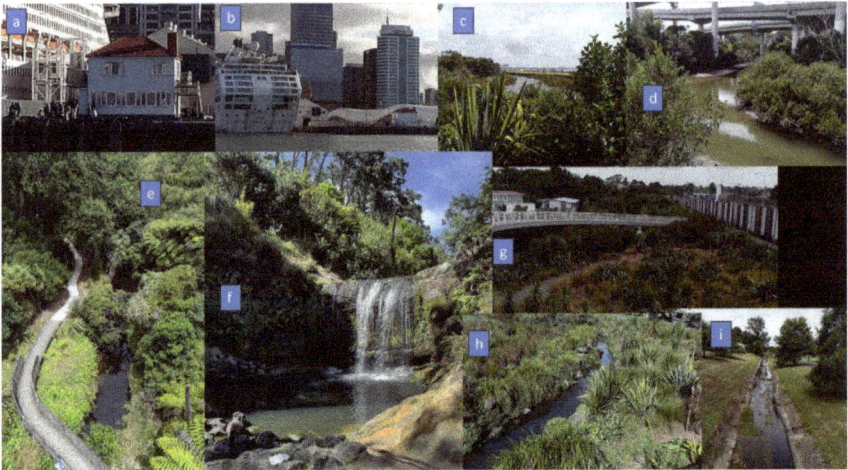

Fig. 6.2 Gary Brierley, 2024. A, B. Michael Parekowhai's sculpture entitled 'The Lighthouse' places Captain Cook inside a State House at Auckland Harbour Ferry terminal. If reflects upon Māori relations to the land, conceived 'From the Sea to the Mountains'. It reflects upon the making and shaping of New Zealand society—the waterfront in Sydney Australia has an Opera House, Auckland (Tamaki Makaurau) has a state house. C, D. Long a site of cultural connections and relations, alongside early colonial 'developments', a recently constructed motorway cuts across the mouth of Oakley Creek – Te Auaunga at the entrance to Waitemata Harbour. E. Recent restoration interventions have greatly enhanced ecological and recreational activities along lower-middle reaches of Oakley Creek, including wonderful walking trails and bike access. F. A 6 m high waterfall divides lower and middle-upper reaches of Oakley Creek. G. Reconstructed wetlands are part of a major restoration and flood protection scheme, tied in part to a recently completed motorway construction programme. H. Restored river reaches have a strong emphasis on ecological and recreational values. I. Channelised reach of the upper catchment that is yet to be restored.

Urbanisation, development, historical management and control of the catchment

Major urban expansion, consolidation and infrastructure development took place in the middle-upper catchment following the Second World War. Over half of the contemporary catchment residents were born overseas. Levels of socio-economic deprivation in the area are high to very high (Fig. 6.1C). A major motorway link that cuts across the mouth of the catchment to facilitate transport between western Auckland and the city centre has recently been extended by a new motorway that connects to Auckland airport, with a major tunnel at Waterview adjacent to the road intersection (Fig. 6.2F). Excavation of a ditch sought to address drainage concerns in swampland areas upstream of convergent lava flows, but significant flood management concerns continue to this day in the middle-upper catchment (Fig. 6.1B).

Tied to the motorway expansion programme, and building on community-led approaches to flood management and river restoration in western Auckland (Project Twin Streams; Gregory and Brierley, 2010; Winz et al., 2011), concerns for flood 'protection' underpinned a major river restoration initiative that has transformed waterways and daylighted various tributary watercourses in the middle-upper catchment (Fig. 6.2G).

Since 2004, the Friends of Oakley Creek has been critical to the scope and success of such programmes (see Oakley Creek | Te Auaunga). High profile community engagement includes participatory approaches to revegetation, local festival events including a celebration of the native shining cuckoo (pipiwharauroa) and multiple activities that emphasise healthy socio-cultural relations to healthy rivers (and vice versa; see Warne, 2013).

A recently completed project in Walmsley and Underwood Reserves in the upper catchment restored 1.5 km of formerly channelised urban stream along Te Auaunga, daylighted seven piped tributaries, restored 8 ha of open space, and treated the water quality of the contributing catchment. A complementary package of interventions prevents flooding from nearly 200 homes in three Local Board areas and enables housing intensification in a brownfield site. The bioengineering project was facilitated through design workshops, a community liaison group, governance meetings, public open days, and school workshops. Alongside creation and enhancement of various recreational facilities, environmental education interventions work with ngā taonga tākaro to interpret the environmental and cultural narratives of various sites. Among desired outcomes of the iwi Management Plan developed by Ngāti Whātua o Ōrākei in 2018 (Ngāti Whātua Ōrākei, 2018), it is explicitly stated that "Water should be managed, and where necessarily restored, to maintain or enhance mauri (life force) and to protect ecosystem, amenity, and man whenua values."

This area now contains an important marine reserve that supports key biodiversity values including migratory bird species. The largest waterfall in the Auckland region (6 m high) separates incised lower reaches from the middle and upper catchment (Fig. 6.2D). This area now incorporates a major urban park and walkway/bikeway (Fig. 6.2E). Ongoing/future concerns for wastewater management in the region are being addressed through a 14.7 km long, 4.5 m diameter Central Interceptor tunnel that cuts across upper Oakley Creek (see www. watercare.co.nz).

Other site considerations:

The site is easily accessible using public transport, supported by first-rate urban walkways and bike paths through the catchment. It is considered a 'safe' site that students could readily revisit in a self-guided capacity as they wished. Recognising that this would not be possible for all students, we developed a 'virtual' field trip and a moderately comprehensive resource kit to enhance engagement with the catchment. Site selection was supported by a well-developed Council and Government resource base and an extensive grey literature of sufficient breadth and depth to support informed engagement/enquiry on local issues (e.g., Auckland Council. 2022a; 2016). Access to a diverse range of information from multiple sources, including local Iwi, community groups and cultural exhibitions alongside field experiences, enabled familiarity with multiple values, datasets and realities/possibilities in the conduct of this experimental, non-prescriptive learning exercise.

i. Rivers exercise

Box 6.1: Catchment case study context—context, location and history of Oakley Creek (Te Auaunga), Auckland (Tamaki Makaurau), Aotearoa New Zealand

Physical geography context:

Oakley Creek lies in the heart of the City of Sails (Auckland City (Tamaki Makaurau)) which lies on an isthmus between Waitematā and Manukau Harbours (Fig. 6.1). The former is connected to Hauraki Gulf and the Pacific Ocean on the East Coast, while the latter flows into the Tasman Sea to the west. The region has a temperate climate with average winter temperatures of 1.9°C to 17.6°C and average summer temperatures of 10.5°C to 27.6°C. Mean monthly rainfall peaks at 134.6 mm in July and is at its lowest in February at 61.4 mm. Characteristic of the more than fifty scoria cone volcanoes in the Auckland region, lava flows exerted a key influence upon the topography and soils of Oakley Creek catchment as swamplands used to lie upstream of convergent lava flows in the middle-upper catchment.

The CPG rivers exercise was designed as a future-focused (proactive, precautionary, preventative) exercise, requiring students to assess river health and develop a coherent catchment management plan (CMP), but provided no fixed solutions or expectations (see Box 6.2). Prior to the exercise the students were taught that CMPs have historically been envisaged and framed in instrumentalist (reductionist) terms in Aotearoa New Zealand. They were challenged to "do much better than this"! Aspirations extend beyond traditional 'command and control' technocentric practices and 'solutions', seeking to establish locally-situated practices that work with nature, work with the river (Brierley and Fryirs 2022; Fryirs and Brierley 2021). An emphasis upon relations between 'process' and 'product' required critical reflection on socio-cultural values about rivers (each reach, and collectively) and concern for *Te Mana o Te Wai* and ancestral connections (see Ruru 2018; Stewart-Harawira 2020; Te Aho 2019). Students were also pointed towards literature on more-than-human (MTH) values and geoethics (Brierley et al. 2019; Sharp et al. 2022) and their relevance as part of postcolonial co-design processes (e.g., Parsons et al. 2021). Importantly, they were also taught that CMPs should incorporate understandings of the evolutionary trajectory of a river, as this shapes determination of 'what is realistically achievable' into the future, typically over a timeframe of 10-50 years (Brierley and Fryirs 2022).

Box 6.2. Rivers exercise assignments

Group-based development of a Catchment Management Plan (CMP) and individual CPG-based rational and justification for Oakley Creek – Te Auaunga, central Auckland

Premise

Perhaps inevitably, co-development and application of a CMP in ANZ is often highly contentious. Contested deliberations incorporate concerns for differing worldviews, values and the politics thereof (see Boelens et al., 2023). Profound repercussions ensue—now and into the future. As such, this is perfect terrain for academic discourse in CPG! What could a CMP look like for river systems in the Auckland region? How could/ should it be designed, applied, communicated, updated? Working in a group of 5 or 6 students, each group is asked to document their approach to the design, implementation and adaptation of a proposed CMP for Oakley Creek – Te Auaunga.

Your task

Framed as an exercise in Critical Physical Geography, each group is tasked to demonstrate:

a. How place-based approaches to design and implementation of a CMP incorporate efforts to describe, explain and predict river systems

b. How biophysical considerations are considered alongside socio-economic, cultural and institutional (governance) arrangements in the design, implementation and updating of within CMPs.

c. How, under terms of Te Tiriti o Waitangi, an emphasis upon Mātauranga Māori is a central pivot of a CMP, embracing and incorporating principles of Te Mana o Te Wai (Ruru, 2019; Te Aho, 2019)

d. How scientific assertions of river condition (health) relate to Māori notions of ora (collective wellbeing) (see Blue, 2018). What attributes should be monitored and measured, how (technique; where, when, how often, etc), what do we measure against, and how do we analyse, interpret, synthesise and present findings (Blue and Brierley, 2016)?

Guidance for this open-ended assignment urged due regard for process and product, framed with explicit recognition of the situatedness and positionality through which the work is conducted. This entails consideration of what the plan looks like and how it is to be (co) developed and (co)enacted.

Individual task: Independent essay write-up: Academic (intellectual) grounding and critical reflection upon your proposed Catchment Management Plan for Oakley Creek – Te Auaunga

Task: Make careful use the academic literature to document the intellectual rationale and justification for your proposed Catchment Management Plan for Oakley Creek – Te Auaunga. This will include a summary of concerns for values, selected information base and approach to use/analysis, and reporting/communication procedures. You are to demonstrate how and why your approach to design and implementation of a CMP builds upon and reflects principles of Critical Physical Geography, presenting a clear evidence-based justification for your chosen approach to decision-making (product and process).

ii. Climate exercise

This exercise was designed to engage students in the collection and analysis of large volumes of data along the same route as the rivers exercise. Students were already familiar with the site and challenges for managing floods and river restoration. The exercise encouraged them to think more holistically about the environmental setting. They learned through the use of real examples about the impact of the little, apparently inconsequential decisions made when collecting environmental data (such as when, where, and how the data are collected). This highlights the nuanced differences between the resulting data sets collected by different groups. Students were challenged to analyse the data in different ways, using different tools and at different spatial and temporal resolutions and to evaluate the results (Box 6.3). They began with a group exercise to collect and analyse data using standard biophysical approaches but then were asked to individually reflect on the limitations of this approach when using the data to make decisions about air quality management or making recommendations to individuals as to how they can be proactive about reducing their personal exposure to air pollution.

In the final exam students reflected on the whole course. This required them to draw parallels across river and atmospheric management approaches and to make compelling arguments as to the added value mixing methodologies brought to environmental management when conceived and considered through a CPG lens.

Box 6.3. Climate exercise Assignments

Group-based of individual exposure to air pollution along Oakley Creek – Te Auaunga, central Auckland and individual CPG reflection on the task.

Group task: Bio-physical assessment of exposure to air pollution

This exercise requires you to examine the value of data and its limitations including the uncertainties when analysing population exposure to air pollution. It also prompts you to consider the need for, and challenges of, delivering data for evidence-based decision making. You need to collect and analysis data to compare and contrast exposure to air pollution in different commuter microenvironments and determine the implications of these results for commuter route choice.

Overview of assignment:

a. Describe and explain the variability in pollutant exposure observed in your data set

b. Provide a quantitative analysis of the data using either R or GIS based on the lab activities

c. Use this information to make general recommendations for active commuters wishing to choose low exposure routes during their commute to work. Provide examples of commuter routes in Auckland to support your argument. State and justify any assumptions made.

d. Use figures, maps, photographs and other forms of visual communication to support your assessment.

e. Your report should take the form of an academic journal article and include an introduction, literature review, methodology, results & discussion and conclusion (as well as references and acknowledgements). It will be marked holistically, so you can choose to give weight to individual sections.

Note: This is a group exercise. Divide activities equally between group members and include a clear description of who did what in the acknowledgements at the end.

Individual task: Reflect on your biophysical assessment

This exercise requires you to consider the need for, and challenges of, delivering data for evidence-based decision making. It also asks you to examine the value of incorporating multiple types of evidence and need for different types of knowledge in decision making. Reflecting on your group report on exposure to air quality consider the limitations of the methodologies chosen and the implications of these for the results and conclusions drawn. Suggest how the study could have been done differently to result in different outcomes. Use this reflection to make recommendations for the data set we need to design low-exposure commuter routes.

The format of your report is not prescribed, but you MUST draw on the literature to support your reflections and observations. The report will be marked holistically. Recognising that the list below is merely indicative and far from exhaustive—it is designed to get you thinking!— your reflection should explicitly consider the following issues:

Methodological choices

a. How does the size of the data set affect its value?

b. How many route repetitions do you need to be confident in your results?

c. How easy is it to extrapolate individual data to population scales? What are the limitations of doing so?

d. What other data did you need to be able to make recommendations about route choice more accurately?

e. Would modelling techniques/ data from fixed air quality monitors help you extrapolate the results?

Analytical tools and communication methods

a. Show how the same data set could have been analysed in different ways to support a different set conclusions.

b. You may choose to include other types of analyses (such as analysis of significance or qualitative approaches).

c. How can you express the limitations of the data analysis effectively?

Management implications
What data are needed to design better commuter routeways in urban areas?

Challenges

Teaching mixed methods is inherently challenging as students, even geography students, typically do not have all the necessary foundational skills. By design, the course encourages explorations in new and challenging spaces on complex tasks, and not all students want or are well prepared for this. Instructors may also be teaching outside of their dominant research area. We overcame these issues through the use of group assignments in which students with different backgrounds were encouraged to work together, and through the careful choice of teaching assistants (tutors) with a range of different backgrounds. Approaching the course with a shared sense of learning, where instructors, lecturers, and students learn together and share their knowledge also helped to ensure that skills were learnt along the way.

Teaching interactively, with students providing feedback to the instructors weekly, also enabled true co-production of knowledge as students engaged with the material and developed truly innovative solutions. This required lecturers and students to be vulnerable in the classroom, and activities (such as those outlined in Sharp et al. 2021) were fundamental in developing a shared sense of safety when working with new ideas and expressing values in group settings.

Marked variability in levels of student engagement was the biggest challenge in running the course. Around one-third of students really grabbed the opportunity to explore mixed methods effectively. Perhaps inevitably, several students adopted a minimalist task-oriented approach that merely focused on passing the course. However, the efforts of key personnel who emerged as group leaders, alongside a 'peer esteem' approach to teamwork, helped to facilitate collective engagement in group exercises. In general terms, framing exercises as open-ended, non-prescriptive assignments works well for some students, but less well for those students who simply like to be told what they have to do to pass. Conversely, as the course design encourages students to explore opportunities to engage with different approaches to learning and expression, it is exceedingly generative in supporting the quest of some individuals to develop their own voice. While working in groups can facilitate and enhance such explorations, the effectiveness with which this happens is greatly influenced by the composition, range of personality types, background skillsets, and dynamics of any given group.

Our limited capacity to support direct engagement with local (community) voices presented a major limitation in the operation of the course. This reflects time constraints on the one hand, and the personal background of the course lecturers on the other. In reflecting upon ethical concerns for positionality and situatedness, direct experience and training with Māori colleagues presents a key constraint in the conduct of the course to date.

Marketing such a mixed method course was also challenging, as students were initially put off by the course description and amount of required reading. Word-of-mouth promotion of the course by students who completed it ensured higher participation in later years, along with the draw of fieldwork which could fit around other commitments (such as employment). Time limitations constrained prospects for critical reading and reflection; as a result everything seemed a little rushed. However, running the course was greatly aided by the design and conduct of the core feeder courses in the second year at UoA, which are built around group work.

Discussion and conclusions

The approach to teaching critical mixed methods using a CPG framework outlined in this chapter promotes a conscious, ethically-aware commitment to co-development and use of integrated 'knowledges-in-place'. Our course teaches students the basics of the design and enactment of scientifically-informed decision-making, thereby addressing a key educational gap. Activities are appropriately 'grounded-in-place', recognising and acknowledging multiple knowledges on the one hand, while giving careful consideration to appropriate contextualisation of understandings and practices on the other. We feel that we are currently at a critical juncture where students need these skills but are seldom taught how to combine and value different sets and types of information.

Contemplating what is required in the design and enactment of a safe and just space for humanity is an inherently aspirational and generative task. Enactive engagement with place and a commitment to sharing and listening facilitates effective communication of differing perspectives and reveals the value of different approaches to studying biophysical worlds in cultural settings. This helps efforts to encourage more careful consideration of the multiple but incomplete sources of information and insight that shape possible solutions to environmental issues and focusses attention on deliberations between idealism and pragmatism (Blue 2018). Engagement in the 'dark art of interpretation' (Brierley et al. 2021) incorporates understandings derived from remote sensing (Braun, Chapter 39) and modelling applications, field data (observations, measurement and processing), historical sources, government documents, oral narratives, and local knowledge (including grey literature). Juxtaposing conventional scientific principles and practices alongside mātauranga Māori relates data-driven approaches to enquiry to sensory relations and connections that express intuitive or instinctive understandings. Critically, appropriately contextualised understandings explicitly articulate the perspective (positionality/situatedness) from which they were derived in relation to alternatives, formally acknowledging and expressing inherent assumptions and associate limitations and uncertainties in understandings.

Experience gained in this teaching exercise indicates that students are keen to engage actively with positive, proactive, ethical encounters that "do" something in hopeful ways, envisaging and seeking to enact better futures that move beyond the negativity of the global post-COVID malaise. Creativity, experimentation, and innovation come to the fore in scoping better, more hopeful, more caring futures. The future-focused exercises outlined in this chapter explicitly present opportunities to 'share dreams', empowering students to envisage and seek to enact imaginaries of a 'better world' (see Agyeman et al. 2003). An open-ended learning experience purposefully moves beyond fixed notions of foundational pedagogic building blocks, recognising that comprehensive, coherent solutions are unlikely to be envisaged and implemented through approaches and practices that created the problems in the first place.

Ultimately, coping with emergent and uncertain futures entails 'asking the right questions', rather than using existing practices to scope particular answers or solutions (Fryirs et al. 2019a). Efforts to create a safe space for engagement and interaction in such experimental ventures embrace principles that underpin a kaupapa Māori approach to pedagogy, striving to facilitate openness, humility and trust in approaches to sharing and listening (Bishop and Glynn 2000). Conducted effectively, group dynamics and the pedagogic experience in its entirety seek to build confidence in expressing ideas and scoping new ways of being, thinking, and doing, moving beyond limitations of existing approaches and practices.

Acknowledgements

We thank our fantastic tutors, Hamesh, Elliot, and Darrell, who played a fundamental role in developing the resource base. We also thank various students for their contributions in sharing perspectives and associated discussions and interactions in delivery of this course.

References cited

Auckland Council. 2016. *Te Auaunga Awa Oakley Creek Upper Catchment Strategy 2016-2019*, http://oakleycreek.org.nz/wp-content/uploads/2014/05/Te-Auaunga-Awa-Oakley-Creek-Upper-Catchment-Strategy-2016-2019.pdf

Auckland Council. 2022a. *Environmental Monitoring Data for Oakley Creek*, https://www.lawa.org.nz/explore-data/auckland-region/river-quality/oakley-creek

Agyeman, J., R.D. Bullard, and B. Evans. 2003. *Just Sustainabilities: Development in an Unequal World* (Routledge). https://doi.org/10.4324/9781849771771

Bishop, R. and T. Glynn. 2000. 'Kaupapa Maori messages for the mainstream', *Set*, 1, pp. 4–7. https://doi.org/10.18296/set.0785

Blue, B., C. Gregory, K. McFarlane, M. Tadaki, P. van Limburg-Meijer, and N. Lewis. 2012. 'Freshwater geographies: Experimenting with knowing and doing geography differently', *New Zealand Geographer*, 68.1, pp. 62–66. https://doi.org/10.1111/j.1745-7939.2012.01223.x

Blue, B. 2018. 'What's wrong with healthy rivers? Promise and practice in the search for a guiding ideal for freshwater management', *Progress in*

Physical Geography: Earth and Environment, 42.4, pp. 462–477. https://doi.org/10.1177/0309133318783148

Blue, B. and G. Brierley. 2016. '"But what do you measure?" Prospects for a constructive critical physical geography', *Area*, 48.2, pp. 190–197. https://doi.org/10.1111/area.12249

Boelens, R., A. Escobar, K. Bakker, L. Hommes, E. Swyngedouw, B. Hogenboom, and K.M. Wantzen. 2023. 'Riverhood: Political ecologies of socionature commoning and translocal struggles for water justice', *The Journal of Peasant Studies*, 50.3, pp. 1125–1156 https://doi.org/10.1080/03066 150.2022.2120810.

Braun, A.C. 2021. 'More accurate less meaningful? A critical physical geographer's reflection on interpreting remote sensing land-use analyses', *Progress in Physical Geography: Earth and Environment*, 45.5, pp. 706–735. https://doi.org/10.1177/0309133321991814

Braun, A., Chapter 39, this volume. '(Critical) Satellite remote sensing'.

Breznau, N., E.M. Rinke, A. Wuttke, H.H. Nguyen, M. Adem, J. Adriaans, and J. Van Assche. 2022. 'Observing many researchers using the same data and hypothesis reveals a hidden universe of uncertainty', *Proceedings of the National Academy of Sciences*, 119.44. https://doi.org/10.31235/osf.io/j7nc8

Brierley, G.J. 2020. *Finding the Voice of the River: Beyond Restoration and Management* (Springer Nature). https://doi.org/10.1007/978-3-030-27068-1

Brierley, G., M. Hillman, and L. Devonshire. 2002. 'Learning to participate: Responding to changes in Australian land and water management policy and practice', *Australian Journal of Environmental Education*, 18, pp. 7–13. https://doi.org/10.1017/s0814062600001063

Brierley, G. and K. Fryirs. 2014. 'Reading the landscape in field-based fluvial geomorphology', in *Developments in Earth Surface Processes*, ed. by J.F. Schroder Jr. (Springer), pp. 231–257. https://doi.org/10.1016/b978-0-444-63402-3.00013-3

Brierley, G. and K. Fryirs. 2022. 'Truths of the Riverscape: Moving beyond command-and-control to geomorphologically informed nature-based river management', *Geoscience Letters*, 9.1, pp. 14. https://doi.org/10.1186/s40562-022-00223-0

Brierley, G. and K. Fryirs. 2024. 'Geomorphic meanings of a resilient river', in *Resilience and Riverine Landscapes*, ed. by M.C. Thoms and I.I. Fuller (Elsevier), pp. 117–134. https://doi.org/10.1016/b978-0-323-91716-2.00001-7

Brierley, G., K. Fryirs, H. Reid, and R. Williams. 2021. 'The dark art of interpretation in geomorphology', *Geomorphology*, 390. https://doi.org/10.1016/j.geomorph.2021.107870

Brierley, G., M. Tadaki, D. Hikuroa, B. Blue, C. Šunde, J. Tunnicliffe, and A. Salmond, 2019. 'A geomorphic perspective on the rights of the river in Aotearoa New Zealand', *River Research and Applications*, 35.10, pp. 1640–1651. https://doi.org/10.1002/rra.3343

Chan, K.M., P. Balvanera, K. Benessaiah, M. Chapman, S. Díaz, E. Gómez-Baggethun, and N. Turner. 2016. 'Why protect nature? Rethinking values and the environment', *Proceedings of the National Academy of Sciences*, 113.6, pp. 1462–1465. https://doi.org/10.1073/pnas.1525002113

Couper, P. R.2023. 'Interpretive field geomorphology as cognitive, social, embodied and affective epistemic practice', *The Canadian Geographer/Le Géographe canadien*, 67.3, pp. 430–441. https://doi.org/10.1111/cag.12821

Fryirs, K. and G. Brierley. 2021. 'How far have management practices come in 'working with the river'?', *Earth Surface Processes and Landforms*, 46.15, pp. 3004–3010. https://doi.org/10.1002/esp.5279

Fryirs, K.A., J.M. Wheaton, S. Bizzi, R. Williams, and G. Brierley. 2019a. 'To plug-in or not to plug-in? Geomorphic analysis of rivers using the River Styles Framework in an era of big data acquisition and automation', *Wiley Interdisciplinary Reviews: Water*, 6.5. https://doi.org/10.1002/wat2.1372

Fryirs, K., G. Brierley, M. dos Santos Marçal, M.N. Peixoto, and R. Lima. 2019b. 'Learning, doing and professional development – the River Styles Framework as a tool to support the development of coherent and strategic approaches for land and water management in Brazil', *Revista Brasileira de Geomorfologia*, 20.4. https://doi.org/10.20502/rbg.v20i4.1560

Fuller, I.C., G.J. Brierley, J. Tunnicliffe, M. Marden, J. McCord, B. Rosser, and M. Thomas. 2023. 'Managing at source and at scale: The use of geomorphic river stories to support rehabilitation of Anthropocene riverscapes in the East Coast Region of Aotearoa New Zealand', *Frontiers in Environmental Science*, 11. https://doi.org/10.3389/fenvs.2023.1162099

Gibson-Graham, J.K. 2011. 'A feminist project of belonging for the anthropocene', *Gender Place and Culture: A Journal of Feminist Geography*, 18.1, pp. 1–21. https://doi.org/10.1080/0966369x.2011.535295

Gibbons, M. 2013. 'Mode 1, Mode 2, and Innovation', in *Encyclopedia of Ceativity, Invention, Innovation and Entrepreneurship*, ed. by E.G. Carayannis (Springer). https://doi.org/10.1007/978-1-4614-3858-8_451

Gillet, N., E. Vogel, S. Noah, and C. Hatch. 2018. 'Proliferating a new generation of critical physical geographers: Graduate education in UMass's Riversmart Communities Project', in *The Palgrave Handbook of Critical Physical Geography*, ed. by R. Lave, C. Biermann,and S.N. Lane (Palgrave Macmillan), pp. 515–536. https://doi.org/10.1007/978-3-319-71461-5_24

Gregory, C.E. and G.J. Brierley. 2010. 'Development and application of vision statements in river rehabilitation: the experience of Project

Twin Streams, New Zealand', *Area*, 42.4, pp. 468-478. https://doi.org/10.1111/j.1475-4762.2010.00946.x

Hikuroa, D., G.J. Brierley, M. Tadaki, B. Blue, and A. Salmond. 2021. 'Restoring sociocultural relationships with rivers: Experiments in fluvial pluralism', in *River Restoration: Political, Social, and Economic Perspectives*, ed. by B. Morandi, M. Cottet, and H. Piégay (Wiley-Blackwell), pp. 66–88. https://doi.org/10.1002/9781119410010.ch3

Hoskins, T.K. and A. Jones. 2020. 'Māori, Pākehā, critical theory and relationality: A talk by Te Kawehau Hoskins and Alison Jones', *New Zealand Journal of Educational Studies*, 55.2, pp. 423–429. https://doi.org/10.1007/s40841-020-00174-0

Jackson, T. and J. Dixon. 2007. 'The New Zealand Resource Management Act: an exercise in delivering sustainable development through an ecological modernisation agenda', *Environment and Planning B: Planning and Design*, 34.1, pp. 107–120. https://doi.org/10.1068/b32089

Knight, C. 2016. *New Zealand's Rivers: An Environmental History* (Canterbury University Press).

Koppes, M.N. 2022. 'Braiding knowledges of braided rivers–the need for place-based perspectives and lived experience in the science of landscapes', *Earth Surface Processes and Landforms*, 47.7, pp. 1680–1685. https://doi.org/10.1002/esp.5380

Lane, S.N., C. Biermann, and R. Lave. 2018. 'Towards a genealogy of critical physical geography', in *The Palgrave Handbook of Critical Physical Geography*, ed. by R. Lave, C. Biermann, and S.N. Lane (Palgrave Macmillan), pp. 23–48. https://doi.org/10.1007/978-3-319-71461-5_2

Lave, R., M. Wilson, E. Barron, C. Biermann, M. Carey, C. Duvall, L. Johnson et al. 2014. 'Intervention: critical physical geography', *The Canadian Geographer/Le Géographe canadien*, 58.1, pp. 1–10. https://doi.org/10.1111/cag.12061

Lave, R., C. Biermann, and S.N. Lane. 2018. 'Introducing critical physical geography', in *The Palgrave Handbook of Critical Physical Geography*, ed. by R. Lave, C. Biermann, and S.N. Lane (Palgrave Macmillan), pp. 3–22. https://doi.org/10.1007/978-3-319-71461-5_1

Malone, M. 2021. 'Teaching critical physical geography', *Journal of Geography in Higher Education*, 45.3, pp. 465–478. https://doi.org/10.1080/03098265.2020.1847051

Mitchell, D., E.W. Laurie, R.D. Williams, K.A. Fryirs, G.J. Brierley, and P.L. Tolentino. 2024. 'Developing an equitable agenda for international capacity strengthening courses: environmental pedagogies and knowledge co-production in the Philippines', *Journal of Geography in Higher Education*, 48.2, pp. 281–311. https://doi.org/10.1080/03098265.2023.2235668

Ngāti Whātua Ōrākei. 2018. *TE POU O KĀHU PŌKERE. Iwi Management Plan for Ngāti Whātua Ōrākei* (Ngāti Whātua Ōrākei). https:// knowledgeauckland.org.nz/media/1400/te-pou-o-k%C4%81hu-p%C5%8Dkere-iwi-management-plan-ngati-whatua-orakei-2018.pdf

Parsons, M., K. Fisher, and R.P. Crease. 2021. *Decolonising Blue Spaces in the Anthropocene: Freshwater Management in Aotearoa New Zealand* (Springer Nature). https://doi.org/10.1007/978-3-030-61071-5

Ruru, J. 2018. 'Listening to Papatūānuku: a call to reform water law', *Journal of the Royal Society of New Zealand*, 48.2–3, pp. 215–224. https://doi.org/10.1080 /03036758.2018.1442358

Samuelson, L., B. Blue, and A. Thomas. 2023. 'Restoration as reconnection: A relational approach to urban stream repair', *New Zealand Geographer*, 79.2, pp. 107–120. https://doi.org/10.1111/nzg.12372

Salmond, A. 2014. 'Tears of Rangi: Water, power, and people in New Zealand', *HAU: Journal of Ethnographic Theory*, 4.3, pp. 285–309. https://doi. org/10.14318/hau4.3.017

Salmond, A. 2017. *Tears of Rangi: Experiments Across Worlds* (Auckland University Press). https://doi.org/10.1017/s0165115319000627

Sharp, E., J. Fagan, M. Kah, M. McEntee, and J. Salmond. 2021. 'Hopeful approaches to teaching and learning environmental "wicked problems"', *Journal of Geography in Higher Education*, 45.4, pp. 621–639. https://doi.org/1 0.1080/03098265.2021.1900081

Sharp, E.L., G.J. Brierley, J. Salmond, and N. Lewis. 2022. 'Geoethical futures: a call for more-than-human physical geography', *Environment and Planning F: Philosophy, Theory, Models, Methods and Practice*, 1.1, pp. 66–81. https://doi. org/10.1177/26349825221082168

Stewart-Harawira, M.W. 2020. 'Troubled waters: Maori values and ethics for freshwater management and New Zealand's fresh water crisis', *Wiley Interdisciplinary Reviews: Water*, 7.5. https://doi.org/10.1002/wat2.1464

Tadaki, M., K. McFarlane, J. Salmond, and G.J. Brierley. 2011. 'Theorizing "crisis" as performative politics: a view from physical/environmental geography', *Dialogues in Human Geography*, 1.3, pp. 355–360. https://doi. org/10.1177/2043820611421557

Tadaki, M, J. Salmond, R. Le Heron, and G.J. Brierley. 2012. 'Nature, culture and the work of physical geography', *Transactions of the Institute of British Geographers*, 37.4, pp. 547–562. https://doi. org/10.1111/j.1475-5661.2011.00495.x

Tadaki, M., G.J. Brierley, M. Dickson, R. Le Heron, and J. Salmond. 2015. 'Cultivating critical practices in physical geography', *The Geographical Journal*, 181.2, pp. 160–171. https://doi.org/10.1111/geoj.12082

Tadaki, M. and J. Sinner. 2014. 'Measure, model, optimise: understanding reductionist concepts of value in freshwater governance', *Geoforum*, 51, pp. 140–151. https://doi.org/10.1016/j.geoforum.2013.11.001

Te Aho, L. 2019. 'Te Mana o te Wai: An indigenous perspective on rivers and river management', *River Research and Applications*, 35.10, pp. 1615–1621. https://doi.org/10.1002/rra.3365

Thomas, A.C. 2015. 'Indigenous more-than-humanisms: Relational ethics with the Hurunui River in Aotearoa New Zealand', *Social and Cultural Geography*, 16.8, pp. 974–990. https://doi.org/10.1080/14649365.2015.1042399

Tuhiwai Smith, L. 1999. *Decolonizing Methodologies: Research and Indigenous People* (University of Otago Press).

Warne, K. 2013. 'Pilgrim at Oakley Creek', *New Zealand Geographic*.

Wilkinson, C., D.C. Hikuroa, A.H. Macfarlane, and M.W. Hughes. 2020. 'Mātauranga Māori in geomorphology: existing frameworks, case studies, and recommendations for incorporating Indigenous knowledge in Earth science', *Earth Surface Dynamics*, 8.3, pp. 595–618. https://doi.org/10.5194/esurf-8-595-2020

Winz, I., G.J. Brierley, and S. Trowsdale. 2011. 'Dominant perspectives and the shape of urban stormwater futures', *Urban Water Journal*, 8.6, pp. 337–349. https://doi.org/10.1080/1573062x.2011.617828

7. Integrating ethnographic and physical science methods in interdisciplinary research projects: Reflections on pedagogy and practice for 'deep interdisciplinary' engagement within the Sajag-Nepal Project

Amy Johnson, Katie Oven, Nick Rosser, Dipak Basnet, Nyima Dorjee Bhotia, Tek Bahadur Dong, Anuradha Puri, Sunil Tamang, and Mark Kincey

Introduction

Funded through the UK Research and Innovation's Global Challenges Research Fund (GCRF), the Sajag-Nepal Project: Preparedness and Planning for the Mountain Hazard and Risk Chain is one of a number of interdisciplinary projects within the broad field of disaster studies. Co-developed with partners in Nepal, the project brings together physical and social scientists and humanities scholars, policy makers, and practitioners across four countries (Nepal, the UK, Canada, and New Zealand). Central to its success is team working with "researchers trained in different academic disciplines com[ing] together to complement the skills and perspectives of each to address 'difficult' research questions" (Shanker et. al. 2013: 2). Indeed, as a team, we have all committed to bringing our expertise to bear with the aim of reducing the impacts of cascading mountain hazards in Nepal's Himalaya. To achieve this, we recognise the need for 'deep interdisciplinarity' which requires engagement across the physical and social sciences and humanities, but are under no illusion as to just how difficult this can be. Such projects require team members "to be open to alternative methods, willing to embrace different working definitions and committed not just to basic science, but also applied science, to advance solutions" (Rigg and Mason 2018: 1031). All too often researchers within interdisciplinary projects can be seen to retreat into their own disciplinary silos resulting in a series of discipline-focused outputs and attempts at a 'crossover piece' at best. One reason for this, we argue, is a lack of recognition of the need to develop a shared approach to, and training to support, 'deep' interdisciplinary working. This takes time and commitment to ensure that all members of the team feel equipped to 'work deeply'. The emerging field of Critical Physical Geography (CPG), which combines "critical attention to relations of social power with deep knowledge of a particular field of biophysical science or technology in the service of social and environmental transformation" (Lave et al. 2014: 3), provides a broad framework to guide our shared approach.

A team member often overlooked in the design and delivery of interdisciplinary research projects is the 'in-country' research associate (RA). The positionality of RAs within project teams is complicated in part because of the international political economy of academic labour and because the role of RAs in project teams often transcends categories, being neither a student or a project lead, such as a recognised academic/ Co-Investigator (Co-I) (Sukarieh and Tannock 2019; Middleton and Cons 2014). For the majority of RAs with discipline specific training, working on projects such as Sajag-Nepal requires a willingness to step outside disciplinary comfort zones (see Lane and Lave, Chapter 3), but to do so requires significant training and support from the wider project team. Perhaps surprisingly, given the important role of the RA, training in the context of interdisciplinary research projects is rarely discussed in the literature despite a recognition within CPG of the importance of pedagogy as we look to develop "the frameworks to create and encourage new cultures of epistemic pluralism (Castree 2012)" (Lave et al. 2014: 6).

In this chapter we discuss how interdisciplinary pedagogy and training has been at the forefront of the Sajag-Nepal project. As authors, we include project Co-Investigators (Co-Is) (KO and NR) and the Research Fellow (AJ), responsible for coordinating training and research from the UK and undertaking collaborative research in Nepal, as well as a team of social science RAs in Nepal (AP, NDB, TBD, DB, ST) who have been undertaking much of the research locally.[1, 2] Specifically, the chapter reflects on our differently positioned experiences in training, learning, and engaging interdisciplinary styles of working within the context of CPG-inflected fieldwork on mountain hazards and risks, with a specific focus on the RAs' experiences of implementing slope and landslide monitoring, which has brought together different disciplines in important and necessary ways. This action research project involved foundational ethnographic work (see Sayre, Chapter 34) with communities across multiple locations in Central Nepal that began six months before the slope monitoring work described below, collaborative decision-making with our local partners in selecting the sites where we are working, exploring different knowledges and ways of understanding the subsurface, getting comfortable with the technology, and much more besides. We reflect here on the mixed methods that we are using and tease out our experiences preparing our team to undertake the research through a deep interdisciplinary pedagogy that joined knowledge, sensibility, and practice. AP, NDB, TBD, DB, and ST offer their own reflections on their training and research experience, as well as their research background before joining Sajag-Nepal in 2021. Their reflections are quoted in the chapter and show that while methods training is important for the RAs on the project, cultivating sensitivities and sensibilities that strengthen confidence in working

---

1   This chapter was the initiative of anthropologist Johnson, interdisciplinary geographer Oven, and physical geographer Rosser. With interdisciplinary projects such as Sajag-Nepal often led by scholars in the Global North, a point that we elaborate upon below, we were keen to critically reflect upon project relations and the implications for interdisciplinary practice within the project, with a particular focus on our landslide monitoring work. The author list reflects the contribution to the framing and writing of this chapter.

2   The social science Research Associates (RAs) are based at Social Science Baha, a leading Nepali social science research centre located in Nepal's capital, Kathmandu. In addition to the social science RAs at Social Science Baha, a team of GIS, geoscience, and disaster management RAs are based at the National Society for Earthquake Technology-Nepal (NSET). The NSET-based RAs joined Sajag-Nepal at the same time as the social science RAs and underwent a similar interdisciplinary training programme, however they had limited involvement with the landslide monitoring research which we describe in this chapter. Two geologists, teaching faculty at Tribhuvan University, joined the project as RAs in summer 2022 and had a significant role in the landslide monitoring research we discuss. Unlike the other RAs on the project, they worked part time only on Sajag-Nepal. This had implications for the training they received and were able to participate in which we reflect on further below.

within a CPG research paradigm has been essential to realising the deep interdisciplinary approach envisioned for the project.

Critical Physical Geography: a framework for deep interdisciplinary engagement

Oughton and Bracken (2009) define interdisciplinary research as research that "encompasses all forms of scientific collaboration where the field of a single discipline is transgressed" (p. 385) (see also Lane and Lave, Chapter 3). They emphasise the importance of framing and reframing the research problem, linking project success to the communication of clear conceptual ideas at the outset of research (Oughton and Bracken 2009). For Rigg and Mason (2018: 1031), the productive challenge of deep interdisciplinary research is found in the reconceptualisation of methods, working relationships, and research goals. Revisiting methods, goals, and working relationships therefore has the potential to advance interdisciplinary research by "transcend[ing] the boundaries between disciplines to create new perspectives and new ways of thinking about the world that can be sustained and developed over time" (Clark et al. 2017: 244).

CPG—which brings together the various sub disciplines of geography from geomorphology and climatology to political, social, and cultural geographies (Demeritt 2008)—provides a conceptual framework for interdisciplinary working. CPG is concerned with "studying material landscapes, social dynamics and knowledge politics together" (Lave et al. 2018: 3). It advocates "careful integrative work that addresses both crucial geoscience questions while taking seriously the power relations, economic systems, and socio-cultural and philosophical presumptions upon which modern society has been built" (Lave et al. 2018: 4). In doing so, CPG offers a framework for 'deep' interdisciplinary engagement— the very kind articulated by Rigg and Mason (2018). As a sub-discipline within geography itself, CPG builds on political ecology and other allied disciplines in trying to explain how the political shapes environmental change. For some critics of political ecology, however, the ecology is rarely an equal partner to the political (see, for example, Walker 2005), an issue CPG seeks to address. Indeed, for Lave et al. (2014), in their intervention in *The Canadian Geographer*, "[t]he integrative holism of CPG requires critical human geographers to engage substantively with the physical sciences and the importance of the material environment in shaping social relations, while expanding physical geographers' exposure to and understanding of power relations and human practices that shape physical systems and their own research practices" (p. 4). Engagement is a two-way street.

A particular aspect of CPG that is relevant to Sajag-Nepal is the focus not only on bridging expertise across the human-physical divide, but also the participation of publics in the research process (see, for example, collaborative research by Lane et al. (2011) on flooding and Whitman et al.'s (2015) work with farmers on farm slurry pollution, both in the UK). Our approach within Sajag-Nepal centres local knowledges and expertise in the areas where we are working, something that Lane et al. (2011) refer to as "uncertified expertise". We take this further, however, recognising the plural ways in which the Earth is understood through different Indigenous and local knowledge systems. Here our conceptual framework has also benefited from ideas from political geology, which we argue complement CPG and explicitly ask what knowledge about the Earth is rendered "knowable and sensible" and what is excluded (Bobette and Donovan 2018: 13). The idea of geology as the "view from nowhere" reflecting scientific objectivity is in direct contrast to our perspective which, like political geology, recognises multiple, grounded ways of knowing and "'takes seriously' non-Western traditions of geological thought and experience...dislodging 'our' own categories and terms" (Bobette and Donovan 2018: 19). In this regard, we also align our work with disaster justice scholarship (Shrestha et. al. 2019) and decolonial approaches to knowledge production in disaster studies (Cadag 2022) through epistemic experimentation and collaboration with team members and participants in Nepal.

In terms of 'doing' interdisciplinarity, Clark et al. (2017: 255) highlight the importance of "shaping (voluntary) opportunities to provide landscapes of practice, shared experience of sites and spaces of expertise, without fixed objectives and with an open, critical mind, accompanied by opportunities for dialogue, and developing relationships, [which] can result in a move beyond identification of difference to hybridisation through the notion of boundary experiences." They stress that the coproduction of research through these boundary experiences in shared practice contexts generates new approaches to research methodology in interdisciplinary research. By orienting scholars of different disciplinary backgrounds to a specific socio-physical problem, CPG gives licence for team members to break through knowledge barriers and transcend disciplinary methodologies in the pursuit of solutions and the advancement of research goals.

Generating a shared field of practice, however, involves socialising team members into a critical and self-reflexive stance where boundaries of knowledge are identified and breached. In university settings in the Global North, it has become more commonplace to teach and demonstrate interdisciplinary styles of working as evidenced in the rise of interdisciplinary departments (e.g., Environmental Studies) and grant funding trends (Sukarieh and Tannock 2019). As a consequence, researchers on interdisciplinary projects generated in the context of Global North university settings are more likely to have been exposed to interdisciplinary research and collaborative teamwork before designing and proposing interdisciplinary projects. It should be noted however, that many of the academics leading interdisciplinary projects—at least in the UK—have often only felt comfortable engaging in interdisciplinary working later in their careers and once established within their own field of expertise. The foundations of an academic career remain disciplinary for the great majority of scholars (see Lane and Lave, Chapter 3). One consequence of this pathway is that interdisciplinarity has been largely discovered and self-taught, and so passing on and sharing this learning is not straightforward. In Nepal, the contributing disciplines to the Sajag-Nepal project of anthropology, geography, environmental science, and geology are typically traditionally researched and taught, meaning that forays into interdisciplinarity remain rare. Extending an ethos of interdisciplinary research to all team members equally is therefore a challenge that underlies the implementation of interdisciplinary projects.

Recognising and valuing RAs as interdisciplinary team members

Just as CPG projects bring together different fields of knowledge and methodologies across the social and physical sciences, they also draw upon different traditions in the training and incorporation of RAs in research projects. These disciplinary legacies are also in need of interrogation in CPG projects, which we discovered in the course of implementing research for Sajag-Nepal. For example, anthropology, which prides itself on the image of the solo field researcher, is often silent around the importance of RAs in the coproduction of knowledge. This can be taken as an example of "anthropology's hidden colonialism" (Cons 2014; Middleton and Cons 2014; Middleton and Pradhan 2014; Sanjek 1993). In the physical sciences, where multi-author publishing is more common, RA positions are viewed largely as a stepping stone for the RAs who receive training and gain research experience in the early stages of their careers. In both contexts there are undeniable power dynamics between researcher and assistant (researcher–trainee, employer–employee) that require acknowledgement and negotiation within any research project (Deane and Stevano 2016). Within the Sajag-Nepal project, we sought to tackle this from the outset through the development of a shared set of project principles and a mentorship scheme that guided the way we worked as a team and that we were all committed to following. There was also a need to acknowledge the ways in which intersecting subjectivities (researcher–assistant–participant) shape knowledge production, and hence recognise the critical role RAs play in the unfolding of research, findings, and impacts (Anwar and Vikar 2017; Caretta 2014).

Surprisingly, given the strong reliance on RAs within research projects—including interdisciplinary projects such as Sajag-Nepal—there is little available literature on training in the core conceptual and theoretical ideas driving research objectives, or in research methods. Where literature does exist, for example in the social sciences, the focus tends to be on enumerating surveys and delivering interview schedules (Stevano and Deane 2019). In the physical sciences, there is little reflective writing on the training process, which is often specific to a given project—for example, training a team in landslide mapping for assessing risk. This paucity of literature may reflect a discomfort by some academics about the dependency on RAs in field-based research, especially in contexts where the researcher does not speak the language or have much familiarity with the socio-political context of their research site. For others, the discrepancy may reflect an interest in communicating research findings rather than the process of 'getting there'.

It is precisely the process of 'getting there' that interests us—in particular, how teams are trained in preparation for interdisciplinary research as well as during the research process. This is particularly important as RA training is thought to become more complex when research is expected to be directed more independently by RAs themselves who often act as mediators of fieldwork, gatekeepers, interpreters, the project representatives, interviewers, and much more. Training can also become more challenging when the research is carried out at a distance from principal and co-investigators who are not always physically present in research sites (Nguyen 2022). This scenario is increasingly the reality for North-South collaborative grant-funded projects, such as Sajag-Nepal. Distance in time and space between researchers, which was exacerbated during the COVID-19 pandemic with international travel restrictions (Nguyen 2022), underscored the need for strong training and preparation before and during research so that RAs who are closest to the day-to-day fieldwork are confident to lead research in line with the project's aims, methods, and ethics. The interdisciplinary nature of Sajag-Nepal, with its added complexities, further compounds the need for training.

Within interdisciplinary research, emphasis is placed on the reciprocal learning that emerges from being part of an interdisciplinary research team, which itself is a positive reason for involvement (Lorenzetti et. al. 2022). However, when it comes to preparing research teams for distanced and independent work, there is a need to consider interdisciplinarity as a *method*. Devising conceptual frames and fostering "boundary experiences" in shared practice contexts (in our case, as we explore below, the monitoring of slopes and landslides that is one area of focus of the Sajag-Nepal project) is key to effective interdisciplinary training and working, and this requires a receptive hands-on approach by all.

When we searched for literatures about interdisciplinary training we were struck by the paucity of material on the topic. Pedagogical literature on interdisciplinarity focuses on classroom settings, introducing students to interdisciplinary research approaches and applying them in limited settings outside the classroom. Examples include Malone's (2021) reflections on teaching a CPG class on contamination in community gardens. Pedagogies for the field, especially in the context of interdisciplinary research projects, require a different approach. In line with calls "to strengthen institutional spaces for cross-training" (Lave et al. 2014: 6), we argue that projects such as Sajag-Nepal provide an important training space. Stakes are high for researchers and participants, and the research team relationships are structured through employer-employee dynamics, with the need to deliver skills and mentorship for career progression alongside methods training. In the context of Sajag-Nepal and other international interdisciplinary research projects, we can think of interdisciplinary pedagogy as consisting of three interrelated parts: 1) educating all researchers in the disciplinary orientations and methodological interests featured in the project (knowledge); 2) engaging in boundary experiences that cross disciplinary divides (sensibility); 3) fostering opportunities for experimentation and leadership in field research (practice).

Knowledge: learning across disciplines

Sajag-Nepal set out to recruit four Nepal-based RAs to the project team with experience of performing qualitative social science research but not necessarily familiarity with all the expected methods and disciplines incorporated within the project. RAs were to be engaged primarily in community level research in four predetermined case study municipalities. A further six RAs were recruited with expertise in geology, GIS and computer modelling, and disaster management and were based at different institutions. (see footnote 2). We focus explicitly here on the experiences of the social science RAs. The induction process involved an intense period of training where RAs and other team members collectively gained some fluency in the different disciplinary perspectives represented by researchers in the project. There was thus at the outset a recognition that all of us, RAs included, had different strengths. By teaching from our individual strengths but collectively working across strengths, we wished to level the field of practice in the project so that everyone on our team was empowered to participate as equal members of the research team.

Our Nepal-based social science RAs came to the project from anthropology, environmental science, and sustainable development backgrounds. The three anthropologists (AP, NDB, TBD) had master's degrees, but varying levels of experience conducting ethnographic fieldwork independently. Two had worked at various points as RAs with foreign doctoral students in anthropology. The three anthropologists also had experience conducting field- and desk-based research on consultancy contracts and were familiar with the academic labour market that facilitates international project-based research employment in Nepal. The environmental scientist (ST) had a master's degree and experience conducting field research for his thesis. ST exited the project after the first year to begin a master's degree programme in Europe. DB, with a master's in sustainable development, joined the project in 2022 to replace ST but did not partake in the same introductory training as his peers. All the RAs were eager to share their experiences with the team and were motivated to learn new modes of working both for the benefit of the project as well as for their professional development, with all envisioning some level of further higher education or research (management) roles after their time with Sajag-Nepal.

Sajag-Nepal officially began in January 2021. The Nepal-based RAs were hired in late Spring 2021. Due to the evolving COVID-19 situation, it was not safe or ethically appropriate for the RAs to travel to the rural Himalayan locations where the research was to be conducted. What was initially planned as an intensive few weeks of methods and background training in the early summer expanded into three months. Over June, July, and August 2021 we undertook a virtual induction programme for team members using Zoom as our learning platform.

The first month served as an introduction to the team and the basis for the project, reflecting on prior (interdisciplinary) research that shaped the development of Sajag-Nepal and developing a shared understanding of the research problem through accessible discussions across disciplines. We devoted significant time to outlining our approach to interdisciplinary working, emphasising knowledge sharing amongst team members. We held knowledge sharing forums every week consisting of presentations by team members on their research background and expertise, reading groups on project themes, and group discussions connecting project themes and goals. Each week the RAs gave presentations on their individual backgrounds and research interests to the team, introducing the wider project team to their prior experience and the kinds of research they were interested in pursuing as part of Sajag-Nepal. More senior researchers on the team gave talks about their research background and organised readings related to their research and project themes. Readings were discussed as a group, opening conversation on how our project engaged literatures and methodologies in different disciplines and how we wished to move forward in our interdisciplinary research. Given that the RAs came from anthropology and environmental science backgrounds, there was keen interdisciplinary interest in the research talks and the subsequent reading groups that more senior project researchers created to delve deeper into key subject areas. AP reflected,

> The best part for me during the training was the literature review on various issues connected with disaster like geopolitics, social and physical aspects of disaster, and so on written by authors from various disciplines. To be honest, being used to anthropological literature, I used to get lost reading some of the more technical articles. But I got a lot of insight during discussions in the larger team. This helped us think about disaster issues through multiple lenses. The courses and process of training was very knowledgeable and productive.

As AP described, the process of becoming familiar with concepts and writing from different disciplines was sometimes difficult. Team members depended on each other to help translate concepts and ideas across disciplines. ST explained that,

> as part of the training, we had to read many scientific publications and reflect on our understandings during virtual group sessions. Either before or after the virtual session, the RAs would share if they had any difficulty in understanding any theme or topic in the papers. In many cases, I found myself [an environmental scientist] as a 'bridge' connecting them to pure science literature.

The role of "bridge" rotated amongst team members regularly throughout the induction programme.

The value of gaining perspectives from beyond one's home discipline was noted by TBD, who summarised,

> While social sciences help to understand the cultural practice of disaster, natural science strengthens knowledge of how different landscapes originated in geological time, the formation of rock, stone, sand, and soil as well as the weathering features of these elements. Teamwork with research associates and senior researchers helped to strengthen confidence before heading to fieldwork.

For the first month we facilitated regular interdisciplinary team discussions on core topics of the project, including the mountain hazard chain, vulnerability and resilience, local knowledge, governing disaster, and geopolitics. We ended the month with a team reflection on our model of interdisciplinary thinking and working, highlighting the ways we anticipated joining the dots to frame questions and develop methods suitable for studying the socio-political and physical dynamics of hazards and risks in the Nepal Himalaya.

Research methodologies, fieldworker safety, and ethics of disaster research were the focus of the second month of the induction programme, which was designed to prepare the RAs to conduct and lead the day-to-day field research. We held two sessions each week with the first session devoted to fieldworker safety and ethics training and the second to methodologies. Prior to each session, team members were provided readings and resources to review. Different members of the project team coordinated sessions based on their experience and interest. For example, the methodologies sessions began by focusing on ethnographic and participatory methods (see Sayre, Chapter 34). Anthropologists on the team led discussions about positionality in ethnographic research, ethnographic interview techniques, and participant observation. These discussions were partnered with interactive activities and workshops that provided opportunities to experiment with methods. In the workshops, we co-created and practised the delivery of a person-centred ethnographic interview schedule. We also individually researched examples of participatory visual methods (see Arce-Nazario, Chapter 17; Ingram, Chapter 23) and shared them with each other to spark ideas about different ways to incorporate participatory visual methods in the project that connected to a photography exhibition that had been earlier identified as a public-facing project output. The interactive sessions were also chances to engage critically with project goals and think through how we would put these goals into practice. In this context, the RAs helped deconstruct the meaning of participation in the project by mapping out the uneven power relations amongst actors included as project participants and devising methods for working equitably and ethically amongst participants with different ethnic, caste, and economic backgrounds. Fieldworker safety and ethics sessions were thus closely linked to the methodology sessions such that by the final week, we were able to facilitate a workshop on fieldwork dilemmas in which team members candidly discussed the ethical dimensions of field research and how situations in the field shape research experience and outcomes. Central to our ethics discussions were the unique position of the in-country RAs, who held dual roles as national insiders but cultural outsiders in three of the four research sites, with only one RA (TBD) working in their home area and in the community's first language, Tamang.

In the final phase of the induction programme, we prepared for our first period of fieldwork scheduled for September 2021. It was only at this point that we established which RA would be working in which of the four predetermined project field sites. The field sites had been selected during the initial proposal submission because they were locations where Co-Is on the project had conducted long term research or had long standing personal connections. In assigning RAs to a field site, we were therefore assigning them a research partner and advisor (one of the project Co-Is) who would collaborate with and support the research carried out in the field, while also providing direction in reference to the overall aims of the project. During the third month of the induction programme, we were concerned with building the RAs' knowledge as burgeoning "experts" in their field sites and as interdisciplinary-minded ethnographic fieldworkers. We facilitated weekly check-in meetings where RAs shared what they were learning about the socio-cultural, political-legal, and geographical profiles of their field sites and the locale's disaster history. We also began a series of workshops to create research tools collaboratively. We worked together to design informed consent forms, project information sheets, and interview guides in English and Nepali languages. Our practice of collaborative working has continued throughout the project as we develop new interview schedules and coordinate fieldwork to focus on different dimensions of our interdisciplinary research, such as monitoring landslides, which will be discussed below in detail. We ended our induction programme by holding several sessions about fieldwork pragmatics and logistics. These meetings covered how to develop a daily field note writing practice, how to store data in line with institutional and governmental guidelines, how to record and anonymise participant data, how to begin preliminary data analysis during fieldwork, and how we planned to share and organise information across the project.

The three-month induction programme was ambitious. We were able to execute it in part because of the unique context of COVID-19. Although at the time we felt the pressure to begin field research, in hindsight we credit the induction programme for providing a foundation for our deep interdisciplinary style of working and for bringing the RAs into the project more fully as team members. As we moved forward in the project and needed to hire additional RAs, we found it difficult to facilitate the same kind of interdisciplinary interactive training for newer arrivals given that we were at a different stage of the research process. This difference in training was felt in the field when we went to undertake slope monitoring, as discussed below.

Sensibility: Ethnography as a steppingstone for interdisciplinary exploration

The temporal and spatial distance between the Nepal-based RAs and Co-Is and internationally based Co-Is and Research Fellows, made greater during the pandemic, meant that the Nepal-based social science RAs were largely self-directed day-to-day during their individual fieldwork time, with periods of RA-led fieldwork set within a wider collaborative programme with the Co-Is joining in the field. The aim of the induction programme was therefore not only to train the team in interdisciplinary methods but to cultivate a sensibility for interdisciplinary working that was sufficiently resilient to persist over time and space. That is, we sought to develop a willingness to pursue "boundary experiences" (Clarke et. al. 2017) to make sense of the socio-cultural, political, and environmental complexities of interconnected mountain hazards and risks in the project's core field sites. As NDB articulated,

> [interdisciplinary working] helps transcend the binary of science of us and them, and also challenges the prevalent dominant knowledge system. Hence, interdisciplinary thinking and working can bring in the layers of truths to a subject of research [landslide monitoring] without denying and undermining diverse knowledge systems.

Approaching boundaries was, and is, not always comfortable. Reaching the edge of our disciplinary limits elicited excitement. New understanding seemed just on the horizon. But it also unleashed anxiety as we individually felt our disciplinary grounds for knowing slip away and leaned on others in the team for guidance and reassurance. Thus, mixing methods to us required more than combining different disciplinary tools to apply to a research problem. It entailed encouraging an attitude of experimentation that normalised false starts and celebrated insights as they occurred in the course of research. For team members accustomed to short-term, time sensitive projects with no scope for second-guessing, the level of experimentation at the beginning of fieldwork was challenging. We had to assure everyone that a slower pace would deliver deeper and more nuanced knowledge in the long run.

We began our fieldwork by using ethnographic methods (Sayre, Chapter 34) as our primary orientation to people, place, and landscape. Methodologically, ethnographic research builds toward emic knowledge of people and place through an ethical and self-reflective process that involves establishing relationships of trust and reciprocity with participants (see Meadow et al., Chapter 5), slowing down and reassessing aims and objectives, and reflecting critically on researcher positionality, power, and intersubjective knowledge production. Ethnography provided the foundation for our deep interdisciplinary working as it helped cultivate a mode of attention about how knowledge and social relationships are produced in practice amongst people and environment (Sayre, Chapter 34), including amongst team members within the Sajag-Nepal project.

In the beginning, the RAs engaged ethnographic methods to gain a sense of the communities where they were working and how they were experiencing hazard and risk in their particular environments. This provided an important foundation for the interdisciplinary landslide monitoring work that we discuss below. The decisions about which communities to approach for participation in the research were informed by the ethnographic research as well as the landslide mapping work. For example, as shared by TBD: "I also benefited from the landslide scientists especially because it helped in the selection of the core and secondary field sites when I received a hazard map." The time spent participating in daily routines, observing village life, conversing with participants, and documenting landscape changes helped us build a more comprehensive understanding of overlapping project themes, such as local knowledges, disaster governance, exposure and vulnerability, and infrastructure. Moreover, we were able to begin developing interpersonal relations with participants as the RAs repeatedly visited the field sites and built rapport with communities. DB, new to ethnographic fieldwork when he joined the project in June 2022, explained how "the training in social science helped to build field relations which during the monitoring work were very useful." As these relationships deepened, we were able to better communicate the more novel interdisciplinary ambitions of the project and ethically address the scope of possibility for collaborative interdisciplinary working with research participants.

Practice: Sensing landslides

In the context of the slope monitoring work, our starting point was to learn more about landsliding from the perspective of rural residents across multiple sites in Central-Western Nepal. Specifically, we were interested in learning about the locations and causes of landsliding; how landslide hazards are perceived and understood by those impacted; the influence of landslide hazard and risk on settlement patterns and the implications for exposure and vulnerability; and how landslides and landslide risk are managed, as well as documenting the questions and concerns in relation to landsliding of rural residents themselves. This involved multiple periods of ethnographic work including semi-structured interviews (Johnston and Longhurst, Chapter 32) and participant observation (Sayre, Chapter 34) undertaken by the RAs in Nepal.

Building relationships locally was key and these relationships facilitated the next phase of the research. Here we brought together different data sets: landslide hazard maps produced by the physical scientists on the team (Rosser et al. 2021; Kincey et al. 2022) and the findings from the ethnographic work to begin to identify possible slopes for monitoring. This was an iterative process involving—as Lave et al. (2018) describe—a back and forth between the physical and social scientists, as well as our interlocutors in the case study municipalities, including rural residents and local government officials, as we sought to identify sites appropriate for slope monitoring.

For the geomorphologists on the team such as NR, their starting point was a lack of understanding of how, when, and why slopes fail in the Middle Hills—questions that mirrored those of community members, who, as NDB described in relation to his fieldwork in Annapurna Rural Municipality, "know that their land has been continuously subsiding, yet they don't know at what rate or under what circumstances." Responding to calls for the development of more informed planning and early warning systems but recognising that the science is some way from being able to produce a reliable alarm, members of the team were interested in approaches which focused on understanding deformation below the surface where material strength, water pressure, and gravity interact to define where and when landslides occur. The funding call also set aspirations for scientific novelty, which led to the adoption of an innovative acoustic emissions (AE) sensor as the focus of our instrumentation that effectively 'listens' to subsurface deformation associated with ongoing landsliding across a number of sites. Our aim of directly monitoring extremely small movements (sub-mm) was also a means of providing a tangible, relatable indicator of landslide risk by articulating the local conditions that cause landslides to start and to stop. This intentionally avoided a requirement to appreciate often complex process linkages from, for example, rainfall to landsliding. Thinking through how to approach this raised questions around local understandings of what is below the ground surface and what it would mean to disturb the ground to monitor movement in this way.

Taking this work forward was a team effort and required flexibility from team members as plans for how and at what speed to implement slope monitoring were adapted over time. While slope monitoring was always a focus of our research, we had not decided at the beginning of the project what method we would use to do so. Holding back on methodological specificity was intentional as we wanted conditions in the field sites to direct us toward a methodology, rather than choosing a methodology before learning more about the places and communities involved. However, the iterative nature of our decision making created some tensions between the RAs, Co-Is, and community members. As AP shared, "I think slope monitoring is one of the most important steps that Sajag has taken which is beneficial for both the community and the Sajag team. I think it would be easier for us to communicate with the locals and stakeholders if the plan of installing the [slope monitoring equipment] was shared from the beginning." NDB added, "I also felt we (RAs) were not involved and informed from the beginning of the [design of the] landslide monitoring process; hence, it came to me in a parachuted way—when the team mentioned our [social science RAs] involvement in the installation and communication of the landslide monitoring equipment." Contemplating these reflections, the team recognised the intrinsic challenges of coordinating research across different parts of a large interdisciplinary project, particularly when all project components are co-evolving, and deliberating the most appropriate path for the research to follow. These reflections highlight the pressures placed on RAs when supporting and facilitating experiential methods and obtaining community consent for what can be 'blue skies' research, a topic we visited in our induction training, but which should have been raised again during the research process as we headed into the logistical implementation of our chosen slope monitoring method.

Implementing slope monitoring in the case study municipalities required us to revisit training across the physical and social divide, and what Lave et al. (2014) refer to as a "methodological retooling" (p. 6). As members of a team there was certainly no expectation that the social science RAs would become expert geomorphologists or field technicians, rather that they would develop through their training "a basic competence in—and mutual respect for—the methodological frameworks of CPG collaborators" (Lave et al. 2014: 6). Nonetheless, as the main interlocutors for the project locally, the RAs expressed feelings of inadequacy over the course of the project, as they were often asked specific scientific questions by community members that they did not necessarily have the knowledge to answer. DB illustrated the depths of this feeling when he reflected:

> Although I was from social science background, while talking about the monitoring work locals usually misjudged me as an engineer—which I was not. People asked about what the rock and soil was like in their place? Was their place vulnerable to landslides? How can the landslides be mitigated? I did not have answers to these questions, and I always

used to clarify to locals that I come from a social science background. Since we had geologists and geographers in the team, I encouraged the locals to ask their questions to them. This was one of the advantages of working in an interdisciplinary project. However, inside I felt that [a more in-depth] understanding of the geography and landslides would have been helpful for my own understanding of the landscape.

As DB's reflections revealed, local people involved in the research project were less keen on distinguishing between expertise amongst team members or the fine separation of disciplines and skills that academics care about. They did, however, pick up on other forms of distinction which were deemed more important than disciplinary ones, such as age, gender, nationality, race, caste, or ethnicity. Consequently, when engaged as the expert for the project, the RAs were expected by community members to be able to describe in geological terms the changes they were observing in their landscapes. Thinking through to what depth physical and social science training was needed by the RAs was therefore a subject of negotiation as the fieldwork progressed. This retooling provided an opportunity to develop a deeper knowledge of landsliding and landslide processes. It also precipitated a series of discussions with geomorphologists on the Sajag-Nepal project about the mechanics of the AE sensors, both for the understanding of the technology and science by the RAs and other team members.

For the RAs, interactions with geomorphologists about the AE sensors were especially important because they anticipated the need to answer questions from local communities who were interested in knowing the practicalities of installation, including how the equipment would be installed, the nature of the data that would be generated, and how data could be shared and discussed with local partners in useful and productive ways. As AP reflected:

The training we received online regarding the slope monitoring equipment was really very useful to me. When the team was talking about the slope monitoring equipment initially, I was puzzled as I didn't know about that. But later when Nick [Rosser] and team clarified about that then I learned a lot and became clearer about it. Of course, yours [Katie Oven/Nick Rosser/Amy Johnson] and the team visit to Dolakha was very insightful. I learned a lot from the team about the geography, risk, and so on. In return I could communicate with the local people accordingly regarding the causes of risk and their geological condition. The debrief session was wonderful as we were able to share our thoughts, what locals wanted, and what local government wanted. I believe because of that we became able to address the concern of both locals and government officials.

The other project RAs, based at Tribhuvan University, who were involved more in the technical aspects of the equipment installation and analysis and who came from geology backgrounds, had not received the same interdisciplinary training described above. While it is often assumed that social scientists need to be trained in geology but geologists (or other physical scientists) do not require equivalent social science training, this was far from the case within the Sajag-Nepal project team. The RA positions in this instance were taken up by faculty members rather than more early career researchers. While their geological expertise and field experience was welcome, they were also engaged in teaching within their academic roles and so were unable to engage in the project with the same intensity and continuity as the social science RAs. There was also an uneven power dynamic reflecting career stage which persisted in the field despite the fact that our project principles sought to address such issues.

The difference in approach to interdisciplinary working between the two groups was stark. As DB shared,

> At times, when we had to coordinate with people for support to the installation, the geologists used to ask us to talk with people, since according to the geologists the 'social' side was my work. 'You guys talk to people; I talk to the stone and soil.' This distinction at once highlighted differences in our academic and professional training but also an interdisciplinary approach to understanding the Earth.

Experiencing this difference in the field has made us more confident in the importance of training in interdisciplinarity as a method and as a sensibility, and the need to devote time and energy into training at the outset of interdisciplinary projects to ensure that all team members are invested in interdisciplinary and more egalitarian working. For ST, the experience of interdisciplinary working underlined the power dynamics and ethics negotiated in shared practice contexts. "The collaboration among RAs from Nepal (both social and physical scientists), Co-Investigators and other researchers within the Sajag project and their distinct roles clearly explains dynamics of power sharing and a recognition of its importance in carrying out research activities successfully in an ethical manner."

Reflections

The increasing funding for Global North-South partnered interdisciplinary research collaborations has facilitated a rise in large multi-country and multi-institutional interdisciplinary research projects that are often reliant on in-country RAs to carry out day-to-day research and provide the face for the project locally. This reflects a growing appreciation that researchers local to the place of study are often best placed to conduct the research (see Gaillard and Peek 2019 in the context of natural hazards-focused research); as well as recognition of practical issues, not least the time and distance between researchers based in different countries and institutions. Pedagogy of interdisciplinary research has not caught up with this trend. Indeed, in reflecting on our experience with Sajag-Nepal, we strongly argue for the need for more sustained attention to pedagogy in interdisciplinary research, particularly in the training of team members for deep interdisciplinary styles of working. More attention to training applies for all researchers involved in such projects but may be felt most acutely by in-country RAs, especially in the Global South, given the unevenness within the political economy of academic labour, with short-term contracts, consultancy work, and limited funding for independent or experimental academic research the norm. With these considerations in mind, we offer some reflections on knowledge building, cultivating an interdisciplinary sensibility, and encouraging experimentation and adaptability in interdisciplinary research practices which, we argue, can be better foregrounded when training researchers to implement CPG-oriented interdisciplinary research projects.

The purpose of interdisciplinary training, as our experience suggests, is less to teach methods than it is to foster an appreciation of and a curiosity about disciplinary boundary experiences in shared practice contexts. The experience of cross-disciplinary knowledge building during training likewise creates a shared conception of the problem or problems being addressed by the research and the various motivations and perspectives that are being brought to bear on that problem. The willingness to experiment, to have false starts, and to learn from experience is crucial to achieving deep interdisciplinarity, and time and energy should be invested to provide researchers the opportunity to grow a sensibility for interdisciplinary working rather than presuming it is inherent amongst researchers involved in interdisciplinary research projects. At the same time, we stress that academic support and mentorship need to be provided early and consistently to enable researchers to become conversant at the interface of multiple disciplinary knowledges and approaches represented in a given research project. Doing so helps to level the field of knowledge amongst researchers and promote equity within the research team. But it also prepares researchers to engage local partners confidently as changes in methods and research direction evolve inevitably over the course of a research project.

For these reasons we feel that training must build from a strong awareness of the broader aims and objectives of the wider project elements. Inevitably, whilst we did do this, there was a lot to take in and some of the training at the very start of the project did not make sense to the RAs until the research was actually underway. An important foundation in our training was discussions around what 'research' actually means to us, with all of its messy, meandering and uneven pathways. This was significant particularly in the context of Nepal where development projects, and the often-formulaic nature of development consultancy work, prescribe distinct project deliverables as the modus operandi. Enquiry-driven research unfolds at a different temporality than the performance indicator-driven pace of consultancy. Training within interdisciplinary research teams can better emphasise the adaptive and open-ended nature of collaborative research processes, preparing team members for the excitement (and sometimes unwieldiness) encountered when implementing a novel method and the need for pause and reflection to ensure that research moves forward ethically in line with community needs and research goals.

Our work, and in particular the interactions of our RAs with communities, made clear the need for a basic level of awareness of the wider project methods amongst team members. But critically the depth of this knowledge needed to be negotiated, treading a fine line between the RAs being conversant in what is being done and knowing everything about doing it. Thus, there was no expectation for our RAs to answer every question that they might encounter. At the same time, we did develop a notion that it was important to have an awareness of different people's skills and expertise and perspectives in the wider project team, such that when questions did arise, the RAs knew where to seek out the answers. A critical means of facilitating this was breaking down power hierarchies amongst team members so that the RAs were comfortable in passing on questions. To ask or convey these questions, we felt that it was important to be conversant in a project-level interdisciplinary language that was intelligible and accessible to all involved, but that maintained respect for each other's disciplines, experience, and training. Finally, given the often-exploratory nature of the research, we recognised the importance of nurturing flexibility to negotiate a research pathway that was convoluted at times, with different pressures waxing and waning as the project evolved. For this reason, we find it essential to invest time and attention to the pedagogy of interdisciplinarity, creating conditions for the unfolding of more equitable fields of knowledge, sensibility, and practice amongst all team members involved in interdisciplinary research.

References cited

Anwar, N.H. and S. Viqar. 2017. 'Research assistants, reflexivity and the politics of fieldwork in urban Pakistan', *Area*, 49.1, pp. 114–121. https://doi.org/10.1111/area.12307.

Arce-Nazario, J., Chapter 17, this volume. 'Space and place in participatory arts-based research'.

Bobbette, A. and A. Donovan. 2018. 'Political geology: an introduction', in *Political Geology: Active Stratigraphies and the Making of Life*, ed. by A. Bobbette and A. Donovan (Springer), pp. 1–34. https://doi.org/10.1007/978-3-319-98189-5_1.

Cadag, J.R. 2022. 'Decolonising disasters', *Disasters*, 46.4, pp. 1121–1126. https://doi.org/10.1111/disa.12550.

Caretta, M.A. 2015. 'Situated knowledge in cross-cultural, cross-language research: A collaborative reflexive analysis of researcher, assistant and participant subjectivities', *Qualitative Research*, 15.4, pp. 489–505. https://doi.org/10.1177/1468794114543404.

Castree, N. 2012. 'Progressing physical geography', *Progress in Physical Geography*, 36.3, pp. 298–304. https://doi.org/10.1177/0309133312436456.

Clark, J., et al. 2017. 'Transformation in interdisciplinary research methodology: the importance of shared experiences in landscapes of practice', *International Journal of Research and Method in Education*, 40.3, pp. 243–256. https://doi.org/10.1080/1743727x.2017.1281902.

Cons, J. 2014. 'Field dependencies: Mediation, addiction and anxious fieldwork at the India-Bangladesh border', *Ethnography*, 15.3, pp. 375–393. https://doi.org/10.1177/1466138114533457.

Deane, K., and S. Stevano. 2016. 'Towards a political economy of the use of research assistants: reflections from fieldwork in Tanzania and Mozambique', *Qualitative Research*, 16.2, pp. 213–228. https://doi.org/10.1177/1468794115578776.

Gailliard, J.C. and L. Peek. 2019. 'Disaster-zone researchers need a code of conduct', *Nature*, 575, pp. 440–442. https://doi.org/10.1038/d41586-019-03534-z.

Ingram, M., Chapter 23, this volume. 'Arts-based environmental research'.

Lane, S.N., N. Odoni, C. Landstrom, S.J. Whatmore, N. Ward, and S. Bradley. 2011. 'Doing flood risk science differently: an experiment in radical scientific method', *Transactions of the Institute of British Geographers*, 36.1, pp. 15–36. https://doi.org/10.1111/j.1475-5661.2010.00410.x.

Lane, S.N. and Lave, R., Chapter 3, this volume. 'Frames, disciplines and mixing methods in environmental research'.

Lave, R., et al. 2014. 'Intervention: Critical physical geography', *The Canadian Geographer/Le Géographe canadien*, 58.1, pp. 1–10.

Lave, R., C. Biermann, and S.N. Lane. 2018. 'Introducing critical physical geography', *The Palgrave Handbook of Critical Physical Geography*, ed. by R. Lave, C. Biermann, and S.N. Lane (Palgrave Macmillan), pp. 3–21. https://doi.org/10.1007/978-3-319-71461-5_1.

Longhurst, R. and Johnston, L., Chapter 27, this volume. 'Focus groups'.

Johnston and Longhurst, Chapter 32, this volume. 'Interviews: Structured, semi-structured and open-ended'.

Lorenzetti, L., et al. 2022. 'Fostering learning and reciprocity in interdisciplinary research', *Small Group Research*, 53.5, pp. 755–777. https://doi.org/10.1177/10464964221089836.

Malone, M. 2021. 'Teaching critical physical geography', *Journal of Geography in Higher Education*, 45.3, pp. 465–478. https://doi.org/10.1080/03098265.2020.1847051.

Middleton, T. and J. Cons. 2014. 'Coming to terms: Reinserting research assistants into ethnography's past and present', *Ethnography*, 15.3, pp. 279–290. https://doi.org/10.1177/1466138114533466.

Middleton, T. and E. Pradhan. 2014. 'Dynamic duos: On partnership and the possibilities of postcolonial ethnography', *Ethnography*, 15.3, pp. 355–374. https://doi.org/10.1177/1466138114533451.

Nguyen, P., et al. 2022. 'From a distance: The "new normal" for researchers and research assistants engaged in remote fieldwork', *International Journal of Qualitative Methods*, 21. https://doi.org/10.1177/16094069221089108.

Oughton, E. and L. Bracken. 2019. 'Interdisciplinary research: framing and reframing', *Area*, 41.4, pp. 385−394. https://doi.org/10.1111/j.1475-4762.2009.00903.x.

Rigg, J. and L.R. Mason. 2018. 'Five dimensions of climate science reductionism', *Nature Climate Change*, 8, pp. 1030−1032. https://doi.org/10.1038/s41558-018-0352-1.

Sanjek, R. 1993. 'Anthropology's hidden colonialism: Assistants and their ethnographers', *Anthropology Today*, 9.2, pp. 13−18. https://doi.org/10.2307/2783170.

Sayre, N.F., Chapter 34, this volume. 'Participant observation and ethnography'.

Shanker, S., et al. 2021. 'The interdisciplinary research team not the interdisciplinarist', *Europasian Journal of Medical Sciences*, 3.2, pp. 104−108. https://doi.org/10.46405/ejms.v3i2.317.

Shrestha, K., et al. 2019. 'Disaster justice in Nepal's earthquake recovery', *International Journal of Disaster Risk Reduction*, 33, pp. 207−216. https://doi.org/10.1016/j.ijdrr.2018.10.006.

Stevano, S. and K. Deane. 2019. 'The role of research assistants in qualitative and cross-cultural social sciences research', in *Handbook of Research Methods in Health Social Sciences*, ed. by P. Liamputtong (Springer). https://doi.org/10.1007/978-981-10-5251-4_39.

Sukarieh, M. and S. Tannock. 2019. 'Subcontracting academia: alienation, exploitation and disillusionment in the UK overseas Syrian refugee research industry', *Antipode*, 51.2, pp. 664−680. https://doi.org/10.1111/anti.12502.

Walker, P. 2005. 'Political ecology: where is the ecology?', *Progress in Human Geography*, 29.1, pp. 73−82. https://doi.org/10.1191/0309132505ph530pr.

Whitman, G.P., R. Pain, and D.G. Milledge. 2015. 'Going with the flow? Using participatory action research in physical geography', *Progress in Physical Geography*, 39.5, pp. 622−639. https://doi.org/10.1177/0309133315589707.

8. The environmental impacts of fieldwork: making an environmental impact statement

Stuart N. Lane

Introduction

In May 2023, I was supporting the fieldwork of a graduate student who was working on a small outlet glacier of the Greenland Ice Sheet. As part of her PhD thesis, she needed to install equipment, plan for UAV (drone) surveys (see Kasvi, Chapter 45), and think through the risks and dangers associated with her project. Her glacier was relatively accessible from Kangerlussuaq, a nearby town which is the major international airport of Greenland and hub for within-Greenland flights, although it has no road connecting it with anywhere else. From Kangerlussuaq, the field site could be reached after a 45-minute drive to the roadhead, a river crossing, and about one hour of walking. Aside from trails for off-road vehicles, the environment is largely devoid of evidence of direct human impacts. The main encounters we had were with musk ox and rock ptarmigan. I was thus surprised when we came across two equipment stores about 100 m apart not far from the glacier (Fig. 8.1). Left behind by scientific expeditions during the mid-2010s, the tarpaulins covering them had decayed to reveal camping equipment, scientific equipment, consumables (including dye for hydrological tracing), and even a scrabble game. This glacier has produced crucially important understanding of the environmental biogeochemistry of the Greenland Ice Sheet faced with human-induced rapid climate warming; yet the researchers left a marked human impact on the environment in which they based their research. Immediately, this raises a question of respect and our ethical position with respect to the non-human world (see Meadow et al., Chapter 5).

 https://doi.org/10.11647/OBP.0418.08

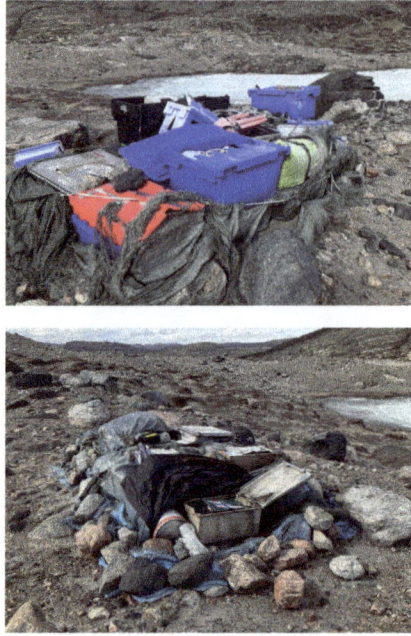

Fig. 8.1 Stuart N. Lane, 2023. Images of two equipment "dumps", 100 m apart, at a glacier in Greenland and left for around 10 years or more. These stores were subsequently removed.

What does this example say about the wider practices we have developed in the name of environmental research? The shock of seeing these equipment dumps opened up, for me, a broader reflection about the environmental costs of field research and how little we reflect about them. This chapter is a practical response to those reflections: a simple environmental impact assessment that allows reflection on environmental questions to be brought into planning for fieldwork.

Replace, reduce, refine as a framework for doing environmental research

The idea that an environmental impact assessment should be an integral part of planning before the decision is taken to do environmental research is not new. For example, the British Antarctic Survey[1] requires

---

1   See https://www.bas.ac.uk/for-staff/polar-predeployment-prep/
     intro-guidelines-and-forms/preliminary-environmental-assessment/

an Environmental Impact Assessment to be completed and approved before research can begin. The aim of such an assessment is not simply to decide what can and can't happen; rather it is to encourage a researcher to ask themselves whether the research needs to be done where it is planned, and whether and how its impact can be reduced.

There are relatively few protocols available for helping to conduct such an assessment for environmental research, but protocols for other research domains can serve as a starting point. One of those is the critical contribution in the last century to how research using animals should be undertaken. Russell and Burch (1959) introduced the 3Rs in relation to testing on animals: can you replace (R1) animals in testing?; can you reduce (R2) the number of animals you test?; and can you refine (R3) how animals are used to improve their welfare? The 3Rs provide a good basis for an environmental impact assessment (and indeed there is a closer parallel for environmental research in the 3Rs of waste management: reduce, reuse, and recycle).

Table 8.1 illustrates the 3Rs as conceived for animal research but in an environmental impact assessment for a planned field data collection campaign on sediment transport and the Greenland Ice Sheet. The assessment follows three broad classes of environmental impact: the scientific activity itself; the local logistics associated with conducting the scientific activity; and transport and travel. Filling in such an assessment encourages a researcher to reflect on whether the activities planned in Greenland could be replaced by a different location with lower transport and travel impacts; or whether it is possible to replace a piece of equipment that is thought to be potentially polluting with another approach, for instance.

If the conclusion is that the planned fieldwork or a particular method cannot be replaced, then the question is whether the impact can be reduced. This might be through limiting the number of participants who are needed; the number of visits required; the duration of the activity; the number of experiments needed, etc. Finally, once reduced as much as possible, if an environmental impact remains, the challenge is whether the fieldwork activity may be refined. For instance, if accessing the field site still requires air travel, then it is possible to limit the environmental impacts by restricting it to segments that can't be done by train. Routinely paying for carbon offsets, notwithstanding debates about their efficiency, could be considered. Developing care in how experiments are done, such as in the manipulation of equipment and in experiments such that environmental impacts are minimised could be considered. "Leave no trace" checks should be routinely planned into fieldwork.

Table 8.1. An environmental impact matrix for implementing a Replace, Reduce, Refine strategy, extended to include Reflection, for a project concerned with monitoring subglacial sediment evacuation from the outlet glacier of the Greenland Ice Sheet.

Environmental impact	Replace – can the activity be replaced to reduce the environmental impact ?	Reduce – how can the magnitude (e.g. duration, frequency) of the activity be reduced ?	Refine – how can the impact of the activity refined to reduce environmental impact ?	Reflect – reflect upon trade offs between activities related to replace/ reduce/refine activities
Section 1. Research Activities : Installation of gauging station (GS)				
Installation of probes and mounting equipment. Main impact is visual.	No replacement possible	Limit gauging station to time period when it is needed; choose a location that is not sensitive for flora and fauna	Design GS to require little modification of installation site; remove installed equipment and infrastructure at end of project	
Section 1. Research Activities : Dye tracing using Rhodamine WT for discharge calibration				
Rhodamine dyes are known to cause risk for aquatic organisms above certain concentrations.	Considered a different tracer; although other approaches may be unsuitable (e.g. salt) or dangerous	Dynamically construct a calibration curve to the point at which no further traces are needed	Estimate minimum dye needed; make sure this is within safe limits. Remove all stored dye at end of project	

Section 1. Research Activities : Suspended sediment sampling				
No environmental impact	Not applicable	Not applicable	Not applicable	

Section 1. Research Activities : Installation of seismic sensors				
Installed away from the river. Only environmental impact is visual	No other method	No reduction possibe	Remove all equipment at the end	

Section 2. Local Logistics: Field camp				
Installation of living and workspace tents, with associated impacts on wildlife, waste etc.	Use local lodgings within driving and walking distance of field site	No local lodgings available/ feasible. Limit number of people involved. Limit spatial footprint of camp.	Export waste for recycling/ disposal Remove all equipment at end of camp. Use walkover survey to remove all traces.	Using local lodgings increases volume of travel between town and field site

Section 3. Transport and Travel: Off road travel between town and field site				
Carbon footprint; disturbance of wildlife	Camp at field site	Limit number of return trips	Consider greener transport (e.g. partial travel by bike if practical)	Camping may have local environmental impacts

Section 3. Transport and Travel: Flights to field site				
Carbon (and other) footprints	Local field sites possible but not for study of an ice sheet outlet-glacier, a poorly studied topic	Limit number of people to max. needed. Focus on 1 long field season rather than multiple visits.	Limit flights to segments that can't be done by train. Carbon offset.	

Reflection, trade-offs and the wider setting

Sometimes 3R analysis reveals the need for trade-offs rather than definitive answers. Trade-offs arise when one or more activities within a project can be realised in different ways, each with different potential environmental impacts. This then requires reflection, where those impacts are compared and contrasted (traded off against one another) in order to shape a final decision as to whether or not to conduct the field campaign and in what form. For this reason, it is important to add a fourth R to the normal 3R study, reflection.

Two considerations should be part of this reflection. First, Table 8.1 illustrates a classic trade-off when undertaking fieldwork: do you camp locally to minimise transport costs when camping has a local environmental impact? Or do you sleep farther away in a town or village where there is accommodation but then you must travel to the field site? Answering this question not only requires reflection about environmental impact but also about other elements of fieldwork planning. For instance, if stopping further away means additional round trips to a field site, is this sustainable in terms of personal welfare, health, safety, and GHG emissions? If the answer is that it is environmentally preferable to camp locally, how is this balanced against, say, health and safety risks associated with polar bears within the region and the potential environmental impacts of setting camp in a relatively untouched area? These specific examples emphasise that an environmental impact assessment should not be designed as an authoritative analysis that allows you to make simple decisions. Rather, it is a tool that helps you to make sure the environmental consequences of research are not forgotten. Environmental sustainability is only one element of sustainability more widely, as the Sustainable Development Goals recognise, and this makes broader reflection particularly important.

Second, beyond the 4Rs, any environmental impact assessment should be subsidiary to local protocols, regulations, customs, and practices that constrain what is a legitimate fieldwork activity. Some of these will be explicitly designed to minimise environmental impact and may affect the trade-offs between different proposed activities. However, consideration of the environmental impacts of research ought to be situated within broader questions regarding what is needed to make research practices more sustainable. Sustainability for example (see the Sustainable Development Goals[2]) does not have only an environmental axis, even though we have a tendency to reduce "sustainability" to "environmental sustainability". The original Brundtland Report of 1987[3] that introduced the notion of "sustainable development" was very clear that development and the environment have to be linked. The SDGs, have well-being, gender equality, decent work, inequalities, justice, and partnerships as themes that all go well beyond environmental ones, but which are bound up with them. In this broader frame, the question of whether we should travel in order to do fieldwork and what we do when we get there is bound up with who is doing that fieldwork and where; why we are doing it and not those who live where it is to be done; and what the consequences might be for those who are perhaps less able to do that fieldwork than we are ourselves (see Miesen and Gevers, Chapter 9). This chapter then has a certain instrumentality in presenting a framework for thinking through the environmental impacts of fieldwork but it should not obscure wider questions regarding the sustainability, broadly defined, of research practice in general (e.g., Wassénius et al. 2023) and fieldwork in particular.

References cited

Miesen, F. and Gevers, M. Chapter 9, this volume. 'Inclusive practices in fieldwork'.

Russell, W.M.S. and R.L. Burch. 1959. *The Principles of Humane Experimental Technique* (Methuen).

Wassénius, E., Bunge, A.C., Scheuermann, M.K. et al. 2023. 'Creative destruction in academia: a time to reimagine practices in alignment with sustainability values', *Sustain Sci*, 18, 2769–2775.

---

2   https://sdgs.un.org/goals
3   https://sustainabledevelopment.un.org/content/documents/5987our-common-future.pdf

9. Inclusive practices in fieldwork

Floreana Miesen and Marjolein Gevers

Introduction

Despite the importance of laboratory-based research and simulation modelling for the environmental and geosciences, fieldwork remains crucial for collecting the samples and data that these approaches need; as well as providing information for challenging and refining established paradigms and developing new ones (Allen 2014). Likewise, fieldwork and field (-based) classes can contribute to personal and career development, fostering practical and transversal skills, personal resilience in the face of challenges, and camaraderie (John and Khan 2018).

However, fieldwork comes with its own set of challenges, exposing individuals, who may be inexperienced Early Career Researchers, to unfamiliar surroundings, uncomfortable conditions, and potential physical and interpersonal hazards (Clancy et al. 2014; Hill 2021; Nordseth et al. 2023). This environment can also be a breeding ground for exclusion, discrimination, and harassment, coinciding with the inherent stresses of fieldwork, including isolation and the challenges of experiencing different scientific cultures (Clancy et al. 2014; John and Khan 2018). The historical portrayal of fieldwork as a male-dominated and physically demanding endeavour has perpetuated biases linked to gender, identity, sexuality, and societal roles, shaping group dynamics during fieldwork, and influencing career trajectories and hiring decisions (Bracken and Mawdsley 2004; Church 2013; Nelson et al. 2017; Bernard and Cooperdock 2018; Bush and Mattox 2020; Chiarella and Vurro 2020; Marín-Spiotta et al. 2020; Berhe et al. 2022; Dance et al. 2024).

Recent efforts towards more inclusive fieldwork practices (Paul and Darke 2004; Vila-Concejo et al. 2018; Iversen et al. 2020; Demery and Pipkin 2021; Hill et al. 2021; Zebracki and Greatrick 2022; Karplus et al. 2022; Osiecka 2022; Coon et al. 2023; Dalrymple and Lane 2024) highlight the potential of diverse groups in enhancing problem-solving and scientific innovation (Amano et al. 2023). But this potential can only be realised if fieldwork itself evolves away from its established practices through broader discussions, encouraging considerations and strategies and implementing proposals for a more equitable and enriched fieldwork experience for everyone involved (see Meadow et al., Chapter 5). The aim of this chapter is to provide some suggestions for how fieldwork and field classes can be conducted differently to realise a more positive experience for fieldwork participants.

Making fieldwork more inclusive for all

1. Physical fitness

Navigating challenging terrains, enduring harsh weather, handling heavy equipment, repetitive movements, and having to adopt ergonomically challenging positions during fieldwork can be difficult for anyone at any physical fitness level. Notably, not all fieldwork demands exceptional fitness or strength, and the impediment may not necessarily be in the inherent physical demands of the tasks (e.g., walking or carrying loads), but rather in a narrative of competitiveness and toughness (e.g., fast pace) that makes these demands more challenging than they need be (Bracken and Mawdsley 2004). As fieldwork is often a significantly costly part of a research project or field class, cost-time efficiency concerns (Williams et al. 1999) create crammed schedules with a focus on data collection to maximise the available field time and little buffer time for rest (Vero 2021; Dance et al. 2024). Fieldwork protective equipment may only be available in standard male sizes, posing safety risks and adding discomfort for those who have to use ill-fitting gear (Arnold 2022). Expensive personal clothing and footwear appropriate for weather and terrain are rarely financed by the research institution and can put a significant financial burden on individuals, especially undergraduate students.

Recommendations

- Integrate transparent communication about physical requirements and limitations during fieldwork planning. A simulation day where planned activities are practised under similar possible environmental conditions may be of value; alternatively, spending the first day familiarising participants with the environment they are working in and how to manage physical demands may be useful.

- Recognise the importance of sufficient recovery time by planning suitable breaks in fieldwork schedules.

- Encourage participants to seek appropriate medical advice and health checks before embarking on fieldwork (Day 2014).

- Consider offering optional adaptive training programmes or resources for individuals aiming to enhance physical fitness prior to fieldwork (Day 2014). For research employees, this should be explicit in their work contract.

- Respect the dietary wishes and needs of participants, while ensuring adequate catering to meet the high energy demands of fieldwork (Gifford 2004).

- Provide personal protective equipment, including dry suits and waders, appropriate for diverse body sizes and suitable for all genders (e.g., drysuits with relief zippers designed for both male and female bodies, correctly fitting life vests accounting for all body weights) that are tested well in advance before the fieldwork.

- Explore means to support team members or students with insufficient means to purchase adequate personal outdoor clothing, e.g., financial support or a stock or rentable gear (raincoats, boots etc.) in a wide range of sizes.

2. Managing privacy needs

Limited access to proper toilet facilities and cultural expectations around toileting practices can be burdensome in remote field settings. Reducing water intake to avoid exposure while urinating has been reported (Greene et al. 2020; Miesen 2020; Nash 2023) and poses the risk of dehydration (Adan 2012; El-Sharkawy et al. 2015). Hiding may put people at risk in challenging terrain or when dangerous wildlife is present. Menstruating individuals may experience temporary declines in physical fitness and comfort, adding to the stigma experienced around menstruation, especially in male-dominated settings (Palinkas 2003; Miesen 2020; Das 2023; Nash 2023; Dalrymple and Lane 2024; Walker 2024). The lack of private disposal facilities for menstruation products like tampons further increases the risk of toxic shock syndrome (Shands et al. 1980).

Toilet breaks, often overlooked in cramped field schedules (Greene et al. 2020), also complicate medical self-care needs, such as insulin injection. Privacy concerns extend beyond the act of exposing oneself to others; they also encompass avoiding being involuntarily pushed into someone else's intimate space (Miesen 2020).

Recommendations

- Lay out the toileting, menstruation, and medical self-care situation during the fieldwork by providing the information needed well in advance of the fieldwork outing. Set the correct expectations with regards to schedule and regular toilet breaks (Greene et al. 2020; Dalrymple and Lane 2024) and communicate about who to talk to in case of any private needs.

- Ask and inform all team members to use private spaces for toilet needs where possible. In an isolated setting, encourage equal toileting routines (e.g., out of sight of others, if safe enough; asking others to turn around if in a setting where hiding is not an option, e.g., when roped up on a glacier—this requires a good level of trust and respect within the team).

- Provide access to private spaces suitable for medical self-care. In remote settings, a makeshift shelter with a signalling system may be useful.

- Address water intake concerns by promoting a supportive environment that encourages proper hydration without compromising privacy or safety.

- Implement flexible work schedules to accommodate varying energy levels and physical ability and comfort during fieldwork, especially for menstruating individuals.

- Ensure unrestricted access to menstrual hygiene products (for example in first aid kits if national regulations allow this; or at a known and common location at the accommodation) and safe and private facilities for disposal.

- Where possible, provide access to gender-neutral toilets in addition to the default female/male facilities.

3. Accessible fieldwork environments

Traditional fieldwork practices commonly pose barriers to individuals with disabilities (Marshall et al. 2022). Disabilities encompass a wide range, including emotional, educational, sensory, communication, or physical impairments, and these may be invisible. The complexity of disabilities means that providing guidance that can cover all possible situations is not possible, and so associated recommendations must remain general.

Recommendations

- If the field team constellation is already known, initiate fieldwork preparation with a focus on accessibility, addressing potential barriers. Otherwise prepare to adapt the plan to specific needs, once requirements of participants are clear.

- Maintain regular, open communication with participants with known disabilities to determine if their requirements are being met or the fieldwork plans need to be modified (Chiarella and Vurro 2020).

- Where specific disabilities have been identified during upcoming fieldwork, undertake a site assessment to identify specific challenges that might be associated with those disabilities, and then take action to compensate for them. This could include, for instance, devising solutions like rerouting access or adding colour coding and tactile markings to equipment for those with visual impairments (Paul and Darke 2004).

4. Caregiving responsibilities

Caregiving responsibilities of any kind (informal caregivers, parents) may affect the willingness to and feelings about participating in fieldwork. Caregiving influences the perceived risk in fieldwork, as the consequences of an incident extend beyond the individual (Jenkins 2020). Policies and expectations of uninterrupted commitment to fieldwork, especially where these are not sensitive to those with care responsibilities, may lead to fears of negative career impacts (Lininger et al. 2021). Visible physical (e.g., breastfeeding, post-partum recovery) and invisible (e.g., sleep deprivation, anxieties) challenges may reinforce bias regarding the ability of caregivers to carry out fieldwork. Individuals often experience mental stress of being away from those for whom they have care responsibilities, such as children, and juggling associated responsibilities in the field (Drozdzewski and Robinson 2015; Jenkins, 2020).

Recommendations

- Provide privacy, flexible schedules and replacements to accommodate physical limitations during pregnancy and post-partum as well as breastfeeding during data collection (Lininger et al. 2021).

- Respect lower risk acceptance for those with care responsibilities by considering alternative field locations, data collection strategies etc.

- Provide financial support for additional caregivers at home or in the field for children (Lininger et al. 2021).

- Account for internet coverage when selecting field sites or provide access to communication means (e.g., satellite phones) to maintain communication with those for whom you have care responsibilities (Jenkins 2020).

5. Addressing prejudice

Prejudice-driven conflicts and discrimination from local communities or within the field team may especially emerge in isolated settings (Coon et al. 2023). Individuals may feel 'othered', facing discomfort and safety concerns (Demery and Pipkin 2021) and are often left alone to manage their own risks (Zebracki and Greatrick 2022). Researchers who have experienced traumatic situations arising from identity-based harassment related to gender identity, sexuality, religion, race, ethnicity, socio-economic status, or ability often chose to alter their behaviour and research career trajectories (Demery and Pipkin 2021). Identity-based cautionary measures that are labelled (e.g., "gender-sensitive") reinforce stereotypes and potentially influence hiring decisions for positions involving fieldwork (Hamilton and Fielding 2018) and omit that individuals can also be members of several minority-identities (Cech 2022).

Recommendations

- Foster a positive team atmosphere by assuming good intentions, promoting collaboration for marginalised and privileged members. Emphasise shared goals for optimal results (Demery and Pipkin 2021; Nordseth et al. 2023).

- Provide private sleeping accommodation, where possible.

- Consider religious and cultural sensitivities around food and alcohol, including fasting times.

- Allocate space and time in the schedule for religious practices.

- Assess risk factors in fieldwork planning, including LGTBQ+ phobia laws and political sensitivities (Zebracki and Greatrick 2022).

- Cultivate relationships with local communities for cultural understanding, preventing misunderstandings, e.g., by engaging them in your field research objectives, plans, and results (Rasch et al. 2019a).

Working as a field team

1. Communication

Poor communication during fieldwork can create unsafe working conditions, an unnecessary sense of urgency, lower productivity, and a decrease in the quality of work, all of which can lead to unsatisfied team members and cause tension and stress (Schneider et al. 2021). This is especially the case if essential information or reasoning for decisions is held back from individuals (Osiecka 2022).

In addition to this, language barriers can significantly impact individuals' sense of inclusion, respect, and safety within the team. Language differences may create feelings of exclusion and hinder clear understanding, potentially leading to misunderstandings or misinterpretations (Amano et al. 2023). Challenges may also arise when the principal (most common) language spoken in the field differs from the principal language back in the research institute or university. This is especially important when it comes to communication on safety instructions and planning.

Recommendations

- Where appropriate, provide documentation and important information in multiple languages to ensure everyone has access to crucial materials (Amano et al. 2023).

- Employ an interpreter/translator for communication with local communities/collaborators for fieldwork abroad where this is needed.

- Use visual aids and hands-on demonstrations to complement verbal communication, making it easier for individuals with varying language proficiencies to understand concepts.

- Pair team members with different language backgrounds as language buddies if possible.

- Emphasise clear and concise communication, avoiding jargon or complex language that may be challenging for non-native speakers to comprehend.

- Set periodic moments during the fieldwork where crucial information is shared.

- Be clear about schedules and required flexibilities and provide unrestricted access to schedules where appropriate (Osiecka 2022).

2. Team morale and well-being

Dangerous situations may evolve from low group morale and a lack of confidence in one's abilities (Tucker and Horton 2018; John and Khan 2018). Climate stressors, geographical and social isolation, perceived danger, or lack of privacy may lead to stress, insomnia, cognitive decline, and interpersonal tension (Mullin 1960; Reed et al. 2001; Wood et al. 2005; Palinkas 2003; John and Khan 2018).

Recommendations

- Use briefing and debriefing techniques to monitor a team's well-being, to set clear expectations and communication norms, and to allow issue resolution (Hill et al. 2021; Karplus et al. 2022). Post-fieldwork debriefings on team dynamics are equally important and a key tool to avoiding recurring negative experiences and patterns of poor teamwork for future fieldtrips (Dance et al. 2024).

- Recognise and praise hard work to enhance team cohesion and offer non-judgmental support to those facing challenges, considering different levels of familiarity and subjective resilience to field hardships (John and Khan 2018).

- Emphasise the positive effects of close-knit groups and recognise personal growth through overcoming challenges (Palinkas 2003).

- Provide private spaces and downtime for team members to decompress and process overwhelming experiences.

- Be transparent about decision-making to keep participants informed. Participative decision-making is recommended for smaller groups (John and Khan 2018; Iversen et al. 2020).

- Acknowledge diverse experiences related to mental health conditions, being mindful of their impact on fieldwork and social interactions (Tucker and Horton 2018).

- View uncomfortable conversations as growth opportunities, listening to team suggestions for cultural and attitudinal improvements (Iversen et al. 2020).

- Follow safety protocols, including a buddy system, emergency devices, and daily meetings for a sense of safety (Rasch et al. 2019a; Rasch et al. 2019b; Iversen et al. 2020).

- Facilitate communication by choosing accommodations with internet coverage or providing satellite phones during prolonged fieldwork to help participants stay connected with trusted individuals outside the team (John and Khan 2018).

3. Addressing conflict and aggression

Uncertainty regarding roles, hierarchies or schedules in isolated group settings can be sources of stress and tensions (Hill 2021; Osiecka 2022). Microaggressions and harassment involve unwanted verbal comments, gestures or physical contact towards individuals based on their identity, causing potentially long-lasting physical or mental harm (Johnson et al. 2018). Women are particularly vulnerable to harassment and assault from senior scientists, whereas male participants are mostly targeted by peers (Clancy et al. 2014).

Recommendations

- Leisure activities during extended fieldwork can alleviate stress and mitigate conflicts (Palinkas and Suedfeld 2008). However, discourage peer pressure and hyper-masculine practices like adventurism, heavy drinking, and inappropriate joking (Nairn 1996; Hanson and Richards 2019).

- Develop strategies to recognise and intervene in group dynamics that may compromise safe spaces (Hill et al. 2021).

- Offer proactive training on anti-discrimination, conflict management, and bystander intervention, while promoting awareness of what microaggressions are, how they can arise, and what can be done to minimise them (Hill et al. 2021; Cronin et al. 2024).

- Inform about reporting systems at the research institute or university prior to the fieldwork.

- Establish transparent mechanisms for reporting and addressing discriminatory behaviour in the field. These should include options that allow victims to speak up without prejudice in situations where there may be substantial asymmetry in power relations (such as when the behaviour is initiated by a senior scientist).

- Include the injured person in the process for resolving the conflict situation (Iversen et al. 2020).

- Use inclusive risk assessments and codes of conduct as effective team-effort tools (see Miesen, Chapter 10).

Key tools for safe and inclusive field teams

In this chapter, we have presented a comprehensive yet not exhaustive list of recommendations for promoting inclusive and safe fieldwork practices. Ensuring safety and inclusion during fieldwork is a complex matter and can be overwhelming for field leaders lacking guidance, competences, or time, but is crucial to avoid persistent challenges. What seems like common sense to one person may not align with another's perspective, underscoring the need to standardise measures of safety and inclusion in a professional context (Kenyon and Hawker 1999; Rinkus et al. 2018). We emphasise three key tools for safe and inclusive fieldwork: inclusive risk assessments, codes of conduct, and training.

Fieldwork safety protocols commonly include filling out a risk assessment (see Miesen, Chapter 10). An inclusive assessment (1) should not only address physical risk, but all factors that undermine the team's or individuals' safety and well-being, including discriminatory behaviour or laws (Prior-Jones et al. 2020; Cronin et al. 2024). The risk of potentially dangerous situations should be evaluated by assigning probabilities and consequences and identifying needs for mitigations measures. Continuous revision based on changing field conditions and team dynamics is recommended (Rasch et al. 2019a).

Additionally, a code of conduct (2) should serve as a set of ethical and practical guidelines outlining expected behaviour and consequences of violation during fieldwork. An inclusive code of conduct should be drafted in a participatory approach where possible and explicitly address issues outlined in this chapter (Nordling 2019; Osiecka 2022;

Cronin et al. 2024). Be sure to inform all participants about the code of conduct.

Lastly, field training (3) about specific practical skills and physical safety (Rasch 2019a; Boon 2024; Jamal 2024) where appropriate will lead to more competent field team members and ensure safe and enjoyable fieldwork for all. Comprehensive proactive training on team safety that includes anti-harassment, anti-discrimination, and bystander intervention practices (Hill et al. 2021; Cronin et al. 2024) into fieldwork preparation will help to ensure that all field team members are not only aware of the importance of inclusivity but are also equipped to actively contribute to a more welcoming and diverse fieldwork culture (Nordling 2019; Osiecka 2022; Dance et al. 2024).

References cited

Adan, A. 2012. 'Cognitive performance and dehydration', *Journal of the American College of Nutrition*, 31, pp. 71–78. https://doi.org/10.1080/073157 24.2012.10720011

Allen, C.D. 2014. 'Why fieldwork?', in *Developments in Earth Surface Processes*, ed. by M.J. Thornbush, C.D. Allen, and F.A. Fitzpatrick (Elsevier). https://doi.org/10.1016/b978-0-444-63402-3.00002-9

Amano, T., V. Ramírez-Castañeda, V. Berdejo-Espinola, I. Borokini, S. Chowdhury, M. Golivets, J.D. González-Trujillo, F. Montaño-Centellas, K. Paudel, R.L. White, and D. Veríssimo. 2023. 'The manifold costs of being a non-native English speaker in science', *PLOS Biology,* 21. https://doi.org/10.1371/journal.pbio.3002184

Arnold, C. 2022. 'The sting of sizeism in the scientific workplace', *Nature*, 606, pp. 421–423. https://doi.org/10.1038/d41586-022-01536-y

Berhe, A.A., R.T. Barnes, M.G. Hastings, A. Mattheis, B. Schneider, B.M. Williams, and E. Marín-Spiotta. 2022. 'Scientists from historically excluded groups face a hostile obstacle course', *Nature Geoscience*, 15, pp. 2–4. https://doi.org/10.1038/s41561-021-00868-0

Bernard, R.E. and E.H.G. Cooperdock. 2018. 'No progress on diversity in 40 years', *Nature Geoscience*, 11, pp. 292–295. https://doi.org/10.1038/s41561-018-0116-6

Boon, S. 2024. 'From crevasse falls to polar bears, train fieldwork leaders for emergencies', *Nature World View*. https://doi.org/10.1038/d41586-024-03155-1

Bracken, L. and E. Mawdsley. 2004. '"Muddy glee": rounding out the picture of women and physical geography fieldwork', *Area*, 36, pp. 280–286. https://doi.org/10.1111/j.0004-0894.2004.00225.x

Bush, P. and S. Mattox. 2020. 'Decadal review: How gender and race of geoscientists are portrayed in physical geology textbooks', *Journal of Geoscience Education*, 68, pp. 2–7. https://doi.org/10.1080/10899995.2019.1621715

Chech, E.A. 2022. 'The intersectional privilege of white able-bodied heterosexual men in STEM', *Science Advances*, 8. https://doi.org/10.1126/sciadv.abo1558

Chiarella, D. and G. Vurro. 2020. 'Fieldwork and disability: an overview for an inclusive experience', *Geological Magazine*, 157, pp. 1933–1938. https://doi.org/10.1017/s0016756820000928

Church, M. 2013. 'Refocusing geomorphology: Field work in four acts', *Geomorphology*, 200, pp. 184–192. https://doi.org/10.1016/j.geomorph.2013.01.014

Clancy, K.B.H., R.G. Nelson, J.N. Rutherford, and K. Hinde. 2014. 'Survey of Academic Field Experiences (SAFE): Trainees report harassment and assault', *PLOS ONE*, 9. https://doi.org/10.1371/journal.pone.0102172

Coon, J.J., N.B. Alexander, E.M. Smith, M. Spellman, I.M. Klimasmith, L.T. Allen-Custodio, T.E. Clarkberg, L. Lynch, D. Knutson, K. Fountain, M. Rivera, M. Scherz, and L.K. Morrow. 2023. 'Best practices for LGBTQ+ inclusion during ecological fieldwork: Considering safety, cis/heteronormativity and structural barriers', *Journal of Applied Ecology*, 60, pp. 393–399. https://doi.org/10.32942/osf.io/kav6m

Cronin, M.R., R.S. Beltran, and E.S. Zavaleta. 2024. 'Beyond reporting: proactive strategies for safer scientific fieldwork', *Trends in Ecology and Evolution*, 39, pp. 213–216. https://doi.org/10.1016/j.tree.2024.01.003

Dalrymple, S.E. and T.P. Lane. 2024. 'Breaking the menstruation taboo to make fieldwork more inclusive', *Nature Career Column*. https://doi.org/10.1038/d41586-024-00044-5

Dance, M., R.J. Duncan, M. Gevers, E.M. Honan, E. Runge, F.R. Schalomon, and D.M.R. Walch. 2024. 'Coming in from the cold: Addressing the challenges experienced by women conducting remote polar fieldwork', *PLOS Climate*, 3.6. https://doi.org/10.1371/journal.pclm.0000393

Das, S. 2023. 'Navigating fieldwork amidst my menstrual cycle. Being a female ethnographer in a remote Indian region', in *Women Practicing Resilience, Self-Care and Wellbeing in Academia: International Stories from Lived Experience*, ed. by I.F. Adi Badiozaman, M.L. Voon, and K. Sandhu (Routledge). https://doi.org/10.4324/9781003341482-10

Day, M. 2014. 'Preparing for Fieldwork', in *Developments in Earth Surface Processes*, ed. M.J. Thornbush, C.D. Allen, and F.A. Fitzpatrick (Elsevier). https://doi.org/10.4135/9781849208949.n3

Demery, A.J.C. and M.A. Pipkin. 2021. 'Safe fieldwork strategies for at-risk individuals, their supervisors and institutions', *Nature Ecology and Evolution*, 5, pp. 5–9. https://doi.org/10.1038/s41559-020-01328-5

Drozdzewski, D. and D.F. Robinson. 2015. 'Care-work on fieldwork: taking your own children into the field', *Children's Geographies*, 13, pp. 372–378. https://doi.org/10.1080/14733285.2015.1026210

El-Sharkawy, A.M., O. Sahota, and D.N. Lobo. 2015. 'Acute and chronic effects of hydration status on health', *Nutrition Reviews*, 73, pp. 97–109. https://doi.org/10.1093/nutrit/nuv038

Gifford, N. 2004. 'Catering for expeditions', in *RGS Expedition Handbook*, ed. by S. Winser (Profile Books).

Greene, S., A. Kate, E. Dunne, K. Edgar, S. Giles, and E. Hanson. 2020. *Toilet Stops in the Field: An Educational Primer and Recommended Best Practices for Field-Based Teaching* (University of Birmingham). https://doi.org/10.31219/osf.io/gnhj2

Hamilton, J. and R. Fielding. 2018. 'Femininities in the field. safety first: The biases of gender and precaution in fieldwork', in *Tourism and Transdisciplinary Research*, ed. by B.A. Porter and H.A. Schänzel (Channel View Publications). https://doi.org/10.21832/9781845416522-004

Hanson, R. and P. Richards. 2019. *Harassed: Gender, Bodies, and Ethnographic Research* (University of California Press). https://doi.org/10.1093/sf/soz166

Hill, A.F., M. Jacquemart, A.U. Gold, and K. Tiampo. 2021. 'Changing the culture of fieldwork in the geosciences', *Eos*, 102. https://doi.org/10.1029/2021eo158013

Iversen, C.M., W.R. Bolton; A. Rogers, C. Wilson, and S.D. Wullschleger. 2020. 'Building a culture of safety and trust in team science', *Eos*, 101. https://doi.org/10.1029/2020eo143064

Jamal, S. 2024. 'Science on the edge: how extreme outdoor skills enhanced our fieldwork', *Nature Career Feature*. https://doi.org/10.1038/d41586-024-02311-x

Jenkins, K. 2020. 'Academic motherhood and fieldwork: Juggling time, emotions, and competing demands', *Transactions of the Institute of British Geographers*, 45, pp. 693–704. https://doi.org/10.1111/tran.12376

John, C.M. and S.B. Khan. 2018. 'Mental health in the field', *Nature Geoscience*, 11, pp. 618–620. https://doi.org/10.1038/s41561-018-0219-0

Johnson, P.A., S.E. Widnall, and F.F. Benya. 2018. *Sexual Harassment of Women: Climate, Culture, and Consequences in Academic Sciences, Engineering, and Medicine* (The National Academies Press). https://doi.org/10.17226/24994

Karplus, M.S., T.J. Young, S. Anandakrishnan, J.N. Bassis, E.H. Case, A.J. Crawford, A. Gold, L. Henry, J. Kingslake, A.A. Lehrmann, P.A. Montaño, E.C. Pettit, T.A. Scambos, E.M. Sheffield, E.C. Smith, M. Turrin, and J.S. Wellner. 2022. 'Strategies to build a positive and inclusive Antarctic field work environment', *Annals of Glaciology*, 63, pp. 125–131. https://doi.org/10.1017/aog.2023.32

Kenyon, E. and S. Hawker. 1999. 'Once would be enough': Some reflections on the issue of safety for the lone researchers. *International Journal of Social Research Methodology*, 2, pp. 313–327. https://doi.org/10.1080/136455799294989

Lininger, K.B., A.V. Rowan, B. Livers, N. Kramer, V. Ruiz-Villanueva, A. Sendrowski, and S. Burrough. 2021. 'Perspectives on being a field-based geomorphologist during pregnancy and early motherhood', *Earth Surface Processes and Landforms*, 46, pp. 2767–2772. https://doi.org/10.1002/esp.5238

Marín-Spiotta, E., R.T. Barnes, A.A. Berhe, M.G. Hastings, A. Mattheis, B. Schneider, and B.M. Williams. 2020. 'Hostile climates are barriers to diversifying the geosciences', *Adv. Geosci.*, 53, pp. 117–127. https://doi.org/10.5194/adgeo-53-117-2020

Marshall, A.M.S., J.L. Piatek, D.A. Williams, E. Gallant, S. Thatcher, S. Elardo, A.J. Williams, T. Collins, and Y. Arroyo. 2022. 'Flexible fieldwork', *Nature Reviews Earth and Environment*, 3, pp. 811–811. https://doi.org/10.1038/s43017-022-00375-9

Miesen, F. 2020. 'Fieldwork and toilets: Pooping with a view—why following nature's call in nature is not always easy-"peezy"', *Feministische Geo-RundMail*, 84, pp. 21–28.

Miesen, F., Chapter 10, this volume. "Fieldwork safety planning and risk management."

Mullin, C.J. 1960. 'Some psychological aspects of isolated Antarctic living', *The American Journal of Psychiatry*, 117, pp. 323–325. https://doi.org/10.1176/ajp.117.4.323

Nairn, K. 1996. 'Parties on geography fieldtrips: Embodied fieldwork?', *Women's Studies Journal*, 12, p. 86.

Nash, M. 2023. 'Breaking the silence around blood: managing menstruation during remote Antarctic fieldwork', *Gender, Place and Culture*, 30, pp. 1083–1103. https://doi.org/10.1080/0966369x.2022.2066635

Nelson, R.G., J.N. Rutherford, K. Hinde, and K.B.H. Clancy. 2017. 'Signaling safety: Characterizing fieldwork experiences and their implications for career trajectories', *American Anthropologist*, 119, pp. 710–722. https://doi.org/10.1111/aman.12929

Nordling, L. 2019. 'Codes of conduct aim to curb harassment at field sites', *Science*, 366, pp. 408–408. https://doi.org/10.1126/science.366.6464.408

Nordseth, A.E., J.R. Gerson, L.K. Aguilar, A.E. Dunham, A. Gentles, Z. Neale, and E. Rebol. 2023. 'The fieldwork wellness framework: a new approach to field research in ecology', *Frontiers in Ecology and the Environment*, 21, pp. 297–303. https://doi.org/10.1002/fee.2649

Osiecka, A. 2022. 'Keep talking to make fieldwork a true team effort', *Nature Career Column*. https://doi.org/10.1038/d41586-022-04416-7

Palinkas, L.A. 2003. 'The psychology of isolated and confined environments. Understanding human behavior in Antarctica', *Am. Psychol.*, 58, pp. 353–63. https://doi.org/10.1037/0003-066x.58.5.353

Palinkas, L.A. and P. Suedfeld. 2008. 'Psychological effects of polar expeditions', *Lancet*, 371, pp. 153–63. https://doi.org/10.1016/s0140-6736(07)61056-3

Paul, S.D. and K. Darke 2004. 'Inclusive expedition practice', in *RGS Expedition Handbook*, ed. by S. Winser (Profile Books).

Prior-Jones, M., J. Pinnion, M.-A. Millet, E. Bagshaw, A. Fagereng, and R. Ballinger. 2020. 'An inclusive risk assessment tool for travel and fieldwork', *EGU General Assembly 2020*. https://doi.org/10.5194/egusphere-egu2020-7678

Rasch, M., E. Topp-Jørgensen, G. Fugmann, F. S. Hansen, F. Tummon, and A. Schneider. 2019a. *INTERACT Fieldwork Planning Handbook* (Danish Centre for Environment and Energy).

Rasch, M., E. Topp-Jørgensen, G. Fugmann, F.S. Hansen, F. Tummon, A. Schneider, and M.F. Arndal. 2019b. *INTERACT Practical Field Guide* (Danish Centre for Environment and Energy).

Reed, H.L., K.R. Reedy, L.A. Palinkas, N. Van Do, N.S. Finney, H.S. Case, H.J. Lemar, J. Wright, and J. Thomas. 2001. 'Impairment in cognitive and exercise performance during prolonged antarctic residence: Effect of thyroxine supplementation in the Polar Triiodothyronine Syndrome1', *The Journal of Clinical Endocrinology and Metabolism*, 86, pp. 110–116. https://doi.org/10.1210/jc.86.1.110

Rinkus, M.A., J.R. Kelly, W. Wright, L. Medina and T. Dobson. 2018. 'Gendered Considerations for Safety in Conservation Fieldwork', *Society & Natural Resources*, 31, pp. 1419–1426. https://doi.org/10.1080/08941920.2018.1471177

Schneider, A.R., E. Topp-Jørgensen, M.F. Arndal, C. Hansen, M.M. Ibáñez, R. Duncan, P. Rajput, and C. Hewitt. 2021. *INTERACT Communication and Navigation Guidebook* (Danish Centre for Environment and Energy).

Shands, K.N., G.P. Schmid, B.B. Dan, D. Blum, R.J. Guidotti, N.T. Hargrett, R.L. Anderson, D.L. Hill, C.V. Broome, J.D. Band, and D.W. Fraser. 1980. 'Toxic-shock syndrome in menstruating women: association with tampon use and Staphylococcus aureus and clinical features in 52 cases', *N. Engl. J. Med.*, 303, pp. 1436–42. https://doi.org/10.1056/nejm198012183032502

Tucker, F. and J. Horton. 2018. '"The show must go on!" Fieldwork, mental health and wellbeing in Geography, Earth and Environmental Sciences', *Area*, 51, pp. 84–93. https://doi.org/10.1111/area.12437

Vero, S. 2021. *Fieldwork Ready: An Introductory Guide to Field Research for Agriculture, Environment, and Soil Scientists* (Wiley). https://doi.org/10.1111/area.12437

Vila-Concejo, A., S.L. Gallop, S.M. Hamylton, L.S. Esteves, K.R. Bryan, I. Delgado-Fernandez, E. Guisado-Pintado, S. Joshi, G.M. Da Silva, A. Ruiz de Alegria-Arzaburu, H.E. Power, N. Senechal, and K. Splinter. 2018. 'Steps to improve gender diversity in coastal geoscience and engineering', *Palgrave Communications*, 4, p. 103. https://doi.org/10.1057/s41599-018-0154-0

Walker, C. 2024. 'It's time to talk about menstruation and fieldwork', *Nature Career Feature*. https://doi.org/10.1038/d41586-024-02021-4

Williams, C., J.S. Griffiths, and B. Chalkley. 1999. *Fieldwork in Science Education* (Ashgate).

Wood, J., L. Schmidt, D. Lugg, J. Ayton, T. Phillips, and M. Shepanek. 2005. 'Life, survival, and behavioral health in small closed communities: 10 years of studying isolated Antarctic groups', *Aviat. Space Environ. Med.*, 76, pp. B89–93.

Zebracki, M. and A. Greatrick. 2022. 'Inclusive LGBTQ+ fieldwork: Advancing spaces of belonging and safety', *Area*, 54, pp. 551–557. https://doi.org/10.1111/area.12828

10. Fieldwork safety planning and risk management

Floreana Miesen

Fieldwork safety challenges

Fieldwork for environmental research may involve activities based in isolated locations and therefore come with a set of complex safety challenges (Fig. 10.1). Environmental and site-specific challenges could include rugged terrain and unstable ground; cliffs with rockfall, landslide or avalanche activity; severe weather conditions; encounters with wildlife; and exposure to harmful plants and pathogens. These challenges may be compounded where fieldwork is undertaken in remote locations with limited or only slow access to help. Operational hazards are also significant, encompassing risks associated with field activities, equipment use, and transportation, with common challenges including fatigue, unfamiliarity with heavy and unwieldy equipment, and the potential for injuries. Social and personal hazards add another layer of complexity, involving barriers to personal comfort and safety, group dynamics and stress during long and strenuous stays outdoors (see Miesen and Gevers, Chapter 9).

Fieldwork safety concerns differ significantly from those in the laboratory, where standard protocols and established prevention and response mechanisms are in place. Managing safety in the field can feel overwhelming due to the need for adaptive and dynamic strategies tailored to the unpredictable nature of the environment. Increasingly, research institutes require their employees to provide written safety plans or risk assessments, yet often lack professional guidance on defining and implementing concrete safety measures. This chapter offers an introduction to addressing fieldwork safety in a structured manner,

  https://doi.org/10.11647/OBP.0418.10

enabling field teams to enhance their preparedness and response capabilities effectively.

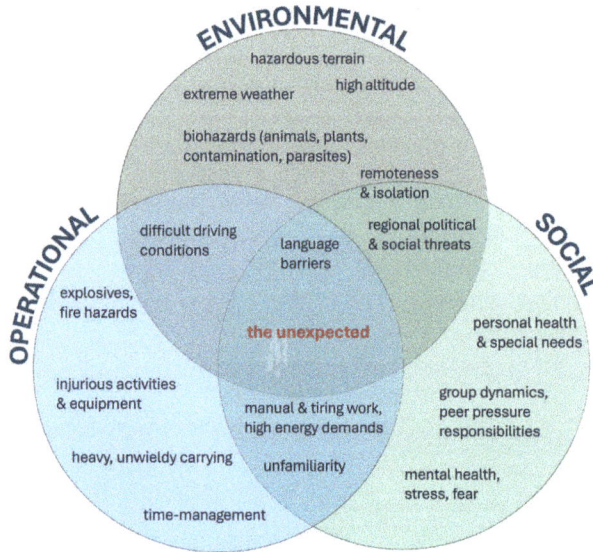

Fig. 10.1 Examples of challenges associated with fieldwork in the context of safety, classified by environmental, operational or social context

Safety planning

Risk assessment

Many research institutes require a formal risk assessment for planned fieldwork activities, where risk is defined as the product of the probability of a hazardous event occurring and the severity or impact of such an event. Typically, a 3x3 matrix is used as the basis of classifying the acceptability of the associated risk (Fig. 10.2). Safety measures must then be defined to mitigate the risk to an acceptable level. In a fieldwork context, risk assessment can be highly subjective due to the lack of reliable statistics or data and non-standardised accident reporting protocols (Cantine 2021), unlike in industrial settings where such information is more readily available. Risk assessments also do not account for the complex conditions that lead to an accident, where hazards or risk factors accumulate and/or interact over various timescales (Lundberg et al. 2009; Vanpoulle et al. 2017).

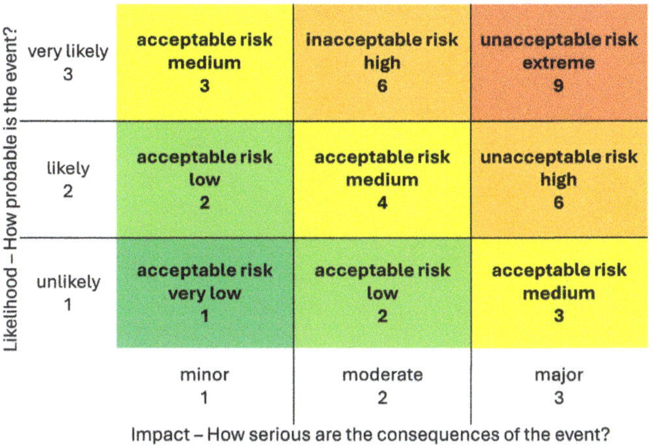

Fig. 10.2 Risk Matrix

Bowtie model for risk management

While risk assessments for environmental fieldwork are often formalised, defining mitigation measures remains challenging due to the unique nature of each campaign. Without standard protocols, fieldworkers often rely on personal judgment and anecdotal experiences from colleagues.

Adapting industrial safety management models, such as the *bowtie model*, to environmental fieldwork can provide useful guidance (e.g., Rasch et al. 2019). The bowtie model, developed in the late 1970s and early 1980s, is widely used in high-risk industries like aviation, material production, and oil and gas exploration and exploitation (de Ruijter and Guldenmund 2014). This model offers a comprehensive, systematic and visually illustrative approach to hazard management, facilitating both accident prevention and consequence mitigation through safety barriers. The model consists of four main components (Fig. 10.3):

1. Hazards—the central point representing the potential source of harm or loss of control.

2. Threats—conditions or activities that could lead to the hazard materialising into an unwanted event (i.e., accident).

3. Consequences—potential outcomes if the hazard is realised, leading to an accident or secondary incident.

4. Safety Barriers—measures which can intervene in the trajectories leading to or from the threat (Sklet 2006). These can be proactive (left side), i.e., preventative measures, detailing the threats and barriers to prevent incidents; or reactive (right side), i.e., mitigation measures to minimise the impact and to regain control.

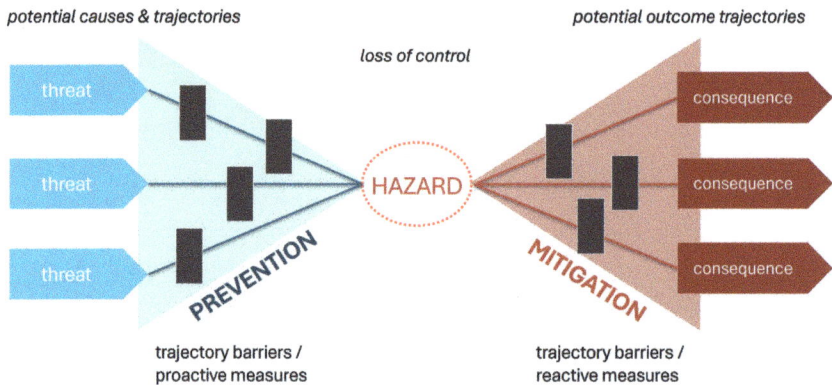

potential causes & trajectories

loss of control

potential outcome trajectories

threat

threat

threat

HAZARD

consequence

consequence

consequence

PREVENTION

MITIGATION

trajectory barriers /
proactive measures

trajectory barriers /
reactive measures

Fig. 10.3 Bowtie model

Proactive and reactive safety barriers in fieldwork

Defining safety barriers for fieldwork can be overwhelming as there are often no standard protocols. A structured approach to safety management is recommended, which classifies safety barriers into (1) strategic (good planning and foresight), (2) skill-based (training and experience), and (3) equipment-based (protective and rescue equipment). Effective hazard management should prioritise these barriers in this order. While essential, relying solely on equipment is inadequate if proper planning and training are lacking. Carrying a high-quality piece of safety equipment, such as a comprehensive first aid kit, can create a false sense of safety, as such equipment is useless if no one on the team know how to use it. Equipment should be viewed as the last resort to keep the team safe, with strategic planning and skill development forming the foundation of a robust safety protocol. See Table 10.1 for examples of proactive and reactive measures related to planning, training, and equipment.

Table 10.1 Examples of proactive and reactive measures in safety planning, training and equipment.

	Proactive measures (examples)	**Reactive measures (examples)**
Planning	Pre-fieldwork site assessment • Consultation of climate charts and seasonal weather records • Natural hazard maps (identify safe access routes, areas to avoid etc.) • Local knowledge and advice, previous field experiences, reconnaissance trip if possible Schedule • Identification of season/ time-dependent hazards, consultation of local event schedules (e.g., road closures, hunting season, hydropower flushing schedule) • Realistic time schedule incl. rest time and unexpected delays • Proper acclimatisation when ascending to high altitudes • Agreement on No-Go Criteria, plan check-out and check-in (e.g., logbook), buddy systems Health and wellbeing • Food supply planning that accounts for high energy demands • Pre-field medical check-up, counselling	• Emergency routine (info on SAR numbers, nearest doctor / emergency room / hospital) • Local distress signal • Back-up plan or alternative route • Nearest shelter (open shelter / warden cabin etc.) • Life or limb emergency card with critical health info and next of kin carried by participants

Training	• Navigation / Orienteering skills • Mountaineering, rope handling, river crossing etc. • Offroad driving • Correct use of safety equipment • Field equipment handling • Harassment, discrimination prevention	• Wilderness first aid • Evacuation protocols (e.g., on ship) • Use of satellite communication devices • Use of rescue equipment • Firearms, wildlife defence • Self defence • Bystander intervention, conflict resolution, emotional first aid
Equipment	• Adequate outdoor clothing (rain gear, shoes with grip / ankle support) • Personal protective equipment (helmets, gloves, goggles, hearing protection, high-vis clothing, personal floatation devices, dry suits, steel-capped boots, sun hat) • Personal technical terrain equipment (crampons, snowshoes, poles) • Local site securing (ropes, barriers, shields, shelter, shading, patting, high-vis marking) • Satellite communication devices for updates on weather, road conditions, etc. • Drinking water purification systems	• First aid kit • Emergency shelter (bivy bag, tent, etc.) • Spare clothes • Fire extinguisher • Avalanche beacon and probe • Satellite communication device to call SAR • Rescue line (throw bag) • Rescue floatation device • Crevasse rescue kit

The description of equipment above relates to personal protection, but a non-negligible risk may also come from poor training in field methods and instrumentation—training that is not only crucial for collecting reliable data, but also avoiding the hazards associated with poor equipment use. For drone flying, for instance, many countries require theoretical and practical training, as well as certification by law (see Kasvi, Chapter 45). Without this training, the risk of accidents and crashes is significantly higher, leading to potential damage to expensive equipment but also personal risk. Such incidents can be dangerous not only to the operator but also to bystanders and uninvolved people and may interfere with aviation. In numerous nature reserves, drone flying is prohibited due to the disturbance it causes to wildlife and the environmental harm from debris and lithium-ion batteries of crashed drones. Proper training ensures safe and responsible drone operation, mitigating these risks effectively.

Example: Bowtie Model for safety planning and risk management when sampling a turbulent mountain stream

Figure 10.4 shows a bowtie analysis for a scenario where fieldworkers need to enter a turbulent, cold and strong mountain stream to undertake sampling. The person may fall in the water for different reasons, with various harmful consequences. Proactive and reactive measures related to planning, training and equipment are defined as safety barriers. Note that some personal protective equipment such as helmets are defined as reactive barriers here. The helmet will not prevent the person from falling, but it will interrupt the trajectory towards more harmful consequences. In a different setting, e.g., when working below a cliff, a helmet would provide a proactive barrier to the hazard of head injuries from falling rock.

potential causes & trajectories   *potential outcome trajectories*

Pre-Fieldwork Planning
Thorough site assessments, check weather forecasts and hydrological data to avoid high-risk periods

Strong Currents
Being swept away by the force of the water

Emergency Response Plans
Position safety personnel with rescue training at strategic points along the sampling area; search and rescue, evacuation plan

Drowning
Being swept away by the current and unable to reach safety

Slippery Rocks
Slipping and falling

Training
Water safety and use of personal protective equipment (PPE); scouting safe entry and exit points along the stream

loss of control

Rescue and First Aid Training
Swift water (self-)rescue techniques, first aid, focus on managing injuries and hypothermia

Injuries
Sustaining injuries from falls, impacts with rocks, or equipment

Falling while sampling in turbulent mountain stream

Sudden Water Level Rise
Flash floods or sudden rise in water level, e.g. after rain

Buddy System
Work in pairs/small groups, monitor and assist each other. Regular check-ins and head counts to confirm everyone's safety

Personal Protective Equipment
Life jackets, helmets, wetsuits, waterproof communication devices for emergency contact

Hypothermia
Prolonged exposure to cold water

Cold Water
Drowsiness / hypothermia from prolonged exposure to cold water

Rescue and First Aid Equipment
Throw bags, rescue boards, floatation devices; first aid kits and equipment for treating hypothermia

Stranding
Becoming trapped or unable to exit water safely

Protective Equipment
Poles, guide ropes, non-slip footwear, insulated clothing or wetsuits

trajectory barriers / proactive measures   *trajectory barriers / reactive measures*

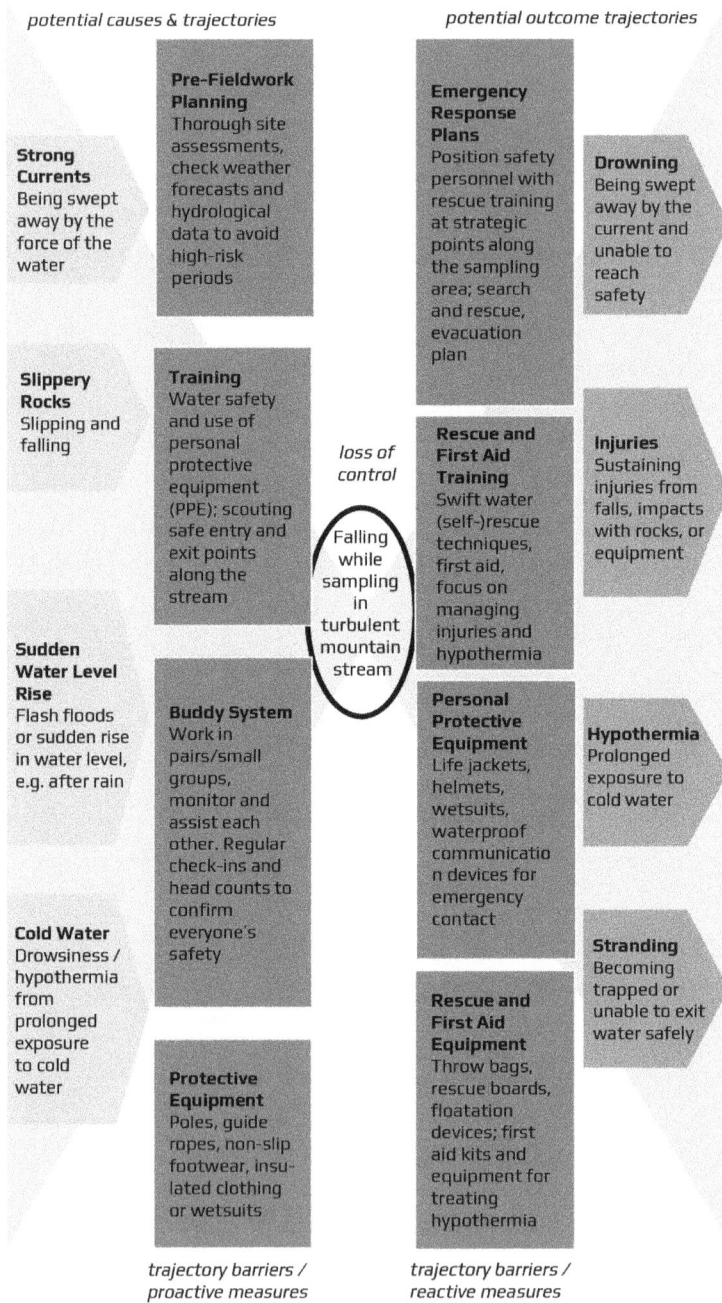

Fig. 10.4 Exemplary Bowtie analysis of water sampling in a turbulent mountain stream

Conclusion

Managing fieldwork safety is inherently complex due to the unpredictable and diverse environments researchers encounter. Addressing these challenges effectively requires a structured and adaptable approach. Emphasising thorough planning, comprehensive training, and proper equipment use is crucial for both preventing and mitigating hazards. By integrating these elements, field workers and field teams can improve their preparedness and response capabilities, ensuring safer and more successful research activities. Like laboratory safety protocols, research institutes should offer guidance on risk management and establish reporting systems to normalise and encourage risk awareness, treating fieldwork as a professional activity where fieldworkers are adequately protected from harm.

References cited

Cantine, M.D. 2021. 'Dying to know: Death during geological fieldwork', *The Sedimentary Record*, 19.3, pp. 5–14. https://doi.org/10.2110/sedred.2021.3.2

Lundberg, L., C. Rollenhagen, and E. Hollnagel. 2009. 'What-You-Look-For-Is-What-You-Find – The consequences of underlying accident models in eight accident investigation manuals', *Safety Science*, 47.10, pp. 1297–1311. https://doi.org/10.1016/j.ssci.2009.01.004

Miesen, F. and Gevers, M., Chapter 9, this volume. "Inclusive practices in fieldwork."

Rasch, M., E. Topp-Jørgensen, G. Fugmann, and F. S. Hansen. 2019. *INTERACT Fieldwork Planning Handbook* (Danish Centre for Environment and Energy).

de Ruijter, A. and F. Guldenmund. 2016. 'The bowtie method: A review', *Safety Science*, 88, pp. 211–218. https://doi.org/10.1016/j.ssci.2016.03.001

Sklet, S. 2006. 'Safety barriers: Definition, classification, and performance', *Journal of Loss Prevention in the Process Industries*, 19.5, pp. 494–506. https://doi.org/10.1016/j.jlp.2005.12.004

Vanpoulle, M., E. Vignac, and B. Soulé. 2017. 'Accidentology of mountain sports: An insight provided by the systemic modelling of accident and near-miss sequences', *Safety Science*, 99, pp. 36–44. https://doi.org/10.1016/j.ssci.2016.11.020

SECTION 2: RESEARCH RECIPES

INSPIRING EXAMPLES OF MIXED-METHODS ENVIRONMENTAL RESEARCH

11. Introduction to the research recipes

Stuart N. Lane and Rebecca Lave

A recent furore has emerged in the news media regarding a series of cookbooks supposedly written by "Teresa J Blair", who is actually "Artificial Intelligence". The books are enticingly entitled *The Ultimate Crockpot Cookbook for Beginners: A Comprehensive Guide to Slow Cooking Success for Novice Chefs, Featuring Mouthwatering Recipes, Time-Saving Tips, and Essential Techniques*; *The Ultimate Anti-Inflammatory Cookbook For Beginners: Nourish Your Body, Soothe Inflammation, and Embark on a Delicious Journey Towards Optimal Health with Flavorful Recipes*; and *Canning and Preserving: A Comprehensive Guide to Canning, Pickling, and Fermenting for Flavorful, Nutrient-Rich Delights that Last—Rediscover the Art and Science of Food Preservation*. Those who have ever experimented with Large Language Models such as ChatGPT will see in these titles the propensity to superlatives like "comprehensive" and "essential" as well as claims that go well beyond what such titles can really deliver. The furore came about[1] because whilst our author "Artificial Intelligence" is able to put together recipes whose ingredients, quantities and preparation appear fine, the resulting dishes are not satisfying. Perhaps because it cannot taste its own work, Artificial Intelligence (AI) can produce plausible recipes but not necessarily good ones.

As with AI cookbooks, we hope you will read the chapters that follow with attention not only to the specific combinations of methods used (the research recipes), but also to the results produced. Further, we strongly encourage you not to take the examples here as templates

---

1   For example, see R. Jones, '"One of the most disgusting meals I've ever eaten": AI recipes tested', *The Guardian*, 31 July 2024, https://www.theguardian.com/food/article/2024/jul/31/one-of-the-most-disgusting-meals-ive-ever-eaten-ai-recipes-tested

    https://doi.org/10.11647/OBP.0418.11

to follow, as they may not produce tasty results when transferred to different field sites, topics and research questions. As we argued in Chapter 1, you should choose the suite of methods you use by listening to your field site, not carefully following an existing template.

All of the chapters in this Section demonstrate both that it is possible to transcend the traditional divides that separate the biophysical and social sciences, and that it is increasingly urgent to do so if we are to address the complex challenges facing the environment. These empirically rich, methodologically innovative case studies address topics including water security and poverty (Lebek and Krueger, Chapter 12), the complexities of modelling water futures (Rusca and Mazzoleni, Chapter 14), flood risk and community engagement (Booth and Druschke, Chapter 15), how the ecology of savannas and forests (Walters et al., Chapter 13) and of land use and land cover are shaped by legacies of colonialism and more recent political economic dynamics (Arce-Nazario, Chapter 17; Kelley, Chapter 19), the ecology of ice free regions in the Antarctic (Chignell et al., Chapter 18), public health and urban contamination (Malone, Chapter 16), and landscape evolution during the Holocene in relation to both climate and societal change (Blond, Chapter 20).

We asked the authors of chapters in this Section to explain how their research recipes evolved: which methods they chose and how they combined them in response to what they studied. We also asked them to reflect upon the lessons that come from their "recipes" for mixing methods in environmental research and it is interesting that so many chose to focus on their engagement with particular places, albeit using quite different methodological mixes. You can see this engagement with place as a means of slowing down, rethinking, and replanning in ways that allow you to mobilise other ways of understanding those places. This is creative and innovative work. As methods are mixed, new kinds of data emerge and new experiences are assimilated, allowing us to make real research progress. It is when initial results don't fit together easily and don't readily combine to create a certain "whole" that we have to think differently, which opens the possibility of new ideas about what interests us (in particular, see Blond, Chapter 20; Chignell et al., Chapter 18; Kelley, Chapter 19; and Lebek and Krueger, Chapter 12; for more general reflections on tensions among results, see Biermann and Gibbes, Chapter 4).

A second characteristic then follows. Many of the chapters show how their slowing down comes from working with people, albeit in different ways. Whether it is building unusual teams of collaborators (e.g., Chignell et al., Chapter 18; Lebek and Krueger, Chapter 12) or working with those whom the research directly concerns, such as local people (e.g., Arce-Nazario, Chapter 17; Booth and Druschke, Chapter 15; Malone, Chapter 16), the chapters show that sensitivity to place is not simply a given, but instead emerges in interdisciplinary (academic collaboration) and transdisciplinary (stakeholder collaboration) settings. Working in teams built by researchers coming from different disciplinary perspectives and/or local perspectives allows a place to be learnt about and understood in different ways. These collaborations are rewarding, but they are also challenging, requiring considerable time, personal effort and, above all, humility (Arce-Nazario, Chapter 17). In some of the chapters, multiple researchers work together in teams (Arce-Nazario, Chapter 17; Blond, Chapter 20); in others, a single author brings training in multiple disciplines to bear (Kelley, Chapter 19 and Malone, Chapter 16). These latter chapters illustrate an important point: it is possible for one person to develop skills in the application of a number of methods, and this often in very different research areas. Whether alone or in a team, mixing methods means working with diverse data types, developing different kinds of skillsets, communicating different ways of working, and recognising different expectations as to how particular methods should be used (see Blond, Chapter 20 for good examples). Many of the chapters show that however advanced your mixing of methods is, however inter- or transdisciplinary your project might be, you will face an underlying challenge around the expectations and frames of the individual disciplines with which you engage (see Lane and Lave, Chapter 3).

Third, the chapters in this Section demonstrate how mixing methods allows you to look at the same question from different perspectives. The chapters include some unusual combinations of methods brought from very different disciplines (e.g., an ethnographer and a hydrologist in Lebek and Krueger, Chapter 12; a geographer, a glaciologist and an environmental historian in Chignell et al., Chapter 18; a hydrologist and a rhetorician in Booth and Druschke, Chapter 15). Embracing these uncommon combinations allows you to go beyond the perspective provided by any one method, illustrated effectively by Arce-Nazario (Chapter 17), Kelley (Chapter 19) and Walters et al. (Chapter 13) in relation to land cover and vegetation. As these authors demonstrate, mixing methods allows you to challenge how things appear from any one perspective and so gain a deeper and better understanding of the concrete realities that are of interest.

Finally, and refreshingly, a sense of ethical responsibility is common to all of the chapters in this Section. If our research has impacts, we are ethically responsible for what those impacts are. The fact that research has impact gives it a certain power; who holds that power (that is, who is involved in the research) and how they execute it makes this firmly a political question. All too often we naïvely assume that power and politics can be excluded from research. Rusca and Mazzoleni (Chapter 14) challenge this view explicitly, arguing that sensitivity to who holds power and how power is (normally unevenly) distributed is central to making research fairer and more equitable. Malone (Chapter 16) shows that research that challenges existing power relations can draw sharp responses. At the same time, a number of chapters show that when methods are mixed in a transdisciplinary setting, existing power relations can be changed (Booth and Druschke, Chapter 15; Malone, Chapter 16).

References cited

Arce-Nazario, J., Chapter 17, this volume. 'Space and place in participatory arts-based research'.

Biermann, C. and Gibbes, C., Chapter 4, this volume. 'Mixed methods in tension: lessons for and from the research process'.

Blond, N., Chapter 20, this volume. 'Mixing geoarchaeology, geohistory and ethnology to reconstruct landscape changes on the longue durée'.

Booth, E. and Gottschalk Druschke, C., Chapter 15, this volume. '"A hydrologist and a rhetorician walk into a workshop," or how we learned to collaborate on a decade of mixed methods river research across the humanities and biophysical Sciences'.

Chignell, S., Howkins, A., and Fountain, A., Chapter 18, this volume. 'Antarctic mosaic: Mixing methods and metaphors in the McMurdo Dry Valleys'.

Kelley, L., Chapter 19, this volume. 'Engaging remote sensing and ethnography to seed alternative landscape stories and scripts'.

Lane, S.N. and Lave, R., Chapter 3, this volume. 'Frames, disciplines and mixing methods in environmental research'.

Lebek, K. and Krueger, T., Chapter 12, this volume. 'On the dialogue between ethnographic field work and statistical modelling'.

Malone, M., Chapter 16, this volume. 'Using mixed methods to confront disparities in public health interventions in urban community gardens'.

Rusca, M. and Mazzoleni, M., Chapter 14, this volume. 'The interface between hydrological modelling and political ecology'.

Walters, G., Hymas, O., Touladjan, S., and Ndong, K., Chapter 13, this volume. 'Revealing the social histories of ancient savannas and intact forests using a historical ecology approach in Central Africa'.

12. On the dialogue between ethnographic field work and statistical modelling

Karen Lebek and Tobias Krueger

At first glance, ethnographic and statistical methods could not be more different: Ethnographic methods aim for interpretations of the particularities of events (see Sayre, Chapter 34), while statistical methods wish to abstract from data to generalities (see Lane, Chapter 42). Ethnographic data are mostly qualitative, while statistical data consist of categories and numerical quantities. The ethnographer benefits from personal immersion in "the field", while the statistician tends to aim for a neutral "outsider" perspective. One could say that ethnographic research is *interpretative*, while statistical research is *generalising* (Krueger and Alba 2022). This is not to say that ethnographies do not use quantitative data and statistics, and that statistical analysis does not benefit from qualitative analysis to inform hypotheses as a precursor to formal quantitative testing. But it seems that in such constellations one approach provides the "master" epistemology while the other performs a service; they do not collaborate at eye-level.[1] Hence, at first glance, these two epistemologies seem to sit uncomfortably together.

We here outline an alternative approach where the different epistemological underpinnings of ethnographic and statistical research can be preserved whilst also brought into conversation with each other so that their different ambitions are taken seriously. Combining the two methods is useful for us primarily as a way of triangulation (see Biermann

---

1   In addition to epistemologies—what and how one can know about the world—the two approaches differ also with respect to underlying ontologies—how the world is or is becoming (Krueger and Alba 2022). So, when we talk about epistemologies for simplicity, one might read these as *onto-epistemologies*.

    https://doi.org/10.11647/OBP.0418.12

and Gibbes, Chapter 4): do participant observations and interview accounts match with survey results? Do we see phenomena observed locally also in larger-scale statistics? The latter points to the generalising ambition of statistical analysis, which will, however, be challenged by the ethnographic epistemology. This tension forces us to engage with what we consider to be a central question of any empirical analysis: to what extent can we generalise from concrete observer situations? Neither ethnographic nor statistical research raises this question on its own.

Our approach eschews integration at all costs (epistemologically speaking) and will instead thrive on the contrast and friction of analytical positions.[2] Because neither of us has extensive disciplinary training in either of these methods, the research we present here was not a particularly agonistic experience (see Lane and Lave, Chapter 3). Should antagonism arise, however, we would not wish to brush this under the carpet, as is often the case, but instead would want to explore its generative momentum. We admit that this approach sits more comfortably with the reflexive attitude of ethnography, which attends to the conditions for and contingencies of producing knowledge about people and places and the consequences of the research process *for* these people and places, than with the objective outlook of statistics.

So, is our approach just another case of subordination, this time of statistics? We argue no for two reasons. First, statistics has a long tradition of problematising its own assumptions, particularly the Bayesian variant of statistics we champion here. Embedding statistical analysis in a larger interpretative frame thus makes sense from the perspective of statistics, too. We refer here to the "small world/large world" distinction in the Bayesian statistics textbook of McElreath (2020), for example: in the small world of data and statistical models, both can be shown to be compatible (or not), i.e., models can serve as interpretations of data. But what this means for the large "real" world requires additional qualifiers beyond statistics. We argue that ethnography provides just such an interpretative frame. Second, we connect with the methodological ambitions of Bayesian statistics in a way that tries to push the boundaries of the field. We aim at statistical analysis that will be recognised as going

---

2   Exploiting the generative potential of antagonism between research traditions has been called *agonistic* mode of interdisciplinarity by Barry and Born (2013).

beyond the state-of-the-art according to the discipline's *own* standards, not just as a service to another discipline.

As noted above, we have practiced this mix of methods as "outsiders" to the disciplines: two researchers with biophysical science backgrounds (geoecology and hydrology), an experience and fascination with statistical methods, and an interest in and appreciation of ethnographic research. The ethnographic fieldwork was made possible through guidance from social anthropologists and human geographers at our interdisciplinary institute. We care deeply about the philosophical underpinnings of both ethnography and statistics, but we have not been brought up in either of these disciplines. It would be interesting to engage researchers more fully immersed in the respective research traditions in such a collaboration, especially to bring the agonistic potential of the confrontation of approaches to the fore more strongly than we were able to do.

Our reflexive attitude to ethnography and statistics is in line with the wider project of critical mixed methods. Mixing methods for us encourages the reflexivity implicit in critical research as one method challenges the other. The juxtaposition of analytical positions requires us to reflect on our own knowledge practices, as well as our epistemological and ontological commitments. From there it seems a logical next step to scrutinise the politics of that knowledge and recognise our own ethical responsibility as researchers, which for us has meant engaging with water injustices, for example (Lebek et al. 2021). A subsequent step to truly transformative research that tries to do something about these injustices is not far away (Krueger et al. 2016). Rusca and Di Baldassarre (2019) have highlighted the transformative potential of quantitative research given the currency that numbers have today. A critical analysis that engages with statistics and numbers can mobilise them for normative purposes, e.g., to reveal patterns of uneven power and related uneven access to water resources.

Case study of rainwater harvesting in rural KwaZulu-Natal, South Africa

We here re-interpret a case study of rainwater harvesting (RWH) in rural KwaZulu-Natal, South Africa that we previously reported on in Lebek and Krueger (2023). The story begins during ethnographic field work by the first author (KL) on water insecurity in this rural setting which involved participant observations, interviews and a household survey (see Johnston and Longhurst, Chapter 32; Sayre, Chapter 34; Winata and McLafferty, Chapter 43). By chance, KL noticed practices of "makeshift" rainwater harvesting next to "conventional" RWH systems. RWH is especially useful in the widely dispersed settlements of rural South Africa, where water infrastructure and services are only partially developed and, even where available, are often dysfunctional and unreliable. Surprisingly, according to previous studies and data from the South African General Household Survey (GHS), only 1 to 3 % of South African households practice rainwater harvesting. However, these studies and surveys only included conventional RWH systems, i.e., straight, off-the-shelf gutters connected to large 4–6 kl storage tanks. In the case study in Kwazulu-Natal, by contrast, over 90% of households practiced RWH, yet only 25% harvested rainwater in a conventional way. The majority of households collected rainwater in what we call a "makeshift mode". Makeshift RWH systems consisted of drums or dishes of different volumes for the collection of rainwater. In some cases, the water was routed towards these containers through homemade gutters made of bent metal sheets, plastic bottles or hollow tree trunks. Makeshift RWH was mostly associated with the traditional round huts in the region, while the conventional systems were usually associated with brick houses with straight walls which some households owned (Fig. 12.1).

A household usually consisted of more than one building and could be a mix of round huts and brick houses. The conventional RWH systems with straight gutters and large tanks evidently could not be installed on the round huts. The brick houses, however, where the conventional systems fitted, seemed more expensive to build than round huts. Observations also suggested that brickmaking required access to water; KL saw a number of unfinished brick houses whose building had stalled allegedly for lack of water. So, could it be that income, water access and type of housing all contributed in interrelated ways to the mode of rainwater harvesting that rural households could practice, which in turn contributed to the volumes of water they had at their disposal? Field work suggested that this was the case, but we wanted to see whether this phenomenon was also apparent in a household survey we had conducted previously, as a way of triangulating the field observations and our assumptions about the factors influencing RWH mode. In particular, we assumed income would play a large role for the RWH mode. While triangulation (see Biermann and Gibbes, Chapter 4) is good ethnographic practice in any case, we used statistics not as a service but with the aim of conducting an analysis that would be considered novel also by the statistical community.

Fig. 12.1 Karen Lebek. (a) Round hut with makeshift gutters made of metal sheet and a drum for harvesting rainwater. (b) House with makeshift rainwater harvesting. (c) Conventional rainwater harvesting system with straight gutters and tank. (d) Brick house with conventional rainwater harvesting. In the background there is another tank for storage of water and a round hut without a rainwater harvesting system.

South African General Household Survey and Kwazulu-Natal household survey

While our household survey in Kwazulu-Natal (KZN) was restricted to 67 households in a small study area, the GHS covers private households in all nine provinces of South Africa. The GHS aims to measure the level of development and performance of government programmes and projects and compile indicators of education, living standards and service delivery. It is not related to a specific season (wet or dry). The GHS comprises 213 questions in 10 different sections on, *inter alia*, health and general functioning; household information, including water and sanitation; health, welfare and food security; and household livelihoods. Information on RWH in the GHS is limited to a question on the main source of drinking water, where "RWH tank" is one of the possible answers. It does not hold information on RWH for other purposes, or on makeshift RWH.

In order to investigate household water insecurity in greater detail, we designed a household survey drawing from the core questions on drinking water and sanitation for household surveys developed by the WHO/UNICEF Joint Monitoring Programme (WHO and UNICEF 2006). The KZN survey comprises 52 questions on different aspects of household water security, with a focus on water collection, domestic water use, water treatment, sanitation and hygiene, and water-related health. The KZN survey provides data on the presence of any kind of RWH system, regardless of whether it is the main source of water and what the rainwater is used for. The household survey was subsequently carried out with 67 households, working in two teams of one researcher and one translator each.

Ethnographic field work

Based on the responses to the household survey, KL asked the interviewees from rural households further in-depth questions and noted any field observations. Through participant observation, she and her research assistants noticed the modes of RWH in different households and what they meant for water availability and quality. They then integrated questions on RWH mode into the household survey. Beyond the survey, KL conducted further interviews and attended meetings. In particular, she participated in a large community meeting on conflicts over water within the study area. The meeting was attended by women and men from the community and surrounding communities, the mayor of the district municipality, and the king and traditional leaders from different villages.

Household survey and hypotheses

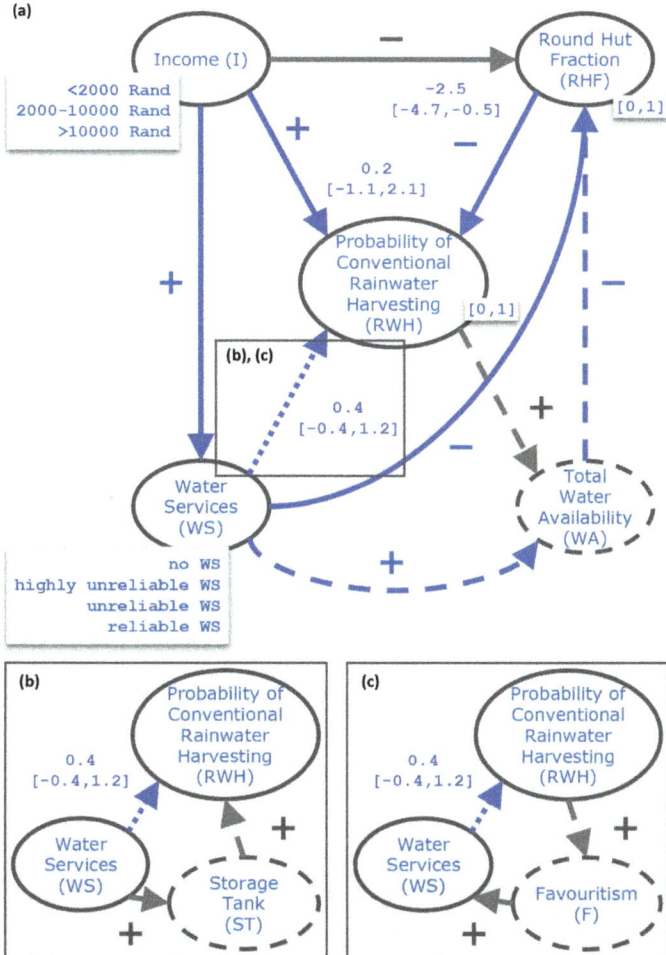

Fig. 12.2 (a) Relations between rainwater harvesting mode (*RWH*) and predictors. The ranges of variables and the categories, respectively, from the household survey are noted alongside. Total water availability (*WA*) is an unobserved variable marked with a dashed line together with its corresponding relations. Hypothesised relations and corresponding signs that were not confirmed by the statistical model are in grey, confirmed ones are in blue together with the estimated coefficients (mean and central 95% uncertainty interval). The relation between water services (*WS*) and RWH emerging from the statistical analysis is considered spurious, marked with a dotted line. (b) Hypothesised explanation of the spurious relation with storage tank (*ST*) as mediating variable ("pipe"). (c) Hypothesised explanation of the spurious relation with favouritism (F) as mediating variable ("fork").

The household survey used for triangulation had previously been reported by Lebek et al. (2021). From all variables of the survey, round hut fraction (*RHF*)—the proportion of round huts belonging to a household—income (I) and quality of water services (*WS*) were deemed relevant for explaining the probability that a household would practice conventional rather than makeshift rainwater harvesting (*RWH*). The hypothesised relations are depicted in Figure 12.2a: a greater round hut fraction (*RHF*) would decrease the probability of conventional rainwater harvesting (*RWH*), while a greater income (I) would increase *RWH*. These relations were already visible in bivariate regression models (Lebek and Krueger 2023). The quality of water services (*WS*) also showed a positive relation with *RWH* in bivariate regression, but this was considered a spurious effect which we thought might vanish when income and round hut fraction were adjusted for. Hence, the reason to include *WS* in the subsequent multiple regression was to disprove its effect. We also hypothesised that the three predictors would co-vary (Fig. 12.2a): a greater income would afford more brick houses and hence decrease the round hut fraction, while also allowing households to afford better water services. Better water services, in turn, would facilitate more brick houses via the unobserved variable total water availability (*WA*) and hence decrease *RHF*.

Statistical modelling and causal inference

The nature of the response variable (*RWH*) being bounded by 0 and 1 demanded a logistic regression, which we cast in a Bayesian framework (see Table 12.1 for mathematical detail). The first technical challenge that made the statistical modelling interesting in its own right was the formulation of the predictor variables income (I) and quality of water services (*WS*) as ordered categories after Bürkner and Charpentier (2020). The second challenge was the causal interpretation of the multiple regression, which is a research frontier in statistics (Pearl et al. 2016). Following McElreath (2020), we ran models with all possible combinations of predictors and interpreted any changes in effect estimates in terms of confounding to get a handle on causality. Since causality is a strong claim for both ethnographers and statisticians, we note that we talk about causality in the "small world" of the data and models at hand, where we merely claim that our analysis is consistent with a causal interpretation of the relations in the data as quantified by the models. The main problem with this approach is that we will have unobserved confounders that bias the effect estimates in unknown ways, some of which we hypothesise about. Nevertheless, used in a triangulation with field observations, the statistical analysis is one piece of evidence to be considered.

Triangulation of relations and new hypotheses

When comparing the effect estimates of all model variants (Table 12.1), we look for estimates that are stable across the model variants and for those that change. Changing effect estimates suggest confounding effects, either spurious or masked relations. We believe that we have not created spurious effects by introducing "colliders" into the model (compare Fig. 12.2a).[3] So, by including all predictors at hand we adjusted for confounding effects as much as possible (model 5 in Table 12.1). In that model, round hut fraction (RHF) had the most important effect on conventional rainwater harvesting probability (RWH) with a stable estimate of -2.5 and an uncertainty interval well below zero (though considerably large). This confirms the hypothesised negative relation between RHF and RWH (Fig. 12.2a). Income (I) had the least important effect on RWH with an estimate of 0.2 whose uncertainty interval spanned zero by a considerable margin. If anything, the income effect is positive as hypothesised, but not as great as we expected. The quality of water services (WS) had a slightly larger effect on RWH of 0.4, with an uncertainty interval barely on the positive side of zero. This means that what we considered to be a spurious effect of WS on RWH in the bivariate models did not vanish when we adjusted for the two other predictors.

The WS effect persisted even though we confirmed the hypothesised negative relation between WS and RHF and the hypothesised positive relation between WS and I when comparing the model variants (Table 12.1); the WS effect becomes slightly more positive when RHF is absent

---

3   Colliders are potential predictors that causally arise from two or more other
     potential predictors (which thus "collide"). Including a collider in the model
     introduces a spurious relation between those other potential predictors
     (McElreath 2020).

(models 1 and 4), while *RHF* becomes slightly more negative when *WS* is absent (models 3 and 6). This can be explained by a negative relation between *WS* and *RHF*, which is compensated for by a more negative *RHF* effect when *WS* is absent and vice versa. The *I* effect, in turn, becomes more positive when *WS* is absent (models 6 and 7), and switches signs when *WS* is the only other variable (model 4). Hence, much of the income effect seems to be channelled through better water services. When *RHF* is present on top of *WS* and *I* (model 5), then the negative *I* effect is adjusted again to a barely positive value through the aforementioned negative relation between *WS* and *RHF* because the *WS* effect is less positive when adjusted for by *RHF*. The hypothesised negative relation between *I* and *RHF* (Fig. 12.2a) is not apparent from the statistical analysis because the *RHF* effect did not change, whether *I* was present or not (models 3 and 6). Note, we here update our interpretation in Lebek and Krueger (2023) where we mistakenly diagnosed a negative relation between *I* and *RHF*. Other than this, the two interpretations yield the same conclusions. The hypothesised positive relation between *RWH* and *WA* could not be assessed because cyclic graphs cannot be handled by statistical inference (in addition, *WA* was unobserved). We expect this positive feedback between *RWH* and *RHF* via *WA* to be accounted for implicitly in the large negative effect of *RHF* on *RWH*. From a statistical perspective, it would be an interesting next step to try and isolate this "endogeneity" (Daniel et al. 2022).

Table 12.1 Effect estimates (mean and central 95% uncertainty interval) of the three predictors income (*I*), round hut fraction (*RHF*) and quality of water services (*WS*) for all seven model variants compared.

Model	β_I	β_{RHF}	β_{WS}
1: WS			0.5 [-0.1,1.2]
2: RHF+WS		-2.5 [-4.6,-0.6]	0.4 [-0.2,1.1]
3: RHF		-2.6 [-4.7,-0.7]	
4: I+WS	-0.2 [-1.4,1.7]		0.6 [-0.2,1.4]
5: I+RHF+WS	0.2 [-1.1,2.1]	-2.5 [-4.7,-0.5]	0.4 [-0.4,1.2]
6: I+RHF	0.7 [-0.7,2.3]	-2.6 [-4.7,-0.7]	
7: I	0.5 [-0.9,2.3]		

One distinct advantage of the statistical analysis of the hypothesised relations is that we could quantify effect strengths: the existence of round huts has a much more important effect than household income, though income works partly through affording better water services and—via total water availability—more brick houses. Even more valuable for the research process are the unconfirmed and new hypotheses: household income is apparently not a direct factor in the choice of housing (round huts or brick houses). This should be probed further by the next round of ethnographic field work.

To explain the spurious relation between quality of water services (*WS*) and conventional rainwater harvesting probability (*RWH*), we posit two new hypotheses that again provide novel ethnographic questions: the positive relation between *WS* and *RWH* might be explained by a mediating variable that acts either as a "pipe" between *WS* and *RWH* (Fig. 12.2b) or as a "fork" influencing both *WS* and *RWH* simultaneously (Fig. 12.2c).[4] Both types of confounders would generate a spurious relation between *WS* and *RWH* (Pearl et al. 2016). A possible pipe would be a variable we might call storage tank (*ST*): field observations suggest that users with reliable water services (yard taps) commonly buy a rainwater harvesting tank to store municipal water. This they can then also use for rainwater harvesting (Fig. 12.2b). A possible fork would be favouritism of the water provider (*F*): in Lebek et al. (2021) we reported that households are frequently disadvantaged by the water provider for specific (sometimes political) reasons and thus end up with highly unreliable water services or none at all. We expect that the water provider might also influence the provision of rainwater harvesting systems to households (Fig. 12.2c).

Differences in rainwater harvesting mode in the context of household water insecurity and uneven power relations

The results from this study and the new hypotheses need to be understood in the wider context of prevalent household water insecurity

---

4   In causal inference, a pipe describes two variables linked causally via a third variable, which induces a spurious direct relation between the two variables (the two "ends" of the pipe) when not adjusted for. A fork, in contrast, describes two variables that are both causally influenced by a third variable, which again creates a spurious relation between the two variables (the two "prongs" of the fork) when not adjusted for. The third variable in question may be unobserved.

in the study area that we have reported in Lebek et al. (2021). The District Municipality responsible for water services has largely failed to improve household water security in the form of adequate, accessible, and reliable water, with adverse effects on health and productivity. Where households do have reliable and adequate water access it is through self-supply by illegal connections. Public standpipes, where water is available at no cost, are often dysfunctional and highly unreliable and 50% of households collect water from surface sources that are often polluted. Some households have the necessary financial means to install illegal taps in their yards and divert water from public standpipes, further exacerbating the unreliability of the standpipes. Important secondary water sources for households are rainwater harvesting and the *waterkan*, which fills public storage tanks along the road and acts as an interim water service. However, our research shows that the *waterkan* is unreliable and characterised by mismanagement. For households without an illegal yard connection, rainwater is the only on-site water source, which can reduce the strain of having to collect water in steep terrain where a roundtrip for water collection exceeded 30 minutes for half of the households surveyed. In light of the importance of rainwater for reducing household water insecurity, the capacity of rainwater harvesting tanks was an initial part of our household survey and the RWH mode was integrated into the survey when its importance became apparent.

The ethnographic field work yielded observations related to the quality of rainwater that a household survey alone would not have been able to provide. If it rained during the day, household members took the rainwater inside where it could be used for 'inside' purposes, such as drinking, cooking, and bathing. If it rained at night or in the dark, the rainwater was said to be of low quality and stayed outside for irrigation and laundry. Water quality and health concerns for conventional RWH arise from bacterial growth inside tanks, insect breeding, and low maintenance of tanks. These risks are likely lower where rainwater is collected in drums because users of drums have more control over the quality of their rainwater. They can move around their RWH container and thus prevent it from being polluted (e.g., by taking it inside the house during the night). Moreover, drums can be easily closed with a lid and cleaned.

RWH in our study area is thus embedded in a context of uneven power relations and related uneven access to water resources. With our statistical analysis we further quantified these uneven relations in the spirit of harnessing the power of numbers (Rusca and Di Baldassarre 2019): the interrelations of income, housing, total water availability, and RWH mode result in the accumulation of water resources in the form of reliable water services plus large RWH tanks in some households, while others lack reliable water services and conventional RWH systems altogether. The two hypotheses we generated from the model (storage tanks and favouritism) are in line with our observations on uneven water access in the study area.

Ethnographic field work and statistical modelling in a reflexive dialogue

Mixing ethnographic field work with statistical modelling for us meant using one method to critically reflect on and challenge the other, lending robustness to the research process and the results in a form of triangulation. This became apparent at every stage of the research process as illustrated next. The mixing of methods has, however, not been as contradictory or uncomfortable as it could have been had we been more invested in the research traditions of ethnography or statistics. In cases where contradictions or discomfort are greater, we would welcome these for creating generative momentum that could move understanding of a field or a methodology forward.

In our study, when KL discovered makeshift RWH practices by chance during her field work, this immediately put into question the assumption of previous studies, relying on the General Household Survey, that RWH is not prevalent in rural South Africa. In the General Household Survey, respondents are asked whether they own a RWH tank. Only a local case study (see Lane, Chapter 24) could reveal that there is more to RWH than conventional RWH tanks and that RWH is presumably far more prevalent in rural South Africa if makeshift RWH is included, which has implications for household water security and RWH dissemination efforts. This is an example where a statistical analysis of a national-scale household survey would miss important information.

From the ethnographic fieldwork alone KL could conceptualise hypothetical relations between income, housing, water availability, and RWH mode. Quantifying these relations statistically and estimating the strengths of the effects was a way of triangulating (see Biermann and

Gibbes, Chapter 4) and thus reflecting critically on the ethnographic data. For example, we assumed that the effect of income on RWH mode was much bigger than what the statistical analysis of the household survey suggested. The statistical model, in turn, reached its limits when it came to allegedly spurious relations it cannot explain on its own. More generally, a statistical model is always limited by what quantitative data are at hand, and it is unidirectional and thus cannot handle feedbacks. Ethnographic field work is one way of providing richer context, identifying feedbacks, and evaluating new hypotheses that could explain spurious relations. As much as the statistical model benefits from the ethnographic information, the interpretation of the ethnographic observations benefits from the statistical model: the spurious relations pointed to something that had previously been missing in the qualitative conceptualisation. This is the point in the research process where the mixing of the two methods was most productive.

Conclusions

We propose the iteration of ethnographic field work and statistical modelling as a useful research process for learning about particular places. Importantly, we do not see statistical modelling as the end point that ethnographies may provide hypotheses for, but as a recurring step of quantification that also generates valuable questions for subsequent ethnographic field work. The reflexive frame of ethnography, we argue, sits well with the scrutiny of statisticians towards their own assumptions. That said, we have observed that the scrutiny of assumptions often does not carry over to statistical analyses performed by other disciplines (and not statisticians). In such cases, collaboration with ethnographers might force the quantitative analyst into a deeper engagement with the statistical underpinnings of their work and hence engender reflexivity. The resulting research process has the potential to be a collaboration at eye-level where ontological and epistemological positions are being challenged—but in a generative (agonistic) spirit that keeps asking what one can actually say about the study case at hand.

For such a collaboration to be fully symmetrical we believe that it helps to choose statistical methods that are stimulating in their own right. We did this here through a Bayesian framing in general and by engaging with ordered regression and causal inference in particular, which could still be extended to endogeneity considerations. We believe that each discipline needs to get something like this out of the collaboration, something that is valuable according to the discipline's own standards, otherwise the collaboration risks compromising the mutual benefit. Our approach might provide a shared methodology of the kind Lowe et al. (2013) saw missing in interdisciplinary collaborations between social and biophysical scientists. In such a shared methodology, ethnographic and statistical results are considered equally valid, sometimes complementary and at other times conflictual. In any case, they inform each other and provide new insights and questions that go beyond the horizon of each epistemology on its own (Rangecroft et al. 2021).

Coming back to the theme of this book on critical mixed methods, for us, reflecting on one's own ontological and epistemological commitments when doing research is the starting point of "being critical". If an interdisciplinary collaboration or a mix of methods forces this kind of reflexivity, then this is beneficial for the critical project. From this reflexivity then follows an ethical responsibility for the "worlds enacted" through those research practices (Krueger and Alba 2022). At that point, one cannot help but look out for the knowledge politics and questions of power in the field, which are themselves core concerns of critical research.

In this regard, our lessons from the case study of rainwater harvesting in rural KwaZulu-Natal, South Africa are that water security is about more than municipal water services and financial means to install conventional rainwater harvesting systems; it is also about more local and culturally embedded factors like the type of housing. Larger-scale water security policy projects must recognise, support, and build on local interventions such as makeshift rainwater harvesting if these policies are to be effective. At present, makeshift ways of harvesting rainwater are not even captured in national statistics like the annual General Household Survey of Statistics South Africa (Lebek and Krueger 2023), so we do not presently know how important makeshift rainwater harvesting is for rural water security in other areas of South Africa and what its development potential is. In Lebek et al. (2021, we highlighted unequal access to and conflicts over water in rural South Africa in light of the failure of municipalities to supply reliable and equitable water services. Similar questions of power arise from the present study when we hypothesise about the role of favouritism of the water provider in furnishing selected households with rainwater harvesting systems as well as yard taps.

References cited

Barry, A. and G. Born. 2013. *Interdisciplinarity: Reconfigurations of the Social and Natural Sciences* (Routledge).

Biermann, C. and Gibbes, C., Chapter 4, this volume. 'Mixed methods in tension: lessons for and from the research process'.

Bürkner, P.-C. 2017. 'brms: An R package for Bayesian multilevel models using Stan', *Journal of Statistical Software*, 80.1, pp. 1–28.

Bürkner, P.-C. and E. Charpentier. 2020. 'Modelling monotonic effects of ordinal predictors in Bayesian regression models', *British Journal of Mathematical and Statistical Psychology*, 73.3, pp. 420–451.

Daniel, D., S. Pande, and L. Rietveld. 2022. 'Endogeneity in water use behaviour across case studies of household water treatment adoption in developing countries', *World Development Perspectives*, 25.

Krueger, T. and R. Alba. 2022. 'Ontological and epistemological commitments in interdisciplinary water research: Uncertainty as an entry point for reflexion', *Frontiers in Water*, 4.

Krueger, T., C. Maynard, G. Carr, A. Bruns, E.N. Mueller, and S. Lane. 2016. 'A transdisciplinary account of water research', *Wiley Interdisciplinary Reviews: Water*, 3, pp. 369–389.

Lane, S.N. and Lave, R., Chapter 3, this volume. 'Frames, disciplines and mixing methods in environmental research'.

Lane, S.N., Chapter 24, this volume. 'Case studies'.

Lane, S.N., Chapter 42, this volume. 'Statistical inference'.

Lebek, K. and T. Krueger. 2023. 'Conventional and makeshift rainwater harvesting in rural South Africa: exploring determinants for rainwater harvesting mode', *International Journal of Water Resources Development*, 39.1, pp. 113–132.

Lebek, K., M. Twomey, and T. Krueger. 2021. 'Municipal failure, unequal access and conflicts over water: A hydrosocial perspective on water insecurity of rural households in KwaZulu-Natal, South Africa', *Water Alternatives*, 14.1, pp. 271–292.

Johnston and Longhurst, Chapter 32, this volume. 'Interviews: Structured, semi-structured and open-ended'.

Lowe, P., J. Phillipson, and K. Wilkinson. 2013. 'Why social scientists should engage with natural scientists', *Contemporary Social Science*, 8.3, pp. 207–222.

McElreath, R. 2020. *Statistical Rethinking: A Bayesian Course with Examples in R and Stan* (CRC Press).

Pearl, J., M. Glymour, and N. P. Jewell. 2016. *Causal Inference in Statistics: A Primer* (Wiley).

Rangecroft, S., M. Rohse, E.W. Banks, R. Day, G. Di Baldassarre, T. Frommen, Y. Hayashi, B. Höllermann, K. Lebek, E. Mondino, M. Rusca, M. Wens, and A.F. Van Loon. 2021. 'Guiding principles for hydrologists conducting interdisciplinary research and fieldwork with participants', *Hydrological Sciences Journal*, 66.2, pp. 214–225.

Rusca, M. and G. Di Baldassarre. 2019. 'Interdisciplinary critical geographies of water: Capturing the mutual shaping of society and hydrological flows', *Water*, 11.10, pp. 1973.

Stan Development Team. 2022. "Stan User's Guide (Version 2.31)", https:// mc-stan.org/.

13. Revealing the social histories of ancient savannas and intact forests using a historical ecology approach in Central Africa

Gretchen Walters, Olivier Hymas, Stevens Touladjan, and Kevin Ndong

Ecosystem histories: from received wisdom to testing assumptions

The state of ecosystems and people's impact on them raise instant questions, and lots of assumptions. I (GW) was consulting aerial photographs in Paris in 2006, at the Institut Géographique National (IGN). A series of black and white images from the 1950s of Gabon's forest-savanna mosaic sat in front of me. A staff member looked at them and raised a seemingly innocent question, "What happened to all the forest?" Their question was based on a stereotype of savanna, assuming that the whole area had once been entirely forested and that the savanna was the result of relatively recent anthropogenic degradation. In reality, these are ancient savannas that have been kept open by customary fire use, as we will see later. These are some of the most ancient savannas in sub-Saharan Africa, with thriving human and ecological communities. So, why do people suppose they are not only anthropogenic, but degraded?

These persistent, misleading stories about ecosystems are "received wisdom", permitting researchers, decision-makers, NGOs and many others, to leave histories of ecosystems unquestioned. Well-known examples include unsupported national deforestation statistics (Fairhead and Leach 1998), assuming that tropical savannas

https://doi.org/10.11647/OBP.0418.13

were previously forest and are degraded by fire (Fairhead and Leach 1996a; Bond 2016a), or that intact forests have little human influence, being an "unbroken expanse of natural ecosystems" (Heino et al. 2015).

To move beyond assumptions about ecosystems (Kelley, Chapter 19), it becomes important to look below the canopy of these forests (Haurez et al. 2017), where the legacy of past and present societies becomes visible (Biwolé et al. 2015; Morin-Rivat et al. 2017; Roberts 2019). When researchers use interdisciplinary methods, they discover exciting things, such as combining land use and archaeological evidence to understand that most of the Earth has been inhabited by humans for over 12,000 years (Ellis 2021). To understand ecosystems, one must ask questions beyond disciplinary boundaries, and overcome assumption drag (see Lane and Lave, Chapter 3; Ascher 1979). Why is this particular ecosystem here? How has it been shaped over time by people? Without asking these questions, we risk "misreading" the landscape and encouraging erroneous policy pathways supported by degradation narratives about the West African forest-savanna mosaic (Fairhead and Leach 1996b), Sami herding practices in the Arctic (Benjaminsen et al. 2015), or desertification in North Africa (Davis 2007). Understanding the history of landscapes is one of the "Grand Challenges" for conservation in the 21st century (Gillson et al. 2020). Addressing this challenge requires understanding both ecological and social "keystone processes" (Marcucci 2000: 72) that shape landscapes over time, such as those impacting intact forest and ancient savanna.

Intact forests and ancient savannas have a descriptor ("intact" and "ancient") in their titles, potentially presupposing little human history. Such assumptions come from the colonial period. In the case of savannas, fire use was seen as destructive (Kull 2004; Laris and Wardell 2006) despite being necessary for fire-dependent species. Over time, debates have shifted from viewing savannas as degraded ex-forest to insisting that many are old-growth or ancient and maintained by anthropogenic fire (Bond 2016b). Old-growth savannas are defined as "ancient ecosystems characterized by high herbaceous species richness, high endemism, and unique species compositions" (Veldman et al.

2015). However, many of these old-growth savannas are regularly burned by people (Walters et al. 2022), and thus anthropogenic. Such smallholder fire use is rarely reported, yet remains an important part of these ecosystems (Smith et al. 2022). So, what does fire use look like in an ancient grassland?

Extensive tropical forests suffer from a different issue: in many African countries, colonial policies emptied forests of people (Cinnamon 2003; 2010) and claimed that empty spaces were uninhabited or pristine (Hymas et al. 2021a). Many of these forests are now called "intact". These intact forest landscapes (IFL) are defined as "a seamless mosaic of forest and naturally treeless ecosystems with no remotely detected signs of human activity and a minimum area of 500 km^2" (Potapov et al. 2017), and are often associated with an absence of roads (Kleinschroth et al. 2017). IFLs are identified by remote sensing methods (see Braun, Chapter 39), but such methods can only detect limited kinds of activity and will likely lack a long-term, historical view. These views from above miss the fact that forests have histories, linked to the societies that inhabit them (Walters et al. 2019).

What methods can we use to test our assumptions about the occupation and histories of these ecosystems? In the past, problems were successfully confronted in a stepwise fashion where a single problem was investigated with a single set of methodologies through the prism of one academic discipline, heavily influenced by the assumptions of one's culture, resulting in a single solution that could be applied globally and at any time. However, with the current multi-faceted global crises such as climate change or changing forest extent, this stepwise process no longer works. Furthermore, unlike Newton's apple that would fall in exactly the same way, regardless of where you are in space and time, studying the current multi-faceted crises with a stepwise, recipe-like methodology results in different outcomes in both space and time (Reiners and Lockwood 2010; Reiners et al. 2019). Instead, today's researchers facing multi-faceted problems need to move beyond a siloed approach and (Tett 2015) a) take a case-by-case approach, b) be agile in their methodology, and c) be open to other scientific disciplines (Fig. 13.1a and b).

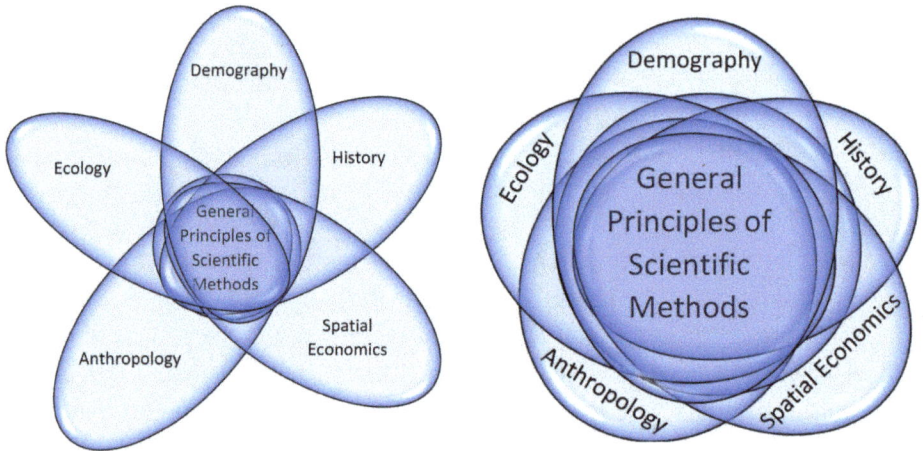

Fig. 13.1 Gretchen Walters. Example of disciples used in Case 1. A. Siloed approach used to study simple problems, where there is little need to integrate other silos. B. Approach needed when investigating multi-faceted problems, where we need to expand our general principles of scientific methods to other disciplines. Note however, we will always have biases towards certain disciplines and there will always be methods that are very specific. Inspired from (Gauch Jr 2002)

Mixed methods have become essential to understanding ecosystems, producing works which have become standard references in the field, such as the high-impact research by anthropologists James Fairhead and Melissa Leach who went to study the forest islands in the tropical savannas of Guinea Conakry. They had originally assumed that they were studying degraded tropical savannas. Only by setting aside their cultural and methodological assumptions, by working with local knowledge-holders about landscape history, and by consulting archival information and aerial photos were they able to understand that these forest islands had been created by people rather than being forest remnants within degraded savannas (Fairhead and Leach 1996). What they were seeing was not deforestation, but reforestation. Although such a combination of methods has since become more common, it remains marginalised in the face of more traditional methods of studying ecosystems. Such work across disciplines requires moving beyond disciplinary boundaries and cultures (Táíwò 2019) and understanding the "conceptual worlds of our colleagues"(Darbellay 2015: 167).

Historical ecology methods: what and why

In the early 2000s, many ecologists believed that societies had not greatly shaped ecosystems (Foster et al. 2003).[1] The long-term impact of historical human disturbance on forests is now more accepted (Ellis et al. 2013) thanks to ground-breaking work in environmental anthropology, geography, and historical ecology (Denevan 1992; Fairhead and Leach 1996a; Balée and Erickson 2006) that considered the influences of humans on forests.

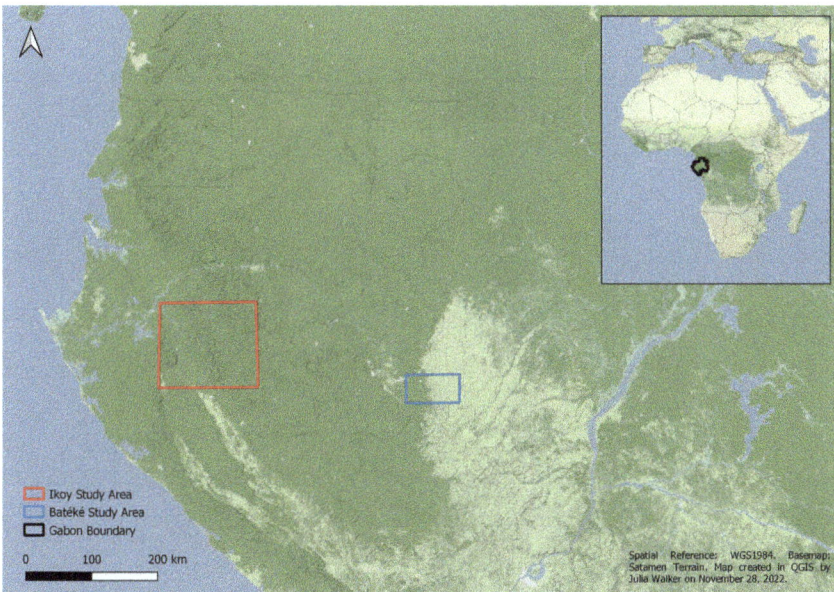

Fig. 13.2 Gretchen Walters. The Ikoy study area in red in an Intact Forest Landscape and the Batéké study area in blue in an ancient savanna

Historical ecology (see Davis, Chapter 29) is an interdisciplinary approach that can be used to unravel the interaction between humans and the environment at different spatio-temporal scales (Szabó 2015), bringing together the ecological and social sciences (see Davis, Chapter 29; Balée 2018). It places people at the centre of investigation, as a "keystone species" (Erickson 2021) having an impact at the landscape scale (Dodaro and Reuther 2016). Historical ecology research has been conducted in Amazonia

---

1   There are exceptions, e.g., Leopold (Leopold 2013: 375).

(Balée and Erickson 2006) since the 1980s, but has only recently become more common in Central Africa (de Saulieu et al. 2016), where research suggests that forests have been heavily impacted by past anthropogenic disturbances (Oslisly et al. 2013; Garcin et al. 2018). Historical ecology comprises many methods, with a 2021 manual proposing 17 (Odonne and Molino 2020), and can be summed up as requiring the analysis of "two types of archives: ecological....and human" (Decocq 2022: 3). Here, we focus on methods that help discern how intact forests and ancient savannas have been created by societies over time. We look at migration patterns, the history of settlements as derived from oral and archival sources (see Cope, Chapter 22; Chakov et al., Chapter 33; Grenand and Davy 2021), and customary fire use and indicator species derived from botanical data and interviews (Ekblom et al. 2019). We focus on cases from the Ikoy forest and the Batéké savanna of Gabon (Fig. 13.2).

Case 1: The intact forest of the Upper Ikoy

The Upper Ikoy Valley of Gabon is classified as an Intact Forest Landscape (IFL) (Potapov et al. 2017). These forests are called "seamless". But are they? The vegetation suggests that previous agricultural fields and village settlements have been resettled by a pioneer tree species, while two palm species suggest opposing landscape histories of presence and absence of people. What is happening under the canopy of this IFL? Who were the inhabitants over time? How was the forest composition and its people impacted by colonisation, disease, and trade? To answer these questions, one must combine several methods from the social, historical, and ecological sciences.

In this study we bring together oral histories (Chakov et al., Chapter 33) and two frameworks to understand the historical events that affected the people and the environment of this area: one from trade and the other from disease. The first framework focuses on trade in the colonial era (Chamberlin 1977). The second framework focuses on the spread of disease (Headrick 1994) (Hymas et al. 2021b). Using these two frameworks and oral histories we explore how the spread of disease and trade impacted this IFL (Fig. 13.3).

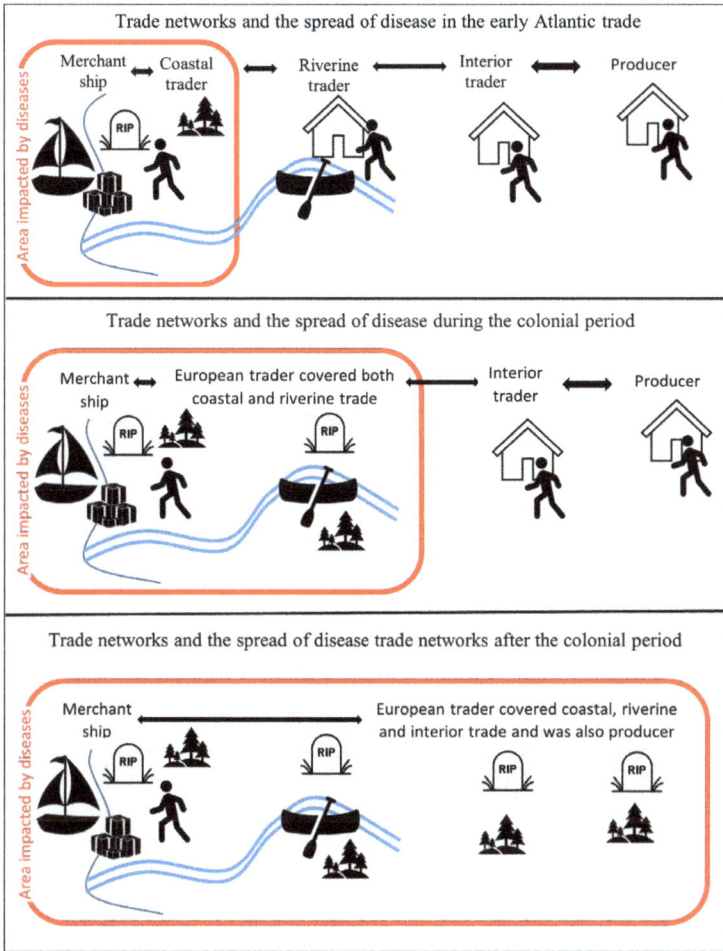

Fig. 13.3 European traders appropriated local trade networks while also aiding the spread of disease. Traders are split into three tiers: first tier traders dealt directly with merchant ships, second tier traders were middlemen, and third tier traders operated the stages between the producers and second tier middlemen (based on Chamberlin 1977: 6–7)

The upper Ikoy study area and methods

In order to understand the forest ecology of the Ikoy, one must move beyond a more traditional method of using forest plots or transects (and so limiting oneself to only biological information) (Walters et al. 2019) or only using interviews, and so limiting one's understanding to what is remembered. In using an ethnographic approach (Sayre, Chapter 34) to unravel migration stories in combination with archival photos and maps, one gains a historic depth to the research and a view of what has happened under the canopy of an intact forest landscape over time.

The upper Ikoy study area is located in Gabon's Du Chaillu Massif (Fig. 13.2) and contains a road that runs up the Ikoy valley along which there are villages populated by Bantu-speaking Mitsogho and Akele peoples. The other villages are mostly populated by the Babongo, a hunter-gatherer group.

We developed our understanding of the historical and epidemiological impacts of the Atlantic trade period mainly from authoritative secondary sources and trading accounts dating from this time. We also used online archives (Cope, Chapter 22) such as the Internet Archive (www.archive. org), the British Library (www.bl.uk), the British Newspaper Archive (www.britishnewspaperarchive.co.uk), Bibliothèque Nationale de France (gallica.bnf.fr), Persée (www.persee.fr), and Horizon Pleins Textes (horizon.documentation.ird.fr).

To place the findings in the local context, we recorded anonymised village histories. Due to the small size of Gabonese villages no sampling strategy was used. The oral migration histories were recorded first with a group of adults and elders and then in separate smaller groups, some of which were based on ethnicity. In each of the group oral history interviews we drew a map on the ground consisting of the villages that people had created during their migration. Then, we asked open-ended questions about each of the villages mentioned, including why they left the previous village, why they moved to the village in question and the routes they used. This information was supplemented with the current locations of named places and a local event calendar we constructed from dates of major, local, historically salient events.

These village oral histories (Chakov et al., Chapter 33) were carried out in French and Mitsogho. We triangulated this information with oral histories from other villages, foresters' accounts, and historical accounts. The University College London's ethics committee reviewed and approved the methods used. Oral consent was requested from study participants, with the understanding that all oral histories would be anonymised. The study was carried out using the Free, Prior and Informed Consent (FPIC) process.

Care was taken not to conduct oral history interviews during busy parts of the year, i.e., when people are in their plantations[2] To encourage participation and to compensate participants for their time in a locally meaningful way, we distributed locally purchased drinks and food items.

---

2   Plantations are large forest gardens created using slash and burn agriculture to cultivate large crops of manioc or plantains; typically, each adult married woman

Results

Period	Event	Impacts on people	Impacts on environment
ca. 1860	Fang migration south	Akele displaced taking over trade at Samba falls	Increased exploitation of natural resources
		Retreat of Mitsogho and Babongo	Creation of "dead zone" between Akele and the Mitsogho/Babongo (ca. 10,000 km²)
			Recovery of environment within "dead zone"
ca. 1900	Creation of the SHO	French take over trade from Akele	Increased exploitation of natural resources
			Disappearance of dead zone
1920 to 1930		Spread of disease	Creation of new "dead zone" along upper reaches of Ikoy valley (ca. 5,000 km²)
			Recovery of environment within enlarged "dead zone"
1940 to 1970	*Regroupement*	Akele, Mitsogho and Babongo either migrate towards urban areas or into the forest zone"	Reduction of "dead zone"
			Increased exploitation of natural resources on edges of "dead zone
ca. 1960	Arrival of the timber company in the upper reaches of the Ikoy valley	Akele, Mitsogho and Babongo migrate towards the logging roads	Further reduction of edge of "dead zone" with renewed exploitation of natural resources
At various times from 1990	Collapse of timber company	Degradation of roads	
		Akele, Mitsogho and Babongo migrate down the road	Creation of a new "dead zone" in the upper Ikoy valley

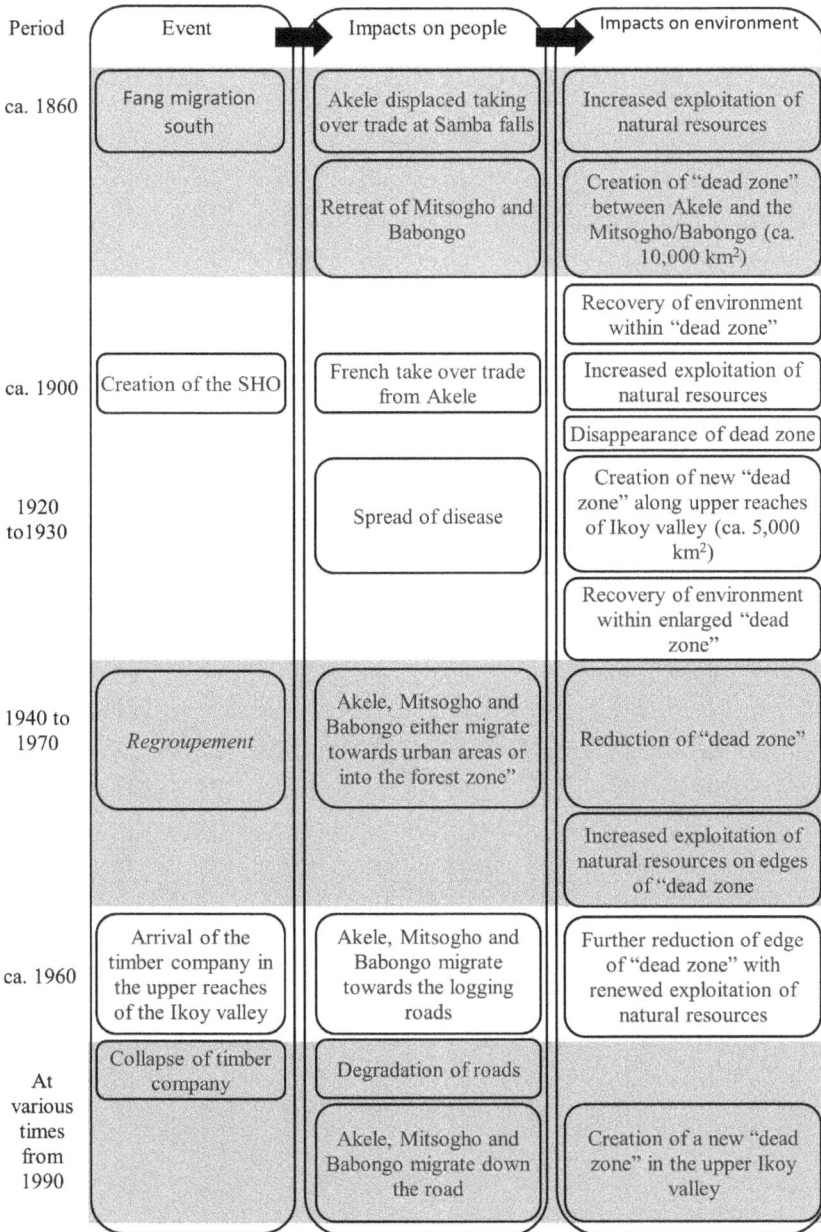

Fig. 13.4 Gretchen Walters. Summary of events and the impact that they had on both the people and the environment of the villages of the Ikoy area

in a household will have her own plantation. These plantations are rotated over time, allowing the forest to recover in between plantings.

To understand the migration of people into the Ikoy area, multiple strands of information must be brought together. This account starts with the European migration from the coast into the interior of Gabon. It continues with how this migration brought disease and, with resettlement policy[3], resulted in the depopulation of the Ikoy area. We then describe the migration histories of the three ethnicities currently living in the area (Fig. 13.4) and discuss how these events influenced the landscape of the area.

Colonisation and migration history

National context

During the 1470s the Portuguese were the first Europeans to enter into contact with the Mpongwe coastal people (Patterson 1975; Aicardi de Saint-Paul 1987: 6; Gaulme 1988: 63; Merlet 1990: 19; Knight 2003; Gardinier and Yates 2006). The Atlantic trade of Gabonese natural resources started around the mid-16th century (Patterson 1975: 8; Merlet 1990: 20–21; Pourtier 2010: 2). Trade of forest products and slaves ensued in the 16th century, with a history too detailed to recount here (Rondet-Saint 1911: 101; Sautter 1966: 729; Martin 1972; Patterson 1975; Chamberlin 1977; Vansina 1990; Gray 2002; Gardinier and Yates 2006; Pourtier 2010: 2).

The ensuing surge in coastal trade drew people from the interior towards these new coastal markets (Sautter 1966: 752; Vansina 1990: 234) bringing the Gabonese second- and third-tier traders into contact with the first-tier traders (Patterson 1975; Chamberlin 1977; Gaulme 1988: 58; Vansina 1990; Gray 2002: 26). To strengthen their position in the tier system the French created a concession system and trading routes that favoured French companies (Cuvillier-Fleury 1904: 80; Gaulme 1991: 85; Coquery-Vidrovitch 2001). These French companies covered 70% of Gabon (Vande weghe 2011: 61).

Ikoy area context

In 1893 the Minister of the Colonies awarded Daumas the first commercial concession, which included the Ikoy River valley (Coquery-Vidrovitch

---

3   Colonial and postcolonial policy of regrouping, sometimes forcibly, villages together. In Gabon this usually meant along roads (Sautter 1966).

2001: 380). It comprised eleven million hectares of forest (four times the size of Belgium), and became the Société Commerciale Industrielle et Agricole du Haut-Ogooué (SHO)[4] (Cuvillier-Fleury 1904: 92–93; Rouget 1906: 610–611; Coquery-Vidrovitch 2001: 14, 44). By 1928, the area was crisscrossed with SHO trading posts (Fig. 13.5), caravan routes, and villages (Journal Officiel de L'Afrique Equatoriale Française 1910: 515).

Fig. 13.5 Elephant tusks outside a SHO warehouse, circa 1910 from Moutangou (2013: 95), public domain

The thirty-year monopoly of the SHO in Gabon ended in 1923 (Moutangou 2013: 216–217) by which time the SHO had already left the Ikoy area (Suret-Canale 1987: 130; Gaulme 1988: 116; Coquery-Vidrovitch 2001: 388). Just before independence, in 1960, 100,000 ha of the SHO concession was bought by la Société d'Okoumé de la N'gounié (SONG) (Simon 1953: 26, see maps in Sautter 1966: 762; and Suret-Canale 1987: 235; Lepemangoye-Mouléka 2009), which operated in the Ikoy area for over 30 years (1956–1990s) (Charbonnier 1957; Lepemangoye-Mouléka 2009). By controlling the production of timber, Europeans took over the last tier. La SONG established the modern-day town of Ikobey, where they set up their base and created the area's present-day roads.

---

4   A company which still exists today under the name Tractafric.

Disease and "dead zones"

Disease-related depopulation has a long history in Gabon (Oslisly 2001: 112–113; Spinage 2012: 1194). By breaking down trade barriers, Europeans increased disease transmission and brought diseases previously unknown in Gabon (Sautter 1966: 798). The introduction of quinine[5] helped Europeans to become first-tier traders and to move away from the coast, leading to further disease spread (Gray 2002: 95) from around 1880 until the First World War (Debusman 1993: 40). This was a time when early explorers and traders, such as du Chaillu and de Brazza (Marche 1879: 327–328), were exploring the interior of Gabon (Headrick 1994: 8) and brought coastal diseases such as venereal disease and smallpox (du Chaillu 1867) into places where they were unknown (Spinage 1973: 966). The final phase of disease spread occurred after the First World War with Europeans taking over the second and third trading tiers. The timber concessions and the development of transport infrastructure created a large demand for labour in a country that had already been depopulated by disease. The resulting mixing of people from all over the country in timber camps and administration posts fostered the spread of diseases including during the 1918 influenza pandemic (Headrick 1994: 125, 128; Coquery-Vidrovitch 2001: 456; Hymas 2015a; Hymas et al. 2021b). While French foresters and the colonial administration fought each other over access to labour (Pourtier 1989: 173; Gray and Ngolet 1999; Coquery-Vidrovitch 2001: 455; Rich 2005) villages became depopulated. A forester wrote in 1918:

> In all the exploitable areas of the Gabonese forest, it is becoming more and more rare to find villages in the middle of the forest. Sleeping sickness, alcoholism, venereal diseases have resulted in the disappearance of a large part of the population and the rest, decimated, have slowly come closer to places where they can get easy access to European factories (Quilliard 1920: 645).

Europeans were moving into remote parts of Gabon away from the principal trade routes in search of resources. The SHO's porters from surrounding areas (Martrou 1923; Moutangou 2013: 230) (Fig. 13.6) brought people into contact with new coastal diseases (Hartwig and Patterson 1978: 9–10; Headrick 1994: 42).

---

5   Quinine was one of the earliest treatments for malaria. Malaria was a deadly disease for Europeans who had not previously encountered it.

Fig. 13.6 Porters near Samba falls circa 1923 from Moutangou (2013: 227–228), public domain

Road building was also carried out through the resettlement policy which exacerbated disease spread. People fled as famine and disease spread. During the 1918 Influenza pandemic there was a 16.6% mortality rate in Sindara (Bruel 1935: 338). By the early 1930s, disease and famine resulted in the abandonment of trade routes and the loss of the SHO concession; the depopulation of the area led to it becoming a "dead zone" (Gray 2002: 160; Hymas 2015b).

For around thirty years the area was devoid of people, resulting in the growth of *Okoumé* (*Aucoumea klaineana*), a successional forest tree species, and an increase in animal populations that had previously been hunted (Hymas 2015b). The arrival of La SONG timber company in the late 1960s encouraged people to return. The company entered the dead zone, finding stands of *Okoumé* that had grown in the abandoned village and trade plantations. With the arrival of La SONG and the creation of the town of Ikobey, migration into the area increased (Abitsi and Lepemangoye-Mouleka 2009), finding only forest. One informant noted that "here everything was forest, all that was forest, Nyoe I and Nyoe II, it was La SONG that opened it up" (Makoko, Babongo Ghebondgi 24/02/10), while another said that "there were no old villages" (Nyoe II, Akele 22/05/10). The migration of people into the area culminated in 2000 when a group of Babongo hunter-gatherers from Mount Iboundji followed the Ikoy River downstream. It was in this area that the different ethnicities started to live together. In the next section, we recount three migration histories, using oral history methods, in order to show the complexity of migration into the landscape.

Akele migration history

The Akele migration into the area started in the 1840s due to competition with the Fang traders (du Chaillu 1861: 121; Walker 1870: 142; Sautter 1966: 745; Chamberlin 1977; Cinnamon 1998; Gray 2002).[6] By the 1870s, the Akele took over an important part of the riverine trade at Samba Falls (Gray and Ngolet 1999) and become second-tier traders (Chamberlin 1977), which put them in competition with the Mitsogho and the French. Migration occurred in two directions: one used during the first Akele migration along the Ngounié and along the Ogooué Rivers (Avelot 1905; Gray 2002; Ngolet 2003), and a second (Map 2) along the old trade routes that surround Mimongo and Eteke, villages created by the resettlement policy (1950s–1960s).

During resettlement, a Prefect of the Fang ethnicity named Ekoga asked the Akele, Mitsogho, and Babongo peoples to move towards Mimongo. Some agreed while others refused and went into the forest. Those who refused used the old caravan routes and the Ikoy. As they went down the river they found no villages or signs of old villages. Some of the first villages settled were along the Idemba River, where they met La SONG. By this time, disease and migration had once again decimated populations. Overall, the Akele migrated 40 km through the forest.

---

6   The migration of the Fang from Cameroon triggered the displacement of people throughout northern and central Gabon (Avelot 1905; Gray 2002), including the Akele (Sautter 1966: p. 743; Van der Veen 1991).

Mitsogho migration history

The Mitsogho used two different migration routes to return to the Ikobey area, a migration of approximately 90 km. The migration from Mimongo village also started with the resettlement policy implemented by the Prefect Ekoga. Some accepted, while others refused and started their migration towards Ikobey:

> Since Ngoassa, there was a chief [Prefet], a joker, Ekoga, who asked our parents to leave—at our old village—leave there, go back, go back home, go back home, over there. They said we will not go over there. Ekoga, what did he do, he took all the chiefs, there were six villages...and said 'ok you are being arrogant, it's you who makes it that the population does not go over there, so I will take you, let's go! ....he took them, beat them well. So they got hit, it was the commander Ekoga who took them. But they, when they saw how things were, they said 'as we were hit, we are obliged to go down, we will no longer go to Mimongo, he should not have hit us'. And that is why they went down. Some went down one side of the Oumba River, that was us [people of Nyoe I]. Those who went down the other side of the Oumba River, were the ones who accepted Ekoga's beatings, but after a time they came back, saying that the others, on the other side of the Oumba River, reacted better than we did. Nyoe I, Mitsogho 09/05/10 [recording DS400093].

They first descended the rivers using the colonial trade routes and then followed the Bakounga River, and eventually found La SONG's forestry roads. One informant noted that "it was the road where our parents passed, with the tipoi, they followed it ... and then at rivers we cut trees to cross". Nyoe II, Mitsogho 23/04/10 [*recording DS400080*] (Fig. 13.7).

Fig. 13.7 Tipoi being used during the colonial period in Gabon reproduced from Meyo-Bibang (1975: 51), public domain

Once on the La SONG road, they followed the timber companies, setting up villages in anticipation of the road. As with the other ethnicities in the area, they moved slowly away from badly degraded roads.

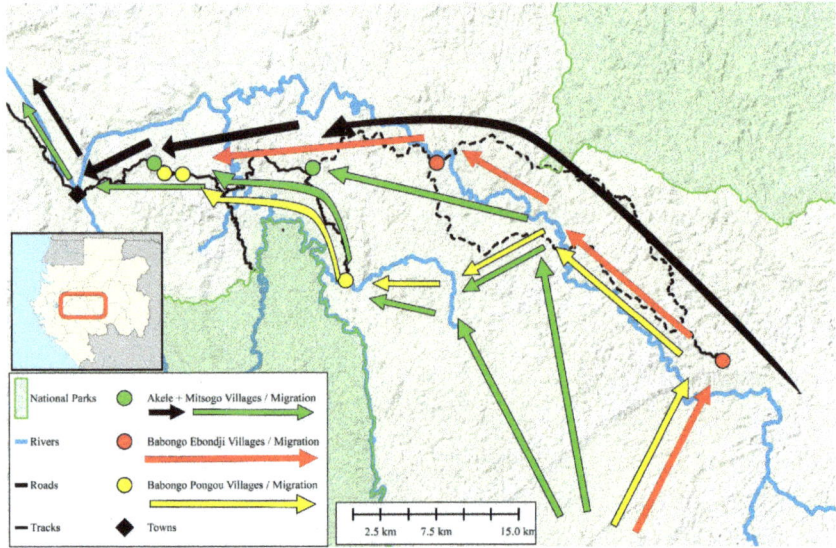

Fig. 13.8 Current villages in the upper Ikoy area with the migration routes of the Akele, Mitsogho and Babongo. Based on the following sources: Basemap (ESRI 2021), park boundaries(Institut Geographique National 2008)

Babongo migration history

As with the Mitsogho and Akele, the Babongo hunter-gatherers were also impacted by resettlement. The same prefect asked them to move towards Mimongo town. Once again some refused and went into the forest. This migration also split into two groups: the Babongo Pongue and the Babongo Ghebondgi.

The Babongo Pongue originally migrated from the village of Pongou to the Oumba River (Fig. 13.8). Currently they live along the road, claiming a common village of origin: Pongou, approximately 130 km away. After refusing to be resettled, they continued their migration down the Ikoy River, sometimes using old trade routes. It was around today's villages of Motombi/Mimongo II where the Babongo Pongue came into contact with La SONG's road. Currently, they are continuing their migration depending on the state of the road and the elephants [recording DS400084]. The second group followed the Ikoy downstream and currently live in the roadless villages of Makoko and Ngondet. They migrated later, having originated from the village of Ebondji. It was at the village of Indamba that the Babongo Ghebondgi came into contact with La SONG's road.

For the Babongo, resettlement is still occurring, with local administrators still asking them to move closer to the road. In 2010, the Babongo Ghebondgi of Makoko and Ngondet were asked to leave the village of Massika and go back towards Mimongo. Instead, some went to Ngondet and others used the old La SONG road to migrate to Makoko or Ossimba.

In 2010, a Chinese logging company, SUNNLY, re-opened part of the La SONG road to a river where one of their old hunting camps was located, and some of the people of Makoko migrated towards this site. By 2014 the village of Makoko had been abandoned with everyone moving down the road towards Nyoe II.

Trade and Okoumé trees of the upper Ikoy

The competition for natural resources, the creation of colonial concessions, and resettlement policy facilitated disease spread. These keystone processes (Marcucci 2000) are "crucial to ensuring we understand the ecological character and effects of long-term human influence" (Gillson et al. 2020: 2) in the Ikoy area. These factors resulted in periods of in- and outmigration, causing a patchwork of forest succession after plantations were abandoned.

The 1930s closure of the SHO in the area meant that large plantations for feeding labourers were abandoned. Combined with disease which created dead zones, the environment was released from anthropogenic disturbance. This resulted in a succession from old plantations to a forest with a high density of *Okoumé* trees, leaving legacies of former land use. The presence of old fruit trees in abandoned villages further contributed to an increase in globally endangered species of animals such as gorillas and chimpanzees and the recovery of elephants that had previously disappeared from the area in the late 1800s (Berton 1895: 214).

The return of logging companies to the Ikoy area in the 1960s resulted in the forest being once more exploited for its natural resources. Ecological surveys undertaken just after the creation of Waka national park indicate a high biodiversity. Maisels et al. (2008) estimated 744–2330 elephants and 2000 great apes in the Lopé-Waka corridor.

The vegetation survey of Balinga et al. (2006) concluded that there was an abundance of large diameter trees and the relatively high number of species made Waka "amongst the most biodiversity-rich of all the Smithsonian Institution's Biodiversity Plot sites" (Balinga et al. 2006: 1, 20). However, they were surprised to find two palm genera species, *Podococus* and *Sclerosperma*, as they usually indicate opposing natural histories: *Podococcus*, "an indicator of long term stability of a habitat" and *Sclerosperma*, "supposedly characteristic of disturbed habitats" (Balinga et al. 2006: 19). While they attributed this contradiction to "unique climatic and topographic features of this landscape" (Balinga et al. 2006: 19), it could also be explained by the history of the area, with waves of settlement and abandonment leaving both deliberately cultivated and successional tree species.

Overall, using the oral history methods and archival work enabled us to understand that the Upper Ikoy forest has been inhabited for a long time, but with substantial impacts from trade and disease brought by colonisation. Using migration histories, we learned the history of this supposedly intact forest landscape and how this history created *Aucoumea* forests, showing how the social history and its vegetation are intertwined. If this case study had been undertaken with a disciplinary approach, we would have ended up with an incomplete history, heavily influenced by assumptions based on the cultural baggage of the researcher, which in turn is susceptible to, among other things, colonial ideologies (Trisos et al. 2021). These mistaken assumptions can have important consequences. Conservation projects have been designed using previous studies of the history of the area that were based on assumption drag and a single methodological approach that resulted in project failure, including misunderstanding which peoples were Indigenous inhabitants of the forest.

Case 2: Fire use in ancient savanna of the Plateaux Batéké

The savannas of the Batéké Plateaux were often considered to be degraded or secondary by colonial-era foresters and botanists (Aubréville 1949; White 1983) and thus of little interest for biodiversity. However, their "underground forests" tell a different story: shrub species have large, often woody root systems, but comparatively little above-ground biomass, which is burned back every year and so resembles herbs (Bond and Zaloumis 2016). Recent work shows that these endemic species are fire-dependent and are an indicator of the ancient status of these savannas (Walters et al. 2022), which are estimated to be at least 30,000 years old (Giresse 1978). Located in the forest-savanna mosaic transition zone of the Guineo-Congolian forest (White 1983), the savannas sit on the Kalahari sands, one of the deepest sand deposits in the world (Haddon 2000). These savannas are part of the greater Batéké Plateaux, spanning more than 120,000 sq. km across south-eastern Gabon, central Republic of Congo, and southwestern Democratic Republic of Congo.

Local fire use plays a role in keeping these ancient savannas open. However, the old-growth definitions say little about the human dimensions of these ancient landscapes. Returning to the savannas we encountered in the aerial photographs of the introduction, where a photographer assumed that the savannas were degraded rather than ancient, we ask: what does fire use look like in an ancient savanna? And what is the link with species diversity? Employing a historical ecology approach, we use interviews and archival research to unravel the history of fire use in this ancient tropical habitat and its links to some savanna species. Using a combination of methods is critical for understanding ecosystem change over time and how people have shaped these landscapes.

Methods and study area

We conducted fieldwork in Gabon's Plateaux Batéké over 19 months (2006–2010, 2022). We triangulated the historical literature with oral histories of elder men and women. We collected site-based information on current fire use via participant observation and semi-structured interviews (see Johnston and Longhurst, Chapter 32; Sayre, Chapter 34). The first and third authors lived in the community, participating in gathering, fishing, agricultural, and ceremonial activities, and helping the community with medical care, education, and transport. Formal, semi-structured, recorded oral history (see Chakov et al., Chapter 33) interviews were conducted with 38 elders. Interviews were conducted in French or Batéké, the latter with the help of a translator.

All informants' names have been changed. A survey on savanna fire use, hunting and gathering was conducted with 122 people in four villages (see Winata and McLafferty, Chapter 43). Botanical specimens were collected, identified, and deposited in the Herbier National du Gabon, the Missouri Botanical Garden, and other herbaria. Samples of the insects and plants were identified by specialists. In the case of birds and mammals, no samples were taken, but the species was identified by biologists who worked in the study area.

Fire use in the pre-colonial and colonial eras

Anthropogenic savanna fires in the Plateaux Batéké have occurred since at least 2,100 years BP (Schwartz 1988b), with lightning fires having occurred for much longer. Early large-scale Bantu migrations into the Plateaux Batéké ended around 1,000 AD; smaller migrations continued, influenced by politics and natural resources (Vansina 1990; Dupré and Pinçon 1997). The Batéké-Alima people who currently live in these savannas have occupied them for hundreds of years (Papy 1949). The savannas were once governed by supreme land chiefs, who used fire largely for communal hunting in the long dry season. Remnants of this system remain today, influencing the ecosystem structure and diversity (Walters 2012; Walters et al. 2014).

Fire use was frequently observed by European explorers in the late 1800s, showing the utility of fire for hunting and gathering (which we call fire foraging). In de Brazza's dry-season crossing of the plateaus in 1880, he repeatedly noted fire use (Brunschwig 1972). These accounts are summarised elsewhere (Walters et al. 2014), but here we provide some examples. Guiral, one of de Brazza's travel companions, observed how fire was used in the hunt,

> Towards the month of September, when the prairies are dried by the sun, they burn all the grass, letting the wind carry the flames forward. In a small space, they install their nets, supported every so often by stakes in the earth. A group of small rodents who live in the prairies, flushed out by the fire, find themselves in the area of the nets.
>
> Guiral 1889: 154

These ancient savannas were governed by people. The Batéké territories, or *ntsé*, were governed by a land-chief (Ebouli 2001), who amongst other duties, was responsible for organising the annual collective fire drive. *Ntsé* were subdivided into permanent burn units called *ewa*. After rituals for a productive hunt, the fire specialists (*otiugui*) would burn fires towards nets, which extended for kilometres (Dusseljé 1910), with some 50 hunters joining their nets end to end. Animals were driven by the fire into the nets. Women would follow, gathering small animals that had been killed. This hunt is further detailed elsewhere (Walters et al. 2014). These ancient savannas were also important for the gathering of many species of caterpillar and insects. Guiral talks about the savanna foods, as well as the link of the *nkieli* caterpillars harvested on the *ololo* tree (a dominant savanna shrub, Annona senegalensis):

> ... the Batéké were extremely fond of these foods. Thus, they eat with gusto grilled rats..., toads that were smoked alive, sun-dried grasshoppers, and fat yellow caterpillars that they abundantly gather on a special tree...
>
> Guiral: 160

Based on these historic sources, we see that fire use was important for livelihoods, but also linked to species, which we will explore in the next section.

In 1960 the Gabonese state gained independence from France. New state laws removed land tenure and burning privileges from the local people and customary authorities. At the same time, gun possession increased, and a rural exodus caused a reduction in communal hunting. Today, communal fire drives are non-existent and land fertility rituals are becoming increasingly rare. Hunters set fires that cross former hunting territory boundaries; these fires are set to attract game to pasture later, rather than using fire to actively drive animals to nets (Walters 2015).

Post-colonial fire use in the ancient savannas

Today, fire foraging continues in these ancient savannas. In our 2007 survey about present day and past fire practices, almost all respondents indicated that they still burn the savanna today. Reasons for burning can be categorised into subsistence activities, safety (visibility, reptile habitat removal, path clearing), and for pleasure (Table 13.1). Answers related to subsistence comprise more than half of the reasons for pre-independence burning and today. As stated by Tricia, "If the savanna is burned, we can easily find *olu*, caterpillars or grasshoppers" (Tricia, Age 55). Many respondents indicated that burning and foraging were linked, particularly in the past. Gorgie noted, "In the past, the women would burn the savanna to look for grasshoppers and the men for the hunt. After the grasshoppers, the new grass shoots would call the *ntsienstiele* caterpillar" (Georgie, age 67). This quote demonstrates the link between fire and the creation of new shoots needed by caterpillars for food.

Table 13.1 Reasons, in order of importance, for past and present
savanna burning.

Pre-independence burning (respondents born pre-1960)	Current burning (all respondents)
1. Grasshopper gathering	1. Clearing paths
2. Hunting (fire drive)	2. In disorder (negative)
3. Savanna plantation	3. Hunting (creating pasture)
4. Rodent gathering	4. Grasshopper gathering
5. Clearing paths	5. Savanna regeneration
6. To eat	6. Savanna plantation
7. Caterpillar gathering	7. To eat
8. Bird hunting	8. Caterpillar gathering
	9. Visibility
	10. Fun
	11. Protection from reptiles
	12. Mushroom gathering
	13. Dead grass removal
	14. Rodent gathering
	15. Bird hunting

A major portion of Batéké rural livelihoods is derived from savanna resources. According to our survey, the list included more than 25 items.[7] These responses ranged from bush meat to insects and fruits, representing a range of seasons and fire-foraging methods (Table 13.2).

---

7   More than 25 species are represented, with "bushmeat" representing several species.
     Spellings correspond to a Batéké-Alima lexicon where possible (Linton 2009).

Table 13.2. A list of savanna derived foods in order of importance. The letter "I" denotes fire used indirectly to obtain food; "D-I" means "direct-immediate"; "D-D" means "direct-delayed".

Category	Lateghe name	scientific name (or order/family)	Direct or indirect burning	Number of respondents (N = 122)
Insect-larva	*enkele*	*Bunaeopsis licharbas* (Saturniidae)	D-D	99
Insect-juvenile/adults	*ampari*	Several species	D-I	98
Insect-larva	**Ntsienstiele ntsintsili**	*Spodoptera exempta* (Noctuidae)	D-D	62
Insect-adults	*enginiña*	Cetoine (Phoenomeanidae)	I	56
Insect-larva	*evura*	*Anthuea insignata* (Notodontidae)	I	42
insect	**akuraku**	*Antheua* sp. (Notodontidae)	I	17
fungi	**tutsa**	*Termitomyces striatus* (Tricholomataceae)	I	15
plant	*olu*	*Albizia adianthifolia* (Fabaceae)	D-D, I	15
plant	*kura*	*Hymenocardia acida* (Phyllanthaceae)	D-D	6
insect	*evatu*	Unknown (Coleoptera)	I	5
animal	rats	Several species	D-I	4
animal	*nyama*	Several species	I	4
plant	*eburi/ mfulugu*	*Landolphia owariensis* (Apocynaceae)	I	3
Animal (bird)	*ankumbi*	*Ciconia abdimii* (Ciconiidae)	D-I	3
Insect-adults	*kayie*	Unknown (Homoptera)	I	3
Insect-larva	*entsaaba*	Unknown (Psychidae)	I	3

Insect-adults	antsama	Macrotermes bellicosus (Termitidae)	I	3
plant	mbaama	Cogniauxia podolaena (Cucurbitaceae)	I	2
plant	bli ntsa	Parinari capensis (Chrysobalanaceae)	I	2
plant	ntunu e ntsege	Afromomum alboviolaceum (Zingiberaceae)	I	1
Insect-juvenile/ adults	eŋeye	Ruspolia differens (Tettigoniidae)	I	1
plant	mfuu	Anisophyllea quangensis (Anisophyllaceae)	I	1
other			I	4

Eighty-five percent of the respondents spoke directly of or implied links between foraging and fire. For these foods, foraging and fire are linked in three ways according to the application of fire and the timing of the resource's harvest post-fire:[8]

- *Direct-immediate benefits of burning*: fire is intentionally applied to acquire a resource the same day. Examples: the historic *ntsa* fire drive, attracting Abdim's stork with smoke and singed grasshoppers, attraction of game to cinders, and grasshopper gathering. 12% of the savanna resources are in this category.

- *Direct-delayed benefits of burning*: fire is intentionally applied to gain a resource that appears weeks or months after application. Examples: creation of forage for caterpillar species several months post-fire; creation of forage for grazers days and weeks post-fire. 15% of the savanna resources are in this category.

- *Indirect benefits of burning*: fire maintains the savanna habitat in which many gathered and hunted organisms are found. 73% of the foods listed are in this category. During the survey, fire was never stated as a tool to harvest these species.

---

8   Categories were determined by GW, not by informants and so do not reflect folk taxonomy.

Here we see that fire's relation to livelihoods in this ancient savanna is linked to maintaining an open habitat, but also to the life cycles of some species.

Direct-immediate benefits of burning: Grasshopper gathering

In the mid-2000s, every long dry season, when the stands of dry grass are full of nymph grasshoppers, Batéké women would burn small patches of the savanna to gather grasshoppers. This form of gathering was done historically either during the communal *ntsa* hunt or in peri-village savannas. Women limit the fire surface area with the idea of conserving a grasshopper-rich spot of savanna for the following day's gathering. Groups of women depart early in the morning to an area where they have recently estimated high grasshopper abundance. Once the conditions are judged right for burning (low humidity, low wind), *ntseli* (long stems of dry grass to use as an ignition torch) are gathered. Lighting is performed by two or more people, either walking along the perimeter of the burn or walking up the middle of the area to be burned (this latter burn type is called *onya*). As the wind takes the fire away from the ignition point in a head-fire (*mba ya olumi*), leafy branches of *ongalaga* (*Hymenocardia acida*) trees are used to beat out the advancing flames. Once the flames are extinguished, the women gather the dead nymph-stage grasshoppers. There are several varieties in Batéké terminology such as *tsara, ngokolo, kadula, jele, oyara, ambali, kafuuyi*, and *anai* according to one informant (interviewed 25 September 2007).

Direct-delayed burning: Caterpillar gathering

The Batéké gather several caterpillars, but here we will focus on the main savanna one: *kankele*. For years a fire specialist, Antoine Mbia, lit savannas in July for *kankele* gathering in November, something that he was taught by his father. Even though fires no longer have a controlled calendar, many people surveyed in 2007 indicated that burning was essential in order to gather *kankele*. *Kankele* is a Saturniidae family African emperor type moth whose major food plant at the larval stage is *ololo* (*Annona senegalensis*). *Ololo* is a co-dominant, fire-resistant savanna shrub. Not a dense wood, these shrubs are not used for firewood. Batéké view them primarily as the host plant for *kankele*; when burned, the shrub resprouts tender leaves which the larvae consume.

In order to gather *kankele*, groups of women leave on all-day expeditions. Some women return to the places where they have gathered since they were young, near former village sites. Hunters discover *kankele* sites. After gathering, upon return to the village, some women begin the time-consuming task of cleaning, cooking, and smoking *kankele*. They can be stored for months and can also be sold in nearby markets.

Caterpillars were proposed as a cash crop in the 1950s (Merle 1958). When the declining abundance of some species was noted in the Kwango area of the Democratic Republic of Congo, investigators linked this to the demise of the traditional authorities who once controlled fires (Leleup and Daems: 19). Unseasonal fires were thought to injure the subterranean larvae of the Saturniidae in the Kwango area, as well as failing to provide tender leaf forage at the right time, linking fire, tender leaf production, and feeding of Saturniidae species (Leleup and Daems 1969).

Indirect burning and ancient savanna indicator species

Another contribution that fire makes to Batéké gathering is through enhancing the growth of gathered fruits and leaves. *Olu* (*Albizia adianthifolia*) is a widespread fire-resistant savanna tree used for medicine, subsistence, and as a host plant of edible caterpillars (Latham 2004: 20). The leaves of *olu* contain as much as 10.81 g of protein per 100 g; this is twice that of *Gnetum africanum*, the most commonly consumed wild leaf in the study area (Mbemba and Remacle 1992: 23) and one of the most common leaf dishes in several Central African countries. Gathering of *olu* occurs primarily after fire, when leaves are resprouting.

Several species are indicator species of ancient savannas, which form underground forests of roots due to the repeated passage of fire that favours constant resprouting (Bond and Zaloumis 2016). Several of the taxa in the survey comprise these indicator species, notably *mufulugu* (*Landolphia owariensis*), *mfuu* (*Anisophyllea quangensis*), and *Parinari capensis*. *P. capensis* is primarily consumed by the *ntsa* (*Sylvicapra grimmia*), the prey of the fire drive and the object of conservation concern, requiring regular burning, which in this case comes largely from customary fire use.

In these examples, we see that fire use in this ancient savanna is diversified and has changed over time. Burning remains a central strategy for obtaining savanna foods, many of which are fire-dependent. When these historic and modern accounts of fire use are taken into account, we see that endemic, fire-dependent species of these ancient savannas co-exist with customary fire use. If we had only applied botanical methods, we would have understood plant diversity, but not its linkages with livelihoods and fire regimes. If we had only applied ethnographic methods, we would have understood fire use but not its connection to the vegetation. Bringing these methods together, we can understand the ties between vegetation, fire use, and local livelihoods.

Discussion and conclusion

These case studies use historical ecology methods (see Davis, Chapter 29) to test assumptions about two ecosystems: intact forests and ancient savannas. In our analysis, we show that the processes behind these landscapes are complex. Bringing together a variety of methods, we are able to understand that in the first case, intact forest landscapes may appear to be "seamless" when viewed from the canopy today, but in the past, the area was crossed with colonial trading routes and impacted by disease and village resettlement politics. The tree canopy bears legacies of these movements through its present day *Aucoumea* populations, which are testimonies of past disturbance (sensu Morin-Rivat et al. 2017). Furthermore, the presence of fruit trees in the abandoned village sites attracts elephants. The curious mixture of a successional timber species and a threatened animal drew the interest of both loggers and conservationists to the area. In the second case, we look at how species and people are linked in hunting and gathering traditions where fire is used as a primary tool for livelihoods but is also necessary for maintaining the biodiversity of species and habitats.

When the first and second authors were university students working on our bachelor's and master's degrees in biology (United Kingdom) and botany (United States) in the 1990s, our courses were siloed by discipline. However, our interdisciplinary doctoral programme in environmental anthropology in the late 2000s allowed us to combine environmental anthropology with ecology to understand complex environmental problems. Such training has helped us to research and understand ecosystems through different angles and to work with other disciplines and forms of data.

Using a historical ecology approach in both cases permitted us to explore the histories of an intact forest landscape and an ancient savanna, using archival work, interviews, and participant observation, but in light of ecological data on species ecology and diversity. Integrating data from multiple disciplines brings these landscapes alive, providing insight into the complex processes that create them and understanding the legacy effects of former land uses (Bürgi et al. 2017). By putting people at the centre of inquiry, historical ecology enables us to analyse how people have historically crafted landscapes. This requires asking social questions of habitats and ecosystems.

Working with diverse types of data can be challenging. We may not have enough time to conduct all the work necessary to answer a question with statistical satisfaction. We may have to translate historical documents, oral histories (see Chakov et al., Chapter 33), and maps into workable data, which requires us to develop different skillsets or collaborate with colleagues from other disciplines. Combining different disciplines that use diverse lexicons, timeframes, epistemologies and methodologies is one of the biggest challenges of conducting interdisciplinary research (Lele and Kurien 2011). Another is publishing such interdisciplinary work, which requires a review process that values the contributions of different disciplines (Campbell 2005).

Intact forest and ancient savannas form parts of scientific and policy debates in different ways. On the one hand, intact forest extent was first defined through satellite imagery without factoring in what was happening under the forest canopy or its history (Potapov et al. 2008; 2017). Later research challenged this, showing the importance of such areas for Indigenous peoples (Fa et al. 2020). However, intact forests form part of forestry policy in some countries (Brouwer 2021) and so understanding how these forests are influenced by societies over time, how this impacts timber stocks species such as *Aucoumea* (Walters et al. 2019), and how local people claim these forest territories (Evine-Binet 2022) should inform policies that recognise that people created these valued forests and in many cases may protect them (Zanjani et al. 2023). On the other hand, ancient savannas are also part of policy debates which focus on the extent to which forests have been favoured over open habitats, like in large-scale restoration (Veldman et al. 2015a, b) fire suppression, and exclusion of megafaunal herbivores (native or domestic. It has been argued that there is a predisposition for forests over savanna leading to "Biome Awareness Disparity" (Silveira et al. 2021). Although an ancient savanna classification is better than an assumption of a savanna being degraded, understanding how ancient savannas are influenced over time by practices such as fire use is often not at the forefront of debates. If we want to ensure that such ecosystems remain functional, we should also ensure that the practices influencing their biodiversity, such as customary fire management, are supported (Walters et al. 2014; 2022).

When considering the state of ecosystems, we should always test our assumptions and look what is going on underneath the canopy (Nguyen 2022; Kelley, Chapter 19) including understanding resource politics, societal change, and how this relates to species in the landscape. The approach we show here is but one way to challenge assumptions about tropical ecosystems. Our best chance of understanding these complex systems is by combining diverse methods and working in interdisciplinary teams to raise critical questions when tropical ecosystems are presented without a historical or social context.

Research and writing process and credits

GW conceived the chapter and lead the writing process. OH, KN, and TS contributed the Ikoy case study. GW and TS contributed the Batéké case study.

Acknowledgements

For the Ikoy case, we thank the people of the Ikoy River Valley. For the Batéké case, we thank the people of the Djouya River valley for participating in the research in the Batéké case. This work was funded by the Rufford Foundation, the Parkes Foundation, and University College London. This work was permitted by the Centre National de Recherche Scientifique et Technique, Gabon. Thanks to P. Laris, J. Fairhead, K. Homewood, P. Burnham, R. Poligui, R. Lave, and S. Lane who read earlier versions of this manuscript. R. Poligui and G. van de Weghe are thanked for entomological identifications; P. Latham is thanked for access to his personal library.

References cited

Abitsi, G. 2006. *Inventaires de reconnaissance des grands mammifères et de l'impact des activités anthropiques: Parc National de Waka, Gabon* (WCS Gabon).

Abitsi, G. and F. Lepemangoye-Mouleka. 2009. *Evaluation des zones de hautes priorités pour la conservation dans le corridor Lopé-Waka, Gabon* (WCS Gabon).

Aicardi de Saint-Paul, M. 1987. *Le Gabon. Du roi Denis a Omar Bongo* (Albatros).

Anon. 1996. *Second nature* (Institute of Development Studies).

Ascher, W. 1979. 'Problems of forecasting and technology assessment', *Technological Forecasting and Social Change*, 13.2, pp. 149–156. https://doi.org/10.1016/0040-1625(79)90109-4

Aubréville, A. 1949. *Climats, forêts et désertification de l'Afrique tropicale* (Société d'éditions géographiques, maritimes et coloniales).

Avelot, M.R. 1905. *Recherches sur l'histoire des migrations dans le bassin de l'Ogooue et la région littorale adjacente* (Imprimerie Nationale).

Balée, W. 2018. 'Brief review of historical ecology', *Les Nouvelles de l'Archéologie*, 152, pp. 7–10. https://doi.org/10.4000/nda.4150

Balée, W. and C.L. Erickson. 2006. 'Time, complexity, and historical ecology', in *Time and Complexity in Historical Ecology*, ed. by W. Balée and C.L. Erickson (Columbia University Press), pp. 1–17.

Balinga, M., T. Sunderland, G. Walters, Y. Issembe, E. Fombod, and S. Asaha. 2006. *A Vegetation Assessment of the Waka National Park, Gabon* (Smithsonian Institute).

Benjaminsen, T.A., H. Reinert, E. Sjaastad and M.N. Sara. 2015. 'Misreading the Arctic landscape: A political ecology of reindeer, carrying capacities, and overstocking in Finnmark, Norway', *Norsk Geografisk Tidsskrift—Norwegian Journal of Geography*, 69.4, pp. 219–229. https://doi.org/10.1080/00291951.2015.1031274

Berton, J. 1895. 'De Lastourville sur L'Ogooué à Samba sur le N'gounië (septembre et octobre 1890)', *Bulletin de la société géographie*, 7.16, pp. 211–218.

Biwolé, A.B., J. Morin-Rivat, A. Fayolle, D. Bitondo, L. Dedry, K. Dainou, O.J. Hardy, and J.-L. Doucet. 2015. 'New data on the recent history of the littoral forests of southern Cameroon: an insight into the role of historical human disturbances on the current forest composition', *Plant Ecology and Evolution*, 148.1, pp. 19–28. https://doi.org/10.5091/plecevo.2015.1011

Bond, W. and N.P. Zaloumis. 2016. 'The deforestation story: testing for anthropogenic origins of Africa's flammable grassy biomes', *Philosophical Transactions of the Royal Society B: Biological Sciences*, 371.1696. https://doi.org/10.1098/rstb.2015.0170

Bond, W.J. 2016. 'Ancient grasslands at risk', *Science*, 351.6269, pp. 120–122. https://doi.org/10.1126/science.aad5132

Brouwer, M. 2021. 'Certification is the future of the industry', in *Central African Forests Forever*, by M. Bouwer (CNPIEC Digital Printing), pp. 36–51.

Bruel, G. 1935. *La France Equatoriale Africaine: le pays, les habitants, la colonisation, les pouvoirs publics* (Larose).

Bürgi, M., L. Östlund and D.J. Mladenoff. 2017. 'Legacy effects of human land use: ecosystems as time-lagged systems', *Ecosystems*, 20.1, pp. 94–103. https://doi.org/10.1007/s10021-016-0051-6

Campbell, L.M. 2005. 'Overcoming obstacles to interdisciplinary research', *Conservation Biology*, 19.2, pp. 574–577. https://doi.org/10.1111/j.1523-1739.2005.00058.x

du Chaillu, P. 1867. *A Journey to Ashango-Land : And Further Penetration into Equatorial Africa* (D. Appleton and Company).

du Chaillu, P.B. 1861. *Explorations and Adventures in Equatorial Africa; with accounts of the manners and customs of the people, and of the chase of the gorilla, crocodile, leopard, elephant, hippopotamus, and other animals* (John Murray).

Chakov, A., Chang, T., Covey, H., Dickson, T., Goggins, S., Harris, N., Purna, S., Widell, S. and Druschke, C.G., Chapter 33, this volume. 'Oral history'.

Chamberlin, C. 1977. *Competition and Conflict: The Development of the Bulk Export Trade in Central Gabon During the Nineteenth Century* (University of California, Los Angeles).

Charbonnier, F. 1957. *Gabon, terre d'Avenir* (Encyclopédie d'Outre-Mer).

Cinnamon, J. 2003. 'Narrating equatorial African landscapes: conservation, histories, and endangered forests in northern Gabon', *Journal of Colonialism and Colonial History*, 4.2. https://doi.org/10.1353/cch.2003.0038

Cinnamon, J. 2010. 'Counting and recounting: dislocation, colonial demography and historical memory in northern Gabon', in *The Demographics of Empire: The Colonial Order and the Creation of Knowledge*, ed. by K. Ittman, D.D. Cordell, and G. Maddox (Ohio University Press), pp. 130–156.

Cinnamon, J.M. 1998. *The Long March of the Fang: Anthropology and History in Equatorial Africa* (Yale University).

Cope, M., Chapter 22, this volume. 'Archival methods'.

Coquery-Vidrovitch, C. 2001. *Le Congo au temps des grandes compagnies concessionnaires 1898-1930* (Edition de l'EHESS).

Cuvillier-Fleury, H. 1904. *La mise en valeur du Congo français* (Université de Paris).

Darbellay, F. 2015. 'Rethinking inter- and transdisciplinarity: Undisciplined knowledge and the emergence of a new thought style', *Futures*, 65, pp. 163–174. https://doi.org/10.1016/j.futures.2014.10.009

Davis, D.K. 2007. *Resurrecting the Granary of Rome: Environmental History and French Colonial Expansion in North Africa* (Ohio University Press).

Davis, D.K., Chapter 29, this volume. 'Historical ecology'.

Debusman, R. 1993. 'Santé et population sous l'effet de la colonisation en Afrique équatoriale', *Matériaux pour l'histoire de notre temps*, 32.1, pp. 40–46. https://doi.org/10.3406/mat.1993.404115

Decocq, G. 2022. *Historical Ecology: Learning from the Past to Understand the Present and Forecast the Future of Ecosystems* (John Wiley and Sons Inc).

Denevan, W.M. 1992. 'The pristine myth: the landscape of the Americas in 1492', *Association of American Geographers*, 82.3, pp. 369–385.

Dodaro, L. and D. Reuther. 2016. 'Historical ecology: Agency in human–environment interaction', in *Routledge Handbook of Environmental Anthropology*, ed. by H. Kopnina and E. Shoreman-Ouimet (Routledge), pp. 81–89.

Ebouli, J.M. 2001. *Les structures de type féodal en Afrique Centrale le cas des Téké: étude des relations de dépendance personnelle et des rapports de production entre 'A mfumu' et 'Elogo dja Mfumu' (des origines à 1880)* (Université Omar Bongo).

Ekblom, A., P. Lane and P. Sinclair. 2019. 'Reconstructing African landscape historical ecologies: an integrative approach for managing biocultural heritage', in *Historical Ecologies, Heterarchies and Transtemporal Landscapes*, ed. by C. Ray and M. Fernández-Götz (Routledge), pp. 83–100.

Ellis, E.C. 2021. 'People have shaped most of terrestrial nature for at least 12,000 years', *PNAS*, 118.17. https://doi.org/10.1073/pnas.2023483118

Ellis, E.C., J.O. Kaplan, D.Q. Fuller, S. Vavrus, K. Klein Goldewijk, and P.H. Verburg. 2013. 'Used planet: A global history', *Proceedings of the National Academy of Sciences*, 110.20, pp. 7978–7985. https://doi.org/10.1073/pnas.1217241110

Erickson, C.L. 2021. 'Foreword', in *Methods in Historical Ecology: Insights from Amazonia*, ed. by G. Odonne and J.-F. Molino (Routledge), pp. xii–xx.

ESRI 2021. 'World Topographic Map', *MapServer*, http://server.arcgisonline.com/ArcGIS/rest/services/World_Topo_Map/MapServer

Evine-Binet, B. 2022. Ibola Dja Bana Ba Massaha—« la réserve forestière de tous les enfants »—est née au Gabon. *Mongabay*, https://fr.mongabay.com/2022/03/

ibola-dja-bana-ba-massaha-la-reserve-forestiere-de-tous-les-enfants-est-nee-au-gabon/

Fa, J.E., J.E. Watson, I. Leiper, P. Potapov, T.D. Evans, N.D. Burgess, Z. Molnár, Á Fernández-Llamazares, T. Duncan, S. Wang, B.J. Austin, H. Jonas, C.J. Robinson, P. Malmer, K.K. Zander, M.V. Jackson, E. Ellis, E.S. Brondizio, and S.T. Garnett. 2020. 'Importance of Indigenous peoples' lands for the conservation of Intact Forest Landscapes', *Frontiers in Ecology and the Environment*, 18.3, pp. 135–140. https://doi.org/10.1002/fee.2148

Fairhead, J. and M. Leach. 1996. *Misreading the African Landscape: Society and Ecology in a Forest Savanna Mosaic* (Cambridge University Press).

Fairhead, J. and M. Leach. 1998. *Reframing Deforestation: Global Analyses and Local Realities: Studies in West Africa* (Routledge).

Foster, D., F. Swanson, J. Aber, I. Burke, N. Brokaw, D. Tilman, and A. Knapp. 2003. 'The importance of land-use legacies to ecology and conservation', *Bioscience*, 53.1, pp. 77–88. https://doi.org/10.1641/0006-3568(2003)053[0077:tiolul]2.0.co;2

Garcin, Y., P. Deschamps, G. Ménot, G. de Saulieu, E. Schefuß, D. Sebag, L.M. Dupont, R. Oslisly, B. Brademann, K.G. Mbusnum, J.-M. Onana, A.A. Ako, L.S. Epp, R. Tjallingii, M.R. Strecker, A. Brauer, and D. Sachse. 2018. 'Early anthropogenic impact on Western Central African rainforests 2,600 y ago', *Proceedings of the National Academy of Sciences*, 115.13, pp. 3261–3266. https://doi.org/10.1073/pnas.1715336115

Gardinier, D. and D.A. Yates. 2006. *Historical Dictionary of Gabon* (Scarecrow Press).

Gauch, H.G. Jr. 2002. *Scientific Method in Practice* (Cambridge University Press).

Gaulme, F. 1988. *Le Gabon et son ombre* (Karthala).

Gaulme, F. 1991. 'From timber to petroleum', *Review of African Political Economy*, 51, pp. 84–87.

Gillson, L., C.L. Seymour, J.A. Slingsby, and D.W. Inouye. 2020. 'What are the grand challenges for plant conservation in the 21st Century?', *Frontiers in Conservation Science*, 1. https://doi.org/10.3389/fcosc.2020.600943

Giresse, P. 1978. 'Le Contrôle climatique de la sédimentation marine et continentale en afrique centale atlantique à la fin du quaternaire – problèmes de corrélation', *Palaeogeography, Palaeoclimatology, Palaeoecology*, 23, pp. 57–77.

Gray, C. 2002. *Colonial Rule and Crisis in Equatorial Africa: Southern Gabon, c. 1850–1940* (University of Rochester Press).

Gray, C. and F. Ngolet. 1999. 'Lambaréné, Okoumé and the transformation of labor along the Middle Ogooue (Gabon), 1870-1945', *The Journal of African History*, 40.1, pp. 87–107.

Grenand, P. and D. Davy. 2021. 'History and ethnohistory of ancient settlements', in *Methods in Historical Ecology: Insights from Amazonia*, ed. by G. Odonne and J.-F. Molino (Routledge), pp. 133–139.

Guiral, L. 1889. *Le Congo français. Du Gabon à Brazzaville* (Plon).

Hartwig, G.W. and K.D. Patterson. 1978. *Disease in African History: An Introductory Survey and Case Studies* (Duke University Press).

Haurez, B., K. Daïnou, C. Vermeulen, F. Kleinschroth, F. Mortier, S. Gourlet-Fleury, and J.-L. Doucet. 2017. 'A look at Intact Forest Landscapes (IFLs) and their relevance in Central African forest policy', *Forest Policy and Economics*, 80, pp. 192–199. https://doi.org/10.1016/j.forpol.2017.03.021

Headrick, R. 1994. *Colonial Health and Illness in French Equatorial Africa, 1885–1935* (African Studies Association Press).

Heino, M., M. Kummu, M. Makkonen, M. Mulligan, P.H. Verburg, M. Jalava, and T.A. Räsänen. 2015. 'Forest loss in protected areas and Intact Forest Landscapes: A global analysis', *PLOS ONE*, 10.10. https://doi.org/10.1371/journal.pone.0138918

Hymas, O. 2015a. *'L'Okoumé, fils du manioc': Post-logging in remote rural forest areas of Gabon and its long-term impacts on development and the environment* (University College London).

Hymas, O. 2015b. *L'Okoumé, fils du manioc': Post-logging in remote rural forest areas of Gabon and its long-term impacts on development and the environment* (University College London).

Hymas, O., B. Rocha, N. Guerrero, M. Torres, K. Ndong, and G. Walters. 2021a. 'There's nothing new under the sun – lessons conservationists could learn from previous pandemics', *PARKS*, 27, pp. 25–40.

Hymas, O., B. Rocha, N. Guerrero, M. Torres, K. Ndong, and G. Walters. 2021b. 'There's nothing new under the sun – lessons conservationists could learn from previous pandemics', *PARKS*, 27, pp. 25–40. https://doi.org/10.2305/iucn.ch.2021.parks-27-sioh.en

Institut Geographique National 2008 (Carte Touristique).

Journal Officiel de L'Afrique Equatoriale Française 1910. 'Postes et Télégraphes', *Journal Officiel de L'Afrique Equatoriale Française*, 7.19, p. 515.

Kelley, L., Chapter 19, this volume. 'Engaging remote sensing and ethnography to seed alternative landscape stories and scripts'.

Kleinschroth, F., J.R. Healey, S. Gourlet-Fleury, F. Mortier, and R.S. Stoica. 2017. 'Effects of logging on roadless space in intact forest landscapes of the Congo Basin: Roadless Areas', *Conservation Biology*, 31.2, pp. 469–480. https://doi.org/10.1111/cobi.12815

Knight, J. 2003. 'Relocated to the roadside: Preliminary observations on the forest peoples of Gabon', *African Study Monographs Supplement*, 28, pp. 81–121.

Kull, C.A. 2004. *Isle of Fire: The Political Ecology of Landscape Burning in Madagascar* (University of Chicago Press).

Lane, S.N. and Lave, R., Chapter 3, this volume. 'Frames, disciplines and mixing methods in environmental research'.

Laris, P. and D.A. Wardell. 2006. 'Good, bad or "necessary evil"? Reinterpreting the colonial burning experiments in the savanna landscapes of West Africa', *The Geographical Journal*, 172.4, pp. 271–290. https://doi.org/10.1111/j.1475-4959.2006.00215.x

Latham, P. 2004. *Useful plants of Bas-Congo Province, Democratic Republic of Congo* (Paul Latham).

Lele, S. and A. Kurien. 2011. 'Interdisciplinary analysis of the environment: Insights from tropical forest research', *Environmental Conservation*, 38.2, pp. 211–233. https://doi.org/10.1017/s037689291100018x

Leleup, N. and H. Daems. 1969. 'Les chenilles alimentaires du Kwango. Causes de leur rarefaction et mésures preconisées pour y rémedier', *Agric.Trop.Bot. Appl.*, 16, pp. 1–21.

Leopold, A. 2013. *A Sand County Almanac and Other Writings on Ecology and Conservation* (Library of America).

Lepemangoye-Mouléka, F. 2009. *Profil socio-économique de la zone d'extraction de ressources naturelles à l'ouest du Parc National Waka* (WCS Gabon).

Linton, P. 2009. *Non-Verbs* (Latege).

Maisels, F., G. Abitsi, and A. Bezangoye. 2008. *Suivi écologique des grands singes et de l'impact des activités humaines dans l'aire de priorité exceptionnelle Lopé-Waka, Gabon: Rapport finale* (WCS Gabon).

Marche, A. 1879. *Trois voyages en Afrique occidentale Sénégal, Gambie, Casamance, Gabon, Ogowe* (Hachette).

Marcucci, D.J. 2000. 'Landscape history as a planning tool', *Landscape and Urban Planning*, 49.1, pp. 67–81. https://doi.org/10.1016/s0169-2046(00)00054-2

Martin, P. 1972. *The External Trade of the Loango Coast, 1576-1870* (Clarendon Press).

Martrou, L. 1923. 'Le Secret de L'Ofoué', *Annales Apostoliques des PP du Saint-Esprit*, 39.2 and 3, pp. 54–64 and 84–93.

Mbemba, F. and J. Remacle. 1992. *Inventaire et composition chimique des aliments et denrées alimentaires traditionnels du Kwango-Kwilu au Zaire* (Biochimie Alimentarie UNIKIN-Zaire; Biochimie Cellulaire FUNDP-Belgique).

Merle. 1958. 'Les chenilles comestibles', *Notes Africaines*, 77, pp. 20–23.

Merlet, A. 1990. *Le pays des trois estuaires: 1471-1900: quatre siècles de relations extérieures dans les estuaires du Muni, de la Mondah et du Gabon* (Centre culturel français Saint-Exupéry).

Meyo-Bibang, F. 1975. *Le Gabon, le monde: histoire 1er degré* (Hatier).

Morin-Rivat, J., A. Fayolle, C. Favier, L. Bremond, S. Gourlet-Fleury, N. Bayol, P. Lejeune, H. Beeckman, and J.-L. Doucet. 2017. 'Present-day central African forest is a legacy of the 19th century human history', *eLife*, 6. https://doi.org/10.7554/elife.20343

Moutangou, F.A. 2013. *Une entreprise coloniale et ses travailleurs: la Société du Haut-Ogooué et la main d'œuvre africaine (1893-1963)* (Université Toulouse le Mirai l–Toulouse II).

Ngolet, F. 2003. Re-examining Population Decline Along the Gabon Estuary: A Case Study of the Bakèlè. in *Culture, Ecology, and Politics in Gabon's Rainforest*, ed. by M.C. Reed and J.F. Barnes (The Edwin Mellen Press), pp. 137–163.

Nguyen, T.H.V. 2022. T*he Politics of Forest Transition in Contemporary Upland Vietnam Case Study in A Luoi, Thua Thien Hue Province* (University of Lausanne).

Odonne, G. and J.-F. Molino. 2020. *Methods in Historical Ecology: Insights from Amazonia* (Routledge).

Oslisly, R. 2001. 'The history of human settlement in the Middle Ogooue Valley (Gabon). Implications for the environment', in *African Rain Forest Ecology and Conservation. An Interdisciplinary Perspective*, ed. by W. Weber, L.J.T. White, A. Vedder, and L. Naughton-Treves (Yale University Press), pp. 101–118.

Oslisly, R., L. White, I. Bentaleb, C. Favier, M. Fontugne, J.-F. Gillet, and D. Sebag. 2013. 'Climatic and cultural changes in the west Congo Basin forests over the past 5000 years', *Philosophical Transactions of the Royal Society B: Biological Sciences*, 368.1625, pp. 20120304–20120304. https://doi.org/10.1098/rstb.2012.0304

Patterson, K.D. 1975. *The Northern Gabon Coast to 1875* (Clarendon Press).

Potapov, P., M.C. Hansen, L. Laestadius, S. Turubanova, A. Yaroshenko, C. Thies, W. Smith, I. Zhuravleva, A. Komarova, S. Minnemeyer, and E. Esipova. 2017. 'The last frontiers of wilderness: Tracking loss of intact forest landscapes from 2000 to 2013', *Science Advances*, 3.1. https://doi.org/10.1126/sciadv.1600821

Potapov, P., A. Yaroshenko, S. Turubanova, M. Dubinin, L. Laestadius, C. Thies, D. Aksenov, A. Egorov, Y. Yesipova, I. Glushkov, M. Karpachevskiy, A. Kostikova, A. Manisha, E. Tsybikova, and I. Zhuravleva. 2008. 'Mapping the world's Intact Forest Landscapes by remote sensing', *Ecology and Society*, 13.2. https://doi.org/10.5751/es-02670-130251

Pourtier, R. 1989. *Le Gabon, tome 2: Etat et développement* (L'Harmattan).

Pourtier, R. 2010. *Un siècle d'exploitation forestière au Gabon: de l'okoumé-roi à l'exploitation sous aménagement durable* (Saint-Dié-Des-Vosges).

Quilliard, M.C. 1920. 'Les exploitations forestières du Gabon', in *Ongrès d'agriculture coloniale 21-25 Mai 1918. Compte rendu des travaux. Tome IV Agriculture indigène elevage, hygiène, forêt, pêcheries*, ed. by M.J. Chailley and D. Zolla (Augustin Challamel), pp. 642–667.

Reiners, D.S., W.A. Reiners, J.A. Lockwood, and S.D. Prager. 2019. 'The usefulness of ecological concepts: patterns among practitioners', *Ecosphere*, 10.4. https://doi.org/10.1002/ecs2.2652

Reiners, W.A. and J.A. Lockwood. 2010. *Philosophical Foundations for the Practices of Ecology* (Cambridge University Press).

Rich, J. 2005. 'Forging permits and failing hopes: African participation in the Gabonese timber industry, ca. 1920-1940', *African Economic History*, 33, pp. 149–173.

Roberts, P. 2019. *Tropical Forests in Human Prehistory, History, and Modernity* (Oxford University Press).

Rondet-Saint, M. 1911. *L'Afrique Equatoriale française* (Plon).

Rouget, F. 1906. *L'expannsion coloniale au Congo français* (Emile Larose).

de Saulieu, G., M. Elouga, and B. Sonké. 2016. *Pour une écologie historique en Afrique Centrale* (IRD, Yaoundé).

Sautter, G. 1966. *De l'Atlantique au Congo: une géographie du sous-peuplement République du Congo; République Gabonaise* (La Haye, Mouton).

Silveira, F.A.O., C.A. Ordóñez-Parra, L.C. Moura, I.B. Schmidt, A.N. Andersen, W. Bond, E. Buisson, G. Durigan, A. Fidelis, R.S. Oliveira, C. Parr, L. Rowland, J.W. Veldman, and R.T. Pennington. 2021. 'Biome Awareness Disparity is BAD for tropical ecosystem conservation and restoration', *Journal of Applied Ecology*. https://doi.org/10.1111/1365-2664.14060

Simon. 1953. 'L'exploitation de la société du Haut Ogooué (Section Bois) a N'Djolé', *Bois et forêts des Tropiques*, 31, pp. 15–26.

Smith, C., O. Perkins, and J. Mistry. 2022. 'Global decline in subsistence-oriented and smallholder fire use', *Nature Sustainability*, 5, pp. 542–551. https://doi.org/10.1038/s41893-022-00867-y

Spinage, C.A. 1973. 'A review of ivory exploitation and elephant population trends in Africa', *African Journal of Ecology*, 11.3–4, pp. 281–289.

Spinage, C.A. 2012. *African Ecology* (Springer Berlin Heidelberg).

Suret-Canale, J. 1987. *Afrique et capitaux: géographie des capitaux et des investissements en Afrique tropicale d'expression française* (L'Arbre verdoyant, Montreuil-sous-Bois).

Szabó, P. 2015. 'Historical ecology: past, present and future', *Biological Reviews*, 90.4, pp. 997–1014. https://doi.org/10.1111/brv.12141

Táíwò, O. 2019. 'Rethinking the decolonization trope in philosophy', *The Southern Journal of Philosophy*, 57.1, pp. 135–159. https://doi.org/10.1111/sjp.12344

Trisos, C.H., J. Auerbach, and M. Katti. 2021. 'Decoloniality and anti-oppressive practices for a more ethical ecology', *Nature Ecology and Evolution*, 5, pp. 1205–1212. https://doi.org/10.1038/s41559-021-01460-w

Van der Veen, L.J. 1991. *Etude comparée des parlers du Groupe Okani B30* (Université Lumière-Lyon).

Vande Weghe, J.P. 2011. *Lopé, Waka et monts Birougou: le moyen Ogooué et le massif du Chaillu* (Les parcs nationaux du Gabon).

Vansina, J. 1990. *Paths in the Rainforest. Towards a History of Political Tradition in Equatorial Africa* (Wisconsin Press).

Veldman, J.W, E. Buisson, G. Durigan, G.W. Fernandes, S. Le Stradic, G. Mahy, D. Negreiros, G.E. Overbeck, R.G. Veldman, N.P. Zaloumis, F.E. Putz, and W.J. Bond. 2015. 'Toward an old-growth concept for grasslands, savannas, and woodlands', *Frontiers in Ecology and the Environment*, 13.3, pp. 154–162. https://doi.org/10.1890/140270

Veldman, Joseph W., G.E. Overbeck, D. Negreiros, G. Mahy, S. Le Stradic, G.W. Fernandes, G. Durigan, E. Buisson, F.E. Putz, and W.J. Bond. 2015. 'Where tree planting and forest expansion are bad for biodiversity and ecosystem services', *BioScience*, 65.10, pp. 1011–1018. https://doi.org/10.1093/biosci/biv118

Veldman, Joseph W., G.E. Overbeck, D. Negreiros, G. Mahy, S.L. Stradic, G.W. Fernandes, G. Durigan, E. Buisson, F.E. Putz, and W.J. Bond. 2015. 'Tyranny of trees in grassy biomes', *Science*, 347.6221, pp. 484–485. https://doi.org/10.1126/science.347.6221.484-c

Walker, R.B.N. 1870. 'Relation d'une tentative d'exploration en 1866 de la Rivière de L'Ogové et de la recherche d'un Grand Lac Devant se trouver dans l'Afrique Centrale', *Annales des voyages, de la géographie. L'Histoire et de l'archéologie*, 205.6, pp. 59–80; 120–144.

Walters, G. 2012. 'Changing customary fire regimes and vegetation structure in Gabon's Batéké Plateaux', *Human Ecology*, 40, pp. 943–955. https://doi.org/10.1007/s10745-012-9536-x

Walters, G. 2015. 'Changing fire governance in Gabon's Plateaux Batéké savanna landscape', *Conservation and Society*, 13.3, pp. 275–286. https://doi.org/10.4103/0972-4923.170404

Walters, G., J.A. Fraser, N. Picard, O. Hymas, and J. Fairhead. 2019. 'Deciphering Anthropocene African tropical forest dynamics: how social and historical sciences can elucidate forest cover change and inform forest management', *Anthropocene*, 27, pp. 1–7. https://doi.org/10.1016/j.ancene.2019.100214

Walters, G., S. Touladjan, and L. Makouka. 2014. 'Integrating cultural and conservation contexts of hunting: the case of the Plateaux Bateke savannas of Gabon', *African Study Monographs*, 35.2, pp. 99–128.

Walters, G.M., D. Nguema, and R. Niangadouma. 2022. 'Flora and fire in an old-growth Central African forest-savanna mosaic: a checklist of the Parc National des Plateaux Batéké (Gabon)', *Plant Ecology and Evolution*, 155.2, pp. 189–206. https://doi.org/10.5091/plecevo.85954

White, F. 1983. *La végétation de l'Afrique: mémoire accompagnant la carte de végétation de l'Afrique* (ORSTOM, Paris).

Zanjani, L.V., H. Govan, H.C. Jonas, T. Karfakis, D.M. Mwamidi, J. Stewart, G. Walters, and P. Dominguez. 2023. 'Territories of life as key to global environmental sustainability', *Current Opinion in Environmental Sustainability*, 63. https://doi.org/10.1016/j.cosust.2023.101298

14. The interface between hydrological modelling and political ecology

Maria Rusca and Maurizio Mazzoleni

Introduction

Research across hydrology and political ecology has explored various dimensions of what are often referred to as global water challenges. Given the essential role of water in all aspects of development, these challenges are extensive and diverse. They range from increasing drought severity and aridification to agriculture and food insecurity, from pollution and ecological destruction to changing climates, and from access to basic water and sanitation services to the resulting health and social implications (Castro 2007; Giuliano Di Baldassarre et al. 2019; Makarigakis and Jimenez-Cisneros 2019; Srinivasan, Lambin, Gorelick, Thompson, and Rozelle 2012). Much research in the fields of both hydrology and political ecology has been devoted to examining the nature, scale, distribution of and responses to these global water challenges.

Political ecology's conceptualisation of water governance challenges is grounded in the idea that they are political in nature. This scholarship has largely focused on how choreographies of power and economic visions shape water governance processes and the everyday experiences of water challenges across regions, rural–urban populations, intra-urban spaces, identities (e.g., gender, race) and income groups (Boelens, Hoogesteger, Swyngedouw, Vos, and Wester 2016; Collard, Harris, Heynen, and Mehta 2018; Loftus, March, and Purcell 2019; Perreault, Bridge, and McCarthy 2015; Rusca and Cleaver 2022.; Sultana 2020, 2022; Swyngedouw 2004, 2004; Zwarteveen et al. 2017). The field is also committed to apprehending biophysical processes that shape

   https://doi.org/10.11647/OBP.0418.14

or are shaped by power relations and contribute to water governance challenges. However, political ecology has not always succeeded in engaging with biophysical and ecological processes beyond the conceptual (Bakker and Bridge 2006; Gandy 2022; Walker 2005).

Hydrologists, on the other hand, focus on the flow of water through the hydrological cycle and its interactions with both human and natural systems (Wu, Christidis, and Stott 2013). Recent developments within the sub-field of socio-hydrology (Sivapalan, Savenije, and Blöschl 2012) have emphasised the importance of human modification of fluxes of water and focused on the interplay between society and the hydrological cycle by developing quantitative explanations of changes in hydrological flows and risk generated by floods, drought, farming and agriculture, water resources development, water quality degradation, and groundwater exploitation (see Di Baldassarre et al. 2019 for a comprehensive description of these phenomena). However, in exploring these phenomena, society is often described as homogenous, and the politics of water governance processes are largely concealed.

Given that these two very different disciplines share a common object, there is a need to reflect upon the potential of interdisciplinary and transdisciplinary collaborations in researching the interplay of water and society. To date, work that crosses disciplinary boundaries has largely employed an interdisciplinary case study approach (see Lane, Chapter 24), combining political ecology insights with hydrological data and quantitative assessments to explore urban water quality and uneven exposure to waterborne diseases (Rusca et al. 2017; Rusca, Gulamussen et al. 2021), droughts and socioecological crises (Savelli et al. 2022), urban climate impacts (Cousins and Newell 2015), soil contamination (McClintock 2015), and the commodification of shrimp aquaculture and peasants' livelihoods (Hoque et al. 2017). With a few exceptions (Lane et al. 2011a, b; Savelli et al. 2023; Whatmore and Landström 2011), interdisciplinary collaborations in the modelling process have been less explored. Thus, herein we discuss the potential and challenges of placing hydrological modelling (Melsen, Chapter 31) into engagement with political ecology. First, we draw on political ecology and science and technology studies (STS) to reflect on the politics of knowledge production and its implications for hydrological modelling. We then propose two alternative approaches to hydrological modelling, showing how methods can be mixed in ways that make modelling potentially more democratic, reflexive and situated. The first approach concerns participatory modelling (Landström, Chapter 35) and its potential to redistribute power and 'expertise' across different knowledge holders. The second concerns an engagement with political ecology to develop models that are more attuned to power relations and their role in shaping water challenges and the uneven outcomes thereof. We conclude by discussing the explanatory potential of these alternative modelling approaches.

Situating scientific knowledge production: power, history, and positionality

In contrast to mainstream and traditional interpretations of science as neutral, unbiased, and objective, political ecology and STS argue that the production and implementation of environmental knowledge is political and shaped by power relations that determine whose knowledge claims should be deemed more relevant, scientific, and actionable (Budds 2009; Goldman, Nadasdy, and Turner 2019; King and Tadaki 2018; Turner 2011; Zwarteveen et al. 2017). In this perspective, technology and knowledge are never produced in a vacuum. Rather, they "participate in the social world, being shaped by it, and simultaneously shaping it" (Law 2004: 12).

The reframing of scientific knowledge as power-laden carries significant theoretical, epistemological, and methodological implications not only for the knowledge that is produced but also for the way it is produced. The first implication is the reconceptualisation of knowledge as embodied, situated, context specific, and contested by different actors involved in environmental and water governance (see also Yates, Harris, and Wilson 2017; Zwarteveen et al. 2017). Knowledge always comes from a specific viewpoint: it is always situated in a specific context and historical moment and, thus, changes over time and across places. Feminist STS scholar Donna Haraway's influential work on situated knowledges (1988: 582–583) proposes an alternative idea of objectivity, which

> turns out to be about particular and specific embodiment and definitely not about the false vision promising transcendence of all limits and responsibility. The moral is simple: only partial perspective promises objective vision. [...] Feminist objectivity is about limited location and

situated knowledge, not about transcendence and splitting of subject and object. It allows us to become answerable for what we learn how to see.

Thus, situated and partial objectivity calls for apprehending the historical contingency of the knowledge produced, and for considering by whom knowledge is produced. If knowledge is embodied and generated by a particular point of view (rather than from an abstract, objective position), it is essential to consider the positionality of knowledge producers. In very simple terms, positionality is a term used to acknowledge that our personal history, background, and identities shape our understanding and the knowledge we produce (Pratt 2009; Simandan 2019). Positionality is a central concern in qualitative research, where data collection methods such as interviews and observations are grounded on the relation between researchers and participants. However, debates on the politics of knowledge production suggest that the knowledge producer—whether an organisation or an individual, a biophysical or a social scientist—has a particular vision, based on their positionality, that shapes the environmental knowledge they produce.

Moving to the second point, the politicisation of scientific explanations has generated a critical reflection on the way expertise is attributed and whose knowledge claims are valued. Zwarteveen and colleagues (2017) have explored processes through which water expertise is distributed, impacting water governance processes. To illustrate this point, they critically reflect on the construction and transfer of Dutch expertise in delta planning to countries far removed from where that knowledge is produced (e.g., Vietnam, Bangladesh, and Indonesia). As the authors note, claims of Dutch expertise in this field are constructed by promoting the idea of the country's success in dealing with floods in its delta. These claims are supported more by the consistent financial support of the Dutch government for research and development projects on delta planning than by any intrinsic and universal superiority of the Dutch approach. The authors thus conclude that uneven distribution of expertise is shaped by broader systems of cultural, social, and economic dominance (Zwarteveen et al. 2017).

A third implication of the conceptualisation of knowledge as situated and political is a reconsideration of the role of the scientist. As noted by Turner (2011), scientists are the most privileged knowledge producers. A particular idea of science is evoked in mainstream discourses, which generally consists of positivistic approaches and established techniques undertaken by what are generally considered qualified experts (Forsyth 2011). An example of the power of 'hard' scientific knowledge is the way in which it has worked to legitimise political actions on climate change, and the (relatively) quick and wide diplomatic consensus it generated (Demeritt 2001). However, as noted by Demeritt (2001) and others (Castree 2015; Rusca et. al 2023), environmental knowledge generated within 'hard' sciences often focuses on atmospheric changes, overlooking the nature, scale, and distribution of risks, as well as the uneven experiences associated with them.

Sheila Jasanoff has argued that this knowledge is devoid of meaning precisely because of its aspiration to be 'objective'. As she notes (2010: 233), "climate facts arise from impersonal observation whereas meanings emerge from embedded experience". Moreover, when applied, scientific environmental knowledge also produces political and developmental outcomes. Demeritt (2001), for instance, notes that a reductionist and partial construction narrowly focused on scientific language and the physical dimensions of climate change has clear policy implications. Separating the physical forces of climate change from political knowledges generated by critical social scientists, he argues, deflects attention from difficult political questions on the role of colonialism and of the capitalist economy in shaping the climate challenge and in generating uneven impacts across social groups and regions (see also Moore 2017; Sultana 2022). Moreover, scientific knowledge has also been strategically mobilised for economic and political goals. To illustrate, the implementation of Dutch environmental knowledge and technical expertise on delta planning in Vietnam had significant policy and developmental implications (Zwarteveen et al. 2017). Participatory tools promoted by the Dutch were discursively framed as a way to support the beneficiary country. However, they also worked to pave the way for Dutch projects in Vietnam. The Vietnamese government, in turn, used these tools to legitimise delta planning approaches favoured by the political leadership.

To conclude this section, political ecology and STS have questioned the mainstream perspective of science as neutral and shielded from political influences. In so doing, this research has convincingly argued that environmental knowledge is not value-neutral and is always shaped by and shapes particular political, social, and economic agendas. We argue with Forsyth (2003; 2011) that failing to acknowledge these dynamics might prevent scientific knowledge from addressing environmental challenges and effectively informing policy processes. Rather, an uncritical perspective on scientific knowledge might end up fostering greater inequalities, unsustainable development and increased environmental degradation. We, thus, turn to reflect on the implications of the politicisation and situatedness of scientific explanations for hydrological modelling, and how this necessitates a much more effective mixing of epistemologies and methodologies in water research. To this end, we briefly discuss evolutions of hydrological modelling, particularly focusing on the extent to which these models apprehend complex social processes that shape water governance challenges and their outcomes.

The politics of knowledge production in (socio)hydrological modelling

From hydrological to sociohydrolgical models

Numerical models are widely used to analyse and quantify the impacts of climate and hydrological change on humans and the environment. Hydrological models (Melsen, Chapter 31) specifically focus on the physical processes that generate changes in river streamflow and surface runoff due to variation in climate (evapotranspiration, precipitation, temperature) and human activities (e.g., land-use) (Mazzoleni, Alfonso, Chacon-Hurtado, and Solomatine 2015) and may be coupled also to models that allow representation of biogeochemical processes (e.g., water quality) and/or ecosystems. Hydrological models are classified based on their model structure, i.e., how the physical processes are discretised in the basin. There are three approaches to discretise these processes, namely physically-based models, conceptual models, and black box or data-driven models (Solomatine and Wagener 2011). Another way of classifying hydrological models is according to the spatial discretisation of the model itself. Thus, hydrological models can be classified as distributed, semi-distributed or lumped (Solomatine and Wagener 2011).

Conventional hydrological models do not explicitly account for the bidirectional feedback mechanisms that characterise human-water systems (Di Baldassarre et al. 2013). In other words, these models only look at hydrological flows, whilst humans and social processes are conceptualised as external or boundary conditions. As a result, the relation or feedback loop between hydrological and social processes is overlooked. This may also have the unintended consequence of systematically underestimating or overestimating changes and distribution in hydrological risk.

Sociohydrological models address this knowledge gap by reconceptualising hydrological change as human-water dynamics and by predicting the emergence of sociohydrological (rather than hydrological) phenomena through system dynamic or agent-based models (Mazzoleni, Odongo, Mondino, and Di Baldassarre 2021). However, whilst these models account for human influence, they do not apprehend power relations and heterogeneity in society. As a result, they do not critically engage with how power and differential agency determine which economic and development trajectories are chosen, and the implications for hydrological change and distribution of risk. Similarly, little attention is placed on the politics of knowledge production and how this might shape both model and development outcomes. These aspects are further discussed below by drawing on political ecology critiques.

The limitations of (socio)hydrological modelling: a political ecology perspective

Political ecologists have critically examined the knowledge that models generate, arguing that they carry significant power to influence development polices and, in turn, political agendas (Bouleau 2014; Budds 2009; Harris 2022; ter Horst et al. 2023). First, models are powerful producers of scientific knowledge, and they are well positioned to influence policy processes (see Krueger, Chapter 26). To illustrate, in her analysis of water governance in Chile, Budds (2009) unravels the limitations of polices that are predominantly informed by hydrogeological modelling. Her work examines modelling assumptions and their implications for agricultural expansion, allocations of water rights and, in turn, sustainable and inclusive development in the La Lingua valley. Although risks of aquifer depletion were detected in several parts of the valley, only two sections were considered at significant risk and likely to affect surface water rights. Based on these results, allocation of additional water rights was recommended in most of the valley.

Budds notes three limitations of this approach. First, additional water rights were purchased by commercial farmers, who were better placed—both financially and strategically—to claim them. By framing water rights allocation as an administrative and technical matter, only dependent on hydrological flows, the allocation of rights overlooked power differential within the farming community. The allocation of rights on a first come, first served basis further exacerbated inequalities in access and distribution between larger farmers and peasant farmers. In relation to the first point, Budds (2009) demonstrates the vulnerability of peasant farmers was unaccounted for in the model and subsequent decision-making processes. Peasant farmers increased groundwater exploitation also in response to the 1996–1997 drought and to agricultural expansion in the valley. However, this practice was framed as illegal rather than as an adaptation response. Third, models solely grounded on physical data and focusing on environmental aspects can be easily manipulated. Budds (2009) thus concludes that the government privileged a particular interpretation of the data that fostered the political agenda it wanted to pursue and that ultimately reproduced uneven outcomes for local populations.

A second critique, which builds on the first, concerns the risk that models might reproduce, rather than challenge existing injustices and power structures. Concerning this aspect, climate models have received significantly more attention from critical scholars. To illustrate, McLaren and Markusson (2020) have argued that because of the limited engagement with questions of power and equity, models and scenario-based approaches are likely to reproduce and entrench pre-existing power structures and unjust distribution of the 'goods' and bads' of climate and water-related challenges. Rubiano Rivadeneira and Carton (2022) have examined Integrated Assessment Models, which aim to quantitatively assess long-term mitigation options including economic strategies and logics. These models, they argue, carry significant and largely overlooked equity and social justice implications because of the central role they play in informing the IPCC's Assessments and Special

Reports and the resulting policy options. They conclude that there is a need to develop models that are informed by clear ethical assumptions and to critically evaluate the economic assumptions of their models. Without those ethical assumptions, models are likely to reproduce the very systems that have led to climate change and uneven distributions of risks. Importantly, some hydrological modellers have also acknowledged the risks of overlooking these dimensions and have critically engaged with the field's knowledge production practices (Addor and Melsen 2019; Beck and Krueger 2016). Drawing on Actor-Network Theory and STS, Beck and Krueger (2016: 639) have argued that hydrological models can generate "authoritative representations of dominant perceptions of the world". If these representations are not questioned, models might have the unintended consequence of stabilising knowledge by uncritically reproducing hegemonic representations of water-society interplay that overlook questions of power, (in)justice, and uneven development. Building on this work, an analysis of assumptions embedded in different models has shown that these often generate depoliticised explanations of hydrological phenomena, thereby pre-configuring some possibilities and foreclosing others (Krueger and Alba 2022).

A third and related point concerns the privileged knowledge perspectives of modellers, who ultimately decide what assumptions or measurements are included in a model, if the data used in the model and the result can be trusted, and whose interpretation of the results is more accurate. This is particularly the case for sociohydrological models, which often rely on subjective and simplified equations on human behaviour.[1] The selection of the model's assumptions shape the knowledge claims made by a particular model, which are validated as they are published, disseminated and, possibly, mobilised by policy makers to pursue particular development agendas (Lane et al., 2011b). It is, therefore, fundamental for the modeller to recognise their positionality and privilege in the process of producing knowledge, and to critically reflect on how the knowledge produced is circulated and applied. Concurrently, the modeller should also be subject to proper

---

1   As noted above, hydrological models are usually based on a simplified representation of the physical processes observed on the field, and complex differential equations based on physical laws.

democratic accountability (a point we make above), rather than only being accountable to the peer review systems they are a part of.

To conclude, whilst political ecology's critique of hydrological models has clear merits, it is important to also recognise that political ecology has—in a similar fashion—largely overlooked the 'ecology' (Gandy 2022; Walker 2005), here represented by hydrological flows. A recognition of the respective disciplinary focus and limitations thereof is foundational to any collaboration between political ecologists and modellers. We build on these critiques by proposing that if models are world-making tools, alternative models can help to generate alterative worlds. In other words, given the "moral agency" of models (Beck and Krueger 2016: 639), there is a need to reconsider how the assumptions adopted in models reconfigure the knowledge they produce and to democratise the modelling process. In the last part of the paper, we thus propose two strategies aimed at addressing these concerns. The first consists of participatory modelling (Landström, Chapter 35), which has the potential to democratise the modelling process and to pluralise perspectives and knowledge claims included on the model. The second is an engagement with political ecology scholarship on the power relations underlying the uneven distribution of hydrological risk to ensure a focus on justice and sustainability and, in turn, to generate policy responses that are more attuned to these issues (Donaldson, Lane, Ward, and Whatmore 2013; Rusca and Di Baldassarre 2019).

Democratising hydrological modelling: participatory approaches and the reconfiguration of knowledge

Control over how and by whom knowledge is produced and used has significant implications for the equitable distribution of water resources and hydrological risk (Budds 2009; Lemos 2008). This also applies to hydrological modelling, which is increasingly used to inform decision-making processes on water resource management (Dreyer and Renn 2011; Falconi and Palmer 2017). As aptly noted by Lane (2014: 927), "decisions over how the world is modelled may transform the world as much as they represent the world". Not surprisingly, therefore, the proliferation of what Webler and colleagues have termed "models for policy", has generated a critical reflection on implications of hydrological modelling (Webler, Tuler, and Dietz 2011: 474). Conventional models tend to include generalised and politically naïve assumptions on social processes and society, which is often portrayed as homogenous and depoliticised (Rusca et al. 2023a; Budds 2009). The resulting models are unlikely to capture the relations between power, differential agency, equity, and hydrological change.

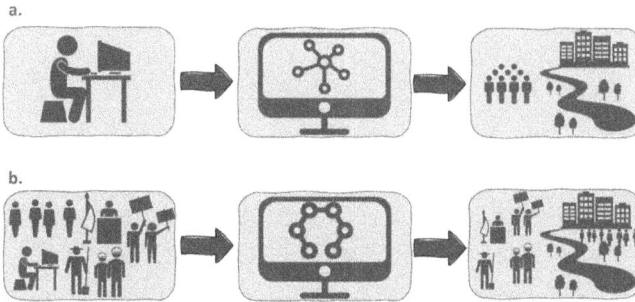

Fig. 14.1 From conventional to participatory hydrological modelling. In conventional approaches to hydrological, the modelller decides what assumptions or measurements are included and if the data in and the results of the model can be trusted. The model will ultimately reflect modellers' knowledge claims and is likely to represent society as depoliticised and homogenous (Panel a). Participatory modelling processes are grounded on the idea of building scientific knowledge with different groups affected by hydrological change in a given location. Assumptions and measurements are defined by bringing together different actors and knowledge claims. This approach can generate a more complex and heterogenous understanding of society, and affected groups' visions, experiences and needs

Some modellers and social scientists have argued for a shift from conventional hydrological modelling practices, in which scientific knowledge production is insulated from affected groups, to participatory approaches in which knowledge is generated with the affected groups to ensure more equitable water management (Basco-Carrera, Warren, van Beek, Jonoski, and Giardino 2017; Falconi and Palmer 2017). Accommodating multiple actors in the modelling process can potentially foster a more critical and democratic reflection on the assumptions that are included or excluded from the model and transform the results the model will generate (Fig. 14.1). For these scholars, opening the modelling process can generate new forms of collective decision-making and social learning (Dreyer and Renn, 2011), as well as more radical processes of knowledge reconfiguration (Lane et al. 2011). This has raised questions about how to effectively engage affected groups and about their role in different phases in the modelling process (Falconi and Palmer 2017; Lane et al., 2011a; Webler et al., 2011; Whatmore and Landström 2011).

The term participatory modelling covers a wide range of approaches that encompass different levels of inclusion and representation of affected groups and their knowledge claims (see Landström, Chapter 35). As noted by Falconi and Palmer (2017), it is essential to consider what forms of participation are being fostered and with what outcomes. Drawing on conceptualisations of participation by the National Resource Council [NRC, 2008], Falconi, and Palmer (2017) present a two-stage framework for evaluating PM based on mechanisms for improving model effectiveness as participatory tools. The five dimensions characterize the "who, when, how, and why" of each participatory effort (stage 1 identify five dimensions of participation in the modelling process:

1. Who is invited to participate and how the identification of interest groups is undertaken—Drawing on an extensive literature review, the authors highlight the risks of further exacerbating inequalities and conflict, as well as reproducing inaccurate results when some affected groups are excluded.

2. The stage of the process at which participants are involved—The authors note that modelling is a complex process involving multiple interrelated tasks. It is thus important to consider at which stage(s) of the process—data collection, definition of the model, construction of the model, verification and validation of the model, or application of the model—participation takes place.

3. The contribution or degree of involvement of each participant, including the organisers—Drawing on different frameworks to assess participation, the authors identify five core practices that generate different degrees of participation: informing, consulting, involving, collaborating, and empowering, which involves placing the decision-making power with the public engaged in the participatory process.

4. The extent to which power asymmetries among participants might shape the ability of different groups or individuals to influence the decision making process—Political ecologists have shown how power relations shape the ability of different actors to control both the construction of knowledge and water

governance processes (see for instance Cornwall 2008; Forsyth 2003; Perreault et al. 2015; Swyngedouw 2004; Zwarteveen et al. 2017). These dynamics extend to the modelling process, where differential agency and power is likely to shape the participants' ability to have their knowledge claims recognised.

5.   The overarching goals of the participation exercise—These are related to the points raised on the degree of involvement of each participant. Importantly, the authors note that different disciplinary perspectives might interpret problem framing, participation, and its overall aim differently. A clear and shared understanding of these dimensions is thus essential to a transparent modelling process.

Participatory modelling experiments that foster close collaborations between modellers, social scientists, and affected residents are still rare. A productive example of the potential of this approach is represented by the collective project developed by Lane et al. (2011a). Here the focus was on flood risks in two towns, one in the north of England, the other in the south. Preliminary research in the areas had shown that rather than a general lack of trust in science, the public was concerned with the ways they had become strategically alienated from scientific knowledge and its production. This was even more frustrating, as recent floods in the areas had increased residents' motivation and claims to be involved in flood risk management.

In both locations, a team of researchers from the social and hydrological sciences (5–7) and volunteer residents (5–8) met six times over a period of one year to collaboratively develop knowledge on flood risk in the region. Competency groups were established as platforms to monitor and at times challenge the knowledge produced by 'the experts', and to generate opportunities to develop new shared competencies and redistribute expertise across academic and non-academic participants (Whatmore and Landström 2011). Key aims included reflecting on the ways scientific knowledge has informed past policy options, making room for volunteer residents to think of and experiment with alternative approaches to flood risk management and, ultimately, producing a model that accounts for these different knowledge claims. As noted by Lane and colleagues (2011a), residents displayed an in-depth qualitative understanding of flood hydrology (also beyond the local context), which was crucial for co-producing the alternative flood risk management solutions that were ultimately implemented in the region. The project thereby served to transform a knowledge controversy on flood risk management into a "generative event in which expert reasoning was forced to 'slow down' and a space for reasoning differently opened up, involving those affected in new political opportunities and associations"(Whatmore and Landström 2011: 604).

Integrating political ecological explanations in hydrological modelling

System dynamic and agent-based models (ABMs) have significantly advanced theorisations of hydrological change and risk grounded on the feedback loops between hydrological processes and society. However, although on paper ABMs are well placed to represent agents with different characteristics, models generally fail to account for the heterogeneity in society and the power and cultural relationships that determine why and how the hydrological regimes are altered by social actors and the uneven outcomes thereof (Rusca et al. 2023b; Budds 2009). In this perspective, political ecologies of water can add to the ways in which society and its relation to the hydrological cycle are represented in hydrological models. Political ecological perspectives have advanced conceptualisations of hydrological flows as reshaped by economic and policy visions, development trajectories, and power relations, thereby questioning the notion that water management is a technical endeavour exclusively the domain of engineers and hydrologists.

Swyngedouw's (1997; 2004) pathbreaking work in Guayaquil, Ecuador, has shown how capital accumulation from oil and agricultural land developments has reshaped the city. Public policies and investments favoured urban elites, who are provided with premium infrastructures and services, whilst the urban poor are marginalised and largely excluded the economic benefits and services of the city. This work demonstrates that water flows internalise power dynamics shaped by capital accumulation. This idea has significant implications for hydrological research, as it suggests a need to reconceptualise scientific problems of hydrological risk in relation to the political economy of water (Lane 2014;

Rusca and Di Baldassarre 2019). To illustrate, interdisciplinary models at the interface between hydrology and political ecology could examine "distinctive hydrological flows (and uneven distribution of risks thereof) produced through capitalist processes in the Anthropocene" (Rusca and Di Baldassarre 2019: 10). Indeed, as argued elsewhere, models that draw on a plurality of epistemological perspectives are likely to be more effective in apprehending and describing the interplay between water and society (Ravera, Hubacek, Reed, and Tarrasón 2011).

In the sections below, we describe two alternative approaches to account for political ecological explanations in hydrological modelling, showing how methods can be mixed in ways that make modelling potentially more sensitive to power relations. We focus on the urban scale, and present two examples to engage with both floods and droughts, whilst also exploring different social dynamics that might shape the hydrological risk profile of a city.

Example 1: Modelling flood risk and vulnerability in the uneven city

Models are only as good as the assumptions about human behaviour that they include (Lane 2014). In this model, we develop assumptions of human behaviour in response to urban floods by drawing on theoretical explanations developed by urban political ecologists (Rusca et al. 2023). Importantly, we do not argue that this approach can generate an objective understanding of social dynamics. As discussed earlier in the chapter, knowledge always comes from a specific viewpoint, and here the viewpoint is the one of political ecology research on natural hazards, which has argued that differential vulnerability to and uneven recovery trajectories from natural hazards is socially constructed. Thus, the aim is not objectivity, but rather pluralising knowledge claims and accounting for power dynamics in hydrological modelling.

A review on the political economy of vulnerability and adaptation to natural hazards (Rusca, Messori, and Di Baldassarre 2021) has identified three interrelated social dynamics that constitute the fundamental assumptions about human behaviour in our model:

> <u>Assumption 1:</u> intersectional dimensions of inequality—including gender, race, status, class, ethnicity, disability and more—shape vulnerability (see for instance Adger 2006; Bolin 2007; Cutter, Boruff,

and Shirley 2003; Enarson 1998). Scholars emphasise that intersecting dimensions of inequality are not neutral and do not simply represent apolitical demographics. Rather, they argue, these multiple dimensions of inequality are socially constructed and, thus, shaped by power relations.

Assumption 2: intersectional dimensions of inequalities are generated by wider patterns of uneven development. Simon (2014: 1199) has defined this dynamic "vulnerability-in-production" to highlight that it is tied to policy and development process that make some more vulnerable than others to natural hazards. This includes, among others, processes that marginalise some social groups and allow others to benefit from resources and services (Collins 2009; 2010), as well as public disinvestment in and gentrification of marginalised areas both before and in the aftermath of a climate-related disaster (Bullard and Wright 2009; Hamideh and Rongerude 2018; Keegan 2020; Kimmelman 2017; Klein 2007). This literature also highlights that inequalities are likely to polarise in the aftermath of a disaster: those who are most vulnerable before a disaster are likely to take longer to recover from it, and government responses and recovery measures may benefit higher income households the most.

Assumption 3: disasters can trigger the potential for transformative change. In this perspective, disasters are "tipping points" (Pelling and Dill 2010), "creative destruction" (Rozario 2019) or "critical junctures" (Olson and Gawronski 2003) that catalyse mobilisation, empowerment of civil society and, ultimately, transformative change towards more equitable modes of governance and, in turn, reduced vulnerability (see also Cretney 2019; Luft 2009). They simultaneously represent the outcome of hegemonic and inequitable development trajectories, and a window of opportunity to subvert the order that has generated uneven vulnerability in the first place (Donovan 2017; Pelling and Dill 2010; Shaw 2012).

The model draws on these theoretical explanations to explore the interplay between floods, vulnerability and socio-political transformations in a city characterised by highly uneven development (Fig. 14.2). While models often assume economic growth as a leverage point for addressing poverty, unemployment, and urban decline (see Meadows, 1999 for a critique), this model acknowledges both the costs of growth and their uneven distribution. To reflect the uneven development of the city (Assumption 1 and 2), the model simulates 1000 neighbourhoods that have different socio-economic characteristics. To represent the floods in the city and modulate their intensity over a period of 200 years, physically-based hydraulic models (Lane, Chapter 30) can be used to convert discharge information into water level. Water level can then be input into a depth-damage curve to assess the floods' impacts based on specific land use information. Rather than modelling the impacts of these floods on the city as a single unit, the model can analyse how these floods affect different neighbourhoods across the city. A core assumption of the model, based on well-established research in political ecology discussed above, is that entrenched inequalities that engendered vulnerability in the first place are likely to polarise following a flood (Assumption 2). Increased inequality and vulnerability generate greater opposition (or political misalignment) to the government, which in turn can generate socio-political transformation in the city (Assumption 3).

In the model, the government is the entity responsible for the development trajectory of the city, through planning strategies, urban policies, and the allocation of limited financial resources. Decision-making by the government is represented by a set of policies which include:

- **Business as usual pro-growth policy,** in which decision making by the government is focused on stimulating economic growth, which in turn leads to processes of capital accumulation and increased inequalities across neighbourhoods. When political alignment is high, this policy is predominant.

- **Flood infrastructure** is a structural measure (e.g., levees, retention basins), which aims at reducing flood intensity across urban areas. In the model, we assumed that in the city everybody equally benefits from this adaptation action.

- **Affordable housing** is a non-structural measure aimed at supporting lower income households by increasing the ratio between rental costs and income, thereby reducing poverty within neighbourhoods.

- **Building back better** is a structural measure, grounded on the idea that a flood related disaster is a window of opportunity to improve the building quality in neighbourhoods in which they are less likely to withstand floods.

- **Social protection** is a non-structural measure that aims at reducing economic inequalities in the city.

Transformative policies that aim to reduce urban inequalities and vulnerability across urban spaces (Affordable housing; Flood infrastructure; Building back better; Social protection) are only prioritised when political alignment (i.e., support for the government) decreases. Increasing and decreasing alignment with the government is determined by several variables that represent uneven development, such as differentiated flood-protection infrastructures and building quality, employment/cost of living, well-being, and damaged housing. When a flood or a series of floods significantly reconfigures these variables, political alignment changes, opening new policy options. However, alternative policy options are always implemented in competition with the business-as-usual pro-growth policy. Specifically, the model implements one transformative option at a time in combination and competition with a pro-growth policy. The degree to which the transformative policy is prioritised over business as usual depends on the political alignment: with greater misalignment with the government the transformative policy becomes more prominent. Figure 14.2 presents a simplified representation of the causal links between the variables included in the model.

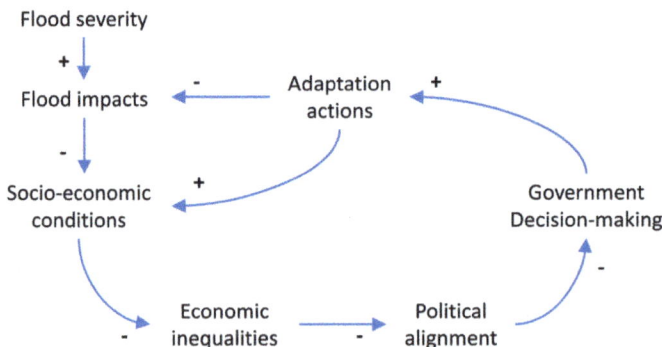

Fig. 14.2 Causal loop diagram of the model showing uneven flood risk and vulnerability, illustrating the links between model variables and their polarity (signs + and -). For instance, a positive polarity indicates that a positive change in one variable (e.g., adaptation actions) leads to a positive change also in the linked variable (e.g., socio-economic conditions). Conversely, a negative polarity means that a positive change in one variable results in a negative change in another

This conceptual model differs from conventional and hydrological ones in three important ways. First, it represents heterogeneity in society by analysing differential impacts of a flood event across different neighbourhoods. Second, the model conceptualises the city as a (potential) site of resistance and contestation (Gawronski and Olson 2013; Harvey 2012; Pelling and Dill 2010). It proposes that flood events may represent a window of opportunity for transformative change. This brings us to our last point. By recognising the possibility of transformative change, the model moves beyond a return to pre-disaster conditions and speculates—through the implementation of a set of alternative policies—on the potential of alternative pathways of urban development. In this way, this model has the potential of generating alternative representations of the world rather than reproducing dominant ones (Rusca et al., 2023b). This approach thereby moves from models that reproduce or reinforce hegemonic representations of the world (see for instance the critique by Beck and Krueger 2016) to approaches that can proliferate alternative worldviews more attuned to questions of justice and equitable development.

Example 2: Modelling uneven vulnerability to droughts in the city

Hydrological models often fail to disaggregate actors and the actions and experiences of different groups affected by hydrological change (Budds 2009). These simplified assumptions of society and human behaviour prevent modellers from understanding equity and justice issues that shape these relations. Political ecology scholarship is grounded on a commitment to understand environmental politics and, more specifically, water governance challenges as contingent and historical. This commitment extends to a methodological emphasis on field research and qualitative methods to explore the meanings, experiences, and everyday practices of dealing with different water challenges. By drawing on this context-specific knowledge, hydrological models can examine, for instance, drought risk in a specific location. These models are well placed to generate a nuanced understanding of the interplay between hydrological and social processes, incorporating multiple perspectives and the everyday experiences of different actors.

In this second example, we develop a conceptual model based on the city of Maputo, Mozambique, which has recently experienced a severe drought (2016–2018). To this aim, we draw on empirical research undertaken in the city to explore the ways in which power, differential agency, and economic and policy visions have shaped responses to the drought across urban spaces (Rusca et al. 2022; 2023). The highlighted multiple interrelated dynamics, combining social, spatial, and economic inequalities that generated varying levels of vulnerability across the city.

Our model includes the following assumptions based on research conducted in Maputo before, during, and after the drought:

- Vulnerability to droughts in Maputo is shaped by social dimensions of inequality such as class and gender. To illustrate, government-enforced water rationing measures to reduce consumption in the city disproportionately affected low-income neighbourhoods, where water shortages were more frequent and longer than in higher-income areas. Indeed, because water distribution centres were placed in proximity of the city centre, those who live in these areas are better served then those who live at the margins of the water supply network. Moreover, higher-income residents rely on storage facilities that have greater capacity (500–1500 L) than those available to residents in low-income areas (250-1.5 L). In the model, the variable "social inequalities" (Fig. 14.3) is a proxy for social vulnerability and accounts for the distance to the distribution centre and income level. Last, lower-income residents are unable to afford tariff increases enforced to reduce water consumption. This uneven distribution across the city affected women the most, as they were tasked with fetching water when taps at home ran dry. Thus, women were exposed to significant psychological and physical stress, as well as to higher risks of experiencing violence as they walked greater distances to fetch water during the drought. In the model, socio-economic inequalities are represented by differential health outcomes and wealth across social groups and by gendered inequalities within lower-income groups, which generate uneven experience of the drought.

- Vulnerability to droughts is socially produced. Uneven levels of vulnerability to the drought are tied to Maputo's urban form and the legacies of colonial and postcolonial development trajectories. First, the systemic prioritisation of the city centre for infrastructure and service development has led to areas with advanced and more reliable water and sanitation services (i.e., lower water insecurity in Fig. 14.3) and other areas facing significant infrastructure and servicedeficits (i.e., higher water insecurity in Fig. 14.3). During the drought, areas with the greatest infrastructure deficits were affected by a cholera outbreak, caused by the combined effects of poor sanitation and drainage systems, extended water shortages, flash floods, and shallowly laid pipes. All together these infrastructure deficits exponentially increased the risk of water contamination in these neighbourhoods, lowering water security. In the model, uneven development is represented by the water security variables, which refers to the type of water infrastructure available for the different communities. Water security is linked to social inequalities as higher inequality will lead to lower water security (see the "-" symbol used in Fig. 14.3). Higher water shortages, i.e., water availability from precipitation lower than water use, can lead to lower water security. Moreover, in the model we assumed that greater water security can lead to improved socio-economic conditions, including better health and income outcomes.

- A large part of the population (34%) is served by approximately 800 small-scale water providers (SSIPs). They primarily serve an emerging middle class in the suburbs through technologically advanced small-scale networked systems drawing groundwater from one of the city's aquifers. The initial investment costs for developing these systems are relatively high, ranging between €4,500 and €10,000 (Schwartz, Tutusaus Luque, Rusca, and Ahlers 2015). Thus, most of the providers themselves are middle class local residents. Because SSIPs rely on groundwater, their customers were unaffected by the drought and were not subject to water rationing measures. Not surprisingly, therefore, some providers were able to capitalise on the drought and expand their market. This, however, has increased the vulnerability of the aquifer, which is increasingly being exploited. In the model, these dynamics are reflected in the private water sources variable.

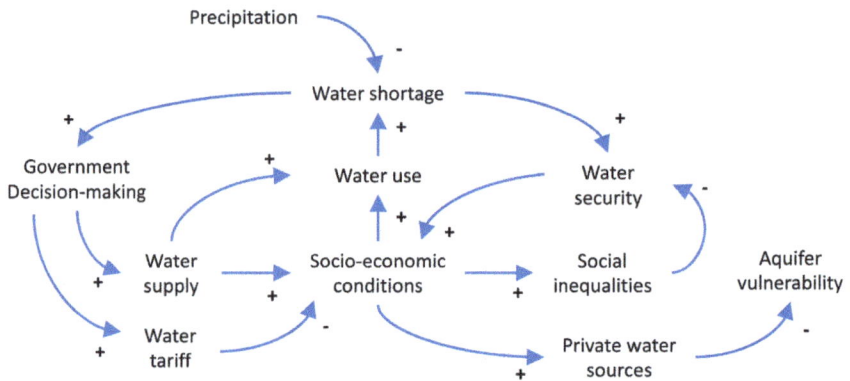

Fig. 14.3 Casual loop diagram representing the conceptual model of uneven vulnerability to drought in Maputo

The model presented above analyses how different social groups within the city were unevenly affected by the drought. As in the previous example, in this model heterogeneity in society is represented by the variable socioeconomic conditions and social inequalities. Rather than modelling the impacts of the drought on the city as a single unit, the model recognises that decisions by the government will unevenly affect residents across urban spaces. It also politicises infrastructure, which is conceptualised as actively shaping the possibilities of accessing water across the city. Indeed, by design the network will favour some citizens over others, thereby generating spaces of drought vulnerability and spaces that are significantly more resilient.

Conclusions

In this chapter, we have argued that knowledge is situated and power-laden, underscoring the need for odelling practices that are more attuned to and reflective of these dynamics. To this aim, we have proposed two alternative approaches to hydrological modelling that combine methods in ways that make modelling potentially more democratic, reflexive, and situated. Participatory modelling (Landström, Chapter 35) can democratise the knowledge process by opening the black box of scientific knowledge production and co-producing understandings of hydrological risks and change with the public. Importantly, participation is inherently political and shaped by power relations. The effectiveness of this approach in democratising the modelling process and in incorporating equity concerns will depend on multiple factors. The type and modality of involvement, which can range from informing to empowering, will determine whether participation serves as an instrumental step in model development or transformative process to empower the public and radically reconfigure knowledge. Secondly, it is crucial that knowledge claims of less powerful actors are recognised and accounted for in the model assumptions. Otherwise, the model is unlikely to generate relevant findings in terms of equity and justice, or to open a debate about political alternatives.

The two conceptual models above show that a political ecology lens can enhance our understanding of the relationship between the generation and distribution of hydrological risks by apprehending processes of social difference, as well as of societal inclusion and exclusion. Specifically, we note three significant contributions of this approach. First, rather than modelling society as an "undifferentiated whole" (Moore 2017: 595), these models account for the uneven outcomes of floods and droughts across intra-urban spaces, identities, and social groups. Second, the models recognise that policy options and development trajectories matter to how risk is generated and distributed. Without acknowledging these dynamics—as we do here—models risk reproducing the hegemonic logics of uneven development that have generated hydrological risk and the uneven distribution thereof in the first place.

Third, whilst the first model draws on abstract political ecological explanations on the social construction of risk and the transformative potential of a disaster, the second is grounded on a context-specific analysis of the 2016–2018 drought in Maputo. They thus provide different approaches to combine hydrological modelling and political ecology analyses. It is important to acknowledge that both strategies are shaped by the modellers and political ecologists involved in the research. It might be argued that the 'tyranny' of one modeller is now simply replaced by the tyranny of the team involved in the model design. However, as noted by STS and political ecology scholarship, knowledge is always generated from a particular and contingent perspective. We argue that the value of these approaches lies in pluralising these perspectives and including assumptions about equity and power that are otherwise overlooked. The aim, therefore, is a partial or incomplete objectivity that—as aptly illustrated by Haraway (1988)—is always embodied (rather than transcendent) and for which knowledge producers are always accountable.

References cited

Addor, N. and L.A. Melsen. 2019. 'Legacy, rather than adequacy, drives the selection of hydrological models', *Water Resources Research*, 55.1, pp. 378–390. https://doi.org/10.1029/2018wr022958

Adger, W.N. 2006. 'Vulnerability', *Global Environmental Change*, 16.3, pp. 268–281. https://doi.org/10.1016/j.gloenvcha.2006.02.006

Arnstein, S.R. 1969. 'A ladder of citizen participation', *Journal of the American Institute of Planners*, 35.4, pp. 216–224.

Bakker, K., and G. Bridge. 2006. 'Material worlds? Resource geographies and the matter of nature', *Progress in Human Geography*, 30.1, pp. 5–27. https://doi.org/10.1191/0309132506ph588oa

Basco-Carrera, L., A. Warren, E. van Beek, A. Jonoski, and A. Giardino. 2017. 'Collaborative modelling or participatory modelling? A framework for water resources management', *Environmental Modelling and Software*, 91, pp. 95–110. https://doi.org/10.1016/j.envsoft.2017.01.014

Beck, M., and T. Krueger. 2016. 'The epistemic, ethical, and political dimensions of uncertainty in integrated assessment modeling', *WIREs Climate Change*, 7.5, pp. 627–645. https://doi.org/10.1002/wcc.415

Boelens, R., J. Hoogesteger, E. Swyngedouw, J. Vos, and P. Wester. 2016. 'Hydrosocial territories: a political ecology perspective', *Water international*, 41(1), 1–14. https://doi.org/10.1080/02508060.2016.1134898

Bolin, B. 2007. 'Race, class, ethnicity, and disaster vulnerability', in *Handbook of Disaster Research*, ed. by H. Rodríguez, E.L. Quarantelli, and R.R. Dynes (Springer), pp. 113–129. https://doi.org/10.1007/978-0-387-32353-4_7

Bouleau, G. 2014. 'The co-production of science and waterscapes: the case of the Seine and the Rhône Rivers, France', *Geoforum*, 57, pp. 248–257. https://doi.org/10.1016/j.geoforum.2013.01.009

Budds, J. 2009. 'Contested H2O: Science, policy and politics in water resources management in Chile', *Geoforum*, 40.3, pp. 418–430. https://doi.org/10.1016/j.geoforum.2008.12.008

Bullard, R.D. and B. Wright. 2009. 'Race, place and the environment in post-Katrina New Orleans', in *Race, Place and Environmental Justice after Hurricane Katrina: Struggles to Reclaim, Rebuild and Revitalize New Orleans and the Gulf Coast*, ed. by R.D. Bullard and B. Wright (Routledge), pp. 19–48. https://doi.org/10.4324/9780429497858-1

Castree, N. 2015. 'Changing the Anthropo(s)cene: Geographers, global environmental change and the politics of knowledge', *Dialogues in Human Geography*, 5.3, pp. 301–316. https://doi.org/10.1177/2043820615613216

Castro, J. E. 2007. 'Water governance in the twentieth-first century', *Ambiente y Sociedade*, 10, pp. 97–118. https://doi.org/10.1590/s1414-753x2007000200007

Collard, R.-C., L.M. Harris, N. Heynen, and L. Mehta. 2018. 'The antinomies of nature and space.' *Environment and Planning E: Nature and Space* 1, no. 1–2 (2018): 3–24. https://doi.org/10.1177/2514848618777162

Collins, T.W. 2009. 'The production of unequal risk in hazardscapes: An explanatory frame applied to disaster at the US–Mexico border', *Geoforum*, 40.4, pp. 589–601. https://doi.org/10.1016/j.geoforum.2009.04.009

Collins, T.W. 2010. 'Marginalization, facilitation, and the production of unequal risk: the 2006 Paso del Norte floods', *Antipode*, 42.2, pp. 258–288. https://doi.org/10.1111/j.1467-8330.2009.00755.x

Cornwall, A. 2008. 'Unpacking "participation": Models, meanings and practices', *Community Development Journal*, 43.3, pp. 269–283. https://doi.org/10.1093/cdj/bsn010

Cousins, J.J. and J.P. Newell. 2015. 'A political–industrial ecology of water supply infrastructure for Los Angeles', *Geoforum*, 58, pp. 38–50. https://doi.org/10.1016/j.geoforum.2014.10.011

Cretney, R. 2019. '"An opportunity to hope and dream": Disaster politics and the emergence of possibility through community-led recovery', *Antipode*, 51.2, pp. 497–516. https://doi.org/10.1111/anti.12431

Cutter, S.L., B.J. Boruff, and W.L. Shirley. 2003. 'Social vulnerability to environmental hazards', *Social Science Quarterly*, 84.2, pp. 242–261. https://doi.org/10.1111/1540-6237.8402002

Demeritt, D. 2001. 'The construction of global warming and the politics of science', *Annals of the Association of American Geographers*, 91.2, pp. 307–337. https://doi.org/10.1111/0004-5608.00245

Di Baldassarre, G., A. Viglione, G. Carr, L. Kuil, J.L. Salinas, and G. Blöschl. 2013. 'Socio-hydrology: Conceptualising human-flood interactions',

Hydrology and Earth System Sciences, 17.8, pp. 3295–3303. https://doi. org/10.5194/hess-17-3295-2013

Di Baldassarre, G., M. Sivapalan, M. Rusca, C. Cudennec, M. Garcia, H. Kreibich, and S. Pande. 2019. 'Sociohydrology: Scientific challenges in addressing the sustainable development goals', *Water Resources Research*, 55.8, pp. 6327–6355. https://doi.org/10.1029/2018wr023901

Donaldson, A., S. Lane, N. Ward, and S. Whatmore. 2013. 'Overflowing with issues: Following the political trajectories of flooding', *Environment and Planning C: Government and Policy*, 31.4, pp. 603–618. https://doi. org/10.1068/c11230

Donovan, A. 2017. 'Geopower: Reflections on the critical geography of disasters', *Progress in Human Geography*, 41.1, pp. 44–67. https://doi. org/10.1177/0309132515627020

Dreyer, M. and O. Renn. 2011. 'Participatory approaches to modelling for improved learning and decision-making in natural resource governance: an editorial', in *Environmental Policy and Governance*, pp. 379–385. https://doi. org/10.1002/eet.584

Enarson, E. 1998. 'Through women's eyes: A gendered research agenda for disaster social science', *Disasters*, 22.2, pp. 157–173.

Falconi, S.M. and R.N. Palmer. 2017. 'An interdisciplinary framework for participatory modeling design and evaluation—What makes models effective participatory decision tools?', *Water Resources Research*, 53.2, pp. 1625–1645. https://doi.org/10.1002/2016wr019373

Forsyth, T. 2004. *Critical Political Ecology: The Politics of Environmental Science* (Routledge). https://doi.org/10.4324/9780203017562

Forsyth, T. 2011. 'Politicizing environmental explanations: What can political ecology learn from sociology and philosophy of science?', in *Knowing Nature: Conversations at the Intersection of Political Ecology and Science Studies*, ed. by M.J. Goldman, P. Nadasdy, and M.D. Turner (University of Chicago Press), pp. 31–46. https://doi.org/10.7208/chicago/9780226301440.001.0001

Gandy, M. 2022. 'Urban political ecology: a critical reconfiguration', *Progress in Human Geography*, 46.1, pp. 21–43. https://doi. org/10.1177/03091325211040553

Gawronski, V.T. and R.S. Olson. 2013. 'Disasters as crisis triggers for critical junctures? The 1976 Guatemala case', *Latin American Politics and Society*, 55.2, pp. 133–149. https://doi.org/10.1111/j.1548-2456.2013.00196.x

Goldman, M.J., Nadasdy, P., and M.D. Turner. 2019. *Knowing Nature: Conversations at the Intersection of Political Ecology and Science Studies* (University of Chicago Press). https://doi.org/10.7208/ chicago/9780226301440.001.0001

Hamideh, S. and J. Rongerude. 2018. 'Social vulnerability and participation in disaster recovery decisions: Public housing in Galveston after Hurricane Ike', *Natural Hazards*, 93.3, pp. 1629–1648. https://doi.org/10.1007/s11069-018-3371-3

Haraway, D. 1988. 'Situated knowledges: the science question in feminism and the privilege of partial perspective', *Feminist Studies*, 14.3, pp. 575–599.

Harris, D.M. 2022. 'The trouble with modeling the human into the future climate', *GeoHumanities*, 8.2, pp. 382–398. https://doi.org/10.1080/2373566x.2022.2043764

Harvey, D. 2012. *Rebel Cities: From the Right to the City to the Urban Revolution* (Verso). https://doi.org/10.4067/s0250-71612014000100013

Hoque, S.F., C.H. Quinn, and S.M. Sallu. 2017. 'Resilience, political ecology, and well-being: An interdisciplinary approach to understanding social-ecological change in coastal Bangladesh', *Ecology and Society*, 22.2, p. 45. https://doi.org/10.5751/es-09422-220245

Jasanoff, S. 2010. 'A new climate for society', *Theory, Culture and Society*, 27.2–3, pp. 233–253. https://doi.org/10.1515/9783839456668-021

Keegan, C. 2020. '"Black Workers Matter": Black labor geographies and uneven redevelopment in post-Katrina New Orleans', *Urban Geography*, 42.3, pp. 340-359. https://doi.org/10.1080/02723638.2020.1712121

Kimmelman, M. 2017. 'Lessons from Hurricane Harvey: Houston's struggle is America's tale', *New York Times*, https://www.nytimes.com/Interactive/2017/11/11/Climate/Houston-Flooding-Climate.html

King, L. and M. Tadaki. 2018. 'A framework for understanding the politics of science (Core Tenet# 2)', in *The Palgrave Handbook of Critical Physical Geography*, ed. by R. Lave, C. Biermann, and S.N. Lane (Palgrave Macmillan), pp. 67–88. https://doi.org/10.1007/978-3-319-71461-5_4

Klein, N. 2007. *The Shock Doctrine: The Rise of Disaster Capitalism* (Macmillan).

Krueger, T., Chapter 26, this volume. 'Environmental modelling'.

Landström, C., Chapter 35, this volume. 'Participatory modelling'.

Landström, C., S.J. Whatmore, S.N. Lane, N.A. Odoni, N. Ward, and S. Bradley. 2011. 'Coproducing flood risk knowledge: Redistributing expertise in critical "participatory modelling"', *Environment and Planning A: Economy and Space*, 43.7, pp. 1617–1633. https://doi.org/10.1068/a43482

Lane, S.N. 2014. 'Acting, predicting and intervening in a socio-hydrological world', *Hydrology and Earth System Sciences*, 18.3, pp. 927–952. https://doi.org/10.5194/hess-18-927-2014

Lane, S.N., Chapter 24, this volume. 'Case studies'.

Lane, S.N., Chapter 30, this volume. 'Hydraulic modelling'.

Lane, S.N., N. Odoni, C. Landström, S.J. Whatmore, N. Ward, and S. Bradley. 2011a. 'Doing flood risk science differently: an experiment in radical scientific method', *Transactions of the Institute of British Geographers*, 36.1, pp. 15–36. https://doi.org/10.1111/j.1475-5661.2010.00410.x

Lane, S.N., C. Landström, and S.J. Whatmore. 2011b. 'Imagining flood futures: Risk assessment and management in practice', *Philosophical Transactions of the Royal Society A: Mathematical, Physical and Engineering Sciences*, 369.1942, pp. 1784–1806. https://doi.org/10.1098/rsta.2010.0346

Law, J. 2004. *After Method: Mess in Social Science Research* (Psychology Press).

Lemos, M.C. 2008. 'Whose water is it anyway? Water management, knowledge and equity in Northeast Brazil', in *Water, Place, and Equity*, ed. by J.M. Whiteley, H.Ingram, and R.W. Perry (MIT Press), pp. 249–270. https://doi.org/10.7551/mitpress/9780262232715.003.0009

Loftus, A., H. March, and T.F. Purcell. 2019. 'The political economy of water infrastructure: an introduction to financialization', *WIREs Water*, 6.1, p. e1326.
https://doi.org/10.1002/wat2.1326

Luft, R.E. 2009. 'Beyond disaster exceptionalism: Social movement developments in New Orleans after Hurricane Katrina', *American Quarterly*, 61.3, pp. 499–527. https://doi.org/10.1353/aq.2009.a317270

Makarigakis, A.K., and B.E. Jimenez-Cisneros. 2019. 'UNESCO's contribution to face global water challenges', *Water*, 11.2, p. 388. https://doi.org/10.3390/w11020388

Mazzoleni, M., L. Alfonso, J. Chacon-Hurtado, and D. Solomatine. 2015. 'Assimilating uncertain, dynamic and intermittent streamflow observations in hydrological models', *Advances in Water Resources*, 83, pp. 323–339. https://doi.org/10.1016/j.advwatres.2015.07.004

Mazzoleni, M., V.O. Odongo, E. Mondino, and G. Di Baldassarre. 2021. 'Water management, hydrological extremes, and society: Modeling interactions and phenomena', *Ecology and Society*, 26.4. https://doi.org/10.5751/es-12643-260404

McClintock, N. 2015. 'A critical physical geography of urban soil contamination', *Geoforum*, 65, pp. 69–85. https://doi.org/10.1016/j.geoforum.2015.07.010

McLaren, D. and N. Markusson. 2020. 'The co-evolution of technological promises, modelling, policies and climate change targets', *Nature Climate Change*, 10.5, pp. 392–397. https://doi.org/10.1038/s41558-020-0740-1

Meadows, D.H., 1999. Leverage points: Places to intervene in a system. https://donellameadows.org/archives/leverage-points-places-to-intervene-in-a-system/

Melsen, L., Chapter 31, this volume. 'Hydrological modelling'.

Moore, J.W. 2017. 'The Capitalocene, Part I: on the nature and origins of our ecological crisis', *The Journal of Peasant Studies*, 44.3, pp. 594–630. https://doi.org/10.1080/03066150.2016.1235036

Olson, R.S. and V.T. Gawronski. 2003. 'Disasters as critical junctures? Managua, Nicaragua 1972 and Mexico City 1985', *International Journal of Mass Emergencies and Disasters*, 21.1, pp. 3–35. https://doi.org/10.1177/028072700302100101

Pelling, M. and K. Dill. 2010. 'Disaster politics: Tipping points for change in the adaptation of sociopolitical regimes', *Progress in Human Geography*, 34.1, pp. 21–37. https://doi.org/10.1177/0309132509105004

Perreault, T., G. Bridge, and J. McCarthy. 2015. *The Routledge Handbook of Political Ecology* (Routledge).

Pratt, G. 2009. 'Positionality', in *The Dictionary of Human Geography*, ed. by A. Rogers, N. Castree, and R. Kitchin (Oxford University Press), pp. 556–557.

Ravera, F., K. Hubacek, M. Reed, and D. Tarrasón. 2011. 'Learning from experiences in adaptive action research: A critical comparison of two case studies applying participatory scenario development and modelling approaches', *Environmental Policy and Governance*, 21.6, pp. 433–453. https://doi.org/10.1002/eet.585

Rozario, K. 2019. *The Culture of Calamity: Disaster and the Making of Modern America*. (University of Chicago Press). https://doi.org/10.7208/chicago/9780226230214.001.0001

Rubiano Rivadeneira, N. and W. Carton. 2022. '(In)justice in modelled climate futures: A review of integrated assessment modelling critiques through a justice lens', *Energy Research and Social Science*, 92, 102781. https://doi.org/10.1016/j.erss.2022.102781

Rusca, M., A.S. Boakye-Ansah, A. Loftus, G. Ferrero, and P. Van Der Zaag. 2017. 'An interdisciplinary political ecology of drinking water quality. Exploring socio-ecological inequalities in Lilongwe's water supply network', *Geoforum*, 84, pp. 138–146. https://doi.org/10.1016/j.geoforum.2017.06.013

Rusca, M. and F. Cleaver. 2022. 'Unpacking everyday urbanism: Practices and the making of (un) even urban waterscapes', *Wiley Interdisciplinary Reviews: Water*, e1581. https://doi.org/10.1002/wat2.1581

Rusca, M. and G. Di Baldassarre. 2019. 'Interdisciplinary critical geographies of water: Capturing the mutual shaping of society and hydrological flows', *Water*, 11.10, p. 1973. https://doi.org/10.3390/w11101973

Rusca, M., N.J. Gulamussen, J. Weststrate, E.I. Nguluve, E.M. Salvador, P. Paron, and G. Ferrero. 2021. 'The urban metabolism of waterborne diseases: Variegated citizenship, (waste) water flows, and climatic variability in

Maputo, Mozambique', *Annals of the American Association of Geographers*, 112.4, pp. 1159–1178. https://doi.org/10.1080/24694452.2021.1956875

Rusca, M., G. Messori, and G. Di Baldassarre. 2021. 'Scenarios of human responses to unprecedented social-environmental extreme events', *Earth's Future*, 9.4. https://doi.org/10.1029/2020ef001911

Rusca, M., E. Savelli, G. Di Baldassarre, A. Biza, and G. Messori. 2023. 'Unprecedented droughts are expected to exacerbate urban inequalities in Southern Africa', *Nature Climate Change*, 13.1, pp. 98–105. https://doi.org/10.1038/s41558-022-01546-8

Savelli, E., M. Mazzoleni, G. Di Baldassarre, H. Cloke, and M. Rusca. 2023. 'Urban water crises driven by elites' unsustainable consumption', *Nature Sustainability*, 6, pp. 929–940. https://doi.org/10.1038/s41893-023-01100-0

Savelli, E., M. Rusca, H. Cloke, T.J. Flügel, A. Karriem, and G. Di Baldassarre. 2022. 'All dried up: The materiality of drought in Ladismith, South Africa', *Environment and Planning E: Nature and Space.* https://doi.org/10.5194/egusphere-egu21-1991

Schwartz, K., M. Tutusaus Luque, M. Rusca, and R. Ahlers. 2015. '(In)formality: The meshwork of water service provisioning', *Wiley Interdisciplinary Reviews: Water*, 2.1, pp. 31–36. https://doi.org/10.1002/wat2.1056

Shaw, I.G.R. 2012. 'Towards an evental geography', *Progress in Human Geography*, 36.5, pp. 613–627. https://doi.org/10.1177/0309132511435002

Simandan, D. 2019. 'Revisiting positionality and the thesis of situated knowledge', *Dialogues in Human Geography*, 9.2, pp. 129–149. https://doi.org/10.1177/2043820619850013

Simon, G.L. 2014. 'Vulnerability-in-production: A spatial history of nature, affluence, and fire in Oakland, California', *Annals of the Association of American Geographers*, 104.6, pp. 1199–1221. https://doi.org/10.1080/00045608.2014.941736

Sivapalan, M., H.H.G. Savenije, and G. Blöschl. 2012. 'Socio-hydrology: A new science of people and water', *Hydrological Processes*, 26.8, pp. 1270–1276. https://doi.org/10.1002/hyp.8426

Solomatine, D.P. and T. Wagener. 2011. 'Hydrological modelling', in *The Science of Hydrology. Treatise on Water Science*, ed. by P.A. Wilderer (Elsevier), pp. 435–457. https://doi.org/10.1016/b978-0-444-53199-5.00044-0

Srinivasan, V., E.F. Lambin, S.M. Gorelick, B.H. Thompson, and S. Rozelle. 2012. 'The nature and causes of the global water crisis: Syndromes from a meta-analysis of coupled human-water studies', *Water Resources Research*, 48.10, pp. 1–16. https://doi.org/10.1029/2011wr011087

Sultana, F. 2020. 'Embodied intersectionalities of urban citizenship: Water, infrastructure, and gender in the global South', *Annals of the American*

Association of Geographers, 110.5, pp. 1407–1424. https://doi.org/10.1080/24 694452.2020.1715193

Sultana, F. 2022. 'The unbearable heaviness of climate coloniality', *Political Geography*, 99, 102638. https://doi.org/10.1016/j.polgeo.2022.102638

Swyngedouw, E. 1997. 'Power, nature, and the city. The conquest of water and the political ecology of urbanization in Guayaquil, Ecuador: 1880–1990', *Environment and Planning A*, 29.2, pp. 311–332.

Swyngedouw, E. 2004. *Social Power and the Urbanization of Water: Flows of Power.* (Oxford University Press). ://doi.org/10.1093/oso/9780198233916.001.0001

ter Horst, R., Alba, R., Vos, J., Rusca, M., Godinez-Madrigal, J., Babel, L. V., ... & Krueger, T. (2023). Making a case for power-sensitive water modelling: a literature review. *Hydrology and Earth System Sciences Discussions*, 2023, 1–31. https://doi.org/10.5194/hess-28-4157-2024

Turner, M.D. 2011. 'Production of environmental knowledge: Scientists, complex natures, and the question of agency', in *Knowing Nature: Conversations at the Intersection of Political Ecology and Science Studies*, ed. by M.J. Goldman, P. Nadasdy, and M.D. Turner (University of Chicago Press), pp. 25–29. https://doi.org/10.7208/chicago/9780226301440.001.0001

Walker, P.A. 2005. 'Political ecology: Where is the ecology?', *Progress in Human Geography*, 29.1, pp. 73–82. https://doi.org/10.1191/0309132505ph530pr

Webler, T., S. Tuler, and T. Dietz. 2011. 'Modellers' and outreach professionals' views on the role of models in watershed management', *Environmental Policy and Governance*, 21.6, pp. 472–486. https://doi.org/10.1002/eet.587

Whatmore, S.J. and C. Landström. 2011. 'Flood apprentices: an exercise in making things public', *Economy and Society*, 40.4, pp. 582–610. https://doi.org/10.1080/03085147.2011.602540

Wu, P., N. Christidis, and P. Stott. 2013. 'Anthropogenic impact on Earth's hydrological cycle', *Nature Climate Change*, 3.9, pp. 807–810. https://doi.org/10.1038/nclimate1932

Yates, J.S., L.M. Harris, and N.J. Wilson. 2017. 'Multiple ontologies of water: Politics, conflict and implications for governance', *Environment and Planning D: Society and Space*, 35.5, pp. 797–815. https://doi.org/10.1177/0263775817700395

Zwarteveen, M., J.S. Kemerink-Seyoum, M. Kooy, J. Evers, T.A. Guerrero, B. Batubara, and A. Wesselink. 2017. 'Engaging with the politics of water governance', *WIREs Water*, 4.6, e1245. https://doi.org/10.1002/wat2.1245

15. 'A hydrologist and a rhetorician walk into a workshop,' or How we learned to collaborate on a decade of mixed-methods river research across the humanities and biophysical sciences

Eric G. Booth and Caroline Gottschalk Druschke

Introduction

On the morning of Tuesday, August 28, 2018, the two of us were at our respective homes in Madison, Wisconsin, obsessively refreshing our browsers for the latest weather news and trying to make contact with a graduate student we co-advised. She was alone in the field, measuring stream cross-sections on a proposed stream restoration site we were studying in southwestern Wisconsin's rural Kickapoo River Watershed. She was eager to get through the final few cross-sections before the start of classes the following week, even though it had started raining in the area the night before and the forecast became worryingly worse as the day progressed. We texted to check in on her, increasingly anxious about the possibilities of flash flooding as she stood waist deep in the creek in an area with poor cell coverage. At noon, Caroline suggested she head to a local hotel, rather than trying to make it back to Madison, and she texted back: "I'm going to finish this transect then work on clearing the lines for surveying to at least keep working until it gets bad!" Eric texted a few minutes later to suggest she had 30 minutes at most, and by 1pm sent an urgent, "Time to pack it up. Lots of storms coming." The rest, as they say, is history. Unprecedented storms swept through the area that week and again the following week, with up to a foot of rain at some locations in under 24 hours. Heavy, non-stop rain pummelling the steep slopes of this unglaciated corner of Wisconsin created flood conditions

  https://doi.org/10.11647/OBP.0418.15

worse than anyone had yet seen in this flood-prone region. The deluge flooded local creeks and rivers, creating the flood of record on the Kickapoo River in 80-plus years of chronicled history. This accelerating reality of flooding—its impacts on the region's streams and watersheds, communities, and stream restoration practices—sits at the centre of our collaborative, mixed-methods research.

We share this story at the outset because we think it illustrates a number of points that are central to our mixed methodological collaboration. First, the anecdote demonstrates our emphasis on open communication. Just as we were in touch with each other and our student during this flood—as well as checking in on her at the hotel that afternoon and evening, and again as she tried to navigate the network of road closures across the region the following day—we make it an explicit habit to be in touch with one another: to share research updates, to explain and demonstrate new methodologies, to ask questions, to collaborate on shared drafts, to share critiques and frustrations and hopes.

Second, it illustrates the ways we draw on complementary but not duplicative skill sets. We bring different disciplinary strengths to our work—Eric is trained in hydrology, engineering, and ecology, and Caroline in rhetorical studies, critical theory, and freshwater science—but we also share overlapping interests. By problem solving in collaboration, we continue to find ourselves better positioned to come up with creative solutions: to a student stuck in a flooding creek, or to community flood response that demands mixing qualitative interviews and hydrologic and hydraulic modelling (see Johnston and Longhurst, Chapter 32; Melsen, Chapter 31; Lane, Chapter 30).

Third, the anecdote above reveals our shared interest in ethics and care (see Meadow et al., Chapter 5). We care for each other and our shared students, try to support them in research and life, and try to develop their leadership capacities and independence, while emphasising that they are humans first and foremost. And we try to extend that care to the communities we work with and alongside. Fourth, the opening story illustrates that our shared work is interested in taking risks—insomuch as any innovative, mixed methodology, interdisciplinary research agenda must—but not in taking risks that would be unwise (see Miesen, Chapter 10). The risks we take are taken with thoughtfulness and intention.

Fifth and finally, this story highlights our ongoing interest in responding flexibly and creatively to changing conditions in the landscape, an interest that necessitates a mixed methodological approach. Just as we spent that day and the next navigating how to respond to our student's changing needs, we have spent the five years since then shaping and re-shaping our shared work—which had been focused squarely on stream restoration in the region—to respond to this new reality of accelerating and worsening flooding. In light of this urgent need, we are working to find ways of mixing field monitoring of geomorphic change, using ground- and drone-based methods (see Kasvi, Chapter 45), with interviews (Johnston and Longhurst, Chapter 32) and oral histories (see Chakov et al., Chapter 33) from community members and stream and land managers, hydrologic and hydraulic modelling, and public engagement activities, and to collaborate with a suite of local organisations to try to understand the impacts of flooding and to help support conversations about how to respond.

Our mixed-methods approach to understanding the impacts of flooding and supporting conversations about flood response is grounded in our shared goal of doing good work to benefit towns, villages, counties, watersheds, and community members across southwestern Wisconsin's Driftless Area, work that is done *with* the members of those communities. This work consists of years of listening to and learning from community residents and stream managers to develop research questions and methods that help to address questions emerging from community members, which has meant, in this case, monitoring stream changes and modelling future flood scenarios for watershed change. We are committed to continually returning what we hear and find and model back to community members, supporting—but not leading—ongoing deliberations about how best to move forward in these increasingly flood prone watersheds. This listening, responding, gathering, synthesising, returning, and refining is at the heart of our shared work, and it also requires diving deeply into the historical and biophysical context of the region to do it well.

We wanted to find ways to address the questions we have heard emerging from members of the Driftless Area communities where we work: about the impacts of land use decisions and riparian vegetation on downstream flooding; about how to respond as homeowners, towns, villages, counties, and watersheds to increasing rainfall and flooding; and about available tools that might help to support these discussions. To answer these questions, we are working to stitch together our various methodological approaches to integrate qualitative conversations with field monitoring and hydrologic and hydraulic models to tease apart process drivers that can often be modified for the worse or for the better by human management.

The scenarios that result from this work—scenarios that area watershed councils, county conservationists, farmers, and business owners have urgently asked for—can serve as boundary objects for difficult conversations about more liveable futures, where boundary objects are entities and ideas that straddle the boundaries of various communities, and force challenging and consequential conversations around common problems (Star and Griesemer 1989; Wilson and Herndl 2007). This doesn't mean that these scenarios provide easy paths forward, or even points of consensus. But they do offer a shared object of focus, allowing community decision makers, from all their different perspectives, to engage with area flooding not as a "matter of fact", but as a "highly complex, historically situated, richly diverse" "matter of concern" (Latour 2003; Lane et al., 2011). Our point here is that our work is committed to responding to concerns emerging in complex communities, and that the only way for us to do that has been to find creative ways to work across these disciplinary perspectives, with deep attention to the broader contexts from which these questions are emerging, in collaboration with community members.

What we do in this chapter, then, is detail the accelerating reality of flooding in southwestern Wisconsin's Driftless Area and explain how we came together to do this work. We then turn to a description of the specific methodologies we are weaving together in our current flood-focused research, and close with a reflection on what this collaborative, community-engaged, mixed methodological work gets us and why it matters. We offer this specific methodological example to make a few key points. First, we argue that interdisciplinary work—which we define via Stock and Burton (2011) as an effort towards "addressing specific 'real world' system problems" that "forces participants (from a variety of unrelated disciplines) to cross boundaries to create new knowledge"—is essential to working towards better community outcomes. Second, we acknowledge that this kind of interdisciplinarity involves, honours, and emerges from not just different disciplines but different epistemologies. In other words, we recognise that our own perspectives on complex problems in stream-floodplain systems, and the methodologies we use individually to explore them, represent dramatically different ways of understanding the world around us, and that there is important labour in highlighting and translating those epistemologies to one another to consider what comes out the other side (see Lane and Lave, Chapter 3).

Third, and relatedly, we insist that the best mixed methodological work is synthetic. It is not about using the work of multiple disciplines and methodologies in an additive manner, where, for instance, an economist weighs in, followed by an ecologist, followed by a hydrologist. Instead, we continue to work to learn new approaches, consider problems from multiple perspectives, work to flexibly adapt to the problems that emerge in front of us, and emphasise ethical connections to community members through it all. We draw upon and are inspired by practitioners of co-production (Norstrom et al. 2020; Schuttenberg et al. 2015), community-based (Strand et al. 2003; Stoecker 2003), and participatory action (Kindon et al. 2007; Kemmis et al. 2014) research. Our interdisciplinary approach shapes the very questions we ask and drives the ways we approach our work. We hope our chapter inspires and equips readers to take a similar approach.

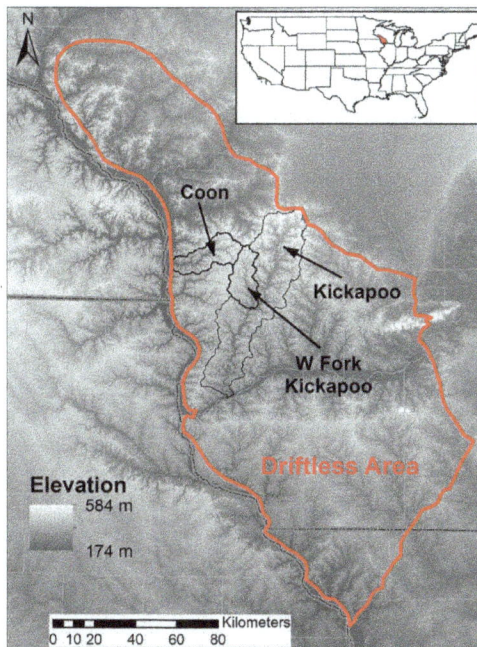

Fig. 15.1. Map of the Kickapoo River and Coon Creek watersheds, including the West Fork Kickapoo subwatershed, within the Upper Mississippi River Basin's Driftless Area. The unglaciated Driftless Area stretches throughout southwestern Wisconsin, dipping into the northwestern edge of Illinois

Working together in place

The Upper Mississippi River Basin (UMRB) has seen increasingly frequent flooding in recent decades, damaging cities, towns, farms, and rural infrastructure in a trend that climate change will almost certainly worsen (Jha et al. 2004; Jha et al. 2006; Frans et al. 2013; Rajib and Merwade 2017). Southwestern Wisconsin's Kickapoo River and Coon Creek watersheds—in the UMRB's Driftless region, a highly dissected landscape with broad, loess-covered uplands, steep hillslopes, and deeply incised valleys (Knox 2019)—have experienced at least one 100-year and two 50-year floods in the last decade, and climate forecasts predict this pattern will intensify (Fig. 15.1; WICCI 2021). The worst of these floods, in August 2018, caused an estimated $29 million in damage to businesses, homes, and public infrastructure in Vernon County alone, almost $1,000 per person in a county with a 15.9% poverty rate (Lu 2019; Couleecap 2016). Four years later, communities throughout the Kickapoo and Coon Creek watersheds—like communities across the world living with climate change-amplified flood events—are struggling to clean out buildings and repair damaged infrastructure, and, with the leadership of local community groups, begin an increasingly public discussion about what flood resilience might look like. Across the Upper Mississippi River Basin, and many other parts of the world, the major question is no longer whether the next catastrophic flood will happen, but how soon.

Chronic and accelerated flooding across the Driftless is a result of feedbacks at historic and current time scales. Early Euro-American agriculture and logging beginning in the 1840s (Fig. 15.2a) accelerated soil erosion across the region's steep hillslopes, contributing over four meters of post-settlement alluvium to valley floors in some places and creating a flume-like riparian corridor that efficiently conveys flood peaks downstream (Fig. 15.2b) (Trimble and Lund 1982; Knox 2006; Belby et al. 2019). More intense and more frequent storms associated with climate change (Fig. 15.2d) are exacerbating a legacy of poor land management (Fig. 15.2a, b, c), negatively impacting water quality, and degrading world-class trout fishing, a major economic driver in the region, contributing $957 million in direct spending to the Driftless economy in 2015 (Anderson 2016). In response, local NGOs and farmers have promoted agricultural best management practices in the uplands (Fig. 15.2c) and in-channel stream restoration projects (Fig. 15.2g), but efforts are complicated by a legacy of anthropogenic damage (Fig. 15.2a, b, c), increasing flood frequency (Fig. 15.2f) and magnitude (Fig. 15.2e), and uncertainty about which approaches are likely to produce the largest impacts.

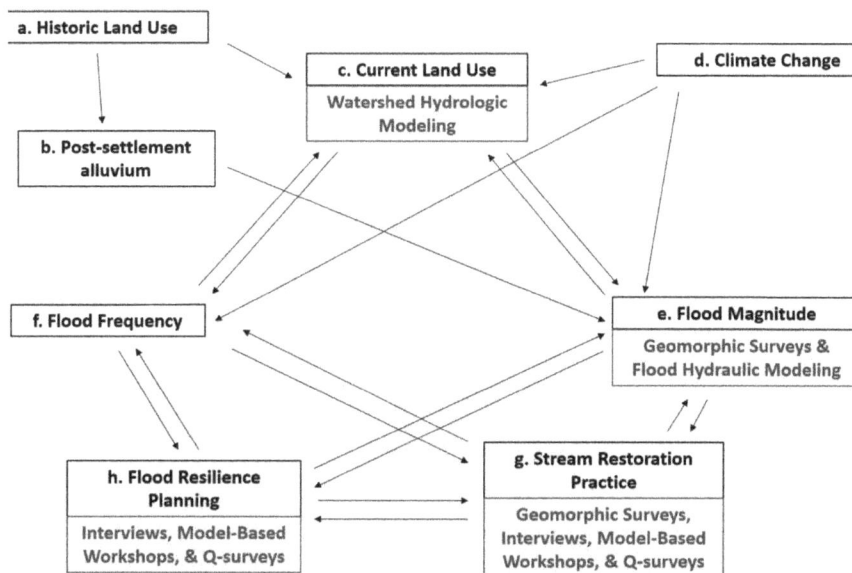

Fig. 15.2. Dynamics of historic and current land use (a, c), soil movement (b), climate change (d), flood magnitude and frequency (e, f), stream restoration practice (g), and flood resilience planning (h) in southwestern Wisconsin's Kickapoo River and Coon Creek watersheds with project methodologies for addressing the interactions of each

We understand flooding as a complex problem that demands complex, interdisciplinary, mixed methodological strategies. And so, we have been working together on the issue since 2017 to build a multi-pronged research agenda to address accelerating flooding through a wide suite of diverse but surprisingly complementary approaches.

Our work together actually has its roots several years earlier, when, as our title suggests, a hydrologist and a rhetorician walked into a workshop. To be more precise, Eric refers to himself as a hydroecologist, trained across environmental engineering, hydrology, fluvial geomorphology, and ecology. Caroline refers to herself as a fluvial rhetorician, to indicate her varied disciplinary training and ongoing interest in borrowing rhetorical approaches to focus on river systems, where rhetoric is an academic discipline focused on "the capacity of language to persuade audiences, connect individuals, and affect the biophysical world" (Gottschalk Druschke and McGreavy 2016). The workshop was a meeting of the Scenarios, Services, and Linkages to Society Research Coordination Network (S3RCN), a National Science Foundation-funded network led by researchers from Harvard Forest, focused on working with land-use stakeholders from across the region to identify drivers of landscape change, develop scenario narratives based on those drivers, and map the scenarios as land-use through the year 2060. Eric and Caroline were brought in as guest consultants for the meeting, invited to share Eric's work building futures scenarios in southern Wisconsin on the Yahara 2070 project (Carpenter et al. 2015; Booth et al. 2016) and Caroline's work supporting the development of a community-driven decision support tool aimed at prioritising wetland restoration projects as a faculty member in Rhode Island and research fellow at the USEPA Atlantic Ecology Division (Mazzotta et al. 2016; Mazzotta et al. 2019). We began talking with each other at the workshop after being asked to draw a picture of something we loved as part of an introductory activity. Caroline's watershed network drawing and Eric's hand drawn canoe initiated what is now an almost decade-long conversation.

We talked once or twice a year after that, recognising that we were both intensely curious people and both extremely interested in roughly the same thing: how to link rich, qualitative, human-centred methodologies with data intensive field work and modelling to support community-driven conversations about how to deal with challenging water issues. Eric's training came first in hydrology, engineering, and ecology, but he was fortunate to have started working with other researchers in the social sciences and humanities through the Yahara 2070 project and came to appreciate strongly the unique insights they offered to human and environmental wicked problems. Rittel and Webber (1973) describe these as, unlike problems in the biophysical sciences, ones that are "ill-defined", and that "rely upon elusive political judgment for resolution. (Not 'solution.' Social problems are never solved. At best they are only re-solved—over and over again.)" (p. 160). Caroline had been trained in rhetorical studies, supplemented with restoration ecology and freshwater sciences, and she had always been interested in how those things fit together.

When Caroline was offered a faculty position in the Department of English at University of Wisconsin-Madison in 2017, Eric's home institution (he is housed in the Departments of Plant and Agroecosystem Sciences and Civil and Environmental Engineering), she got in touch with Eric to consider proactively ways how we might work together. An internal funding opportunity gave us that chance, when we were awarded funding in 2018 from the Kickapoo Valley Reforestation Fund (KVRF) in UW-Madison's College of Agricultural and Life Sciences, a fund meant to support projects that "enhance the ecological, economic and social well-being of the Kickapoo Valley and its residents" and "originate with the citizens of the Valley".

That first project, which kicked off in summer 2018, emerged from planning discussions with local partners including the Trout Unlimited Driftless Area Restoration Effort (TUDARE), the Valley Stewardship Network (VSN), and the Wisconsin Department of Natural Resources (WDNR). We focused on disagreement among managers, anglers, and landowners about the use of riparian forests as a stream restoration tool to potentially enhance habitat, to provide shade to mediate extreme stream temperatures, and to stabilise banks to lessen flood impacts in the Kickapoo Valley (Gottschalk Druschke et al., 2023). To address this stakeholder-identified disagreement, we proposed an integrated approach that combined biophysical, humanistic, and spatial methods to provide a holistic view synthesising knowledge related to forested stream-riparian ecosystems and human perspectives on how best to manage these multifunctional systems. Our work involved reviewing the available scientific literature on forested stream buffers; monitoring stream temperature and stream/riparian conditions at active restoration sites; conducting interviews with 16 stream restoration managers; and hosting a Q-method workshop with 35 managers, anglers, and restoration volunteers (Druschke et al. 2019; Lundberg et al. 2022).

We were awarded a second KVRF award in 2020 to extend that work, again in collaboration with Trout Unlimited Driftless Area Restoration Effort (TUDARE), Valley Stewardship Network (VSN), and Wisconsin Department of Natural Resources (WDNR). This time, after the experience of the devastating 2018 flood that opened this chapter, our stream restoration work became explicitly flood-focused. We asked three primary research questions: 1) how does more frequent and more intense flooding in the Kickapoo Valley and across the Driftless Region alter stream restoration practices in the study watersheds, including riparian vegetation?; 2) how does more frequent and more intense flooding shape Valley residents' perceptions about stream management and future flooding?; and 3) how do floods impact stream restoration projects via changes in tree cover and channel geomorphology? To address these questions, we extended our restoration manager interviews to develop an explicit focus on flooding, while continuing to assess changes in geomorphic, habitat, and vegetation conditions at four stream-floodplain sites, with a focus on the intersecting impacts of restorations and floods. We also focused on work that Caroline had gotten deeply involved in the year prior, a community-driven oral history effort called Stories from the Flood.

Stories from the Flood (SFTF), initiated by the not-for-profit Driftless Writing Center in Viroqua, Wisconsin in late 2018, has worked to document extensive flooding in the Kickapoo River and Coon Creek watersheds by gathering oral histories with community members who experienced the 2018 floods and many prior floods in the region (see Chakov et al., Chapter 33; Gottschalk Druschke et al. 2022a; 2022b). The project aimed to help flood-affected residents of the Kickapoo River and Coon Creek watersheds process their trauma, while creating a historical record to inform future planning and support community healing. Oral histories gathered from over 100 participants are now permanently archived at the University of Wisconsin-La Crosse Murphy Library Oral History Program; these oral histories highlight Valley-based perspectives on flood impacts, agricultural practices, stream management, and community-level resilience. The stories gathered in the archive offer an important mechanism for amplifying community-identified paths forward in the face of increasing flooding, working to flag the attention of local, state, and national decision makers to make smarter plans to prevent future flooding that take the interests of local residents into account.

The 2018 flood and its aftermath were also the subject of the Water Resources Management Practicum at UW-Madison from 2019 to 2021 where Eric advised six M.S. students on a shared project focused on opportunities to enhance flood resilience in the Coon Creek watershed. Through the support of a multidisciplinary set of co-advisers including Caroline and local community members, this group of students took a mixed-methods approach that included quantifying the variability in infiltration potential using field measurements and mapping recent land management changes, conducting an economic analysis of agricultural practices that could enhance infiltration, creating a network analysis of all of the institutions involved in flood management and emergency response in the area, and recording interviews with land and water managers and local citizens to gain a better understanding of their perspectives on flooding and their visions of the future (WRM 2022).

We share this rather long history as it helps to situate our current research effort and its realisation following from both patience and time. We have spent many years at this point familiarising ourselves with each other's approaches and methodologies and interests; building relationships with community members; learning more about the history, hydrology, geomorphology, culture, and ecology of the region; experimenting with various combinations of research approaches; and collaborating with a suite of students and partners. That groundwork was central to shaping our current project, a National Science Foundation CNH2-funded award that kicked off in January 2021 with colleague Rebecca Lave at Indiana University focused on investigating the interactive dynamics of stream restoration and flood resilience in a changing climate in the Kickapoo River and Coon Creek watersheds. The project works to support stakeholder-based flood resilience planning by using multiscalar hydrologic and hydraulic modelling iteratively with qualitative and participatory approaches.

Key to this project is that it emerged from—and is intended to support—watershed-based discussions about how to respond to increasingly frequent and severe flooding, particularly in the context of the failures of five large federal flood control structures (authorised under Public Law 566 (PL-566)) in the two watersheds during the August 2018 flood event. We knew from our work on the Stories from the Flood project that these dam failures caused a huge amount of localised physical destruction as floodwaters held back by the dams were released all at once onto the farms and homes below them. We knew also that the dam breaches prompted a huge amount of psychic destruction, as residents began to worry as much about whether the 18 remaining structures in the watersheds might breach as they did about how bad the next flood might become without those structures in place.

The failures of those federal PL-566 flood control structures triggered a Watershed Project Plan-Environmental Impact Statement (PLAN-EIS) process, sponsored by the Natural Resources Conservation Service (NRCS) in partnership with Monroe, Vernon and La Crosse Counties, to offer a recommendation of what to do with the five failed dams and the 18 others remaining in the watersheds, and how to address flooding in the potential absence of these large flood control structures. The process inspired widespread and intense interest from the outset, as community members looked forward to the chance to weigh in on these dam decisions and learn more about recommended strategies to control flooding in the Coon Creek and West Fork Kickapoo watersheds. NRCS and County partners were likewise excited that the PLAN-EIS process could support the production of expensive and labour-intensive hydrologic and hydraulic models of the watersheds, in addition to an economic cost-benefit study of retaining, repairing, removing, or decommissioning the PL-566 dams.

Given our ongoing engagement with watershed partners, we were also excited about the PLAN-EIS process, and tried to consider ways our ongoing community-engaged work might be able to build from and supplement that process, with particular goals of: 1) prioritising meaningful participation from community members; 2) extending the scope of the PLAN-EIS study beyond the impacts of the PL-566 dams on downstream flood peaks to consider flood impacts from upstream land management change; and 3) translating the PLAN-EIS models into forms that could place decision-making support—and a stronger sense of agency over potential pathways to reducing flooding—into the hands of county, watershed, and community decision-makers after the conclusion of the PLAN-EIS study. Our NSF CNH2 proposal was designed around these three goals, hoping to learn from the Stories from the Flood project to highlight community voices, to build from the interactions of watershed-based land management with reach-based flood impacts to think beyond the dams themselves, and to put those PLAN-EIS models into the hands of community decision makers by adapting them into an interactive decision support tool. Crucially, these plans emerged directly from several years of conversations and shared work with local partners, including the Kickapoo Valley Reserve, Trout Unlimited, Valley Stewardship Network, Vernon County (WI) Land and Water Conservation Department, Monroe County (WI) Land Conservation Department, Natural Resources Conservation Service, and the Driftless Writing Center.

We then worked to translate those county- and community-level interests into research-based terms we hoped NSF-CNH2 Dynamics of

Integrated Socio-Environmental Systems reviewers would recognise. What we proposed, then, was to integrate qualitative interviews (Fig. 15.2g, h) with geomorphic surveys (Fig. 15.2e, g), iterative, multiscalar hydrologic (Fig. 15.2c) and hydraulic modelling (Fig. 15.2e), and model-based participatory workshops (Fig. 15.2g, h) to ask a set of nested questions that interrogate the multiscalar dynamic feedbacks between in-stream restoration practices, upland conservation measures, flood peaks, and community-level impacts of and responses to climate change:

Q1. How does more frequent and more intense flooding across the Upper Mississippi River Basin alter stream restoration practices in the study watersheds, linking or breaking the potential connection between restoration and resilience?

Q2. How do stream restoration practices, particularly changes in channel morphology, impact community-level climate vulnerability to flooding?

Q3. How do stream managers, flood decision makers, and landowners balance socio-environmental tradeoffs related to flood vulnerability and resilience across the reach and watershed scales?

Q4. Can iterative, multiscalar hydrologic and hydraulic modelling be incorporated with qualitative and participatory approaches to support stakeholder-based flood resilience planning to enable social-ecological system (SES) transformations?

We understand that these questions may read as though they could have been crafted for any academic research project focused upon flooding in southwestern Wisconsin, but we felt that was needed in order to satisfy the traditional expectations of academic reviewers, and we were absolutely committed to doing whatever was needed to get this work funded in order to support the work of flood recovery and planning in the watershed communities with whom we worked. Almost four years later, we have more confidence that reviewers would have found value in an approach more akin to a co-production where the development of research questions was part of the research process itself. There are some notable exceptions (e.g., Lane et al., 2011), but in the United States in 2019, when we submitted this proposal, we did not assume that would be the case. And whether we were right or wrong in doing so, we saw our role as translating community hopes, interests, and concerns into questions that might be recognised as "good science". But our research questions emerged from community interests and needs, and are meant to create new knowledge that amplifies and responds to those interests and needs, and is delivered back to community members and community decision-makers in forms they have asked for.

Methods in practice: How we are working to study this

Extensive flooding in the Kickapoo River Watershed in August 2018. (Photo courtesy of Emma Lundberg)

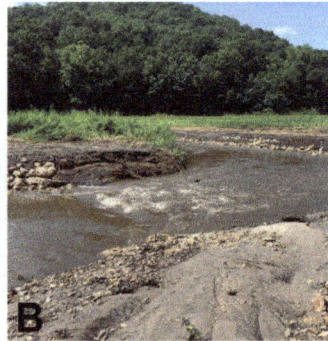

Impact of a flash flood in July 2019 on a recently completed restoration project on Billings Creek

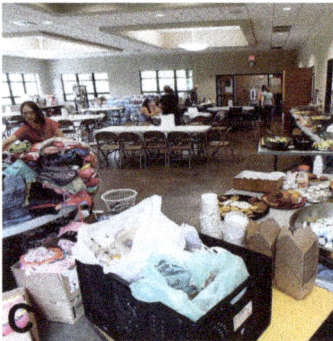

Gays Mills Community Center during August 2018 flood response. (Photo courtesy of Tim Hundt)

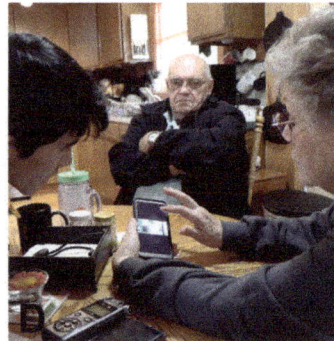

Student volunteers collect a flood story from Angie and Elmer McCauley as part of the Stories from the Flood project. (Photo courtesy of Sydney Widell)

Fig. 15.3. (A) Extensive and persistent flooding impacts floodplain communities (CC BY) and (B) stream restoration efforts across the region (CC BY-NC). (C) Our mixed-methods research explores the biophysical interactions of flooding and stream restoration, integrating perspectives on community resilience (CC BY) and (D)relies on qualitative interviews with community residents and stream and land managers (CC BY-NC-ND 4.0)

Once the NSF-CNH2 proposal was funded, we began by returning to semi-structured interviews (see Johnston and Longhurst, Chapter 32), this time interviewing 16 Kickapoo River and Coon Creek Watershed stream restoration managers and flood decision makers about stream restoration planning and practice and changes in land use in the context of more frequent and more intense flooding. We recruited participants based on a combination of heterogeneity and nonproportional quota sampling to ensure a spectrum of opinions and broad representation (Patton 2002; Lindlof and Taylor 2011; vonHedemann, Chapter 38). This round of interviews built from our previous interviews in the Kickapoo Valley (n = 16), as well as the oral histories gathered in the Stories from the Flood archive (n = 100). We have been supplementing these interviews with ongoing engagement in the watershed: attending monthly watershed council meetings; speaking at flood mitigation alliance meetings; contributing to conservation tours in the watershed; and leading undergraduate students as they develop content on behalf of a local watershed council.

Ongoing visits to the watersheds resulted in the creation of an adjacent, unanticipated collaboration that has dramatically impacted our research: supporting the development, facilitation, and analysis of a community-driven oral narrative project with the newly formed Coon Creek Community Watershed Council (CCCWC). In Fall 2022 and Spring 2023, Caroline and UW-Madison graduate and undergraduate students worked alongside CCCWC on the effort, recording oral narratives about the past, present, and future of flooding and conservation in the watershed from 70 residents, working to archive those with the UW-La Crosse Oral History Program, and reporting out on the ways those oral narratives can shape the work of the council. We never could have anticipated this work when we drafted and submitted the NSF-CNH2 proposal as the watershed council did not then exist. But we have been working to continue supporting their identified efforts and needs by making sure our NSF-CNH2 work responds to those needs. Crucially, we see all of the activities we participate in and collaborate on in these watersheds as central to developing relationships with community members and continuing to tailor our work to the needs of the communities we are working with. This observation is important for any research that is making an intervention into a live project; the parameters surrounding the project will evolve both through and independently from the research being done. This requires a strong sensitivity and reflexivity to changing context to be built into project design.

This relational work offers important context for our research interviews, which allowed us to elicit and analyse flood-focused perspectives on the state of restoration science and management and upland conservation practices in the Kickapoo River and Coon Creek watersheds, specifically, and the Driftless Region more broadly and to create a comprehensive list of common arguments about stream stabilisation and flood impacts on stream restoration planning and construction. These interviews built, in part, from Caroline's past work exploring the complexities of restoration efforts in coastal New England (Druschke and Hychka 2015; Druschke et al. 2016; Hychka and Druschke 2017), and partly from existing conceptual frameworks for incorporating SES resilience into restoration efforts (Krievins et al. 2018) and for evaluating processes of SES transformation (Moore et al. 2014). These interviews are allowing us to consider how more frequent and more intense flooding across the UMRB is altering stream restoration practices in our study watersheds (Q1); letting stakeholders reflect on how flood-inflected stream restoration practices impact community-level climate vulnerability (Q2); and offering insight into how community members balance socio-environmental tradeoffs related to flood vulnerability and resilience across the reach and watershed scales (Q3).

As the project started, in addition to initiating this set of research interviews and ongoing community engagement, we also returned to our biophysical pilot data from 2018 and 2019. This involved pre-restoration geomorphic surveys of four restoration projects in the Kickapoo River watershed using an existing protocol to survey channel cross-sections (Fitzpatrick et al., 1998), supplemented with digital terrain and plant canopy models created using imagery from Uncrewed Airborne Systems (UAS), more commonly known as drones (see Kasvi, Chapter 45). Our ground-based surveys include high-resolution topographic surveys of the stream and floodplain using Real-Time Kinematic Global Positioning Systems (RTK-GPS) equipment and descriptions of geomorphic and habitat characteristics along regularly spaced cross sections. The imagery from the UAS was analysed using structure-from-motion, multi-view-stereo (SfM-MVS) photogrammetry (Kasvi, Chapter 45) which was used to create geo-referenced and geometrically corrected imagery (orthoimagery) and high-resolution, three-dimensional digital terrain and plant canopy models for each survey date. SfM-MVS photogrammetry is an established technique within the geomorphic and river sciences community (Fonstad et al., 2013) and has recently been made widely accessible at reasonable cost by the introductions of consumer grade UASs and open source/affordable software. Our research expands upon previous applications of SfM-MVS photogrammetry to monitor changes in pre- and post-restoration topography (Marteau et al. 2017) and riparian shading (Dugdale et al., 2019).

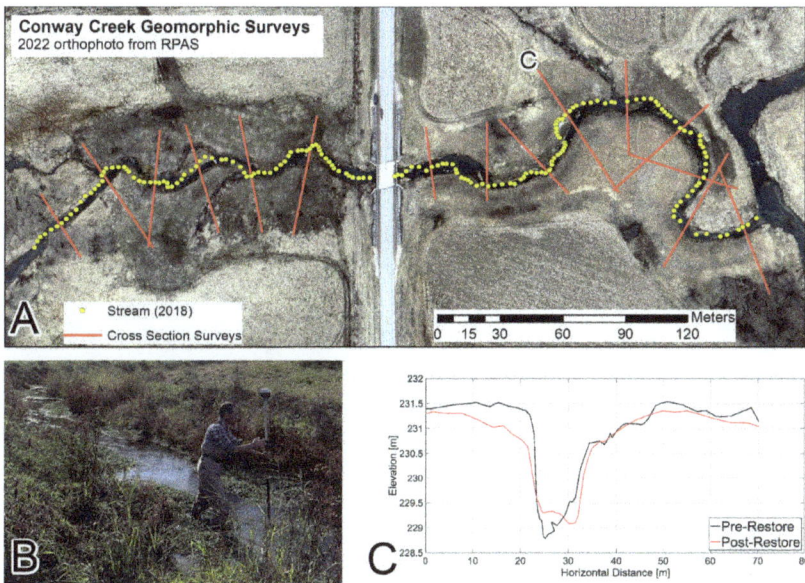

Fig. 15.4. Illustration of biophysical methods. (A) Locations of repeated cross-section surveys (red) and pre-restoration channel position (yellow) shown with post-restoration orthophoto generated from RPAS imagery; (B) Surveying cross-sections with RTK-GPS equipment (CC BY); (C) Comparison of representative cross section before and after restoration

The restoration projects we are monitoring consist of tree removal, bank sloping, and grass re-seeding. Comparison of pre- and post-project surveys allows us to gather precise data about changes in channel geometry resulting from typical stream restoration efforts in project watersheds to inform hydraulic modelling and to add material texture to stream restoration manager interview responses about the ways that more frequent and more intense flooding alters stream restoration practices in study watersheds (Q1). Periodic surveys using repeatable and low-cost methods (e.g., UAS flights) also allow monitoring of geomorphic and ecosystem change associated with large floods that are increasingly likely to occur over our research timeframe.

These pre- and post-restoration geomorphic surveys are also guiding our application of hydrologic and hydraulic models to inform restoration strategies and notably to estimate the impacts of changes in channel-floodplain geomorphology and land use/land management on flood peak attenuation (Q2). As mentioned above, we are making use of the recent watershed hydrologic and hydraulic modelling effort funded by the Natural Resources Conservation Service (NRCS) in our study region that generates and routes flood hydrographs associated with design rainfall events (e.g., 100-year, 24-hr event) through the stream networks of the Coon Creek and West Fork Kickapoo River watersheds using models from the U.S. Army Corps of Engineers' Hydrologic Engineering Center (HEC). This modelling effort is part of a watershed planning process (PLAN-EIS) in response to catastrophic flooding in August 2018 that led to the failure of five major flood control dams.

The first model (HEC-HMS) generated in the PLAN-EIS process is a basic rainfall-runoff algorithm that predicts the storm runoff response to different rainfall events in separate sub-watersheds and routes the storm runoff through the stream network (HEC 2018). The second model generated in the PLAN-EIS process is a widely used one-dimensional river hydraulic analysis (HEC-RAS) that estimates the water surface elevation and inundation extent across the stream network for different peak flows based on channel geometry, slope, and surface roughness (HEC 2016). The PLAN-EIS process created a baseline modelling scenario that represents the current land use/management and channel network characteristics. We are currently at work modifying the baseline conditions and creating potential stream-floodplain restoration scenarios with modified channel properties (geometry, roughness) across varying extents of the watershed's stream network (e.g., all headwater streams) depending on the scenario. Channel modifications are based on the pre- and post-restoration geomorphic surveys we have been conducting over the past several years to ground the model in real-world observations. We are comparing restored and baseline scenarios by determining the ratio in flood peak flows and difference in flood peak elevations at locations of interest across each stream network. Scenarios generated at this stage of the project that modify large extents of the stream network are meant to represent widespread implementation of restoration—something that is likely implausible—to illustrate the maximum impact of various stream-floodplain restoration strategies and inform new, more plausible scenarios generated by stakeholders through participatory workshops that we describe below.

Following this analysis of stream-floodplain flood interventions, we are expanding our focus to watershed-scale land use/land management interventions via watershed hydrologic modelling as an approach to reducing downstream flood peaks that complements stream-floodplain reconnection. We are using and expanding upon an existing decision-support platform (SmartScape™; Tayyebi et al. 2016) to allow community members to design land use and land management scenarios and to estimate their impacts on runoff generation. SmartScape™ is a web-based decision-support system that guides users through the creation of land use/management change scenarios by selecting different areas of the watershed, transforming their land use/management (e.g., from corn to pasture), and then estimating and visualising the changes in a suite of ecosystem services at the watershed scale (Tayyebi et al. 2016).

Our focus on reducing storm runoff and flood peaks will result in a complementary version of SmartScape™, called FloodScape, that includes estimates of runoff generation during large rain events (e.g., 100-year, 24-hour) at a fine spatial resolution (30-meter). The underlying model that estimates changes in storm runoff within the HEC-HMS model is the Curve Number method (SCS 1985), which is a simple empirical model that predicts storm runoff based on storm rainfall and the curve number parameter that represents various land use/management conditions. After a user creates a new scenario, FloodScape then executes the HEC-HMS model (above) with the updated curve number values for each sub-watershed. Outputs from FloodScape consist of total event runoff (overland flow) volume generation and peak event runoff rate; we will map the relative change in these two metrics from the baseline (current) condition to user-generated flood-reduction scenarios as a means of rapidly assessing the scenario's impact on downstream flooding. The creation of FloodScape will allow workshop participants to construct flood-reduction scenarios with various land cover configurations (e.g., pasture, corn-soy rotation) and management practices (e.g., cover crops) during community workshops and, critically, to view the change in flood outcomes.

Our experience with stakeholder interviews and community meetings is shaping our development of FloodScape to guide users in easily creating custom watershed scenarios of land management change and viewing the modelled consequences on downstream flood peaks at points of user interest along the stream network. We plan to trial FloodScape this year at the first of two iterative participatory modelling workshops with community stakeholders. We will begin the workshop with a demonstration of FloodScape and the scenario creation process. Then we will break into smaller groups and allow users to create their own scenarios, assess flood outcomes, and refine additional scenarios based on model results and discussion. Design and usability suggestions will be solicited at this first workshop and incorporated at the second workshop. FloodScape will be offered to local municipal planners and government managers as a tool for fostering conversations about flood vulnerability and resilience and for making challenging decisions about land use and restoration prioritisation throughout study watersheds.

Facilitated conversations at the conclusion of the stakeholder workshops will allow for community feedback on the participatory modelling exercises and will let us consider how community members balance socio-environmental tradeoffs related to flood vulnerability and resilience across the reach and watershed scales (Q3), as well as understanding whether our iterative, participatory, model-based stakeholder workshops impact participants in ways that might enable social-ecological system (SES) transformations (Q4). The idea here is that the results of this work will directly feed back into our study communities, helping to support their ongoing conversations about flood recovery and resilience.

The ongoing influence of community members and community needs on the development of our research questions, on the integration of our methodological approaches, on refining our modelling scenarios, and on providing feedback until our work most closely matches their needs, is central to our mixed methodological approach. This approach represents what our colleagues in the sciences might refer to as "actionable science", what Beier et al. (2017) refer to as "data, analyses, projections, or tools that can support decisions in natural resource management; it includes not only information, but also guidance on the appropriate use of that information". Our colleagues in the humanities and social sciences, in turn, might focus on this work as "community-based" or "community-driven". This is an approach that, as Carlson (2020) describes, tries "to generate possibilities for the region on behalf of community members", and, as Cruz and Bakken (2020) explain, is distinguished as research "done WITH community partners and not ON community members". This means that a researcher centres the community in their work and to the extent possible, engages with the community partner throughout the entire research process so that decisions are made jointly and on behalf of the community. The emphasis on community-driven, actionable work always grounds our collaborative, mixed methodological research.

Where do we go from here?

We conclude this chapter by thinking about where "we"—as in, Eric and Caroline—and "we"—you, the reader, and everyone working on mixed methodological, collaborative research—might go from here.

It is critically important to note that the work we detail above offers a relatively tidy snapshot of the actual messiness of this work. Our approach to this NSF-CNH2 project has already changed from what we had initially envisioned. The federal PLAN-EIS process we noted above has taken on a life of its own, involving more contention and more uncertainty than we imagined, and so we have found it important to let that process take a more central role in our work. So too has the development of the Coon Creek Community Watershed Council, which has become a major player in discussions about the past, present, and future of flooding and conservation in the region and didn't yet exist when our project began. And our timing has shifted as various components of our research depended on particular community decisions, consultant outputs, hiring processes, or simply our own capacity to conduct this work alongside the pressures of the rest of our jobs and lives during a pandemic. Regional land and water managers' concerns about the models generated in the PLAN-EIS process—the models on which our participatory modelling efforts depend—leave us scrambling to think critically about these models while still making our best efforts at building a useful decision support tool from them that can help facilitate urgent community conversations. This emphasises that the kind of mixed-methods research we describe here needs to be flexible, reflexive, and responsive and, as such, be quite different in nature to the traditional pre-planned research question and methodology approach many often expect. The ongoing challenge is to balance that nimble flexibility with some underlying structure to make clear progress within community concerns, academic knowledge-making, graduate degree timelines, and professional demands.

Our engagement with the Coon Creek Community Watershed Council has also revealed the potential utility and value of decision-support tools like FloodScape to help communities create what have recently been described as "conviction narratives". Conviction narrative theory is a psychological theory of decision making in the face of radical uncertainty that involves a mixture of storytelling, information, beliefs, models, and rules of thumb specifically in service of identifying a potential opportunity worth acting on (Thompson 2022; Johnson et al. 2023). In the wake of the dam breaches, and a fraught federal process focused on deciding how to proceed with the breached dams, community members are facing radical uncertainty about how to protect themselves in the face of increasingly frequent heavy rain events. FloodScape is designed to facilitate community-generated creation of land use/management change scenarios that can have meaningful (though highly variable) impacts on downstream flood hazards. Such a tool could serve as a helpful organising framework for community members to envision more flood-resilient futures and spark more positive stories of the future. It could also provide some underlying structure to aid in the kinds of responsive flexibility this work demands.

The need for us to adapt our plans and to retool our methodologies is made all the more urgent by our desire to be accountable to community members who are waiting on the next flood. And so we continue to adapt our work with a self-reflective eye, trying to remain present to the evolving needs of the watershed communities with whom we work. After all, what motivated the project was our investment in shifting resources towards under-resourced communities to support them as they consider how to recover from recent and persistent flooding and prepare for an uncertain future. As we have argued here, our integrated approach is critical to doing that work and trying to do it well. By fitting our various approaches together, we are working to build tools that centre community interests and needs and create processes that make space for community members to make sense of their changing landscapes and advocate for their perspectives, concerns, and needs, all while building community capacity and energy around shared futures. But to do that work, as we have argued above, our multiple disciplinary perspectives and methodologies need to be integrated from the very beginning of a project; they are essential to building its foundation. Such work also needs to be intentionally designed and continually refined to be responsive to community needs.

We feel lucky to have come together as intensely curious people who have been willing to put in the time and hard work to get familiar with each other's approaches, disciplines, methodologies, and areas of expertise. We have built from that work together to co-create new questions and new ways to approach those questions in ways that honour our shared investment in the lands and waters and communities of the Driftless Area. Our shared ethical orientation towards community-driven, justice-oriented, actionable work demands this collaborative, mixed methodological approach, and we hope the work we have shared here equips and inspires others to do the same.

Acknowledgements

The authors wish to thank our wonderful collaborator, Rebecca Lave, the University of Wisconsin-Madison students who have supported so much of this research, especially Emma Lundberg, Ben Sellers, Sydney Widell, Paige Stork, and Julia Buskirk, and all of our collaborators in southwestern Wisconsin. This material is based upon work supported by financial funding from the Kickapoo Valley Reforestation Fund in UW-Madison's College of Agricultural and Life Sciences, UW-Madison's Morgridge Center for Public Service, UW-Madison's Vice Chancellor for Graduate Research and Education, and the National Science Foundation under award #2009353, CNH2-S: Interactive Dynamics of Stream Restoration and Flood Resilience in a Changing Climate.

References cited

Anderson, D. 2016. *Economic Impact of Recreational Trout Angling in the Driftless Area* (Report to Driftless Area Restoration Effort).

Beier, P., L.J. Hansen, L. Helbrecht, and D. Behar. 2017. 'A how-to guide for coproduction of actionable science', *Conservation Letters*, 10.3, pp. 288–296. https://doi.org/10.1111/conl.12300

Belby, C.S., L.J. Spigel, and F.A. Fitzpatrick. 2019. 'Historic changes to floodplain systems in the Driftless Area', *The Physical Geography and Geology of the Driftless Area: The Career and Contributions of James C. Knox*, 543, p. 119. https://doi.org/10.1130/2019.2543(07)

Booth, E.G., J. Qiu, S.R. Carpenter, J. Schatz, X. Chen, C.J. Kucharik, and M.G. Turner. 2016. 'From qualitative to quantitative environmental scenarios: translating storylines into biophysical modeling inputs at the watershed scale', *Environmental Modelling and Software*, 85, pp. 80–97. https://doi.org/10.1016/j.envsoft.2016.08.008

Carlson, E.B. 2020. 'Embracing a metic lens for community-based participatory research in technical communication', *Technical Communication Quarterly*, 29.4, pp. 392–410. https://doi.org/10.1080/10572252.2020.1789745

Carpenter, S.R., E.G. Booth, S. Gillon, C.J. Kucharik, S. Loheide, A.S. Mase, and C.B. Wardropper. 2015. 'Plausible futures of a social-ecological system: Yahara watershed, Wisconsin, USA', *Ecology and Society*, 20.2. http://doi.org/10.5751/ES-07433-200210

Chakov, A., Chang, T., Covey, H., Dickson, T., Goggins, S., Harris, N., Purna, S., Widell, S. and Druschke, C.G., Chapter 33, this volume. 'Oral history'.

Cinderby, S. 2010. 'How to reach the 'hard-to-reach': the development of Participatory Geographic Information Systems (P-GIS) for inclusive urban design in UK cities', *Area*, 42.2, pp. 239–251. https://doi.org/10.1111/j.1475-4762.2009.00912.x

Couleecap. 2016. '50th Anniversary and 2015 Annual Report', https://www.couleecap.org/uploads/1/2/2/5/122572786/arp2015.pdf

Cruz, E. and L. Bakken. 2020. 'Community Guidelines for Engaging with Researchers and Evaluators: A Toolkit for Community Agencies, Organizations and Coalitions', https://ictr.wisc.edu/documents/community-guidelines-for-engaging-with-researchers-and-evaluators/

Dugdale, S.J., I.A. Malcolm, and D.M. Hannah. 2019. 'Drone-based Structure-from-Motion provides accurate forest canopy data to assess shading effects in river temperature models', *Science of the Total Environment*, 678, pp. 326–340. https://doi.org/10.1016/j.scitotenv.2019.04.229

Druschke, C.G., E.G. Booth, and E. Lundberg. 2019. 'Q-rhetoric and controlled equivocation: Revising "the scientific study of subjectivity" for cross-disciplinary collaboration', *Technical Communication Quarterly*, 28.2, pp. 137–151. https://doi.org/10.1080/10572252.2019.1583377

Druschke, C.G., and K.C. Hychka. 2015. 'Manager perspectives on communication and public engagement in ecological restoration project success', *Ecology and Society*, 20.1. http://doi.org/10.5751/ES-07451-200158

Druschke, C.G. and B. McGreavy. 2016. 'Why rhetoric matters for ecology', *Frontiers in Ecology and the Environment*, 14.1, pp. 46–52. https://doi.org/10.1002/16-0113.1

Druschke, C.G., L.A. Meyerson, and K.C. Hychka. 2016. 'From restoration to adaptation: the changing discourse of invasive species management in coastal New England under global environmental change', *Biological Invasions*, 18.9, pp. 2739–2747. https://doi.org/10.1007/s10530-016-1112-7

Fitzpatrick, F.A., I.R. Waite, P.J. D'Arconte, M.R. Meador, M.A. Maupin, and M.E. Gurtz. 1998. 'Revised methods for characterizing stream habitat in the National Water-Quality Assessment Program', *Water-Resources Investigations Report*, 98, p. 4052. https://doi.org/10.3133/wri984052

Fonstad, M.A., J.T. Dietrich, B.C. Courville, J.L. Jensen, and P.E. Carbonneau. 2013. 'Topographic structure from motion: a new development in photogrammetric measurement', *Earth Surface Processes and Landforms*, 38.4, pp. 421–430. http://doi.org/10.1002/esp.3366

Forrester, J., B. Cook, L. Bracken, S. Cinderby, and A. Donaldson. 2015. 'Combining participatory mapping with Q-methodology to map stakeholder perceptions of complex environmental problems', *Applied Geography*, 56, pp. 199–208. https://doi.org/10.1016/j.apgeog.2014.11.019

Frans, C., E. Istanbulluoglu, V. Mishra, F. Munoz-Arriola, and D.P. Lettenmaier. 2013. 'Are climatic or land cover changes the dominant cause of runoff trends in the Upper Mississippi River Basin?', *Geophysical Research Letters*, 40.6, pp. 1104–1110. http://doi.org/10.1002/grl.50262

Gottschalk Druschke, C., E. Booth, R. Lave, S. Widell, E. Lundberg, B. Sellers, and P. Stork. 2023. "Grass versus trees: A proxy debate for deeper anxieties about competing stream worlds." *Environment and Planning E: Nature and Space*, 7.2. http://doi.org/10.1177/25148486231210408

Gottschalk Druschke, C., T. Dean, M. Higgins, M. Beaty, L. Henner, R. Hosemann, and T. Woser. 2022a. 'Stories from the flood: Promoting healing and fostering policy change through storytelling, community literacy, and community-based learning', *Community Literacy Journal*, 16.2, p. 35. http://doi.org/10.25148/CLJ.16.2.010622

Gottschalk Druschke, C., T. Dean, R. Alsbury, J. Buskirk, M. Higgins, E. Johnson, S. Koretskov, B. Steinmetz, E. Waldinger, S. Wood, and C. Zuleger. 2022b. 'Cultivating empathy on the eve of a pandemic', *Reflections: A Journal of Community-Engaged Writing and Rhetoric*, 21.1. https://reflectionsjournal.net/2022/02/cultivating-empathy/

Hychka, K. and C.G. Druschke. 2017. 'Adaptive management of urban ecosystem restoration: Learning from restoration managers in Rhode Island, USA', *Society and Natural Resources*, 30.11, pp. 1358–1373. https://doi.org/10.1080/08941920.2017.1315653

Hydrologic Engineering Center (HEC). 2016. *HEC-RAS River Analysis System user's manual. Version 5.0, February 2016.* (U.S. Army Corps of Engineers, Hydrologic Engineering Center).

Hydrologic Engineering Center (HEC). 2018. *HEC-HMS Hydrologic Modeling System user's manual. Version 4.3, September 2018* (U.S. Army Corps of Engineers, Hydrologic Engineering Center).

Jha, M., Z. Pan, E.S. Takle, and R. Gu. 2004. 'Impacts of climate change on streamflow in the Upper Mississippi River Basin: A regional climate model perspective', *Journal of Geophysical Research: Atmospheres*, 109.D9. http://doi.org/10.1029/2003JD003686

Jha, M., J.G. Arnold, P.W. Gassman, F. Giorgi, and R.R. Gu. 2006. 'Climate Change Sensitivity Assessment on Upper Mississippi River Basin Streamflows Using SWAT', *JAWRA Journal of the American Water Resources Association*, 42.4, pp. 997–1015. http://doi.org/10.1111/j.1752-1688.2006.tb04510.x

Johnson, S.G.B., A. Bilovich, and D. Tuckett. 2023. 'Conviction Narrative Theory: A theory of choice under radical uncertainty', *Behavioral and Brain Sciences*, 46. http://doi.org/10.1017/S0140525X22001157

Johnston and Longhurst, Chapter 32, this volume. 'Interviews: Structured, semi-structured and open-ended'.

Kasvi, E., Chapter 45, this volume. 'Uncrewed Airborne Systems'.

Kemmis, S., R. McTaggart, and R. Nixon. 2014. 'Introducing critical participatory action research', in *The Action Research Planner: Doing Critical Participatory Action Research*, ed. by S. Kemmis, R. McTaggart, and R. Nixon (Springer), pp. 1–31. https://doi.org/10.1007/978-981-4560-67-2_1

Kindon, S., R. Pain, and M. Kesby. 2007. *Participatory Action Research Approaches and Methods: Connecting People, Participation and Place* (Routledge). https://doi.org/10.4324/9780203933671

Knox, J.C. 2006. 'Floodplain sedimentation in the Upper Mississippi Valley: Natural versus human accelerated', *Geomorphology*, 79.3-4, pp. 286–310. https://doi.org/10.1016/j.geomorph.2006.06.031

Knox, J.C. 2019. 'Geology of the Driftless Area', in *The Physical Geography and Geology of the Driftless Area: The Career and Contributions of James C. Knox*, ed. by E.C. Carson, J.E. Rawling, III, J.M. Daniels, and J.W. Attig (Geological Society of America). https://doi.org/10.1130/2019.2543(01)

Krievins, K., R. Plummer, and J. Baird. 2018. 'Building resilience in ecological restoration processes: a social-ecological perspective', *Ecological Restoration*, 36.3, pp. 195–207. https://doi.org/10.3368/er.36.3.195

Lane, S.N. and Lave, R., Chapter 3, this volume. 'Frames, disciplines and mixing methods in environmental research'.

Lane, S.N., Chapter 30, this volume. 'Hydraulic modelling'.

Lane, S.N., N. Odoni, C. Landström, S.J. Whatmore, N. Ward, and S. Bradley. 2011. 'Doing flood risk science differently: an experiment in radical scientific method', *Transactions of the Institute of British Geographers*, 36.1, pp. 15–36. https://doi.org/10.1111/j.1475-5661.2010.00410.x

Latour, B. 2003. 'Why has critique run out of steam? From matters of fact to matters of concern', *Critical Inquiry*, 30, pp. 225–48. https://doi.org/10.1086/421123

Lave, R., C. Biermann, and S.N. Lane. 2018. *The Palgrave Handbook of Critical Physical Geography* (Palgrave Macmillan). https://doi.org/10.1007/978-3-319-71461-5

Lindlof, T.R. and B.C. Taylor. 2011. *Qualitative Communication Research Methods* (Sage Publications).

Lu, J. 2019. 'Kickapoo Valley faces long, winding road to flood recovery', *La Crosse Tribune*. https://lacrossetribune.com/news/local/kickapoo-valley-faces-long-winding-road-to-flood-recovery/article_632e118d-1b7e-5493-81e9-35362d9e07f9.html

Lundberg, E., C. Gottschalk Druschke, and E.G. Booth. 2022. 'A Q-method survey of stream restoration practitioners in the Driftless Area, USA', *River Research and Applications*, 38.6, pp. 1090-1100. http://doi.org/10.1002/rra.3971

Marteau, B., D. Vericat, C. Gibbins, R.J. Batalla, and D.R. Green. 2017. 'Application of Structure-from-Motion photogrammetry to river restoration', *Earth Surface Processes and Landforms*, 42.3, pp. 503–515. http://doi.org/10.1002/esp.4086

Mazzotta, M., J. Bousquin, C. Ojo, K. Hychka, C. Gottschalk Druschke, W. Berry, and R. McKinney. 2016. 'Assessing the benefits of wetland restoration: a rapid benefit indicators approach for decision makers', in *Narragansett, RI: US EPA, Office of Research and Development, National Health and Environmental Effects Research Laboratory, EPA/600/R-16/084* (US EPA Office of Research and Development).

Mazzotta, M., J. Bousquin, W. Berry, C. Ojo, R. McKinney, K. Hyckha, and C.G. Druschke. 2019. 'Evaluating the ecosystem services and benefits of wetland restoration by use of the rapid benefit indicators approach', *Integrated Environmental Assessment and Management*, 15.1, pp. 148–159. https://doi.org/10.1002/ieam.4101

Miesen, F., Chapter 10, this volume, 'Fieldwork safety planning and risk management'.

Moore, M.L., O. Tjornbo, E. Enfors, C. Knapp, J. Hodbod, J.A. Baggio, A. Norström, P. Olsson, and D. Biggs. 2014. 'Studying the complexity of change: toward an analytical framework for understanding deliberate social-ecological transformations', *Ecology and Society*, 19.4, p. 54. http://doi.org/10.5751/ES-06966-190454

Norstrom, A.V., et al. 2020. 'Principles for knowledge co-production in sustainability research', *Nature Sustainability*, 3.3, pp. 182–190. https://doi.org/10.1038/s41893-019-0448-2

Patton, M.Q. 2002. *Qualitative Research and Evaluation Methods* (Sage Publications).

Rajib, A. and V. Merwade. 2017. 'Hydrologic response to future land use change in the Upper Mississippi River Basin by the end of 21st century', *Hydrological Processes*, 31.21, pp. 3645–3661. http://doi.org/10.1002/hyp.11282

Rittel, H.W. and M.M. Webber. 1973. 'Dilemmas in a general theory of planning', *Policy Sciences*, 4.2, pp. 155–169. https://doi.org/10.1007/BF01405730

Schuttenberg, H.Z. and H.K. Guth. 2015. 'Seeking our shared wisdom: A framework for understanding knowledge coproduction and coproductive capacities', *Ecology and Society*, 20.1.http://doi.org/10.5751/ES-07038-200115

SCS National Engineering Handbook. 1985. 'Section 4: Hydrology' (Soil Conservation Service).

Star, S.L. and J.B. Griesemer. 1989. 'Institutional ecology, "translation," and boundary objects: Amateurs and professionals in Berkeley's Museum of Vertebrate Zoology, 1907-39', *Social Studies of Science*, 19.3, pp. 387–420. https://doi.org/10.1177/030631289019003001

Stock, P. and R.J. Burton. 2011. 'Defining terms for integrated (multi-inter-trans-disciplinary) sustainability research', *Sustainability*, 3.8, pp. 1090–1113. https://doi.org/10.3390/su3081090

Stoecker, R. 2003. 'Community-based research: From practice to theory and back again', *Michigan Journal of Community Service Learning*, 9.2. http://hdl.handle.net/2027/spo.3239521.0009.204

Strand, K., S. Marullo, N. Cutforth, R. Stoecker, and P. Donohue. 2003. 'Principles of best practice for community-based research', *Michigan Journal of Community Service Learning*, 9.3. http://hdl.handle.net/2027/spo.3239521.0009.301

Tayyebi, A., J.J. Arsanjani, A.H. Tayyebi, H. Omrani, and H.S. Moghadam. 2016. 'Group-based crop change planning: Application of SmartScape™

spatial decision support system for resolving conflicts', *Ecological Modelling*, 333, pp. 92–100. https://doi.org/10.1016/j.ecolmodel.2016.04.018

Thompson, E. 2022. *Escape from Model Land* (Basic Books).

Trimble, S.W. and S.W. Lund. 1982. *Soil Conservation and the Reduction of Erosion and Sedimentation in the Coon Creek Basin, Wisconsin* (United States Geological Survey). https://doi.org/10.3133/pp1234

vonHedemann, N., Chapter 38, this volume. 'Sampling'.

Wilson, G. and C.G. Herndl. 2007. 'Boundary objects as rhetorical exigence: Knowledge mapping and interdisciplinary cooperation at the Los Alamos National Laboratory', *Journal of Business and Technical Communication*, 21.2, pp. 129–154. http://doi.org/10.1177/1050651906297164

WICCI. 2021. *Wisconsin's Changing Climate: Impacts and Solutions for a Warmer Climate.* (Wisconsin Initiative on Climate Change Impacts Nelson Institute for Environmental Studies). https://wicci.wisc. edu/2021-assessment-report/

16. Using mixed methods to confront disparities in public health interventions in urban community gardens

Melanie Malone

While this chapter is about mixed-methods research, it is also about how to conduct research with a critical lens that identifies how important it is to think through equitable methods (see Meadow et al., Chapter 5). It is also a chapter that illuminates the challenges of doing interdisciplinary public health research in unjust systems. Therefore, my first point in this introduction is about why I use mixed methods. Mixed-methods research is a powerful tool for making interventions in communities impacted by environmental problems that have yet to be investigated thoroughly. Further, when communities are relegated to research subjects, and only one method (e.g., statistics or environmental sampling) is used to assess harm, this usually does not serve communities well. This can be particularly problematic in environmental health studies, which are often wrongly assumed to be objective and without interest, can be extractive, and involve researchers telling the communities what they are experiencing, rather than taking into account methods or questions that the community may request and help formulate (Yen-Kohl and Collective 2016).

I am primarily a physical scientist, and although my research has often been about soils, water, and sediment, my research has also been situated in the U.S., and therefore has always intersected with questions about power and race, even when I did not initially have those questions in mind. And while my research topics have varied over the years and have mostly centred on how contaminants move through the environment, or how someone can trace erosion, my findings are always

   https://doi.org/10.11647/OBP.0418.16

impacted by larger social systems that invariably have to do with who has access to land, manages land, and exerts power over decisions in policy and practice.[1] My research, and many other studies, demonstrates that having a solid grasp of multiple techniques and methodologies can illuminate not just physical environmental factors that contribute to environmental issues, but also how important the social systems are that drive how people interact with and shape the environment.

The second point I want to highlight in this introduction is that methods themselves can enact harm and burden, including on the researcher themselves. Relatedly, it is always difficult to know if one is using the "right" methods in ways that will serve communities best, and ones that do not exacerbate harms. For this reason, being transparent about methodology and its limitations are important, because communities can help identify methods that will serve them best and not replicate problematic ways of doing research. Many challenges can arise when thinking about how to do research that equitably benefits communities. This can be particularly vexing for those of us who are researchers of colour, who often have to not only conduct our research, but also translate our work to those who are not as equitably minded or informed.

Beyond these considerations, mixed-methods research is often very time-consuming and intensive. It requires the researcher to come up with creative solutions for deciding what methods are best for their research and also how to make the research happen. This can be rewarding when all of the pieces come together, but also can create a major burden on the researcher, who is often desperate to do complex research well because they know it is important, but who is also fatigued by the process of mixed-methods research. And circling back to issues of equity in environmental justice research, this can be an even greater burden for marginalised scholars, such as women, people of colour, and transgender individuals who might not have the resources, connections with researchers who are well resourced, or might not even be welcome

---

1   For excellent examples of these dynamics, see Levi Van Sant's research (Van Sant, 2021) on how slavery, access to prime agricultural land, and the development of the U.S. soil classification influences governance to this day, or Megan Ybarra's research (Ybarra, 2021) on how broken treaties with Puyallup Tribe have resulted in the illegal detention of immigrants and inadequate superfund cleanup in a toxic location in the Puget Sound.

or taken seriously because of their background (Ali et al. 2021; Demery and Pipkin 2021; Venton 2020; Wölfle-Hazard 2022).

In this chapter, I will use a project that I have been conducting for multiple years to illustrate some broader points about why mixed methodologies and transparency about methodologies are important for equitable research. I will then discuss some of the challenges in conducting mixed methodology, and then present some of the approaches I have found useful while doing the research.

The project

Seattle, Washington (and the surrounding metropolitan area) has a contamination problem in its urban community gardens (UCGs) (for extensive details see Malone 2021a; 2021b; Malone et al. 2023). I will discuss some of the details of that research, but because this is a methods book, I will mostly elaborate on the methods I chose to use, how I integrated them, and why.

Briefly, I started researching contamination in urban soils in UCGs in Seattle in 2019. After finding widespread and concerning levels of a number of contaminants (arsenic, lead, petroleum products, and glyphosate) in what were supposed to be organic gardens where synthetic pesticides and fertilisers are prohibited, I expanded my study to include more UCGs than I started with. My main goal for this project was to figure out how to inform and protect people from contamination in UCGs where, in many cases, they were not even aware that the contamination was present in their space because of systemic injustices they encountered. Therefore, understanding why people are unaware of contamination in their spaces, and how to do equitable research projects with them to redress the issue, required multiple methods.

These methods include a mix of field work collecting soil and plant samples for analysis of contaminants, interviewing tenants of UCGs about what they add to their soils and composts, and about what they physically do in their gardens to understand contaminant pathways and exposure factors (more on that under conventional risk assessment). It also entails analysing samples in a laboratory and synthesising data to understand what patterns emerge about soils, compost, and plants that uptake contaminants; taking drone photos (Kasvi, Chapter 45) to delineate concentrations of contaminants in soils and plants; community organising and feedback sessions to translate results back to farmers and gardeners; and meetings to advocate for farmers and gardeners when municipal governments are not assisting with contamination concerns. I also have to understand the spatial elements of my research, as some of my research sites are in more polluted areas of Seattle than others (such as the Duwamish River Superfund Site) or have site histories that have left legacy contamination at a site. These are sometimes answered by long-term tenants of gardens, but usually take some investigative detective work, as well as viewing old maps and records about the sites where UCGs are located. Hence, it is a very intensive, mixed-methods research project.

All of my research and my choice of methods have been driven by a number of research gaps and concerns that my community partners have brought to my attention. These are detailed in the following sections.

Conventional risk assessment

In most risk assessment research, risk is defined as the likelihood that something will cause harm. The U.S. Environmental Protection Agency (EPA) defines risk more specifically as "The probability of an adverse effect in an organism, system, or population caused under specified circumstances by exposure to an agent" (EPA 2011: G-9). Most governing agencies that calculate risk do it by basing it on a number of exposure scenarios, meaning scenarios in which a person will interact with a medium (e.g., soil, water, air) carrying specific concentrations of a contaminant. When one looks at screening tables from the EPA, there are different screening levels for residential settings than there are for someone working in a commercial setting, for instance. Residential scenarios usually set the lowest concentrations for screening levels because the ways someone might interact with a soil in a playground or in their backyard, such as through play or digging for yard work, is different than when they might be working on a construction site, where presumably they are wearing personal protective equipment and are limiting their exposure to a contaminant. However, there is no exposure scenario for UCGs, and therefore no implicit assumptions about how someone may be getting exposed to a contaminant in the setting. The EPA defines these assumptions as exposure factors, which "are factors related to human behaviour and characteristics that help determine an individual's exposure to an agent" (EPA 2011: viii).

Risk calculations are based on exposure, duration, and the concentration of a specific contaminant. For example, the EPA calculates what is known as the lifetime cancer risk for a population with an average life expectancy of 70 years as:

$$\Sigma \ (\text{Exposure} \times \text{Duration}/70 \text{ years} \times \text{Potency} \times \text{Age dependent adjustment factor (ADAF))}^2$$

The problems with conventional risk assessment

As my research has evolved, it has become clear that missing exposure factors are at the crux of helping people define what is safe in their gardens. A variety of guidebooks and pamphlets from state and federal agencies as well as peer reviewed papers do have a list of generic ways soil might enter a person's body (e.g., touching or digging in soils, inhaling dust) or enter a person's home after being in a UCG (e.g., tracking dust/soil into homes with shoes), but none of them actually quantify how often this occurs, how long people are interacting with contaminated soil and plants, and what other ways people might be encountering contaminants in their UCGs (EPA 2015; Shayler et al. 2009). Therefore, I have created surveys and interviews to understand what ways people interact with their soils and plants in their gardens. This requires asking very specific questions about how long people are in their UCGs, how long they touch soils with bare skin, whether or not they wash their hands after working with soils, if they wear a mask, if they wash their fruits and vegetables out of their UCGs before eating them, and how much and how often they eat specific foods out of their gardens because each different type of plant stores contaminants differently. And that is not even the full list of questions I ask participants.

Because of my previous years of research, I have the concentrations of contaminants in soils and plants in multiple gardens, but can only apply loose residential screening levels which are not reflective of how a gardener or farmer would interact with a soil or plant in their UCGs. Having more specific information about the ways farmers and gardeners

---

2   Age dependent adjustment factor (ADAF)—The ADAF is applied to account for differences in toxicity and exposure based on a person's age.

in UCGs interact with soils and plants in their spaces helps me calculate risk more definitively, because then I know more about how they come into contact with contamination. And to larger goals, gathering this information will also help inform a major gap in policy on UCGs, as there are currently no federal or state guidelines for contaminants in UCGs because the information that contains exposure factors has not been available for these spaces. At present, the EPA and other governing bodies make assumptions about how long a person breathes in or eats dust or soil, or touches a contaminant in a soil which may be absorbed into their skin. They also assume how much people will eat out of their gardens. But these assumptions are based on a very limited number of studies, and even now people are trying to make new models, which are often based on other models, for how much contamination one might get exposed to by breathing or eating soils (Lupolt et al. 2022). Hence, the surveys I have created intentionally ask questions that will shed light on these matters, but the surveys also ask questions about exposure pathways that might not be on the radar of those studying exposure routes in an effort to understand the many ways people are exposed to contaminants in UCGs. Therefore, I have also included questions in my surveys and interviews that ask people about other ways they think they might be getting exposed to contaminants. Analysing these responses and including them in the study is critical for more equitable ways to redress contamination in UCGs.

But how does one know that even with all of this information, everyone will experience the same adverse effect? The answer is: they will not. Thankfully, the EPA has compiled an "Exposure Factors Handbook" (2011). Although the 1,432-page handbook is intimidating in size, it contains a number of useful tables of how different demographics (e.g., children, adults, and people of various body weights) will interact with specific contaminants, even going so far as to describe how long various body parts will typically be in contact with a contaminant and absorb it. It even defines how much surface area of skin is involved in the calculation. If one flips through the handbook, one will see categories of studies that the EPA has used to figure out exposure for various body parts on different ages of people. While the EPA has been inclusive of a wide range of ages and body weights, notably, the studies that the EPA relies on to make assumptions generally do not account for race, ethnicity, non-binary gender identities, underlying conditions, or genetic pre-dispositions to health issues caused by exposure to contamination. Therefore, even this is limited information.

This is something I wrestle with, and many (CATO Institute 1998; Cram 2016; Fagin 2012; Felter and Dourson 1998) have argued convincingly that risk assessment is not a true science, but rather a quasi-science or trans-science. This is because questions asked in risk assessment can be asked in a scientific way, such as, "how many people will die from exposure to lead in a car factory?", but cannot be answered because most risk assessment relies on known concentrations of contaminants with acute exposures, rather than prolonged exposure to various doses of contaminants (in scientific terms the latter scenario is known as a non-monotonic response). Moreover, questions about how many deaths from exposure are *acceptable* cannot ever be answered with science.

What is acceptable risk?

Yet, even with identification of exposure factors, I am only able to achieve part of my goal in studying risk in UCGs. The EPA itself has acknowledged an issue with a lack of guidelines in UCGs and has published a guide called "Brownfields and Urban Agriculture: Interim Guidelines for Safe Gardening Practices" (2011), but still has not offered definitive guidance for UCGs. Beyond the issues that arise when no guidance is available that I discussed previously, defining what is safe is incredibly difficult because both risk and harm are subjective depending on who one is speaking to. There are actual definitions for risk and harm in risk assessment, but these definitions do not capture whether concentrations of contaminants in UCGs (or any scenario where people are exposed to contamination) is acceptable to the person who is exposed. Having no guidelines for what acceptable concentrations in UCGs is only part of the problem. Understanding what acceptable risk is for the communities where I research is the biggest issue.

For example, if one looks at the image below from the EPA "Lead in Soil Guide for Region III" (EPA 2020), there is a scale of soil lead limits for growing food in soils that contain lead. In my study, there are a number of UCGs that have lead concentrations in soil in the 400 to 600 range, but the majority are in the lower ranges where this diagram indicates it is "safe to garden with children—all food crops safe". This diagram has always piqued my interest for a few reasons. The first is that in the same two-page guide where this diagram is located, there is a statement that says, "There is no single threshold that defines acceptable levels of lead in soil. State and federal regulatory and guidance values may only address specific situations and are mostly focused on cleaning up industrial properties" (p. 1). Therefore, it is interesting and alarming to me that standards for cleaning up industrial properties would be considered appropriate standards for where children can play and where people can grow food. I also am intrigued by this diagram because the EPA has other guidelines for lead in soils that can cause harm, which are much lower and in the range of the category of "safe" soils in this diagram. For example, the EPA's Ecological Soil Screening Level (ECO-SSL) "are concentrations of contaminants in soil that are protective of ecological receptors that commonly come into contact with soil or ingest biota that live in or on soil' (EPA 2003). The EPA ECO-SSL ranges from 11 mg/kg to 120 mg/kg depending on species (avian, mammalian, plant), which is well below the 400 mg/kg listed here (EPA 2005).

U.S. ENVIRONMENTAL PROTECTION AGENCY, REGION III

Recommended soil lead level limits for growing food in gardens

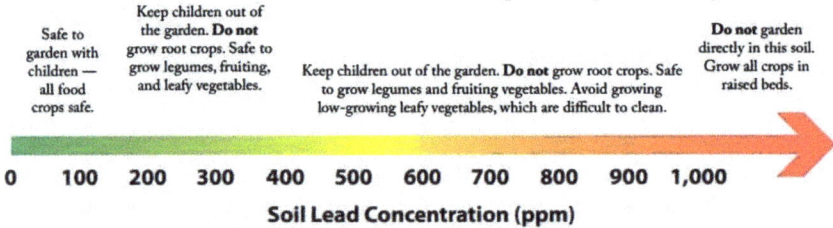

Safe to garden with children — all food crops safe.

Keep children out of the garden. **Do not** grow root crops. Safe to grow legumes, fruiting, and leafy vegetables.

Keep children out of the garden. **Do not** grow root crops. Safe to grow legumes and fruiting vegetables. Avoid growing low-growing leafy vegetables, which are difficult to clean.

Do not garden directly in this soil. Grow all crops in raised beds.

0   100   200   300   400   500   600   700   800   900   1,000

Soil Lead Concentration (ppm)

Assume soil testing for lead with EPA Method 3051A

From Kansas State University Agricultural Experiment Station and Cooperative Extension Service, and adopted by the Penn State Cooperative Extension Service

Fig. 16.1 US EPA Allowable soil lead levels. From https://www.epa.gov/sites/default/files/2020-10/documents/lead-in-soil-aug2020.pdf

Furthermore, the 400 mg/kg national EPA guideline for lead in soil, which I do not use as an acceptable guideline when talking to farmers and gardeners because it is far above levels shown to cause human harm, is widely used and cited as an acceptable level for lead in soils in many pamphlets on safe gardening, including the pamphlet where I took this image from. However, that 400 mg/kg SSL was developed in 1994 by the EPA to protect children from unacceptable exposures to lead in residential scenarios that do not encompass how children would interact with soils in a UCG. Since then, numerous studies have shown that there is no safe concentration of lead for children (meaning no adverse health effect) (Filippelli and Laidlaw 2010; Miranda et al. 2007; Mitchell et al. 2014), and the CDC has reduced the blood lead reference value from 10 µg/dL to 5 µg/dL (CDC 2021). And while only a small (2%) population of children aged 0 to 5 in the U.S. are poisoned with lead, in the Eastern and Midwest regions of the U.S., that figure is 15–20% because of the close contact that children have with lead-contaminated urban soil (Filippelli and Laidlaw 2010; Steffan et al. 2018). Children in UCGs also have close contact with soils and are often breathing in dust, which is one of the most prevalent pathways for lead poisoning in children (CDC 2022), as well as eating soils and unwashed produce that may contain lead.

How mixing methods and transparency about methodology helps with risk assessment

Therefore, in an effort to make an intervention, my surveys, interviews, and community feedback sessions allot time for asking respondents about what their definitions of risk and harm are. I also ask communities what their views are about the lack of guidelines, if they think current guidelines are adequate, and most importantly, what would make them feel safe in their UCGs. To be sure, there is a place for numbers, calculations, and risk assessment. If I did not believe so, I would not have written about it in this chapter and I would not be collecting data to plug into (admittedly flawed) risk calculations. But there should also be a place for the communities impacted by contamination to have a say about what the standards should be, and how conservative they are. In my multiple years of doing research in a variety of contaminated settings, I have seen this theme emerge over and over again. In the Duwamish and Portland Harbor Superfund Sites, I have seen tribes and immigrant populations fight for lower allowable concentrations of PCBs in river sediments and fish because they rely on them for food and sacred connections. In 2021, I saw a local non-profit group fight and question why the limit for cancer-causing polycyclic aromatic hydrocarbons (cPAHs) in waterways in Washington State was being made less conservative, and they rightly pointed out that people of color for whom they were advocating are disproportionately impacted by those same contaminants (Winters 2021).

In UCGs where people are interacting with, breathing, and literally eating soil, and where people are often relying on food from these spaces to supplement their grocery needs, it seems obvious that they should also have a say in what is acceptable risk in their UCGs. As other researchers (Sarewitz 2004; Ybarra 2021) have highlighted, there comes a point when more data is not useful; rather, you need to know and understand what community values are and be transparent about that. In this study, I make transparent what my values are, both to my community partners and to the City, and I ask them to do the same so that we can at least begin to navigate, and hopefully overcome, ideological barriers that ultimately harm people.

Some reasons you should collect your own data: Values in physical research

But now that I have just said that you do not always need more data to prove a point, I will contradict myself a bit and say that there is real value in collecting your own data for certain types of projects. Part of the reason that marginalised groups of people are exposed to contaminants at disproportionate rates is because they never had someone who looked like them collecting data, interpreting it, and sharing it back with the community. Numerous studies have shown (Flynn n.d.; Lewis et al. 2020; Pulido and De Lara 2018; Yen-Kohl and Collective 2016) that communities impacted by contamination often have results interpreted in ways that are flawed and not in their best interest. As Ellen Yen-Kohl writes, communities are tired of experts telling them "what they are experiencing and, more often, what they are not experiencing" (Yen-Kohl and Collective 2016). Furthermore, studies on contaminants in marginalised communities are often limited in their ability to reveal inequity from the start because the research questions asked are not those that are actually useful to the communities impacted. The way risk is calculated is highly subjective as well, and is based on many implicit assumptions that are not made readily apparent, and are generally not set up to protect people in the ways they expect to be protected (Cram 2016). Therefore, collecting your own data means that you should know its limitations, but more importantly its strengths, and what you can defend when advocating for a community.

In my own research project, the physical samples I collect include soil samples and plant samples, and I process some of the samples in my laboratory to understand soil characteristics that might influence contaminant mobility or uptake. I also do preliminary processing of my samples to ensure that they are ready to be sent to laboratories in the best condition possible for contaminant analysis. On that note, I choose laboratories that I trust and have a proven track record with my samples. I know that the labs I use have machinery that can detect very low concentrations of contaminants that can cause harm, whereas many laboratories used by other researchers or state agencies do not. The laboratories I use for contaminant analysis also have National Environmental Laboratory Accreditation Program (NELAP) certification. When I collect samples, I know I am not cross-contaminating my samples because I wear sterilised, nitrile gloves; I properly decontaminant my sampling equipment; I print out field sheets to collect field conditions that might influence results; and I train student interns and community partners on how to properly sample and decontaminant equipment as well.

I also make transparent in my methodology the rationale for what I collect, and where I collect samples. For example, I choose mostly BIPOC UCGs because contamination in locations primarily comprised of BIPOC are usually the most contaminated. Race is the number one indicator of whether you live near contaminated soil, water, or air (Younes et al. 2021). Saikawa and Filippelli (2021) have also argued that BIPOC UCGs should be prioritised for contaminant sampling because these sites are known to be more contaminated than other places and the public health need is greater. I also intentionally sample the plants that take up the most contaminants such as leafy greens and root vegetables, because those are consistently shown in the literature to uptake more contaminants, and they present the greatest danger to community gardeners and farmers.

While it might seem like sample collection methods and rationale for study design are small details, they are actually very important. I have already been questioned about my methodology in this research project and during other research projects as well. For example, in this project, a group of community gardeners that I had not worked with before reached out to me and asked me to sample some recently delivered compost because they had heard about my study and wanted to make sure that the compost they used was safe before spreading it. I collected samples from their compost pile and it had some of the highest levels of petroleum products I had encountered on the project up to that point. The samples even smelled like petroleum, which I can only assume is another reason they asked me to sample the compost batch. After getting the results back, it was not surprising that the contamination concentrations were high.

The community gardeners shared this information with the City of Seattle, because they contract with a large compost manufacturer that supplies most of the publicly owned UCGs in the City with compost. The City then passed this information on to that supplier. After a few weeks, the community gardeners sent me back the responses from the compost supplier, and the comments were bizarre. One of the claims was that the compost might have had high petroleum levels because someone walked into the gated—by the way—UCG and sprayed it with diesel or motor oil. The supplier said this was of no concern, however, because even if there was a high concentration of motor oil and diesel, a cow could biodegrade into the compost and thus become part of the natural system. As far-fetched and odd as these claims were, there were even more insidious claims about my ability to collect samples and do the research. They asked how I had collected the data and suggested that I could be cross-contaminating the samples myself with petroleum because I may not know how to collect soil samples for contamination. Given that I had been collecting environmental samples for approximately 15 years at this point, and that I had been trained to do so for projects that were heavily scrutinised (such as for brownfields cleanup or Superfund cleanup), I found this suggestion particularly offensive, but not surprising.

Ultimately, the compost supplier sent a formal letter (which I suspect was drafted by their legal team, as this particular group is very litigious) to the community gardeners and me explaining that, yes, the compost did have high concentrations of petroleum products, but they were in no way obligated to rectify the situation. They stated that they had followed all the required state guidelines for compost, and so were well within their right to distribute it, which is true. But they omitted that the state guidelines for compost do not involve any requirements for testing contaminants or for revealing those concentrations to those who purchase or use their products. State guidelines for compost generally only require screening out various types of debris (e.g., plastics or materials that do not biodegrade easily). The supplier also used the word "organic" in ambiguous ways—at one point seemingly using the word to indicate that it was made of natural materials, while also simultaneously using it to imply organic certification, which this particular compost was not. They also noted that the screening level for the petroleum products was still below state cleanup levels for Superfund sites, and therefore stated it would be safe.

I had a conversation with the gardeners during this process to help them understand all of these details and to explain the ways in which the compost guidelines the suppliers were referring to were not, in fact, related to contaminants. As I was discussing this, the gardeners revealed that at first, they were unsure who to believe because they had not worked with me before. Yet, the more they interacted with both the City and the supplier of their compost, the more it became apparent to them that something did not seem right, but they were not able to verbalise it. They told me it was helpful to have me there to interpret what they were saying. Notably, after I pushed back against the supplier and they revealed that the compost should be good enough for the UCG because it was clean enough for a Superfund site, the community was fully on board with my interpretation of the concentration levels in their soils, and expressed outrage that the supplier would believe that Superfund screening levels were acceptable for UCGs.

The larger takeaway here is that this is just one of many everyday occurrences of those in power trying to discredit science that both serves and calls out the harm that contamination inflicts on communities. Collecting your own physical data allows you to readily defend the information you collected and can be a powerful way to resist harmful interpretations of data. It is also an opportunity to intervene and question the ways that data have been interpreted or accepted previously, and can provide a model for people doing research in ways that are different from before, which usually do not serve vulnerable communities. In this particular instance, I also gained another community partner who was very vocal about this incident and the way the City forces them to use compost from this supplier even though the City has received other complaints from UCGs partners about the compost. This outcome resulted in the partner getting different compost, more UCGs hearing about the issue, and more pressure on the City to provide alternative free compost that is not from this supplier. Many of those UCGs were already unhappy with this compost before, so having data from this location (as well as other UCGs in my study who have used this compost) was useful information and served as evidence of why the City should at least give gardeners additional choices for better compost with less contamination.

Lessons learned and pathways forward

Through all of the interconnected ways of learning from and supporting each other, I have noticed that my community partners and I are all building new skillsets. I am always learning from the communities that do research with me, and I often find it humorous that I do community research because I am an introvert who generally does not enjoy socialising in large groups of people and feels awkward doing so. I also strongly dislike giving bad news to people, which there is a lot of in this project because of the nature of contamination. And having tense conversations with City officials, who do not want data about contamination in spaces to be widely shared because of fear about liability, are both stressful and infuriating to me because they prioritise liability over the health of people who look like me and are vulnerable to environmental racism. Although I have a lot of expertise in both physical and social science methodologies, no one trained me on how to interact with the public as a scientist, nor with City officials who have agendas that are not always in the interest of public health. Much of how I interact with my community is based on my set of values and what I believe to be equitable research, but also my experience as a woman, person of color, and a scientist who has had to fight to learn methodology and have people explain why they do what they do in scientific research.

Despite these elements of the research, I find that I am getting better and more adept at ways to convey information, and more people invite me into their UCGs each year, so much so that I cannot get to all of them. Even the City now invites me to certain events and now recognises my research. To be clear, there is still much to be desired in how the City could address contamination and systemic racism, but I am encouraged that my way of conducting research has resonated with and served the community, and provided an alternative way to think about doing contamination research in UCGs. Moreover, the communities I work with have been extraordinary in their welcoming of my research, and they often warmly greet me and invite me to events that are not even research related. I am strongly indebted to them, and mention this as encouragement to those who find themselves in frustrating circumstances for research or are experiencing similar dynamics.

I am also fortunate to have colleagues who gave feedback on this chapter, and one of their recommendations was to end this chapter with what I wish I had known as an undergraduate and graduate student who might be embarking on mixed-methods research. So, my first recommendation is to surround yourself with knowledgeable, supportive colleagues and mentors that can help you refine, but also affirm, your research. Beyond that, and without a doubt, I wish someone would have explained in my undergraduate years how epistemologies and ontologies (i.e., ways of knowing and conducting research) influence scientific findings. Given that scientific research has been strongly influenced by racial politics since the beginning of U.S. history, I cannot stress enough that understanding what people know and why they think they know it are essential to understanding and interpreting scientific findings (for a wonderful example of this see Guthman 2011). I give this advice especially to those of us from underrepresented populations who are gaslighted in our research just for existing. Being able to understand the perspectives of others who do research, but also why they might not believe your research, helps to provide context for what you are encountering. It will also help you avoid being overly critical of yourself and your research. A good amount of scrutiny is useful for any research project, but it is also imperative that you are able to contextualise criticism so that it does not undermine research that will serve justice to communities.

References cited

Ali, H.N., S.L. Sheffield, J.E. Bauer, R.P. Caballero-Gill, N.M. Gasparini, J. Libarkin, K.K. Gonzales, J. Willenbring, E. Amir-Lin, J. Cisneros, D. Desai, M. Erwin, E. Gallant, K.J. Gomez, B.A. Keisling, R. Mahon, E. Marín-Spiotta, L. Welcome, and B. Schneider. 2021. 'An actionable anti-racism plan for geoscience organizations', *Nat. Commun.*, 12.

Balotin, L., S. Distler, A. Williams, S.J.W. Peters, C.M. Hunter, C. Theal, G. Frank, T. Alvarado, R. Hernandez, A. Hines, and E. Saikawa. 2020. 'Atlanta residents' knowledge regarding heavy metal exposures and remediation in urban agriculture', *International Journal of Environmental Research and Public Health*, 17.

CATO Institute. 1998. 'Science is badly used in risk assessment', *Cato Institute*, https://www.cato.org/speeches/science-badly-used-risk-assessment

CDC. 2022. 'Lead FAQs', *Lead CDC*, https://www.cdc.gov/nceh/lead/faqs/lead-faqs.htm

CDC. 2021. 'Blood lead reference value', *Lead CDC*, https://www.cdc.gov/nceh/lead/data/blood-lead-reference-value.htm

Cram, S. 2016. 'Living in dose: Nuclear work and the politics of permissible exposure', *Public Culture*, 28, pp. 519–539.

Demery, A.-J.C. and M.A. Pipkin. 2021. 'Safe fieldwork strategies for at-risk individuals, their supervisors and institutions', *Nat. Ecol. Evol.*, 5, pp. 5–9.

Eggen, R.I.L., R. Behra, P. Burkhardt-Holm, B.I. Escher, and N. Schweigert. 2004. 'Challenges in ecotoxicology', *Environ. Sci. Technol.*, 38, pp. 58A–64A.

EPA. 2020. 'Lead in soil', https://www.epa.gov/sites/default/files/2020-10/documents/lead-in-soil-aug2020.pdf

EPA. 2011. *Exposure Factors Handbook 2011 Edition (Final Report)* (EPA).

EPA. 2005. 'Ecological soil screening levels for lead', https://www.epa.gov/sites/default/files/2015-09/documents/eco-ssl_lead.pdf

EPA. 2015. 'Brownfields and urban agriculture: Interim guidelines for safe gardening practices', https://www.epa.gov/brownfields/brownfields-and-urban-agriculture-interim-guidelines-safe-gardening-practices

Fagin, D. 2012. 'Toxicology: the learning curve', *Nature*, 490, pp. 462–465.

Felter, S. and M. Dourson. 1998. 'The inexact science of risk assessment (and implications for risk management', *Human and Ecological Risk Assessment: An International Journal*, 4, pp. 245–251.

Filippelli, G.M. and M.A. Laidlaw. 2010. 'The elephant in the playground: confronting lead-contaminated soils as an important source of lead burdens to urban populations', *Perspect Biol. Med.*, 53, pp. 31–45.

Flynn, L.Y., A. Kofman, A. Shaw, L. Song, M. Miller, and F. Kathleen. n.d. 'Poison in the air', *ProPublica,* https://www.propublica.org/article/toxmap-poison-in-the-air?token=qSckpj4gt77NMS0djk1uIFCImwpRjeg9

Guthman, J. 2011. *Weighing In: Obesity,Food Justice, and the Limits of Capitalism* (University of California Press).

Hunter, C.M., D.H.Z. Williamson, M.O. Gribble, H. Bradshaw, M. Pearson, E. Saikawa, P.B. Ryan, and M. Kegler. 2019. 'Perspectives on heavy metal soil testing among community gardeners in the United States: a mixed methods Approach', *International Journal of Environmental Research and Public Health*, 16.

Kasvi, E., Chapter 45, this volume. 'Uncrewed Airborne Systems'.

Kim, B.F., M.N. Poulsen, J.D. Margulies, K.L. Dix, A.M. Palmer, and K.E. Nachman. 2014. 'Urban community gardeners' knowledge and perceptions of soil contaminant risks', *PLOS ONE*, 9.

Lewis, D., S. Francis, K. Francis-Strickland, H. Castleden, and R. Apostle. 2020. 'If only they had accessed the data: Governmental failure to monitor pulp mill impacts on human health in Pictou Landing First Nation', *Social Science and Medicine*, 288.

Lupolt, S., R. Santo, B. Kim, C. Green, E. Codling, A. Rule, R. Chen, K. Scheckel, M. Strauss, A. Cocke, N. Little, V. Rupp, R. Viqueira, J. Illuminati, A. Schmidt, and K.E. Nachman. 2021. 'The safe urban harvests study: a community-driven cross-sectional assessment of metals in soil, irrigation water, and produce from urban farms and gardens in Baltimore, Maryland', *Environmental Health Perspectives*, 129.

Lupolt, S.N., J. Agnew, G. Ramachandran, T.A. Burke, R.D. Kennedy, and K.E. Nachman. 2022. 'A qualitative characterization of meso-activity factors to estimate soil exposure for agricultural workers', *Journal of Exposure Science and Environmental Epidemiology*, 33, pp. 140–154.

Malone, M. 2021a. 'Seeking justice, eating toxics: overlooked contaminants in urban community gardens', *Agricultural and Human Values*, 39, pp. 165–184.

Malone, M. 2021b. 'Teaching critical physical geography', *Journal of Geography in Higher Education*, 45, pp. 465–478.

Malone, M., S. Hamlin, and S.I. Richard. 2023. 'Uprooting urban garden contamination', *Environmental Science and Policy*, 142, pp. 50–61.

Meadow, A., Wilmer, H., and Ferguson, D., Chapter 5, this volume. 'Expanding research ethics for inclusive and transdisciplinary research'.

Miranda, M.L., D. Kim, M.A.O. Galeano, C.J. Paul, A.P. Hull, and S.P. Morgan. 2007. 'The relationship between early childhood blood lead levels and performance on end-of-grade tests', *Environmental Health Perspectives*, 115, pp. 1242–1247.

Mitchell, R.G., H.M. Spliethoff, L.N. Ribaudo, D.M. Lopp, H.A. Shayler, L.G. Marquez-Bravo, V.T. Lambert, G.S. Ferenz, J.M. Russell-Anelli, E.B. Stone, and M.B. McBride. 2014. 'Lead (Pb) and other metals in New York City community garden soils: Factors influencing contaminant distributions', *Environmental Pollution*, 187, pp. 162–169.

Pulido, L. and J. De Lara. 2018. 'Reimagining "justice" in environmental justice: Radical ecologies, decolonial thought, and the Black Radical Tradition', *Environment and Planning E: Nature and Space*, 1, pp. 76–98.

Ramirez-Andreotta, M.D., M.L. Brusseau, J. Artiola, R.M. Maier, and A.J. Gandolfi. 2015. 'Building a co-created citizen science program with gardeners neighboring a superfund site: The Gardenroots case study', *International Public Health Journal*, 7.

Saikawa, E. and G.M. Filippelli. 2021. 'Invited perspective: Assessing the contaminant exposure risks of urban gardening: Call for updated health guidelines', *Environmental Health Perspectives*, 129.11.

Sarewitz, D. 2004. 'How science makes environmental controversies worse', *Environmental Science and Policy: Science, Policy, and Politics: Learning from Controversy Over The Skeptical Environmentalist*, 7, pp. 385–403.

Shayler, H., M. McBride, and E. Harrison. 2009. *Sources and Impacts of Contaminants in Soils* (Cornell Waste Management Institute).

Steffan, J.J., E.C. Brevik, L.C. Burgess, and A. Cerdà. 2018. 'The effect of soil on human health: an overview', *European Journal of Soil Science*, 69, pp. 159–171.

Van Sant, L. 2021. '"The long-time requirements of the nation": The US Cooperative Soil Survey and the political ecologies of improvement', *Antipode*, 53, pp. 686–704.

Venton, D. 2020. 'Ten simple rules for building an anti-racist research lab', *KQED*, https://www.kqed.org/science/1966972/ten-simple-rules-for-building-an-anti-racist-research-lab

Winters, J. 2021. 'EPA might water down its cleanup standards for Seattle's only river', *Grist*, https://grist.org/science/epa-proposal-environmental-justice-lower-duwamish-superfund-seattle/

Wölfle-Hazard, C. 2022. *Underflows, Feminist Technosciences* (University of Washington Press).

Ybarra, M. 2021. 'Site fight! Toward the abolition of immigrant detention on Tacoma's tar pits (and everywhere else)', *Antipode*, 53, pp. 36–55.

Yen-Kohl, E., Collective, and T.N.F.C.W. 2016. '"We've been studied to death, we ain't gotten anything": (Re)claiming environmental knowledge production through the praxis of writing collectives', *Capitalism Nature Socialism*, 27, pp. 52–67.

Younes, L., A. Kofman, A. Shaw, L. Song, M. Miller, and K. Flynn. 2021. 'Poison in the air', https://www.propublica.org/article/toxmap-poison-in-the-air

17. Space and place in participatory arts-based research

Javier Arce-Nazario

Introduction

This chapter is about the shaping of space as a technique in participatory, arts-based research (Ingram, Chapter 23), and how it has influenced my work in remote sensing and landscape change. My discussion applies to other modes of investigation that use visual imagery or prioritise a high level of bidirectional community engagement. I explore how the attributes of the spaces used for research encounters, including the personal and historical contexts of the spaces and the ways that they invite researchers and participants to move within them, can support us in framing and interpreting this kind of critically oriented mixed-methods research. By narrating several examples from my own experiences in the field, I illustrate why it is valuable to think more frequently and more explicitly about both space and place throughout the research cycle, and to document attributes of the spaces we use in descriptions of our research.

My interest in looking at spaces started with the desire to increase community engagement with an ongoing research project that I was carrying out with mostly undergraduate student collaborators. In searching for ways to recreate the experience of discovering landscape processes through imagery in a way that would be accessible to a broader public, we ended up creating exhibits where we could reimagine our remote sensing image data as art, and where we needed to give thoughtful attention to space and place. Later, I reflected on the choices we made to enhance certain forms of interaction among visitors to these exhibits and started to re-evaluate other phases of the research in

  https://doi.org/10.11647/OBP.0418.17

general, paying attention to how space and place influenced the forms of interaction among researchers and other participants.

Artistic production/co-production is now one of my primary research modalities, so I continue to think about space and place in the context of arts-based research. In this chapter, I will describe projects where visual images and artistic production were central elements of the research and discuss the role of space and place in those projects. However, space is not just relevant in participatory methodologies that use co-mapping or visual art production. Attributes of physical space and the meanings and associations that define places can have direct influences on the most persistent challenges in community participatory research methodologies.

Background

Participatory arts-based research and its challenges

The terms *participatory research* (Mokos, Chapter 36) and *arts-based research* (Ingram, Chapter 23) both represent a continuum of research practices (Cornwall and Jewkes 1995; Bagnoli 2009) which are so frequently linked that "arts-based, creative research has become equated with participatory research" (Brown 2022). In the context of this chapter, participatory research refers to research based on a partnership between the academic investigators and the participants which is fundamentally "bi-directional" or "co-productive", meaning that research questions and answers are not seen as cleanly divided territory between the academic investigator (with the questions) and the participant (providing data). Arts-based modes of research can refer to a variety of creative and crafted ways that information is produced and shared between the investigator and the participant or community. Arts-based methodologies are usually identified either at the stage of gathering data or when results are disseminated (Coemans and Hannes 2017).

Combining these techniques in a participatory, arts-based approach is well-motivated in research that applies cartography. Mapping techniques can be adapted in participatory research and arts-based research, since maps are almost always malleable images, and are usually polyvocal. Participatory, arts-based methodologies are also especially useful for a critically oriented cartography practice that seeks to expose how mapping encodes and exercises power. They can contest the tendencies for cartographers to rely on sensor data in which decisions about data collection were made decades ago, and to apply standardised toolkits for creating images. Making space for creativity and conversation in mapping can make the choices encoded in the resulting images more visible. Inviting non-academic participants to creatively construct maps can also bring forth different choices informed by broader, and often marginalised, perspectives. Finally, projects in which academic researchers present research data artistically often broaden the research collaboration to involve artists, researchers, or students from many disciplines. Engaging more interdisciplinary and diverse perspectives on the academic side of participatory research partnerships can be especially useful in quantitative critical research practices.

The specific participatory arts-based techniques I describe below fit into three categories: co-mapping, community-focused exhibition, and community participatory exhibition. In *co-mapping*, individual participants create map layers by making drawings prompted by researchers, helping integrate new quantitative or qualitative data layers into an image. Co-mapping is similar to "graphic elicitation" interview approaches (Bagnoli 2009) but tends to prioritise participant agency. Co-mapping is often used in settings where Indigenous or local knowledge is being emphasised (Peluso 1995). The other two techniques involve exhibition, which is frequently a part of participatory arts-based research. I use the term *community-focused exhibition* for cases when research results are presented artistically, and participants interact with one another, the results, and the creators in exhibition settings. For example, research through arts-based techniques such as photovoice are likely to include an exhibition of co-produced works or works created by community participants, targeting broader communities or policymakers as an audience (Hergenrather et al. 2009). In *community participatory exhibition*, participants interact with one another in curated spaces (exhibitions) and create new maps and layers by interacting with provided tools. This mode of research aligns with participatory action research principles in that it aims for learning, knowledge production, and community action (McTaggart 1991). The participatory exhibition is a space of discovery, where maps created by investigators and participants can be seen as research results for discussion and analysis. Encouraging network-building and continued community involvement is also an explicit goal of this mode (McTaggart 1991). Community participatory exhibition differs from community-focused exhibition in that the audience-participants are engaged in creation, and the interaction among groups is an important outcome of the method.

One way to think about space and place in participatory arts-based research is to consider how they might improve or facilitate participatory and arts-based methodologies. Among common challenges described in participatory research, two are especially relevant to my examples. The first is that of *establishing and increasing engagement*. In the initial phases of a participatory research project, it is common for participants and academic partners to interact in order to identify the constructs and concepts of the investigation as well as other details of the research. These interactions can have a strong bearing on the scientific validity of the research being conducted and the relevance of the research to the researchers' goals and to the participating community (Balazs and Morello-Frosch 2013). When participants "stand in" for a larger community in framing research, it is useful if they are endorsed by that community or represent the community as fully as possible. However, engaging with a representative group can be difficult, and often proceeds through gatekeepers who create and negotiate various domains of inclusion (De Laine 2000; Lenette et al. 2019).

Another frequently mentioned challenge relevant to the cases described below is ensuring that participants and researchers are prepared to speak and listen to one another as partners in the project. The challenge of shifting and sharing power is usually framed as a consequence of the asymmetry of power afforded by academic affiliations and social positions of the researchers relative to other participants. Rebalancing these asymmetries is an "intrinsic aim" of participatory research (Kara 2017). These asymmetries also arise in mixed-methods participatory research, because the implicitly understood authority carried by quantitative or technical methodologies, usually associated with the academic researchers, can become an obstacle to "bi-directional learning" (Ghanbarpour et al. 2018).

Arts-based research additionally confronts challenges including issues of confidentiality, privacy, and stewardship over the products of research (Brady and Brown 2013), finding external validity for research results within traditional research culture (Simons and McCormack 2007), as well as uncertainty about the final outcome of a project or its potential misreading (Van der Vaart, Van Hoven, and Huigen 2018;

Hodgins, Boydell, et al. 2014). The challenge of goodness can especially be an issue for the researcher producing or co-producing art and worrying about its reception in the community or in the academy, or for participants who must overcome a perceived lack of competence to engage with the project (Hodgins, Boydell, et al. 2014; Simons and McCormack 2007). While not central to the examples below, the challenge of goodness and competence is relevant to me in visually based methods that depend on participants' level of comfort creating or interpreting imagery.

Thinking and writing about space and place

Decades past the "spatial turn" in the social sciences, we still tend to ignore spaces and places of encounter when writing about our participatory research. Consideration of space, place, or setting in mixed-methods research interactions has developed from a simplistic division between the stereotypically "naturalistic" qualitative research setting and the "controlled" quantitative research setting (Philip 1998), to broad theoretical frameworks for integrating the space of interaction into research (Anderson, Adey, and Bevan 2010). Elwood and Martin (2000) frame "the microgeography of the interview" as an important source of information in methodologies that use participant interviews, which should be considered at every stage of the research process (see Johnston and Longhurst, Chapter 32). These latter frameworks fit well with participatory research methodologies in the way they emphasise the reciprocal interactions between researchers and informants. Nevertheless, descriptions of the spaces used for research interactions in participatory research can be surprisingly difficult to find outside of research to explicitly test or describe the influences of spaces of encounter (such as Evans and Jones 2011; Jones et al. 2008; Skovbjerg, Tanggaard, and Sand 2022). Yet even when space is not the focus of the research, simply identifying characteristics of the spaces and places used for encounters can still be informative. For example, in considering how to engage and to build trust with participants from a vulnerable population of people who had experienced violence, the authors of one study briefly noted the attributes of a space (easily accessible exits) which they saw as relevant to the needs of participants from vulnerable populations (Jumarali et al. 2021).

Different explanations for this neglect of the role of space in our methodologies have emerged. Concerns about privacy and anonymity may play a role (Kitchin and Tate 1999; Nespor 2000). For participatory methodologies, there may be additional reasons. Thinking about physical spaces and places for interaction (or the lack of such spaces) is likely to occur, but it might not be recognised as a formal stage of participatory research design because of the fluidity of the process. Details of the research method in participatory approaches are frequently worked out through dialogue between the researchers and participants, and it may also be difficult to identify the boundaries of a site where research "takes place" in participatory methods because not all productive encounters with co-researchers are formalised (Lenette et al. 2019).

In the following sections I narrate ways that the spaces and places of encounter helped us to confront these challenges. The examples are drawn from research projects that varied in how participation and arts were incorporated into the research cycle. The first project, sited in the humid highlands of the Galapagos, relates to the early stages of participatory research, when researchers might face the challenges of community engagement and of establishing an equal footing for conversations between academic researchers and community participants. It involved a mixed method research effort on land cover change and agriculture in the region that prioritised quantitative methodologies, but applied iterative and participatory methodologies to more rigorously investigate the drivers of land cover change and to develop landscape change narratives that were more relevant for farmers, policymakers, local conservation scientists, and other stakeholders. The second example relates to a series of community-focused exhibitions. These involved intentional thinking about space at the dissemination stage, initially in an effort to connect communities in a rural region of Puerto Rico with ongoing research reconstructing the landscape history of the area and its interaction with the water resources there. The research was based on remote sensing (Braun, Chapter 39) to understand landscape processes affecting people's interactions with water quality, and the exhibition spaces were chosen to provoke a critical examination of the quantitative data and analysis used in the research. This project grew to an island-wide project, *geo/visual/isla*, designed to evoke discussion about many themes of landscape and society in Puerto Rico. The final project serves to illustrate the process of choosing place and designing space to enhance the co-production of qualitative data and to facilitate community reflection and action about community concerns. It centres on highly qualitative research into the landscape history of colonisation on the island of Vieques in Puerto Rico, through a community-based exhibition called *Visualizing Vieques*. The island of Vieques reflects the unequal distribution of resources and unequal exposure to environmental risks so common in landscape histories of the Global South. Its future is being written by the long-time residents descended from planters and labourers in the sugar plantations, by newcomers from the main island of Puerto Rico and the US mainland, and from the industries catering to insular, domestic, and international tourists. As in the Galapagos research, this project aimed to incorporate perspectives from several populations on the island that had apparently conflicting stakes in the island's future.

Engagement and co-mapping

Land cover dynamics in the Galapagos are influenced by conservation efforts, by agriculture, and by the ecosystem dynamics of native and introduced species (Laso 2020). The islands' dominant tourism industry and its significance in scientific communities mean that conservation and protection of native species in Galapagos is highly prioritised, so tension has emerged at various pressure points between farmers, the Galapagos National Park, and other communities. The research project I engaged in to understand these dynamics in the Galapagos was anchored by the dissertation work of Francisco Laso (Laso 2021) and included several undergraduate collaborators.

One of the first objectives of the research was a land cover map of the Galapagos highlands which would improve on existing maps via more advanced remote sensing (Braun, Chapter 39) and landscape classification techniques, and a more meaningful set of land cover classes which would be generated through a participatory methodology. Another objective was a spatially explicit understanding of land management practices in the region. Francisco planned a research cycle that included repeated participation by groups of food producers, as well as government officials, scientific experts, and food distributors. We understood that the voices of farmers were often unheard in debates about the economic and ecological sustainability of the islands, so it was especially important for us to understand the dynamics of farming and land cover composition in a way that was relevant to farmers and inclusive of their perspectives. We hoped that broad farmer participation would give us a way to frame and carry out research without imposing constructs or values associated with exotic species from conservation scientists or other interests in the Galapagos so that research results could become a point of common understanding among different stakeholders.

Early in the project, Francisco carried out individual interviews with farmers and other stakeholder groups. To bring the different communities of stakeholders together, Francisco invited farmers to join a meeting to discuss the Galapagos food system, along with students and officials from the Galapagos Ministry of Agriculture. The place designated for the meeting was in the offices of an environmental education organisation. Although the academic researchers, government officials, and others attended, none of the farmers who expressed interest showed up. Their absence disappointed his hopes for inclusive conversation.

In the meantime, we discussed how to collect the land cover data, including inventories of invasive species in the farms. Our initial idea of doing transects would have been very inefficient with our small research team and raised the ethical concern of trampling over participants' crops while doing inventories. Instead, we decided to use drone flights (Kasvi, Chapter 45) to collect high-resolution aerial imagery with which we could later measure the species composition. To obtain permission to fly drones over farmers' landholdings, we contacted them in various ways, including leveraging contacts with the Ministry of Agriculture and visiting the markets where farmers sold produce.

The early data on land management was collected through semi-structured interviews (Johnston and Longhurst, Chapter 32) with the farmers, which took place in markets, in the farmers' homes, and most frequently on the farms where they worked. During the interviews on the farms, we wanted to collect spatially explicit information, so we walked around them while conversing with the farmers about their crops, the plants and wildlife they encountered on the farms, and their experiences managing the farms. We observed that without specific prompting, farmers talked about what was visible nearby as we walked.

At this stage, I introduced a co-mapping methodology into the research to gather a more spatially explicit understanding of land management practices. To create maps with the farmers, we orthorectified the drone imagery and printed the images. We took these images back to the farmers we had visited, asking them to make a map of their farms using tracing paper overlaid on the printed images (Fig. 17.1). Using this technique, the farmers created map layers with their own classification schemes and land management practices, which we could use in tandem with the high-resolution imagery. This co-mapping usually took place in farmers' homes. Some participants who were uncomfortable drawing or orienting themselves while viewing the aerial imagery asked to return to the farm for these interviews. We noted that returning to the farm and identifying landmarks there helped restore these participants' confidence in interpreting the maps (Colloredo-Mansfeld, Laso, and Arce-Nazario 2020).

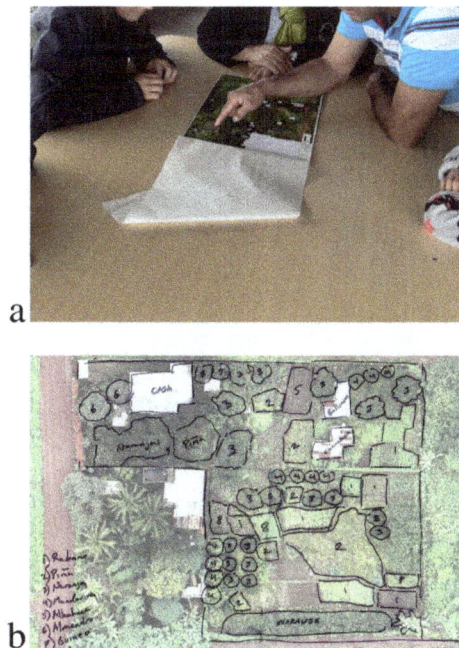

Fig. 17.1, Javier Arce-Nazario, (a) Farmers interpreting their landholding and sketching a map based on their classifications. (b) The result from one of the smaller landholdings in the study

My collaborators and I later reflected on the importance of these farm visits to the participatory research process. We found that when we were in places that were so familiar to farmers but unfamiliar to us, we felt receptive to learning about the farmers' agricultural practices and their view of the different species. In addition, farmers did not have any apparent difficulty assuming the role of instructor or authority and correcting our misconceptions in those settings. The places and landmarks on their farms clearly supported them as they shared information with us. Because we were outside on the farm, with the drone flying overhead, we all had a visible reminder of the position of the farm within a bigger picture and how the farmers' activities had shaped the landscape.

After these farm visits, efforts to recruit farmers to meetings were much more successful. The farmers that we had interacted with in the farm visits had a different idea of the role they could play in this research project than our initial efforts to recruit participants had projected. They helped us determine a better setting for the next series of meetings, which took place in a meeting room belonging to a local agricultural cooperative. These conversations had active and enthusiastic participation from the farming community. Thus, place had a prominent role in our experience in trying to attract a broadly representative group of participants in the Galapagos. Compared to the first meeting venue, the farm conversations took place in settings that were much more easily accessed by farmers, that were rich in features that related to the processes and constructs that our research needed to address, and where the farmers had a history and a material connection to the space. These meetings formed the basis for our future engagement with a more representative group of participants.

Combining the co-mapping methodologies in this project with remote sensing and image analysis (Braun, Chapter 39) provided deeply nuanced narratives of the Galapagos agricultural landscape and insights for promoting sustainable agriculture and protection of native species. As described in (Laso and Arce-Nazario 2023), a fine-grained analysis of remote sensing data combined with our drone imagery highlighted the prevalence of *Psidium guajava*, an introduced species present in about 20% of the agricultural areas. Co-mapping revealed that another introduced species, *Rubus niveus*, was nearly invisible to remote sensing but pervasive under the canopy and considered more challenging than *P. guajava* by some farmers. Mixing these methods allowed us to create map layers that sensors alone could not provide, such as land management and invasive species control practices (Fig. 17.2). It also allowed us to engage with farmers in ways that taught us about the economic coupling between tourism and agriculture, and their combined role in approaches to sustainability and management of introduced species.

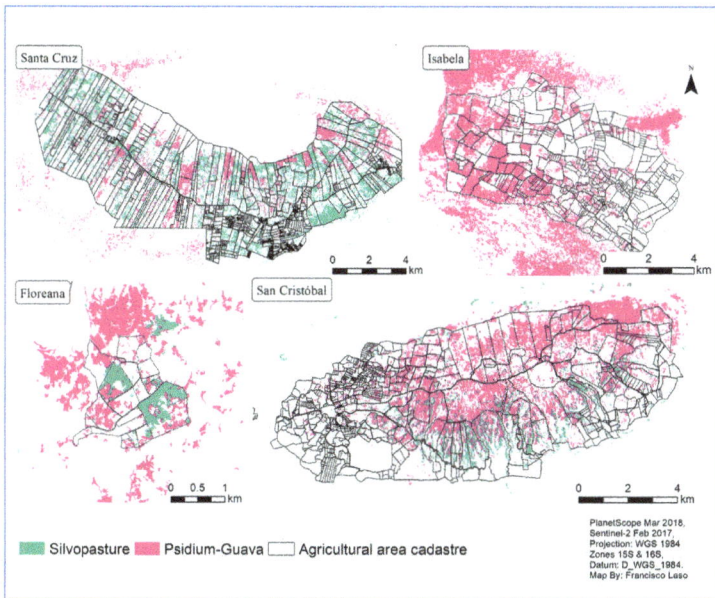

a

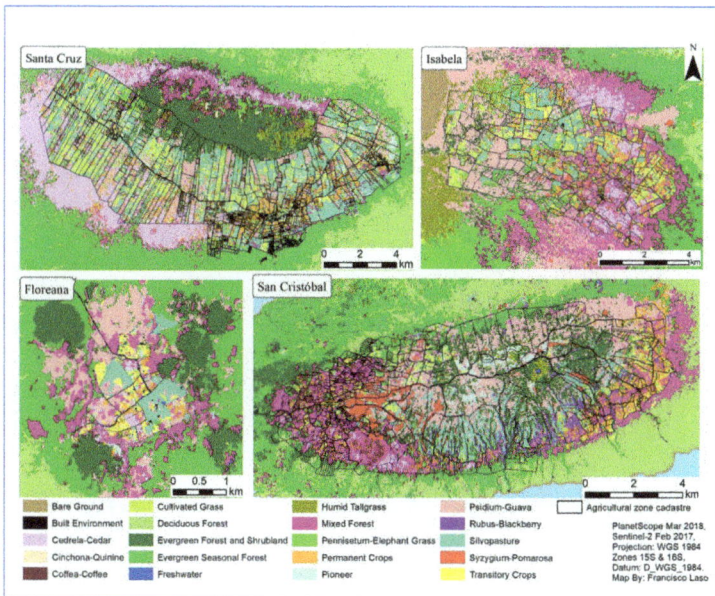

b

Fig. 17.2 Francisco Laso. Remote-sensing classification maps developed through iterative conversations with stakeholders and farm visits. (a) Simplified classification demonstrating the prevalence of *P. guajava* and silvopastures. (b) Classification that includes the other invasive species emphasised by stakeholders

Dissemination, participation, and pedagogy

My next example involves a series of community-focused exhibitions. These were initially designed to disseminate mostly quantitative research carried out with student collaborators at the University of Puerto Rico in Cayey, examining the landscape history and water quality of the surrounding watershed. The students and I took advantage of an invitation to show the history of changes in the area surrounding the university to a group considering a major development project on the last remaining agricultural patch in the region. This event began a process that gradually broadened the dissemination of this work into a new participatory effort with students from the humanities, social sciences and biophysical sciences, and with the input of local environmental activist groups. We created images of landscape change and presented them in community centres, fairs, conferences and other public venues. In 2014 we presented a large set of images in an exhibit at Cayey's Casa De Camineros, a community centre in a restored residence built in the late 1800s for the workers who maintained colonial road networks.

In the Casa de Camineros exhibit, we attempted to provide the public with the chance to interact with landscape change data in the same way we did while rectifying and interpreting historical remote sensing image sequences of nearby landscapes (Arce-Nazario 2016). Along with the effort we spent creating the actual content of the exhibits, I spent significant time thinking of ways to encourage visitors to "read" the landscape, attempting to translate the experience of conducting research. To invite this kind of participation, the quantitatively analysed images had to be experienced as artifacts for interpretation instead of as static and objective reality without obscuring the layers of data they contained. This led me to think about the construction of spaces where images could be experienced as artifacts that invited criticism, revision, and conversation. The student collaborators and I designed the exhibit in a series of galleries with 3-D modelling software while creating the works for display. One of the most rewarding aspects of collaborative research occurs during the act of looking together at data, and the laboratory where this landscape research took place (which I had also designed with 3-D modelling software) allowed all of us to gather around large screens to view maps or learn techniques. Thus, our goal was for the arrangement of interior space to allow users to interact with one another as well as with the themes of the exhibit.

The gallery-like space of the Casa de Camineros was ideal for presenting the images as subjective and interpretable, and its small rooms allowed us to create various spaces for discussions. It was also a good place to encourage participation, as it was known as a cultural centre that welcomed the Cayey community. The history of the building was particularly resonant because roads are such an important driver of landscape change in the Cayey region.

Our remote sensing study of land cover allowed us to quantify the importance of the post-agricultural forests in the region for water quality (Santiago-Rodriguez, Toranzos, and Arce-Nazario 2016). The analysis of historical imagery formed the basis for our method of dissemination, the community-focused exhibition. The exhibition method, in turn, expanded our research in two new directions. First, it raised a new research question about whether our dissemination efforts were effective. We addressed this using entrance and exit surveys (Winata and McLafferty, Chapter 43) administered by students during the exhibit. The surveys used a few questions to understand how people's understanding of the landscape history had changed after visiting the exhibit (Arce-Nazario 2016). Second, we were so excited about the interactions we saw in the halls of the Casa de Camineros that we wanted to create similar spaces for discussion of the broader themes we encountered while working together in the geography laboratory. This motivated the creation of a new exhibit, which we began planning in 2015.

The new exhibit was called *geo/visual/isla*. One of the student collaborators, a visual artist, reached out to the Puerto Rican Institute of Culture, which offered several possible venues. We chose the residence built for Puerto Rico's first colonial governor, Juan Ponce de Leon, which had a gallery space of the appropriate size for the amount of material we planned to produce. In historic Old San Juan, the Casa Blanca museum is a very visible, welcoming, and accessible place for young people, who visit with their schools, as well as for tourists from the island and from overseas. For this exhibit, we developed art pieces made collaboratively from our research data of orthorectified historical maps and aerial imagery of Puerto Rico.

The content of the exhibit was simply the art and a few text prompts to provoke conversation or reflection among visitors. We chose not to include the same kinds of conclusions we would publish in a scientific journal (except for a few results of the land cover classification that were mentioned in relative rather than quantitative terms). Instead, we wanted to share the data itself, because precisely orthorectifying the images took (in 2015) so much labour, and because the results were so beautiful and evocative. We also wanted to reproduce the joyful experience of doing collaborative research in an interdisciplinary group of students, and of inviting individual visitors from the community into our laboratory. Emphasising the imagery allowed us to leave conclusions open to interpretation and revision by the audience.

In contrast to the Casa de Camineros exhibit, we worked with professionals for some aspects of the exhibit, including printing images. We worked with a textbook editor to refine the text so that it was understandable and engaging to school-age children. We advertised the exhibit through postcards, social media posts, and appearances in local media, and planned an official opening with a catered buffet.

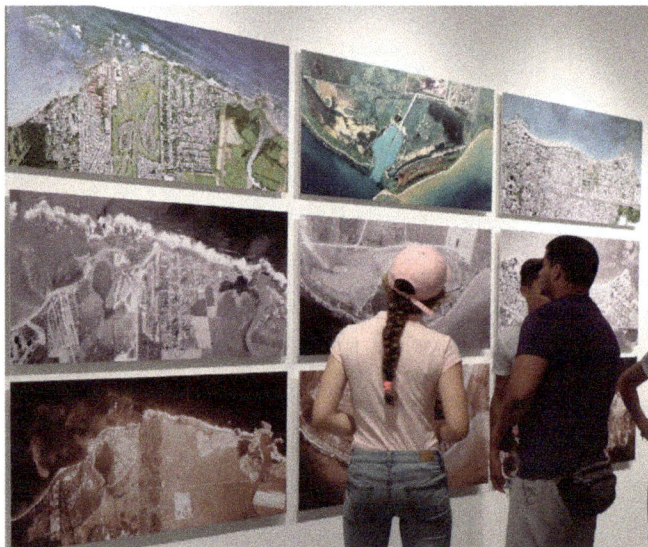

Fig. 17.3 Javier Arce-Nazario, Pieces were arranged in a way that allowed large groups to gather and discuss changes in the landscape

We also made choices about the interior space to facilitate reflection, engagement, and interaction among visitors. The elements of space that we considered were the flow, the sightlines, and the scale. Because the gallery was a long rectangle with an entrance on one of the longer sides, we designed the flow as a counterclockwise circulation around the open gallery space. This was encouraged by printing some of the informational text on arrow-shaped placards. The exhibit pieces were arranged along the walls in order of increasing complexity, so that visitors following the flow could become comfortable reading the landscapes on their own as they progressed through the exhibit. The images were interspersed with the textual prompts which pointed out features in the images and included open-ended questions for discussion. The sightlines allowed visitors entering the exhibit to see participants interacting with anaglyph images positioned towards the end of the exhibit. Anaglyph images, some of which were mounted on the floor of the gallery, required users to view them with 3-D glasses, and their novelty provoked social interactions among strangers (Fig. 17.3). Playful and spontaneous interactions in that space, viewed from a distance, were meant to set a tone for the exhibit that encouraged strangers to share perspectives with one another. Finally, the maps were printed at a scale large enough for several people to gather around each individual image, and for groups of about ten to interact with the larger compositions at the same time. The large format made recognisable features of the images such as houses big enough to point at, and the same scale was used throughout the exhibit, so that viewers did not have to adjust their sense of scale while moving from image to image.

The origin of this project in an undergraduate-serving institution might explain why it was so pedagogically focused, with an aim to teach people skills for reading the landscape, and for reading changes in the landscape in the context of social and ecological processes. We could not explicitly test if our decisions about space supported this aim, but we did take steps to understand the immediate impact of the exhibit by observing visits by individuals and school groups and conducting entrance and exit surveys. Rather than testing for particular readings of the landscape, the surveys asked more general questions about people's views about landscape change science and geography. They demonstrated that visitors to the exhibit developed a broader idea of what could be learned from aerial images. Observing the school visits confirmed our hope that teachers could stimulate conversations among groups of students around one installation. This observation was also supported by our observations of casual visitors conversing and sharing perspectives as they looked at different pieces. We noticed visitors leaning against one another or guiding each other's hands and eyes as they experienced different works. However, determining the longer-term impact of the exhibit was complicated by unforeseen events. *geo/visual/ isla* was initially scheduled to be open from June to July. In reaction to its positive reception, its opening was extended by the Institute of Culture until it was forced to close due to Hurricane Maria's impact on Puerto Rico in September.

Inviting participation

My final example is drawn from my most recent effort in participatory arts-based research. Inspired by the success of the earlier geovisualisation exhibits, a collective including undergraduate students from the University of North Carolina at Chapel Hill worked to create an exhibition that would be a space for artistic and sustained conversation about Vieques and its future. The island of Vieques is known for the victory of civil disobedience over colonial influence, which ended the US Naval occupation of two-thirds of its land in 2003 (Cruz Soto 2008). Nevertheless, Vieques still struggles for sovereignty. This struggle is complicated by the complexity of the ways Viequenses lay claim to their attachment to Vieques, which is the attachment of displaced people and colonisers, and of Puerto Ricans who negotiate a place-identity between the main island, Vieques, and the U.S. mainland (Myers 2021; Duany 2003). Most of the formerly occupied land is now federally managed nature reserves, and housing costs and resource shortfalls for infrastructure are increasingly driven by the dynamics of short-term rentals through multinational agents like AirBnB (Arrojado 2022).

In 2019, the collective became interested in producing digital maps of Vieques and current challenges in its transportation infrastructure based on data from the Puerto Rico Department of Transportation that I had gathered in previous research. Several students visited Vieques to contact local community leaders and better understand the present-day concerns about space and access on the island. These contacts included farmers engaged in land reclamation and the director of the Historical Archive of Vieques (AHV), Robert Rabin Siegal.

Several individuals were interested in hosting installations and activities related to our maps, so we chose venues based on both their physical characteristics and local meaning (space and place). The AHV director had experience supporting and publicising the work of dissident artists during the anticolonial movement in Vieques. He suggested the museum that hosts the archive, el Museo Fuerte Conde de Mirasol (El Fortín), as one site. We planned a second site within an educational space on the land of a participating farmer. In this way, our research was refocused on creating exhibition spaces for participatory cartography about Vieques, with the main activities in El Fortín. The museum, initially a Spanish fortification constructed between 1845 and 1855, is situated on a hill overlooking the main town and transportation port in Vieques. Historian Maria Cruz Soto writes that the fort initially "embodied many things at once," showing Spain's investment in Vieques as a colonial outpost, but also serving as a deterrent in its defence against foreign invasion (Cruz Soto 2008). While it never was used in armed conflict, it functioned as a colonial prison for the unemployed and for political prisoners (Rabín 1991). Now, El Fortín is a highly rated tourist attraction where art, archaeology, colonial history, and the history of civil disobedience in Vieques are mingled in a series of exhibit rooms along with the Vieques Historical Archive. Many see El Fortín as a place reclaimed to tell an activist history of Vieques and to create "a sustainable, and also loving" economy and society (Rabin Siegal 2020).

Hurricane Maria had damaged the gallery space for temporary exhibitions so that it was unusable, so we were granted a wing with broad windows looking towards the port. The space surrounding our exhibit room included the colonial-era brickwork and cannons, archival images of military occupation, and vibrant paintings and photographs created by artists resisting the military occupation. Based on discussions with community leaders, we planned an area for mapping activities centred on land use and tenure. Because of the view of the port, we decided to also include an installation about the marine transportation connecting Vieques to the large island of Puerto Rico. We also planned a section on the issues of military occupation and consent marking the history of the island. Each installation was designed so that visitors would create maps, have discussions, and rearrange and modify what was already present.

Fig. 17.4 Ayana Arce, Visitors interacting with the anaglyph floor graphic installations in *geo/visual/isla*

Planning for the exhibit, which took place during the coronavirus pandemic, involved continuous discussion among the cartography collective, the museum directors, historians, Vieques educators, and farmers via in-person meetings, text message apps, and videoconference meetings. We had so many difficulties accessing the space on the farm and finding appropriate transportation on the island for construction materials that the second exhibit site was never constructed. Ultimately, the spaces for creation and interaction in El Fortín were centred on three areas. The space related to land tenure and the space related to transportation were both in a sunny, bright side of the gallery. Participants were encouraged to modify a large 1883 map of the urban centre of Vieques using string, coloured paper, pushpins, and markers, and a smaller map of the channel between Vieques and the main island of Puerto Rico using hand-folded paper boats. We used low tables and inviting colours in that area to make it accessible to visitors of all ages, and to the groups of students that we would host in the space (Fig. 17.4). It was clear from our conversations that the third space related to military occupation would bring up more sensitive and emotionally affecting themes. One of those maps was a digital interactive map of the bomb impact craters in Vieques that we wanted to project on a wall, and the other was a map juxtaposing craters in a heavily bombarded lagoon with other experiments carried out on people and populations without consent. These installations were made in a darker corner of the gallery, to reflect the more sombre subject matter and allow caregivers to censor the content more easily, and also to allow the digital map to be clearly viewed.

The exhibit was open for six months. Traffic to the museum was reduced because of pandemic travel restrictions to Vieques and because of the limited opening hours, but the exhibit received many groups of casual visitors as well as several dedicated gatherings of participants, including groups of middle school students and university college students. Visitors created works of text, drawings, and folded and cut paper which were added as layers to the cork map and the transportation maps. The cork map alone was modified by more than 200 visitors, showing that participants did not feel reluctant or intimidated about engaging with the map. Children added some of the first map features (Figs. 17.5 and 17.6). The participants' creations included buildings, images of missiles, flowers and trees, hearts, and human figures, as well as messages in Spanish, English, and Japanese. Several messages in English said "I/We love Vieques" and included a location ("AZ", or "From Wisconsin"). Others in Spanish requested "Peace" or "Strength" for Vieques. The map layers even showed evidence of conversations, with an addition asking to "End colonialism—do not become a SETTLER" provoking the responses "Or a gentrifier", "Or a speculator", and "Who is a settler?" (Another conversation was shorter: "follow my TikTok @——" got the response "Don't follow it!"). The layers added to the transportation maps, which were in the form of folded paper boats, included many captions about failures in the ferry service, such as "it left early and we didn't make it", and "it stopped in the middle of the sea". Visitors were moved to tears by the tragedy depicted in the darker section of the exhibit. Some left comments for the museum in a suggestion box by the exit of the gallery. One wrote "create an ecological university", and another wrote "you should make this a centre for organising so others can learn from your experience and lead causes on their fronts". We are still interpreting the mappings and narratives shared by visitors in El Fortín, but the community participatory exhibition significantly enhanced our original image data. The collective's initial focus on transportation challenges was broadened to include historical narratives of ownership, consent, and activism. The conversations that emerged in the mappings show us how diversely Vieques is viewed as a history and how universally it is valued as a place.

The exhibit closed in May 2022 with another tragedy: the passing of Robert Rabin Siegal two months before. After discussion, the collective decided not to move the installation to another venue in Vieques that lacked the characteristics of El Fortín as a space and place, which had helped make the initial exhibit so successful.

Fig. 17.5 Javier Arce-Nazario, (a) Table for crafting boats and map elements for the interactive maps. Users of different ages using the crafting table during (b) a regular visit to the museum and (c) a middle school class visit

Fig. 17.6 Javier Arce-Nazario, Middle school students interacting with the cork map during a class visit

Discussion

Mixing remote sensing and landscape change analysis techniques with participatory, arts-based research seems inevitable in these examples since in each case, the imagery and narratives from the quantitative methods are the starting point for artistic and participatory interactions about the places surveyed. Similarly, the participatory and arts-based methods are necessary in each case discussed, either as a way to incorporate spatially explicit local knowledge into broader scientific and policy discussions, or to disseminate our findings beyond academic venues, bringing data and scientific methodologies for understanding space into local discussions of place. What is more notable is that when these methods were combined, space and place emerged as intentional or unintentional actors in the participatory research interactions. The examples demonstrate that as a research technique, considering space and place can support mixed methods employing participatory or arts-based research by confronting the challenges mentioned previously: representative community engagement, shifting and sharing power, and confidence in artistic production and interpretation.

In the Galapagos, we found that efforts to engage a community that is not already linked to the development of the participatory research creates the first impression, setting the stage for the kinds of interactions that will ensue. Farmers asked to attend a distant meeting in the city, on a workday, might easily imagine that sustained interaction with the project would be a burden and that the participatory relationship will be one in which farmers make sacrifices to satisfy researcher needs. In the three exhibits discussed, participant engagement was clearly shaped by the characteristics of the spaces and places we chose. Because of their colonial histories, each of the historical buildings we used was centrally located near a population centre, making them easily accessible to many. However, the people who saw our advertising chose whether or not to visit based on expectations and evaluations that included the site and the way they valued those places, and so our audience was probably biased towards people who tend to visit museums and felt comfortable in those settings. We needed to counteract this bias in *geo/visual/isla* and *Visualizing Vieques*, so we also invited schools to bring student groups to visit. In *Visualizing Vieques*, we also benefited from the museum's location inside a popular tourist attraction, which allowed us to include the voices of visitors as well as residents.

In a discussion of walking interviews, Jones et al. (2008) explain how power dynamics in participant encounters can be shaped by space and place by contrasting encounters in a university setting where "[p]articipants have to go to, and pass through, a space which can be somewhat exclusionary and intimidating" with interviews in the participant's workplace, where it is possible that the academic researcher might be "transformed into a supplicant". In Galapagos the physical attributes of and the entities found inside of the spaces we used, as well as the characteristics of various places (e.g., meetings at an agricultural cooperative vs. an environmental education NGO), both shaped these dynamics. The plants and other artifacts on the farms played a role in establishing positions of authority during our conversations with farmers. It has been noted that participant authority can sometimes be derived from the fact that participants outnumber the academic researchers: participants feel empowered to challenge academic partners when they are backed up by other participants (Jumarali et al. 2021). In the one-on-one interactions we had on the farms, farmers did not have this recourse. However, place may have offered a form of support similar to like-minded participants, in the form of the farmers' own crops and the larger landscape which was very familiar to them. We, on the other hand, were more likely to ask questions than to answer them when we pointed out objects in the setting.

The way that structural power inequalities, space, and place interacted at the dissemination phase is less clear. In the community-focused exhibitions and the community participatory exhibition, the role of gallery spaces in shifting power relationships among researchers and participants was ambiguous. On the one hand, galleries are not the customary domain of either the academic researchers or (usually) the participants. However, differences in social power that accompany being an academic may correlate with a sense of belonging in galleries and museums. The spaces for exhibition may create another potential division between researchers and participants. Even museums and galleries that specifically aspire to be "participatory cultural institutions" confront challenges in creating genuine collaborations between museums and their audiences (Cruikshanks and Van der Vaart 2019). In the exhibits

I described, the academic researchers and our collaborators expressed authority by determining what was displayed on the walls of the gallery. This might not translate into visitors passively receiving the works as authoritative: Pitts and Price (2021) describe visitors to contemporary art galleries as seeking a "freedom to respond without being wrong, ignorant, or pressured" which can be reinforced or disrupted by the venue. We hoped that our gallery spaces and the attributes of these spaces would connote this kind of freedom and thus invite responses from participants in a way that an academic presentation would not, but this question should be more systematically explored.

Participants' competence and confidence with imagery and artistic production was supported by the space in each of the examples provided. We accidentally explored this interaction in the Galapagos. Farmers' expertise in interpreting and creating images was sometimes challenged by the drone-generated maps we presented, and moving to the actual spaces depicted in the images reframed the exercise of drawing a map into something more concrete, which made the co-mapping with these individuals more successful. In the *Casa de Camineros* and *geo/visual/isla*, we intentionally used space pedagogically to break down the complicated task of landscape image interpretation into smaller parts, and to encourage groups to support each other in exploring those parts. While dissemination via meetings or public presentations could also break down these tasks by presenting them in a time sequence, using space had the advantage that participants could control the pace of their experience. In *Visualizing Vieques*, we created child-friendly, informal, playful spaces for crafting to allow participants of all ages to overcome any reluctance to share their own drawings and art.

Taken together, these examples argue strongly for considering how space affects researcher-participant interactions and the climate in which we communicate through imagery and art in our research. However, we can look beyond the challenge of an academic researcher designing an ideal methodology to see the broader impact of space.

Thinking about place, especially its historical component, can also encourage new perspectives. Because places are persistent, they can create links to historical contexts or futures that might not otherwise be considered. My work derived from postcolonial landscape change research embraced the historical resonances of places linked to road construction and colonial power, but these histories also disrupted our ideas about the research. While we were creating *Visualizing Vieques*, the history of the surrounding places in El Fortín called my attention to the complicated nature of our alliance with Vieques. Even though our goal was to facilitate communities envisioning positive change, while mounting maps and refurnishing the space my thoughts continually returned to the way that the exhibit and the research on Vieques were themselves invasive. A review of contributions to the *Qualitative Inquiry* journal found that in contrast to other empirical and theoretical studies reviewed, "Indigenous intellectual contributions rarely fail to engage in issues of land and place". The authors link this difference to the idea that "Settler societies are designed not to consider place—to do so would require consideration of genocide" (Tuck and McKenzie 2015).

Conclusion

When considering space in different phases of participatory and arts-based research, our task expands beyond considering how interviews are shaped by space and place, and even beyond the additional context they can provide as data for the research. We should also confront the effect that space and place have on who participates, and on the methodologies and overall research design that we employ.

The following seven questions can be a rough guide to considering space and place at various stages of participatory or arts-based research design.

- How is the space accessed?
- Who might feel ownership of the space/place, and what features of the space/place create a sense of ownership?
- How do the features of the space/place "speak for" researchers or participants?
- How does the arrangement of the space encourage or discourage interactions?
- Can the space be arranged to support pedagogical or artistic goals?
- How do the features of the space provoke actions by researchers or participants?

- How do the place and its history or future pertain to the research, and how should they be acknowledged?

The first three questions are broadly useful for any interview-based or elicitation methodology, while the next three are especially relevant for community participatory and arts-based methodologies. The last question, for me, is special in the way it provides a doorway into a critical framing of the research. Answering the question can simply bring serendipitous historical resonances to the fore, but it can also work to problematise the role of the research and the researcher in the community, calling for humility, for a consideration of the project in a broader context, and for a deeper commitment to listening, learning, and collaborative action.

References cited

Anderson, J., P. Adey, and P. Bevan. 2010. 'Positioning lace: Polylogic approaches to research methodology', *Qualitative Research*, 10.5, pp. 589–604. https://doi.org/10.1177/1468794110375796

Arce-Nazario, J.A. 2016. 'Translating land-use science to a museum exhibit', *Journal of Land Use Science*, 11.4, pp. 417–28. https://doi.org/10.1080/1747423x.2016.1172129

Arrojado, J. 2022. *Minding the Gap: Applying a Rent Gap Analysis for Short-Term Rentals in Puerto Rico* (University of North Carolina).

Bagnoli, A. 2009. 'Beyond the standard interview: the use of graphic elicitation and arts-based methods', *Qualitative Research*, 9.5, pp. 547–70. https://doi.org/10.1177/1468794109343625

Balazs, C.L. and R. Morello-Frosch. 2013. 'The three Rs: How community-based participatory research strengthens the rigor, relevance, and reach of science', *Environmental Justice*, 6.1, pp. 9–16. https://doi.org/10.1089/env.2012.0017

Brady, G. and G. Brown. 2013. 'Rewarding but let's talk about the challenges: Using arts based methods in research with young mothers', *Methodological Innovations Online*, 8.1, pp. 99–112. https://doi.org/10.4256/mio.2013.007

Braun, A., Chapter 39, this volume. '(Critical) Satellite remote sensing'.

Brown, N. 2022. 'Scope and continuum of participatory research', *International Journal of Research and Method in Education*, 45.2, pp. 200–211. https://doi.org/10.1080/1743727x.2021.1902980

Coemans, S. and K. Hannes. 2017. 'Researchers under the spell of the arts: Two decades of using arts-based methods in community-based inquiry with vulnerable populations', *Educational Research Review*, 22, pp. 34–49. https://doi.org/10.1016/j.edurev.2017.08.003

Colloredo-Mansfeld, M., F.J. Laso, and J. Arce-Nazario. 2020. 'Drone-based participatory mapping: Examining local agricultural knowledge in the Galapagos', *Drones*, 4.4, pp. 62. https://doi.org/10.3390/drones4040062

Cornwall, A.and R. Jewkes. 1995. 'What is participatory research?', *Social Science and Medicine*, 41.12, pp. 1667–76. https://doi.org/10.1016/0277-9536(95)00127-s

Cruickshanks, L. and M. Van der Vaart. 2019. 'Understanding audience participation through positionality: Agency, authority, and urgency', *Stedelijk Studies Journal*, 8. https://doi.org/10.54533/stedstud.vol008.art02

Cruz Soto, M. 2008. *Inhabiting Isla Nena, 1514-2003: Island Narrations, Imperial Dramas and Vieques, Puerto Rico* (University of Michigan).

De Laine, M. 2000. *Fieldwork, Participation and Practice: Ethics and Dilemmas in Qualitative Research* (Sage Publications).

Duany, J. 2003. 'Nation, migration, identity: the case of Puerto Ricans', *Latino Studies*, 1.3, pp. 424–44. https://doi.org/10.1057/palgrave.lst.8600026

Elwood, S.A. and D.G. Martin. 2000. '"Placing" interviews: Location and scales of power in qualitative research', *The Professional Geographer*, 52.4, pp. 649–57. https://doi.org/10.1111/0033-0124.00253

Evans, J. and P. Jones. 2011. 'The walking interview: Methodology, mobility and place', *Applied Geography*, 31.2, pp. 849–58. https://doi.org/10.1016/j.apgeog.2010.09.005

Ghanbarpour, S., A. Palotai, M.E. Kim, A. Aguilar, J. Flores, A. Hodson, T. Holcomb, et al. 2018. 'An exploratory framework for community-led research to address intimate partner violence: a case study of the Survivor-Centered Advocacy Project', *Journal of Family Violence*, 33.8, pp. 521–35. https://doi.org/10.1007/s10896-018-9987-y

Hergenrather, K.C., S.D. Rhodes, C.A. Cowan, G. Bardhoshi, and S. Pula. 2009. 'Photovoice as community-based participatory research: a qualitative review', *American Journal of Health Behavior*, 33.6, pp. 686–98. https://doi.org/10.5993/ajhb.33.6.6

Hodgins, M.J., K. Boydell, et al. 2014. 'Interrogating ourselves: reflections on arts-based health research', *Forum Qualitative Sozialforschung/Forum: Qualitative Social Research*, 15.1.

Ingram, M., Chapter 23, this volume. 'Arts-based environmental research'.

Jones, P., G. Bunce, J. Evans, H. Gibbs, and J.R. Hein. 2008. 'Exploring space and place with walking interviews', *Journal of Research Practice*, 4.2.

Jumarali, S.N., N. Nnawulezi, S. Royson, C. Lippy, A.N. Rivera, and T. Toopet. 2021. 'Participatory research engagement of vulnerable populations: Employing survivor-centered, trauma-informed approaches', *Journal of Participatory Research Methods*, 2.2. https://doi.org/10.35844/001c.24414

Kara, H. 2017. 'Identity and power in co-produced activist research', *Qualitative Research*, 17.3, pp. 289–301. https://doi.org/10.1177/1468794117696033

Kasvi, E., Chapter 45, this volume. 'Uncrewed Airborne Systems'.

Kitchin, R. and N. Tate. 1999. *Conducting Research in Human Geography* (Prentice Hall). https://doi.org/10.4324/9781315841458

Laso, F. 'Galapagos is a garden', in *Land Cover and Land Use Change on islands: Social and Ecological Threats to Sustainability*, ed. by S.J. Walsh, D. Riveros-Iregui, J. Arce-Nazario, and P.H. Page (Springer), pp. 137–166. https://doi.org/10.1007/978-3-030-43973-6

Laso, F.J. 2021. *Agriculture, Wildlife, and Conservation in the Galapagos Islands* (University of North Carolina).

Laso, F.J. and J.A Arce-Nazario. 2023. 'Mapping narratives of agricultural land-use practices in the Galapagos', in *Island Ecosystems: Challenges to Sustainability*, ed. by S. Walsh, C. Mena, J. Stewart, and J.P. Muñoz (Springer). https://doi.org/10.1007/978-3-031-28089-4_16

Lenette, C., N. Stavropoulou, C. Nunn, S. T. Kong, T. Cook, K. Coddington, and S. Banks. 2019. 'Brushed under the carpet: Examining the complexities of participatory research', *Research for All*, 3.2, pp. 161–79.

Longhurst, R. and Johnston, L., Chapter 27, this volume. 'Focus groups'.

Johnston, L. and Longhurst, R., Chapter 32, this volume. 'Interviews: Structured, semi-structured and open-ended'.

McTaggart, R. 1991. 'Principles for participatory action research', *Adult Education Quarterly*, 41.3, pp. 168–87. https://doi.org/10.1177/0001848191041003003

Mokos, J., Chapter 36, this volume. 'Participatory methods'.

Myers, J.N. 2021. *Narratives of Resilience: Place Attachment in Vieques, Puerto Rico* (Prescott College).

Nespor, J. 2000. 'Anonymity and place in qualitative inquiry', *Qualitative Inquiry*, 6.4, pp. 546–569. https://doi.org/10.1177/107780040000600408

Peluso, N.L. 1995. 'Whose woods are these? Counter-mapping forest territories in Kalimantan, Indonesia', *Antipode*, 27.4, pp. 383–406. https://doi.org/10.1111/j.1467-8330.1995.tb00286.x

Philip, L.J. 1998. 'Combining quantitative and qualitative approaches to social research in human geography—an impossible mixture?', *Environment and Planning A*, 30.2, pp. 261–76. https://doi.org/10.1068/a300261

Pitts, S. and S.M. Price. 2021.'"It's okay not to like it": the appeal and frustrations of the contemporary arts', in *Understanding Audience Engagement in the Contemporary Arts*, ed. by S.E. Pitts and S.M. Price (Taylor and Francis). https://doi.org/10.4324/9780429342455-7

Rabín, R. 1991. *Notas para la historia del fortín conde de mirasol y La Isla de Vieques* (Instituto de Cultura Puertorriqueña).

Siegal, R. and L. Robert. 2020. 'Archivo histórico de Vieques: Memoria histórica de un pueblo en constante lucha y resistencia', *Acceso. Revista Puertorriqueña de Bibliotecología y Documentación*, 1.

Santiago-Rodriguez, T.M., G.A. Toranzos, and J.A. Arce-Nazario. 2016. 'Assessing the microbial quality of a tropical watershed with an urbanization gradient using traditional and alternate fecal indicators', *Journal of Water and Health*, 14.5 pp. 796–807. https://doi.org/10.2166/wh.2016.041

Simons, H. and B. McCormack. 2007. 'Integrating arts-based inquiry in evaluation methodology: Opportunities and challenges', *Qualitative Inquiry*, 13.2, pp. 292–311. https://doi.org/10.1177/1077800406295622

Sand, A.-L., H.M. Skovbjerg, and L. Tanggaard, and. 2022. 'Re-thinking research interview methods through the multisensory constitution of place', *Qualitative Research*, 22. 4. https://doi.org/10.1177/1468794121999009

Tuck, E. and M. McKenzie. 2015. 'Relational validity and the "where" of inquiry: Place and land in qualitative research', *Qualitative Inquiry*, 21.7, pp. 633–638. https://doi.org/10.1177/1077800414563809

Van der Vaart, G., B. Van Hoven, and P.P.P. Huigen. 2018. 'Creative and arts-based research methods in academic research', *Lessons from a Participatory Research Project in the Netherlands*, 19.2, p. 10.   https://doi.org/10.17169/fqs-19.2.2961

Winata, F. and McLafferty, S., Chapter 43, this volume. 'Survey and questionnaire methods'.

18. Antarctic mosaic: Mixing methods and metaphors in the McMurdo Dry Valleys

Stephen M. Chignell, Adrian Howkins, and Andrew G. Fountain

In February 1911, the young Australian geologist Griffith Taylor spent a week exploring what he referred to as "Dry Valley" in East Antarctica, part of a larger region that would turn out to be the largest ice-free area on the continent. As he sought to describe the bare soils and patterned ground in his sledge diary, he drew upon the metaphor of a Roman mosaic:

> The surface of the valley floor hereabout was rather remarkable. The ground resembled a Roman mosaic of closely packed flakes of rock laid down absolutely flat. This I take to be a result of wind action in distributing the frost flakes into which the large boulders disintegrated. The flakes are packed together and there is no agent—such as rain and surface waters—to disturb them. [Geological Report, 4th Feb]

In the years since Taylor's exploration of what would come to be known as Taylor Valley in the McMurdo Dry Valleys, the mosaic metaphor has been expanded to encompass the whole of this remarkable region. In the introduction to a recent environmental management report, for example, limnologist John Priscu writes:

> The McMurdo Dry Valleys (MDV) of southern Victoria Land form the largest ice-free expanse on the continent and represent the coldest and driest desert on our planet. This region consists of a mosaic of glaciers, soils, streams and lakes that are intricately connected to support a fragile ecosystem (Priscu and Howkins 2016).

  https://doi.org/10.11647/OBP.0418.18

In both its early human history and in contemporary science, the mosaic metaphor has been used to help make sense of this other-worldly landscape (Fig. 18.1), albeit in different ways for different people.

Fig. 18.1 Adrian Howkins, Photograph of a contemporary soil wetting experiment in the McMurdo Dry Valleys

Indeed, the metaphors we use often reflect the lens through which we see the world and the values we hold. Taylor's use of the Roman mosaic metaphor, for example, hints at the lingering influence of a classical education in the early 20th century within the British imperial world. While current use of the mosaic metaphor reflects the language of modern landscape ecology (i.e., as a way to describe assemblages or patches of different habitats), its use also reflects something of the cultural values of the scientists who developed ecology as an academic discipline in the late 19th and early 20th centuries. Metaphors also have an important relationship to how we think about a place or problem; certain metaphors can be generative, helping to set the frame by which future research is conducted (Schön 1979; Halffman 2019; Tadaki et al. 2023). Each instance of 'seeing as' enables certain types of metaphors and analyses (and forecloses others). For example, despite the early use of the 'mosaic' metaphor, the MDV are also understood through other metaphors, particularly as a pristine 'wilderness' and a 'natural laboratory'. Glaciers are seen as sources of water and nutrients, streams as 'conduits', and lakes as 'sinks'. Local human activity is seen as 'impact' or 'disturbance', while anthropogenic climate change is seen as an exogenous 'pressure'. Sudden change is viewed as a 'pulse event' while longer term change is described as a 'press event'.

In what follows, we use the metaphor of a mosaic, as well as several others, to help us reflect on our experiences of working together on a decade-long historical research project that emerged from research on the MDV Long Term Ecological Research (LTER) site. Each of us— Chignell (a GIS [geographic information systems] specialist), Howkins (an environmental historian), and Fountain (a glaciologist)—come from different disciplinary backgrounds and career stages, yet share a deep interest in interdisciplinarity. The project began with two overlapping questions: 1) how can a better understanding of MDV history inform the science taking place in the region? 2) how can the results of scientific research be used to deepen our understanding of the history?

In addressing these broad questions and learning to work together collaboratively across very different disciplines, we have found the insights offered by Critical Physical Geography (CPG) particularly helpful. CPG is a new field which brings together social and biophysical sciences to study material landscapes, social dynamics, and knowledge politics together (Lave et al. 2014; Lave et al. 2018). From the outset we wanted to make this a project that was able to "go with the flow", using whichever methods we thought might help produce insights into the past, present, and future of the MDV. In many cases, this included learning new methods in response to questions that emerged as we collaboratively interpreted the results of each analysis. The 'interactional expertise' needed for this kind of work takes time to develop (Gorman 2010), and it helped that our work developed as part of a LTER site, in which the funding model is designed to support long-term research. Being part of a larger scientific project in the MDV while simultaneously writing about the region's history has significantly helped in the analysis, interpretation, and dissemination of our work. Over the years, we have become increasingly reflexive about our roles, and recognise that as three English-speaking white men based at research universities in Canada, the United States, and the United Kingdom, we are in many ways a reflection of the dominant culture of the eco-social system we are studying, and to some extent critiquing.

In this work we have often found the process of mixing methods to be easier than the process of writing and publishing our research. Different methods use different metaphors, and the way we write can reflect our underlying philosophical assumptions, positionalities, and values (Gray 2017). To mix methods, it follows, is also in a very tangible sense to mix metaphors, a point we will revisit in a later section. But while CPG scholars have paid a great deal of attention to the innovative possibilities of mixing methods (Biermann et al. 2021; Biermann and Gibbes, Chapter 4) and the importance of scrutinising the metaphors at work in scientific fields (Kull 2018), the field has paid less attention to the difficulties of 'mixing metaphors', broadly understood, in the actual process of writing up research. In describing our experiences of interdisciplinary collaboration, we highlight the need for CPG practitioners involved in mixed-methods research to pay attention not only to the language used to describe their study sites and methods, but also to the process of writing itself. How should we structure our narratives? Should we be using narrative form at all? What counts as evidence, and how much is enough? What is the role of the writing process in knowledge production? Is there space for speculation, and if so, where? What metaphors do we consciously or unconsciously rely on, and do these mean the same thing for our collaborators and audiences?

Within the wider framing of CPG research, we have found the mosaic a useful metaphor for thinking about the eco-social hybridity of both the MDV landscape as well as our process of researching and writing about it. The rest of this chapter is structured around three main sections. First, we discuss several of the methods undergirding our project, how we came to use them, and how their creative juxtaposition has helped us conceptualise the eco-social hybridity of our study site. In the next section, we reflect on the process of collaboratively writing and publishing the research, as we believe this is a crucial yet under-examined aspect (and natural extension of) the reflexive practice advocated by CPG. We conclude with some key takeaways and reflections which may have broader relevance.

Mixing methods: Researching together

Our project has grown out of the United States National Science Foundation (NSF)-funded MDV LTER Project. Beginning in the early 1990s, the MDV LTER investigates and monitors the changing ecology of this ice-free region as part of a much larger network of LTER sites across the continental United States and beyond. In keeping with a growing recognition towards the end of the 2010s of the roles of humans in ecological change more generally, there was an NSF requirement that all LTER sites should have a 'human dimensions' component as part of their renewal proposals. For most sites this was relatively easy to fulfil since there were plenty of people living, working, and recreating on the research sites. But with no permanent human population, adding a human dimensions component to the MDV LTER site was a little more challenging.

In response to this challenge, Howkins proposed to Fountain, who was lead Principal Investigator of the MDV LTER at the time, that environmental history could usefully contribute to and draw upon the science that was taking place in the MDV. One initial idea, which proved successful, was the use of photographs, diaries, sketch maps, and scientific reports from the so-called 'Heroic Era' of the early 20th century to offer more 'data points' to think about change over time (Howkins 2016). But as the project developed, and as Chignell joined the research group, we increasingly began to ask wider questions about the MDV as an eco-social system, especially in relation to human impacts and environmental management (neither of which can be considered without thinking about history).

As we investigated these issues, we found CPG to be very helpful for offering a constructive framework for doing collaborative interdisciplinary research, which respected and took seriously the contributions that each of us could bring. A major characteristic of CPG is its epistemic flexibility and pragmatic mixed-methods approach (Lave et al. 2018). This freedom allows what you study to define how it should be studied (in other words, to set its own metaphors (see Lane and Lave, Chapter 3)). We can extend the mosaic metaphor introduced previously to include the different approaches we are taking to study the history, understand the present, and think about the future of the MDV. In this context, the idea of a mosaic helps to highlight the fact that the various methods used are mutually informing each other, not just sitting adjacent.

Our work together on the project began with the creation of a digital archive of human activity in the MDV (Howkins et al. 2020). This involved visits to various archives (Cope, Chapter 22) and collecting photos and oral history (Chakov et al., Chapter 33) interviews from scientists who have worked in the region. In reflecting on this process, we realised that we and other historical researchers often use the language of 'digging' through an archive, which suggests a conceptualisation of archival work as an archaeologist looking for evidence buried at their dig sites. The digging metaphor was perhaps particularly fitting for a project that would later involve actual digging in the soils of the MDV to collect a very different type of data about the region's history (see below). Digging can be hard work—even metaphorical digging in archives. An early point of interdisciplinary convergence came after Fountain and Howkins had come away tired after a long day working in the archives of the Byrd Polar Research Institute in Columbus, Ohio, both recognising that most research involves conducting laborious forms of data collection as the price of bringing useful information to the table.

Our chosen method for presenting the early results of our archival research—a web-based relational database—offers a tool for assembling and reassembling the raw material of the history of the MDV (http://mcmurdohistory.lternet.edu/). Photographs, maps, archival documents, publications, and oral history interviews can be brought together in a way that can help to reveal the connections among them. Here the metaphor of a 'web' is useful, as we sought to weave together information and insights across space and time. For example, the academic publications of a scientist working in the MDV can be displayed alongside images of her enjoying a Christmas dinner at a field camp and spoken recollections of her experiences of Antarctic fieldwork. In this way, a web-based relational database can provide a much fuller record of her experience in the eco-social web of relations that comprise the MDV than any single source in isolation. Through the process of creating and interacting with the database and related website, we began to (re)conceptualise the eco-social relations that comprise our study site.

Having collected thousands of photographs, we soon realised that we could use their subject matter as metaphorical 'benchmarks' for locating historic structures and field camps and assessing changes in human activity and environmental impact over time. By locating historic sites of human activity, our digital archive enabled the collection of soil samples that may reveal the long-term effects of various types of human activity on the microorganisms of the MDV. This led to field surveys at six historic (removed) and present (still in use) research camps, where we collected soil samples at former huts, helicopter landing zones, and outhouses. Our colleagues at the MCM LTER then analysed these in the lab to determine nematode abundance and bacterial diversity. We paired this with repeat photography to help identify and contextualise changes at each site over time. With those same colleagues, we are now analysing these data to see whether and how the soil communities (e.g., organism abundance, diversity) at each camp differ from those in the surrounding areas in an attempt to determine how the field-based study of the MDV alters its ecosystems.

In the process of creating the web-based relational database, we collected and digitised numerous sources found in archives. One important example is a three-volume, paperbound *Bibliography of Dry Valleys Publications* containing almost 1,500 records of scientific research from the late 1970s to the early 1990s. This was produced by New Zealand's Department of Scientific and Industrial Research, and the motivations for this work can be related both to New Zealand's scientific interest in the region and to its claim to ownership of the Ross Dependency (the larger region encompassing the MDV) (Templeton 2000). After digitising each record in the paperbound bibliography, we integrated this information with bibliographic data from the Web of Science database (which added approximately the same number of publications again), giving us a fascinating and virtually complete record of the scientific research that has taken place in the MDV since its discovery.

Using the metaphor of a 'network' to think about scientific collaboration, we realised that we could analyse and visualise the various relationships contained within this bibliographic data using the network analysis software *Gephi* (Bastian et al. 2009). Drawing from both quantitative and qualitative traditions of social network analysis (Chignell, Chapter 40), this mixed-methods approach allowed us to understand the social and intellectual structure of MDV research, such as who is publishing with whom, which academic fields are most closely related, and how these have changed over time. Of course, the outputs from such analyses are not flawless (and are only as good as the data that goes into them), and it can be tempting to reify the resulting network graphs as objective or hastily interpret them (Wyatt et al. 2015). However, this can be mitigated by constructing and interpreting the networks together, iteratively tweaking the layouts to identify new patterns and discussing the results (Jacomy et al. 2014; Venturini et al. 2021; Chignell 2023).

Changing scales was also important in this process. Zooming into the network helped us identify important communities and key knowledge brokers (which in turn pointed us to individuals with which to conduct additional oral history interviews). Zooming out and looking at the collaborative relationships among disciplines revealed a different image of the intellectual landscape of the MDV. We found that collaboration among individuals and academic disciplines increased through time, and the most productive scientists in the network are also the most interdisciplinary. Moreover, to our surprise, this zoomed out image of the intellectual landscape strongly resembled the biogeochemical relationships among different features of the biophysical landscape (Fig. 18.2). Oceanography lies at the periphery of the network, similar to the location of the McMurdo Sound at the terminus of each of the valleys. Glaciology interfaces closely with geology, similar to how glaciers scrape and move rock from valley walls. Geochemistry is heavily mixed with Limnology, just as rock and organic material is suspended or dissolved in the region's lakes. Hydrology lies toward the centre of the network, mirroring the connecting role that streams play in transporting sediment and nutrients between the different landscape features of the MDV (Chignell et al. 2022). This raises interesting questions about the role of the material environment in the development of scientific collaboration and field research in the MDV, and their dynamic interaction with socio-cultural and political factors.

Having a near-complete corpus of scientific articles organised in a relational database offered opportunities for integrating GIS technology, which enables the addition of a geospatial dimension to historical data. This included mapping the location and footprint of each structure (historic and existing) to analyse where and how people have interacted with the environment over time. The metaphor of scientific field stations as 'hubs' for research can also be generative of new questions and new analysis. For example, in viewing the camp features on a map, each with their associated photographs and publications, we started to wonder how the placement of field camps might be related to where scientists conduct their field studies. This led us to design a meta-analysis in which we read each publication in the MDV corpus to identify where its author(s) conducted their field work. Using this information, we associated each publication to the nearest geographical feature (e.g., a glacier, stream, lake), and then analysed how the distribution of study sites and their distance from field camps changed over the historic record. We found that research sites have, on average, gotten closer to field camps over time, and that scientific output does not necessarily correspond to the increase in research in the local area (Chignell et al. 2021). This is likely due to the shift from exploratory, descriptive science (primarily based in geosciences) to long-term monitoring and environmental change (primarily ecological sciences). Not only does this have implications for national Antarctic programmes seeking to mitigate the impacts of scientific activity in the MDV, but it also helps to show the contingency of scientific knowledge about the area, as whether and where a camp is established is often decided based on logistical rather than scientific factors.

Thinking geospatially about how the intellectual network affects and is affected by the biophysical landscape caused us to think more deeply about human perceptions of the MDV. This led us to mix in another method to our overall approach—placename analysis. Naming is one of the first activities that human beings do when settling a new environment; names reflect the cultural perception of the landscape, and are highly stable through time (Seidl 2008; Seidl 2019). Placenames (or 'toponyms') can also be thought of as a form of metaphorical 'flag planting' and can reveal much about the motivations, values, and ambitions of the namers (Rose-Redwood et al. 2010). To explore these questions in the MDV, we collated data from official GIS gazetteers to identify and map patterns in placenames across the region. Adapting the approach of placename studies in New Zealand (Atik and Swaffield 2017), we developed a typology with which to classify the placenames of the MDV. This includes the date named, etymology, connotative meaning(s), as well as the namer, namee, and their respective nationalities and genders. Our preliminary results show a diversity of meanings embedded in the landscape, who ascribed them, and how they have changed over time. They also reveal the contested nature of the MDV, as who names what is often related to who claims what (i.e., territorial claims or aspirations of the different national research programmes). This will ultimately provide another lens through which to understand our study site, one that challenges the dominant vision of the MDV (and Antarctica more broadly) as a pristine wilderness that is devoid of culture and politics and reduces human activity to 'impact'.

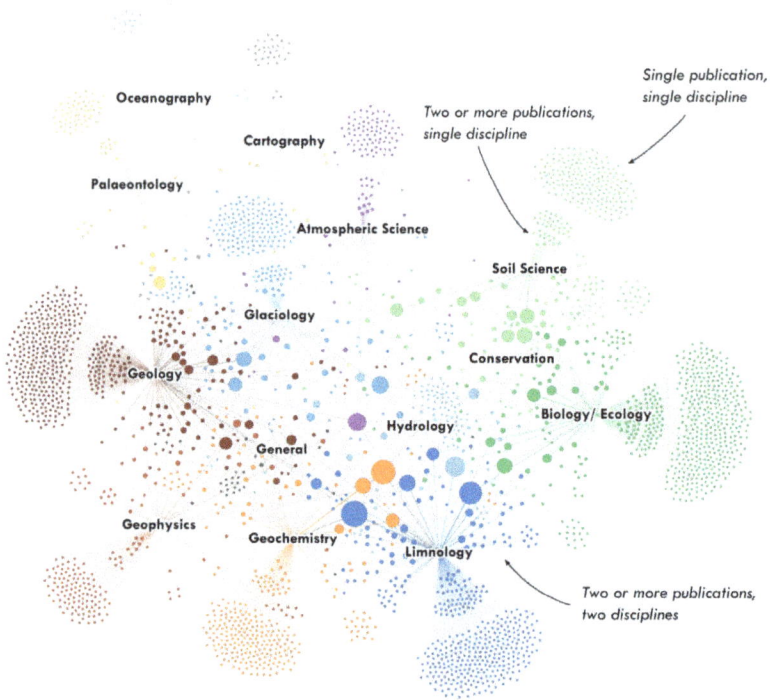

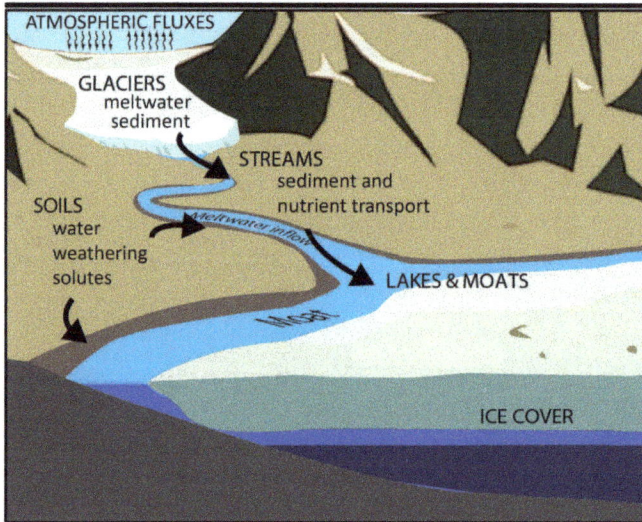

Fig. 18.2 a) Author–discipline network for the MDV showing relationships between individual authors and academic disciplines (Chignell et al. 2022, Fig. 5); b): Conceptual model depicting mass and nutrient flow between landscape components of the MDV. This is a simplified version of the model created by Eric Parrish for the MDV LTER project (McMurdo Dry Valleys LTER 2016, Fig. 7a)

An analysis of MDV placenames or the visualisations produced by the network analysis of scientific collaboration are in many ways a world away from Griffith Taylor's ventures into "Dry Valley" in February 1911, and his observation that the patterned ground he encountered resembled a Roman mosaic. Our project suggests not only that there is room for both in a CPG-informed history of the MDV, but also that there are very real benefits to mixing methods. The images produced by these methods—and the underlying data analysis—can highlight different aspects of the eco-social relations of the MDV, which, when integrated with more traditional forms of historical analysis, offer opportunities for deepening our understanding of change over time. The historical record can, in an almost literal sense, be turned into a visual mosaic. But this visual mosaic still needs to be laid out and explained, and combining these various methods in collaborative writing can be more difficult than doing the analysis.

Mixing metaphors: Writing together

Interdisciplinary research not only involves mixing methods, but also mixing metaphors. By 'mixed metaphors' we are not primarily referring to the classic examples that plague the nightmares of English teachers around the world, and which are often considered something to be avoided (although they may sometimes be present in the MDV: e.g., 'a mosaic that reflects…'). Instead, we are using the idea of mixed metaphors more loosely to describe the different language used by academics in different fields to make sense of their work. Mixing metaphors in this sense is unavoidable in collaborative interdisciplinary writing, but we would argue that the language we use in our own writing—and the values and assumptions it can reveal—is often taken for granted and under-examined. Paying attention to the metaphors we use to think with and the language we use in our writing can help scholars understand each other and communicate to a wider audience. We often take the writing process for granted because it's something familiar to all of us, and one rarely confronts these issues when publishing in one's own field. But the way we write reflects and encapsulates many of the challenges of mixing methods more generally. This is especially true if we hope to meaningfully collaborate across the "two cultures" of the humanities and the sciences (Snow 1959; Snow 2012).

Differences in style

For example, biophysical—and many social—scientists tend to collect and analyse data, create figures, and then fill in the text of a manuscript around a pre-existing structure. The writing is often in a passive voice that implies a certain distance between the subject and the object of study (e.g., "the data were analysed"). Gray (2017) argues that "the prevalence of the distant voice in academic writing can reinforce an unreflective orientation toward objectivist traditions and norms, and potentially obscure the underlying epistemological implications of voice in scholarly writing". This is often paired with a linear roll out of ideas that follows a fairly predictable structure (e.g., methods, results, discussion, conclusion) which is strictly enforced by many scientific journals. This structure effectively extends the underlying epistemological assumptions about the separation of subject and object to the page. In contrast, while historians recognise the value of primary sources and data as the foundation of historical knowledge, they also emphasise the importance of narrative and storytelling in making sense of what happened. The act of writing becomes part of the analytical process, not simply the reporting of analyses that have already taken place. For scientists observing and participating in the process of historical writing, this approach might appear to blur the lines between 'results' and 'discussion', and potentially even seem vague or misleading.

Different approaches to writing have been repeated points of friction and discussion for us as we have worked together. We believe that the emphasis that many humanities fields place on narrative may actually represent a hurdle for interdisciplinary collaboration (see also Alagona et al. 2023). Many biophysical scientists may be put off by a strong focus on narrative, especially if it is perceived to come at the expense of data visualisation and concise writing. For historians, a concern that multiple voices could threaten the integrity of an engaging narrative may underlie their reluctance to collaborate. This argument may be bolstered by the idea that if there is only one author, their biases can more easily be understood by the reader. Thus, articles with multiple voices and authors may be seen by historians as both harder to read and harder to critically assess. While these concerns are warranted, we would also argue that this 'lone wolf' tradition makes it easier for historians to remove themselves from the work. While environmental historians are often reflexive in their practice, unlike critical human (and increasingly, physical) geographers, rarely do they write themselves into their narratives. This can imply a distance or even separation between the author and their object of study; not quite the "gaze from nowhere" common in the biophysical sciences (Haraway 1988), but rather a certain epistemic authority, which is made more potent when paired with a finely crafted narrative.

Developing a shared thought style and 'creole'

Although the above is a somewhat simplistic characterisation, and narrative of course plays a large role in scientific writing, we believe these differences in style can form an obstacle to meaningful collaboration between the "two cultures". In this way, the challenge of writing together blurs the lines among language, epistemology, and practice. We soon became aware that we were each coming with our own perspectives and writing styles—each of which has its own metaphors, prose, conventions, organising structures, and philosophical assumptions. These differences in writing are not simply aesthetic in nature, but often reflect significant—and often unconscious—differences in epistemology (Gray 2017). In the early days of the project we tended to take the writing process for granted, perhaps because it is something that is familiar to all of us. For us, this resulted in a period of what Friedrich Wallner calls 'strangification', the unsettling experience of wading into the unfamiliar waters of another field (Badie and Mahmoudi 2007; Öberg et al. 2013). It was only once we had gone through that process that we were able to begin to affectively (and effectively) think and write *together*, rather than as multi-disciplinary collaborators.

In reflecting on the process of interdisciplinary collaboration in science, Peter Galison observed that scientific subcultures "...form trading zones in which participants create first a shared jargon, then a pidgin, and finally a creole.... [S]cientific fields, like languages, emerge out of this process of hybridization" (Gorman 2010). We have certainly found this to be true in our case. For example, when we first began to write up the results of the project, Howkins would often take the results of our analyses and build a couple of pages of narrative around them, weaving in contextual details, suggesting causal relationships, and blending the discussion of results with uncertainties and future directions (a common approach in historical writing). Chignell, who had previously only written scientific papers, had to first get accustomed to the narrative format, after which he would correct or qualify certain interpretations of the data, adding clarifications without changing the format. However, upon reading this composite draft, Fountain would often want to split the narrative into discrete sections, as he felt uncomfortable with the blurring between results, discussion, and conclusions.

After a few rounds of this, and many discussions, it started to become clear that our back and forth was rooted less in fundamental disagreements, but rather in assumed 'best practices' we had inherited from our respective disciplines (see Lane and Lave, Chapter 3). Howkins' fitting the results to a narrative while heavily caveating the claims made from quantitative analyses reflected his training in historical research, which relies on context and qualification to draw nuanced conclusions from disparate and often uncertain data. Chignell and Fountain were initially uncomfortable about the claims made based on historical context, as these might be considered 'speculation' in scientific writing. This is not just a matter of avoiding confirmation bias, or cherry-picking data, as what one field may consider cherry picking may be commonplace in another. Over time, we each became more aware of our own and each other's disciplinary assumptions and languages to be able to confidently defer to each other's expertise and create a shared voice that reflects a composite thought style.

Challenges in publishing

We found that the review process has similar issues to the writing process, particularly in a publishing system that, despite regular calls for interdisciplinary research, can be surprisingly resistant to articles that read differently than a single discipline's norms (see Lane and Lave, Chapter 3). Deciding on where to publish has involved multiple rounds of deliberation and strategising. It has also fed back into how we write. Do we aim for a science journal, and try to Trojan horse some history into the required article structure? Or do we try to sprinkle some scientific data into a history article? Ultimately, these decisions have tended to come down to the specific point(s) we are trying to make and the specific audience(s) we are trying to reach. This has often meant allowing one disciplinary style to be dominant while trying to bring in a smattering of the creole we developed.

For example, in analysing the spatio-temporal relationships between field camp placement and field site location, we began to think about specific camps in the MDV and their histories. This led us back to the archives and resulted in an article rethinking the history of Vanda Station, New Zealand's largest and only research base on the continent. In the article, we challenge the dominant idea of the station being removed simply as a result of rising lake levels, and instead suggest that the removal was the result of several interconnected factors around geopolitics, science, race, class, and gender (Howkins et al. 2021). While our archival research formed the basis for framing the article on the construction, operation, and removal of Vanda Station as a 'biography', we supplemented this with other methods, most notably the social network analysis (which in turn was based on bibliographic data uncovered from the archival visits). In writing this article, we had candid discussions about the weight and intersection of different types of evidence. What language can (or should) we use to talk about sensitive topics and histories? Do we have enough evidence to call something 'racism'? What counts as 'enough' evidence in history as opposed to science? Ultimately, we decided to publish this work in a humanities journal, and our language reflects the tone and style of historians.

In contrast, our most 'science-y' paper was the GIS paper mentioned previously that analysed the spatio-temporal relationship between field camp placement and field study sites (Chignell et al., 2021). This was a highly technical paper with many moving parts, and we published it in a science journal with a linear format and strict rules about data availability. Although the historical context that we had gained was key in developing our research questions and interpreting the results, the final product looks and reads very much like a scientific article.

Determined to have a manuscript that demonstrated the kind of integrative thinking and 'creole' we had developed, we attempted to combine the social network analysis, bibliometrics, and interview data into a single paper. To do so, we used a mixed style based around an argument of mixing metaphors (i.e., the material landscape resembling the collaboration network). This blended the scientific style with a historical style that attempted to tie together the different datasets into a narrative. Perhaps unsurprisingly, this was the most difficult paper in our project to get published. Starting out, we thought that the ideas in the paper most reflected a geographical approach, and initially submitted to a leading geographical journal. This submission led to mixed reviews. One reviewer praised the large dataset and analytical work we had done, but another thought the methods were 'plodding' and distracted from the narrative. Somewhat ironically, one of the reviewers pointed to a text that one of us had published as an example of a key work that was missing from our analysis. After two rounds of major revisions, during which our work was rejected and referred to a related geographical journal, we were unable to satisfy all the reviewers, and, crucially, the editors. We ultimately decided to step back and publish the distinct disciplinary components separately (e.g., Chignell et al. 2022).

Through this and other publishing experiences, we have found the traditional scientific article to be problematic. Its short word length and rigid style leave little space for the integrative holism that frequently emerges from our collaborative work. A book-length history of the MDV, which we are currently working on, provides more space to explore the nuances and complexities of interdisciplinary research. However, traditional historical monographs—which frequently privilege narrative coherence—have little room for or interest in the detailed descriptions of data processing and technical methods which are integral to our collaborative work. Perhaps appropriately for mixed-methods research, our strategy has involved a mixture of publishing more disciplinary specific research in disciplinary journals, while continuing to develop some genuinely interdisciplinary publications, especially the co-authored book.

Conclusion and takeaways

Like a mosaic, the integrative value of mixed-methods research might not be apparent when looking up close, but at distance, it should generate an image of a place or problem that is both unique and impactful. When working together on a common mosaic, each person has their own style and 'tiles', which can be points of visual (and physical) friction when they run up against each other. These adjacencies, with their discordant colours and shapes—sometimes awkwardly chiselled to fit together—represent areas of epistemic compromise and creativity. The unpredictability of this process—no one knows what the final image will look like even as they co-create it—is different from when you are the sole author, or when you are working on a disciplinary project with a shared language and tighter epistemic bounds. It is a surrendering of control over the project, which is beneficial—and perhaps even necessary—for successfully mixing methods. Moreover, theoretical and practical considerations come together in the writing, and it is not guaranteed that they will ultimately form a cohesive whole. We see the writing process as a key opportunity for communal self-reflection (on each other's positionalities, preferences, and assumptions). This not only helps to blur where one person's 'tilework' ends and another's begins, but can feed back in unexpected and interesting ways on the outcomes of the research project. We believe this reflexive practice is essential for successful interdisciplinary mixed-methods work in CPG and related fields.

Our experience will of course be different to others, as we are a specific group of people working in a very specific context. Still, we offer several practical takeaways that may be useful for others:

- **Mix methods, but don't assume it will be easy.** Each of us has had to learn tools and skills far outside of our disciplinary training and comfort zones. This has required significant time investment. It has also involved reanalysing the data when we realised we did it wrong the first time (and in some cases, the second, and the third!).

- **Give yourselves time.** Time to get to know yourselves, each other, and your data. Lane (2017) draws on Isabelle Stengers' idea of 'slow science' as a potentially valuable approach for CPG scholars (Stengers and Muecke 2018; Lane and Lave, Chapter 3). We couldn't agree more. Several ideas and analyses that we explored in the early years of our project but had to shelve have since become central to our thinking.

- **Mix your skill sets, not just your methods.** Be willing to get your hands dirty in each other's domains. In our case, Howkins was a key part of the soil sampling design and field work.

Fountain learned the challenges and excitement of sifting through archives and conducting interviews. Chignell learned bibliometric analyses and how to design metadata in line with archival standards. This builds a shared understanding and mutual appreciation for the complexity, difficulty, and utility of each other's methods (Greer et al. 2018).

- **Don't take the writing process for granted.** Treat it as a site for constructively critical work. Use it to think deeply about your own and each other's taken-for-granted assumptions, epistemologies, and values. This is a natural extension of the relentless reflexivity advocated by CPG.

- **Be clear but not bland.** Developing a shared language and writing style risks taking the lowest common denominator of what the group can agree on. Try to become comfortable with disagreements when they arise, and explore their epistemic origins together through reflective discussion. Try to develop a 'creole' that you are all conversant in and excited about communicating to others.

- **Not everything has to be integrated.** You may want or need to split your work into disciplinary language and venues. This is perfectly fine, and can serve as published references to cite when communicating the larger integrative project in other venues (e.g., presentations, books, interdisciplinary journals).

- **Take risks, be open, and willing to loosen control.** Don't be too attached to the narrative, voice, direction, or outcomes of the project. This requires a willingness to let go, or at least modify, your favourite theories, terminology, and turns of phrase. When it comes to publishing, be persistent but not pushing; if a journal clearly doesn't like your style, take it somewhere else.

In sum, our project demonstrates that historians (and other humanities scholars) have much to contribute to contemporary scientific research and policymaking, and scientists have much to contribute to historical research. CPG provides a unique and much needed 'trading zone' for this type of work (see also Rader et al. 2023). As the three of us have experienced mutual transformations in our thinking, we have also come to hope that our work will lead to positive eco-social transformation in our study site (Law 2018). To do so, we are working with broader Antarctic groups to provide policy recommendations that challenge dominant narratives and framings (e.g., Hughes et al. 2022)—especially those tied to ideas of wilderness protection which remain prevalent in Antarctica. In this process, it seems that methodological eclecticism serves to draw people in, as it often differs from what they are used to seeing in more disciplinary presentations. This has been facilitated by our use of the creole we developed to communicate these ideas to our colleagues in our respective disciplines. Reflecting on these conversations, we suggest that taking a mixed-methods approach may actually help to preempt knee-jerk reactions to any disciplinary omissions, since it is clear that we are attempting a balanced treatment. We are excited to continue these conversations, and hope these reflections are useful to others seeking to mix their own methods and metaphors.

Acknowledgements

Howkins and Fountain are the leaders of this project, and Chignell is a graduate research assistant. We have also had many key collaborators and co-authors along the way, including Poppie Gullett, Melissa Brett, Evelin Preciado, and Madeline Myers. In addition, we have benefited immensely from the feedback and support of the other members of the McMurdo Dry Valleys LTER. We would like to acknowledge financial and logistical support from NSF Grants #1443475 and #1637708 and British Academy Grant KF3/100152.

References cited

Alagona, P., M. Carey, and A. Howkins. 2023. 'Better together? The values, obstacles, opportunities, and prospects for collaborative research in environmental history', *Environmental History*, 28.2. https://doi.org/10.1086/723784

Atik, M. and S. Swaffield. 2017. 'Place names and landscape character: a case study from Otago Region, New Zealand', *Landscape Research*, 42.5, pp. 455–470. https://doi.org/10.1080/01426397.2017.1283395

Badie, K. and M.T. Mahmoudi. 2007. *Strangification: A New Paradigm in Knowledge Processing and Creation*, ed. by F.G. Wallner. (P. Lang).

Bastian, M., S. Heymann, and M. Jacomy. 2009. 'Gephi: an open source software for exploring and manipulating networks', *Icwsm*, 8, pp. 361–362. https://doi.org/10.1609/icwsm.v3i1.13937

Biermann, C., L.C. Kelley, and R. Lave. 2021. 'Putting the Anthropocene into practice: Methodological implications', *Annals of the American Association of Geographers*, 111.3, pp. 808–818. https://doi.org/10.4324/9781003208211-20

Chakov, A., Chang, T., Covey, H., Dickson, T., Goggins, S., Harris, N., Purna, S., Widell, S., and Druschke, C.G., Chapter 33, this volume. 'Oral history'.

Chignell, S.M. 2023. 'A missing link? Network analysis as an empirical approach for critical physical geography', *The Canadian Geographer / Le Géographe canadien*, 67.1, pp. 52–73. https://doi.org/10.1111/cag.12767

Chignell, S.M., M.E. Myers, A. Howkins, and A.G. Fountain. 2021. 'Research sites get closer to field camps over time: Informing environmental management through a geospatial analysis of science in the McMurdo Dry Valleys, Antarctica', *PLOS ONE*, 16.11. https://doi.org/10.1371/journal.pone.0257950

Chignell, S.M., A. Howkins, P. Gullett, and A.G. Fountain. 2022. 'Patterns of interdisciplinary collaboration resemble biogeochemical relationships in the McMurdo Dry Valleys, Antarctica: a historical social network analysis of science, 1907–2016', *Polar Research*, 41. https://doi.org/10.33265/polar.v41.8037

Cope, M., Chapter 22, this volume. 'Archival methods'.

Gorman, M.E. 2010. *Trading Zones and Interactional Expertise: Creating New Kinds of Collaboration* (MIT Press).

Gray, G. 2017. 'Academic voice in scholarly writing', *The Qualitative Report*, 22.1, pp. 179–196. https://doi.org/10.46743/2160-3715/2017.2362

Greer, K., K. Hemsworth, A. Csank, and K. Calvert. 2018. 'Interdisciplinary research on past environments through the lens of historical-critical physical geographies', *Historical Geography*, 46.1, pp. 32–47. https://doi.org/10.1353/hgo.2018.0024

Halffman, W. 2019. 'Frames: Beyond facts versus values', in *Environmental Expertise: Connecting Science, Policy and Society*, ed. by E. Turnhout, W. Halffman, and W. Tuinstra (Cambridge University Press), pp. 36–57. https://doi.org/10.1017/9781316162514.004

Haraway, D. 1988. 'Situated knowledges: the science question in feminism and the privilege of partial perspective', *Feminist Studies*, 14.3, pp. 575–599. https://doi.org/10.2307/3178066

Howkins, A. 2016. 'Taylor's valley: what the history of Antarctica's "Heroic Era" can contribute to contemporary ecological research in the McMurdo

Dry Valleys', *Environment and History*, 22.1, pp. 3–28. https://doi.org/10.319 7/096734016x14497391602125

Howkins, A., S.M. Chignell, P. Gullett, A.G. Fountain, M. Brett, and E. Preciado. 2020. 'A digital archive of human activity in the McMurdo Dry Valleys, Antarctica', *Earth System Science Data*, 12.2, pp. 1117–1122. https:// doi.org/10.5194/essd-12-1117-2020

Howkins, A., S. Chignell, and A.G. Fountain. 2021. 'Vanda Station, Antarctica: a biography of the Anthropocene', *Journal of the British Academy*, 9.6, pp. 61–89. https://doi.org/10.5871/jba/009s6.061

Hughes, K.A., M. Santos, J.A. Caccavo, S.M. Chignell, N.B. Gardiner, N. Gilbert, A. Howkins, B.J.V. Vuuren, J.R. Lee, D. Liggett, A. Lowther, H. Lynch, A. Quesada, H.C. Shin, A. Soutullo, and A. Terauds. 2022. 'Ant-ICON– "Integrated science to inform Antarctic and Southern Ocean conservation": a new SCAR Scientific Research Programme', *Antarctic Science*, 34.6, pp. 446–455. https://doi.org/10.1017/s0954102022000402

Jacomy, M., T. Venturini, S. Heymann, and M. Bastian. 2014. 'ForceAtlas2, a continuous graph layout algorithm for handy network visualization designed for the Gephi software', *PLOS ONE*, 9.6. https://doi.org/10.1371/ journal.pone.0098679

Kull, C.A. 2018. 'Critical invasion science: Weeds, pests, and aliens', in *The Palgrave Handbook of Critical Physical Geography*, ed. by R. Lave, C. Biermann, and S.N. Lane (Springer), pp. 249–272. https://doi. org/10.1007/978-3-319-71461-5_12

Lane, S.N. and Lave, R., Chapter 3, this volume. 'Frames, disciplines and mixing methods in environmental research'.

Lane, S.N. 2017. 'Slow science, the geographical expedition, and critical physical geography: Slow science and critical physical geography', *The Canadian Geographer / Le Géographe canadien*, 61.1, pp. 84–101. https://doi. org/10.1111/cag.12329

Lave, R., M.W. Wilson, E.S. Barron, C. Biermann, M.A. Carey, C.S. Duvall, L. Johnson, K.M. Lane, N. McClintock, D. Munroe, R. Pain, J. Proctor, B.L. Rhoads, M.M. Robertson, J. Rossi, N.F. Sayre, G. Simon, M. Tadaki, and C.V. Dyke. 2014. 'Intervention: Critical physical geography', *The Canadian Geographer / Le Géographe canadien*, 58.1, pp. 1–10. https://doi.org/10.1111/ cag.12061

Lave, R., C. Biermann, and S.N. Lane. 2018. 'Introducing critical physical geography', in *The Palgrave Handbook of Critical Physical Geography*, ed. by R. Lave, C. Biermann, and S.N. Lane (Springer), pp. 3–21. https://doi. org/10.1007/978-3-319-71461-5_1

McMurdo Dry Valleys LTER. 2016. *LTER: Ecosystem Response to Amplified Landscape Connectivity in the McMurdo Dry Valleys, Antarctica Project*

Overview (McMurdo Dry Valleys LTER). https://mcm.lternet.edu/sites/default/files/proposal_2016.pdf

Öberg, G., L. Fortmann, and T. Gray. 2013. ‹Is interdisciplinary research a mashup?›, IRES Working Paper Series 2013.2.

Priscu, J.C. and A. Howkins. 2016. *Environmental Assessment of the McMurdo Dry Valleys: Witness to the Past and Guide to the Future* (Department of Land Resources and Environmental Sciences, College of Agriculture, Montana State University).

Rader, A.M., C. Biermann, S.M. Chignell, K.R. Clifford, L.C. Kelley, and R. Lave. 2023. 'Practicing critical physical geography: New trading zones and interactional expertise in an expanding field', *The Canadian Geographer / Le Géographe canadien*, 67.1, pp. 10–16. https://doi.org/10.1111/cag.12828

Rose-Redwood, R., D. Alderman, and M. Azaryahu. 2010. 'Geographies of toponymic inscription: New directions in critical place-name studies', *Progress in Human Geography*, 34.4, pp. 453–470. https://doi.org/10.1177/0309132509351042

Schön, D.A. 1979. 'Generative metaphor: a perspective on problem-setting in social policy', *Metaphor and Thought*, 2, pp. 137–163.

Seidl, N.P. 2008. 'Significance of toponyms, with emphasis on field names, for studying cultural landscape', *Acta Geographica Slovenica*, 48.1, pp. 33–56. https://doi.org/10.3986/ags48102

Seidl, N.P. 2019. 'Engraved in the landscape: the study of spatial and temporal characteristics of field names in the changing landscape', *Names*, 67.1, pp. 16–29. https://doi.org/10.1080/00277738.2017.1415539

Snow, C.P. 1959. 'Two cultures', *Science*, 130.3373, pp. 419–419.

Snow, C.P. 2012. *The Two Cultures* (Cambridge University Press).

Stengers, I. and S. Muecke. 2018. *Another Science is Possible: A Manifesto for Slow Science* (Polity).

Tadaki, M., J. Clapcott, R. Holmes, C. MacNeil, and R. Young. 2023. 'Transforming freshwater politics through metaphors: Struggles over ecosystem health, legal personhood, and invasive species in Aotearoa New Zealand', *People and Nature*, 3. https://doi.org/10.1002/pan3.10430

Venturini, T., M. Jacomy, and P. Jensen. 2021. 'What do we see when we look at networks: visual network analysis, relational ambiguity, and force-directed layouts', *Big Data and Society*, 8.1. https://doi.org/10.1177/20539517211018488

Wyatt, S., S. Milojević, H. Park, and L. Leydesdorff. 2015. *Quantitative and Qualitative STS: The Intellectual and Practical Contributions of Scientometrics* (Social Science Research Network). https://doi.org/10.2139/ssrn.2588336

19. Engaging remote sensing and ethnography to seed alternative landscape stories and scripts

Lisa C. Kelley

Introduction

This chapter reflects on the expanding influence of a rapidly evolving remote sensing (RS) science—and on the opportunities, contradictions, and challenges of engaging RS within a critical and ethnographically-oriented account of human-environment change.

I ground these reflections in roughly 10 years' experience engaging ethnography (Sayre, Chapter 34) and RS (Braun, Chapter 39) to study land and labour relations in eastern Sulawesi, Indonesia. Lands and land systems in this area have long been managed by the Indigenous Tolaki people through practices of in-situ landscape mobility across diversified forests, agroforests, and wetland complexes. Extra-local labour migration is also now common in a landscape context marked by histories of colonial and state-led development, in-migration, and state, corporate, and small-scale commodity productions. My work traces these transformations, from a spectacular boom and bust in smallholder cacao production beginning in the late 1970s to recent intensifications in agro-industrial development and extreme flood events.

Below, I begin by reviewing the literature that informs my approach while making a case for practices of integration that go beyond identifying the deficiencies with dominant RS science to engaging both RS and ethnography with a view to what they reveal and how they can be reimagined. I then discuss three linked observations shaped by ongoing work in Sulawesi. The first reflects on how staying with place and process can shape intuitive approaches to integration where

  https://doi.org/10.11647/OBP.0418.19

integration is necessary to deepen understanding of a particular socio-spatial relation. The second focuses on how staying with the tensions between RS and ethnographic findings can enhance knowledge reflexivity. The third reflects on how embracing the epistemological continuities between RS and ethnography can inspire approaches that embed integrative instincts, whether in spatialising ethnography to map scale-making processes or leveraging RS to fashion an ethnographically thick description of place-based relations.

My overarching argument is that mixing RS and ethnographic methods is less about eliminating or overcoming barriers to integration than it is about facilitating conversation between researcher, process, and place. I argue that fostering such conversation can enable modes of integration that are intuitive, reflexive, and generative of questions and methods more attentive to place, positionality, and interpretive possibility.

Starting points

Diverse bodies of work have reflected on the promise and peril of integrating geospatial science and qualitative data (Kwan 2016; 2002; Schuurman and Pratt 2002; Turner 2003). Critical and participatory geographical information systems (GIS) and radical cartographic traditions have leveraged ethnography to look beyond the Cartesian grid, revealing other ways of sensing and knowing nature (Bryan 2021; Elwood 2006; Fujikane 2021; Kitchin et al. 2013). Political geographers have engaged qualitative insights to "further inscribe the vertical" into overly horizontal notions of power (Graham 2004: 16), illustrating how air and space power foster geopolitical control over volumetric territory (Adey 2010; Elden 2013). Accounts marrying political ecology and land change science have mixed RS and ethnographic data to re-read conflictual and contested processes of landscape change (Fairhead and Leach 1996; Lukas 2014); document methodological omissions within RS (Bennett and Faxon 2021; Kelley et al. 2017); and fashion 'multiperspectival' RS practices (Arce-Nazario 2016; Bennett et al. 2022; Ferring and Hausermann 2019; Gleason and Hamdan 2017).

These insightful reflections on integrating geospatial and qualitative data cannot be the last word on the subject, though. This is particularly true given rapid shifts within RS science. These include expanding public access to large satellite data, as from Landsat, MODIS, and Sentinel (Zhu et al. 2019), the development of cloud-based computing platforms (Gorelick et al. 2017), and algorithmic leaps in training and classifying satellite data (Belgiu and Drăguț 2016), including in the application of artificial intelligence methods to automate RS monitoring (Yang et al. 2022). RS capacities have also expanded over the past decade with improvements in the capacity of sensors to capture data across different wavelengths and amid the rapid development and launch of nanosatellite constellations (Nagel et al. 2020). Large swathes of the Earth's surface, for instance, are now being mapped at 3 m resolution daily (*Planet | Homepage* 2022).

Such developments rework the politics of knowing, sensing, and governing human-environmental relations (Havice et al. 2022), making it possible to measure, monitor, and detect eco-social relations in ways that transmute them, for instance, into calculable risks (Hind and Lammes 2016). RS data and methods are deployed for climate adaptation planning along the US Gulf Coast, groundwater monitoring in California's Central Valley, and deforestation monitoring in countries such as Indonesia and Brazil (Goldstein and Nost 2022). "Big" RS data are also being instrumentalised by states and other powerful actors, including the private sector entities financing, guiding, and benefiting from many of the above developments (Nagel et al. 2020). These tendencies shape important concerns over how RS tools are being applied to produce 'surveillable space' (Archer 2021), including through the "collection, harmonization, and automated analysis of big (environmental) data from a variety of sources" (Rothe 2017: 334) in ways that feed into violent forms of exclusion, territorialisation, and border securitisation (Nost and Goldstein 2022).

These tendencies highlight the power-laden ways governments, corporations and other powerful groups engage dominant scientific methodologies, begetting exclusionary structures for studying, knowing, and managing nature. As Robbins (2015: 90) notes, knowledge politics such as these have encouraged critical nature-society scholars to adopt an "ambivalent, if not wary, relationship to the simple models of materiality and measurability that underlie expert practices and methods". While such wariness is merited, this chapter rejects the corresponding conclusion that it should result in "'having it both ways': advancing arguments by mobilising diverse branches of nature/society research while simultaneously undermining these same branches of science." Using dominant science to undermine dominant science may advance critique, but it does little to sow the seeds of a more deeply relational and transformative scientific praxis.

This is problematic because integrating RS and ethnography can also enable more inclusive, just, and transformative methods-and-ethics of mapping. Developing ways of doing and being, or "looking for the cracks" (Harvey and Haraway 1995: 514) that already exist within RS, for example, has been a means of reworking dominant environmental discourses, elevating situated knowledges of change, and empowering marginalised actors (Bennett et al. 2022). As Mei-Po Kwan notes (2016: 281), "[a]s the algorithmic mediation of the knowledge production process increases, the precise ways in which data are generated, processed, and analysed tend to become increasingly invisible to and detached from researchers". Deep engagements with RS are necessary to become intimate with such invisibilities.

This chapter thus begins from the premise that while RS and ethnography tend to be engaged within distinct scalar, epistemological, and disciplinary paradigms, the two practices are not necessarily antagonistic. Rather, RS and ethnography can be convivial partners in telling a multiplicity of landscape stories and writing new mapping scripts—despite, and in some cases, because of the tensions between such stories and scripts (see Biermann and Gibbes, Chapter 4). In centring both RS and ethnography, I also begin from the assumption that fostering such an intimate dialogue requires holding the 'guts' of both methods in one's hands (so to speak). As Liboiron (2021: 119) writes of their efforts to run an anti-colonial marine biology lab, for example:

> Through our handling of dead birds, little red organic or plastic bits, specimen bag, tweezers, and the statistics of outliers, we find that we must be accountable to these things and their worlds in ways that don't always show themselves when we are theorizing at our desks and handling keyboards and books. Thinking at desks is still a way of doing, of course. But when your hands are in someone's guts unanticipated issues tend to present themselves.

Holding the guts in hand, with respect to RS and ethnography, refers to working directly with both methods with a "care for the subject" (Schuurman and Pratt 2002) and by "standing in faith" (TallBear 2014), i.e., in ways that commit to critiquing a given practice and investing in its transformation. It also refers to the unanticipated ways "placing oneself in a situation where that situation can speak back" (Lane 2017: 95; Lane and Lave, Chapter 3) can shift one's sense of the dynamics being studied and the modes of engagement required of the researcher (cf. Bunge 1979). Such orientations make it possible to engage both RS and ethnography as living methods—living in that they are not finished protocols or products but still possible to strategically reimagine and repurpose. It also makes it possible to develop approaches that better engage the material and discursive dimensions of human-environmental change, or the ways in which landscapes are "as much the product of unequal power relations, histories of colonialism, and racial and gender disparities as they are of hydrology, ecology, and climate change" (Lave et al. 2014: 3).

Staying with the conversation

When I began working in eastern Sulawesi, Indonesia in 2011, I did not plan to integrate RS and ethnography. I had initial research questions that turned out to be less interesting than the stories told by the land itself and by the people I met in fields, at kiosks, and on porches. In this section, I reflect on how staying with such stories—and the emergent research conversation more generally—guided a more flexible and generative approach to integrating RS and ethnographic methods than if I had focused on how best to achieve integration in abstract. Integration can be intuitive, I suggest, where it is necessary to respond to questions posed by place and process.

I was in Sulawesi to study accelerating public-private investments in the smallholder cacao sector, interventions initiated in the early 2000s and expanding by the 2010s under the rubric of multi-year, multi-million-dollar corporate sustainability commitments. Associated programmes generally aimed to increase cacao yields by providing participants with grafts and fertiliser, and by training them in farm management techniques designed to address yield losses due to pests and pathogens. Programmatic claims to sustainability were rooted in an understanding of cacao as a deforestation crop; an assumption predicated on the yield advantage derived from planting cacao in forested land. According to some observers, this Ricardian "rent" shaped an inexorable linkage between cacao and deforestation. As Ruf et al. (2004: 108) write, for instance:

> Where [cacao boom and bust cycles] started, they led to the opening up of new forests, sometimes at a tremendous speed. Where they ended, they left behind, in the best cases, disease-infested groves of low productivity in a secondary forest environment but often only poor fallows and pastures.

Programme sponsors claimed that increasing cacao yields in people's existing fields would "spare" further forest from clearance while simultaneously sustaining the livelihoods of smallholders presumed dependent on the crop.

People's stories in Sulawesi complicated these propositions. Many people recounted planting cacao in former swidden rice fields through the 1980s and 1990s following decades of political violence and armed conflict, using the crop to physically inscribe their claims to land. Such inscription was vital following the rise of Suharto's authoritarian regime (1966–1998) and the introduction of the 1967 Basic Forestry Law, which established the Indonesian government's de jure rights over all lands perceived to be 'un-used' or 'under-utilised' (Peluso and Vandergeest 2001). Cacao, as a state-sanctioned tree crop commodity, comprised a legible form of use in a context of rampant state enclosures for rattan and timber production, transmigration settlements, and wet-rice expansion. Other people recounted planting cacao in Imperata grasslands because such lands were seen to be more 'marginal' for state-led production, and were thus less likely to be expropriated by the state than were mixed agroforests, even had these been planted in cacao.

More generally, by the 2010s, people's dependency on cacao had begun to decline. Many of the individuals targeted by intensification initiatives could be better understood as 'flexible frontier-makers' (Zhu and Peluso 2021) than as full-time cacao farmers given their routine circular migration to mining, logging, and construction work sites. Long-term deteriorations in grower value capture and price volatility in the cacao sector were also making the crop unattractive relative to alternative commodity crops like peppercorn—undermining the viability of yield intensification initiatives predicated on increasing the labour people invested in cacao. In this context, most people were trying to transition out of cacao and into crops more amenable to intermittent labour availability, often within their existing fields.

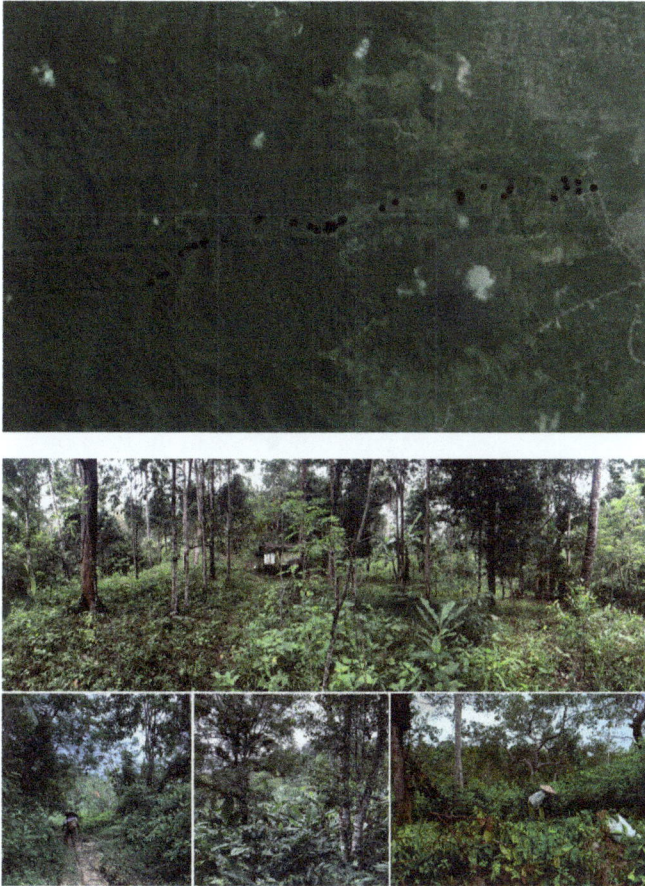

Fig. 19.1 Lisa C. Kelley, Former swidden landscapes within the state forest:
a) the view from above; b) the view from below

Though I did not arrive at the above understandings over an initial six weeks of ethnography, I did start to ask the questions that led to my eventual engagement with RS. Most centrally, I wanted to explain how the forests and farmers so centred in dominant discourses of cacao expansion (Clough et al. 2009; Ruf et al. 2004; Ruf and Yoddang 1999), and in imaginaries of tropical nature more generally (Slater, 2003), were more complex than a single vantage point could reveal. Figure 19.1 illustrates this point, depicting lands within protected forests better understood as regrown swidden rice fields. Figure 19.1a overlays GPS point data depicted in black over a satellite image encapsulating a lowland settlement (far right) and an adjacent state forest (centre and left). Each point marks a tree at least 50 years old that had been planted at the time a given rice field was either cultivated or fallowed. Figure 19.1b illustrates how such trees (e.g., teak, rambutan, durian, coffee, and sago palms), among other things, seed the forest as food and timber gardens still accessed for such purposes.

Returning to graduate coursework in the fall, I was interested in further exploring the cacao-forest questions initial ethnographic observations had raised. Though ethnographic data is often used to ground the satellite gaze, I hoped to use satellite data to ground the ethnography, specifically, to reveal the spatiality and spatial extent of tree cover regrowth. I had admittedly modest goals. I intended to download a few images and take a quick look at the how the landscapes I had visited might have looked in the periods prior to widespread cacao adoption.

My initial understandings of RS were overly simplistic. It took me a couple weeks to do even the most basic work of selecting, downloading, and rendering satellite images for visual interpretation. For me, this began the process of attending to the RS field's disciplinary "vernacular" norms, and conventions (Johnston 1986), whether to identify coding and processing platforms that enabled imagery to be opened on a budget laptop or to select bands that resulted in a "true" or "false" color composite. Reworking my a priori assumptions about RS science through this process of "strangification" (Öberg et al. 2013), wherein common research foci took on new dimension when viewed through other categories, concepts, and models, also required that I suspend doubts about where I was headed or my capacity to develop associated research skills. By the semester's end, I had compiled a wall-to-wall composite image of Southeast Sulawesi province, one which suggested that large parts of the landscape had few if any trees as of 1972, the year of the first publicly available satellite imagery in the region.

My interest in cacao-forest relations deepened alongside a deepening engagement with the RS literature, where I saw that the handful of studies exploring cacao expansion in Sulawesi had exclusively examined forest loss rather than tree gain (Clough et al. 2011; Erasmi et al. 2004; Steffan-Dewenter et al. 2007). Moreover, these studies focused on known periods of cacao expansion and sites of known forest clearance, selection biases that likely precluded the possibility of 'seeing' tree cover gain even had it been incorporated into the analysis. Tree cover gain takes years and, unlike tree clearance, is not visible in back-to-back satellite images. Whilst tree cover loss is often concentrated on the landscape, tree cover gain in this area was likely to be more dispersed given rotational histories of swidden farming. Rather than present a more 'objective' or 'accurate' understanding of land change in Sulawesi, then, existing RS analyses largely reproduced existing understandings of cacao as a deforestation crop that would lead to further forest loss if production was not sustained within existing fields (e.g., Clough et al. 2009).

The systematic erasure of tropical tree cover gain was not unique to cacao. It reflected a longstanding political and epistemic fixation on tropical forest loss, one that continues to motivate an overwhelming focus on developing ever-more resolved and real-time techniques for monitoring forest loss within RS science. Erasure of tree gain also reflected the multiple methodological factors then limiting long-term, spatially extensive analyses, including the high cost of acquiring satellite imagery, heavy cloud coverage characteristic of images from the humid tropics, and limitations in the computing power available to most analyses.

Fortunately, at least three transformations in RS science had begun to address these limitations by the early 2010s. First, the release of the entire Landsat satellite archive for public use in 2008 created the world's largest free library of geospatial imagery (Woodcock et al. 2008; Zhu et al. 2019). Second, the National Aeronautics and Space Administration (NASA)'s developed a Global Land Survey (GLS) database for the reference year 1975, correcting for radiometric and geometric distortions specific to the first Landsat satellite platform and sensor, and making multi-decadal analyses more accessible than previously (Hansen and Loveland 2012). Third, the development of cloud-based computing platforms made it possible to summon, to manipulate, and to analyse petabytes of data in near-real time. This enabled hundreds of individual Landsat scenes to be mosaicked into wall-to-wall maps of relatively cloud-free satellite imagery for a given reference year (Gorelick et al. 2017).

The RS analyses I published several years later drew on these developments to map tree cover loss and gain for all Southeast Sulawesi province spanning the then full 42-year extent of the Landsat archive, 1972–2014 (Kelley et al. 2017). These analyses illustrated that gross rates of tree cover gain had been three times those of tree cover loss over the period 1972–1995, years in which cacao expansion had been particularly rapid. Combining maps of tree gain and loss with maps of contemporary cacao production also demonstrated that 48% of tree cover loss associated with ongoing cacao production had been concentrated in 10 villages, eight of which belonged to a single administrative district. More generally, my analyses revealed that tree clearance, much of it unrelated to cacao, had been common along lowland roads where state claims to land were likely strongest, especially between 1995–2000 and since 2010: periods encompassing the collapse of Suharto's authoritarian regime, resource decentralisation, and the precipitous increase in land allocation to corporate palm and mining concessionaires (Kelley 2018; Kelley 2020). My findings recast the smallholder-deforestation relation as a smallholder-state-deforestation relation. They also resituated smallholders as agents of extensive 'woodland resurgence' (Hecht 2014).

From one perspective, this approach reflects classic strategies of integrating qualitative and quantitative data. My ethnographic analyses sequentially guided my RS analyses while providing a basis for triangulating interpretations of observed land cover changes (see Biermann and Gibbes, Chapter 4). My RS analyses, in turn, set the stage for subsequent ethnographic work. Why had the landscape changes associated with cacao expansion been so variable, and why had ostensibly similar lowland regions experienced such distinct trajectories of change? Why had a dramatic smallholder cacao boom given way to corporate commodity concessions by the early 2010s? Such questions, emergent from the analysis, guided my ethnographic site selection and shaped a corresponding focus on understanding the localised politics of forest access and histories of mobility and resistance critical in shaping variable trajectories of cacao expansion and forest change. For instance, the decision of an ex-army general to remove land from the state estate through a Surat Keterangan Tanah (letter of land clarification) to settle in-migrants perceived to be more commodity savvy and amenable to state rule in a site marked by histories of insurgency and counter-insurgency (enter: rapid deforestation for cacao in those 10 villages though not others).

While it would be possible to conceive of this research as comprising distinct phases, my experience was that integration emerged intuitively in relation to deepening engagements with place, process, and my own orientations and abilities. Questions of technological capacity or methodological difference did not go away, but they could be situated from the outset as "contingent, relational, contested, and context-dependent", allowing the process of integration to become "a process of seeking to solve a set of relational problems" (Kitchin et al. 2013: 482). Integration, in other words, became a practice of staying with the research conversation as it unfolded, a necessary response to questions posed by place and process.

4. Staying with the contradictions

Laura Ogden discusses landscape ethnography as a practice of following a tangle of overlapping and unfolding territorial claims and land relations where it goes, depicting what it might mean to stay with the research conversation. Resonant of my experience above, Ogden also notes how such a practice can enable iterative "invention, experimentation, and play" that builds towards understanding rather than (necessarily, or simply) breaking down existing ideas (2011: 26). As I discuss in this section, however, it is not always clear that RS and ethnography are building towards shared understandings or even engaged in the same conversation.

Two tensions I encountered during a year of ethnographic work in Sulawesi between 2014–2015 illustrate this. First, classifying land in terms of percentage tree cover within RS had made it possible to render distinct places comparable, situating cacao expansion in broader agrarian and forest transformations, and countering regional discourses of land change on their own terms. Yet RS analyses had almost nothing to say about the localised ecologies in which trees thrived; how or why the fruits of a given tree might be shared, guarded, sold, or divided; or how trees comprised not only 'cover' but claims on land, living histories, and sites of memory and cultural and spiritual meaning. In other words, while RS findings had purchase in certain epistemic, social, and political contexts, their insights seemed hollow in the landscapes they purportedly described.

Second, while ethnography was well suited to revealing such on-the-ground relations, ethnographic findings did not just fill in the RS blanks. They also confused and contradicted RS findings at times. For example, while RS analyses suggested certain lands had been characterised by dense and sustained tree cover for more than four decades, these same lands might be described as a fishing site by one individual, a long-standing swidden and vegetable garden by another, and a fallowed forest by another.

Arguably, such tensions simply point to other ways of staying with the conversation. Above, for instance, I argued that integrating RS and ethnography sequentially, or by means of triangulation (see Biermann and Gibbes, Chapter 4), allowed me to stay in dialogue with questions posed by place and process. Perhaps the above tensions simply illustrate the complementarity between RS and ethnography in illuminating distinct facets of a given research puzzle, shaping a more 'holistic', 'interdisciplinary', or 'socio-biophysical' research conversation. Whereas RS speaks to 'how much' material land change has taken place, for instance, ethnography reveals 'how' and 'why' such change has taken place with reference to how material shifts are socially perceived, represented, or contested (see Sayre, Chapter 34). From this perspective, the conversation is not breaking down, but expanding.

An alternative perspective might see the above tensions as a byproduct of engrained analytical, scalar, and epistemological distinctions between RS and ethnography. For example, while RS is an act of subjective interpretation, it is often engaged within a logically positivist orientation that situates spectral data as a 'neutral', 'objective', and 'reproducible' empirical observation of the Earth's land cover. Ethnographic work, in contrast, is typically situated within a research paradigm that conceives of reality as multi-faceted, contingent, and socially constructed. Similarly, whereas RS analyses often measure biophysical land cover change based on generalised typologies and relationships between spectral reflectance and surface vegetation, ethnographic analyses are typically rooted in an inductive and open-ended understanding of how landscapes come to be socially constructed through context-specific, contingent, and ultimately idiosyncratic behaviours, histories, ideologies, and so forth. The former draws conclusions by aggregating generalities; the latter by attending to the irreducibility of the specific. Perhaps there are simply limits on the degree to which RS and ethnographic findings can or should be reconciled.

Notably, however, both ways of engaging the apparent tensions between RS and ethnographic findings ultimately explain them away. But what if RS and ethnography not only complement and confound one another, they also reflect and reveal one another?[1] In this section, I suggest how staying with the apparent contradictions in the research conversation can expand analytical insight and enhance research reflexivity. I do so by reflecting on what the two tensions sketched above reveal about the shared knowledge politics and partialities embedded within my initial approaches to both RS and ethnography.

To begin, although the abstracted and remote landscape views I had become so acquainted with through RS did not seem to say much about on-the-ground relations, they attuned me to the landscape complexity they obscured. Visiting cacao fields and state forests, for instance, I noticed how some long-fallowed cacao fields had begun to resemble forests populated with dense secondary regrowth, or how seemingly forested lands resembled vegetable fields in their understories. Walking to the main road to find a cell phone signal, waiting out a rainstorm on someone's porch, or washing clothes by the river, I also noticed the diverse abundances I had classified in terms of tree cover absence: marshes replete with fishing sites, watery mud plains doubling as sago palm groves, and sandy riverbank lands that turned green with pumpkin, green bean, peanut, and corn crops in early dry season months.

Such observations can be seen as reiterating the distinctions between RS and ethnography. In contrast with the generalised landscape views I had produced within RS, flexible and open-ended ethnographic encounters expanded a sense of land's 'double infinity', or its infinite physiographic and social and institutional variability (Fowler and Gray 1988; Sayre 2004). My point, however, is more about how having engaged with RS oriented my ethnographic attention to such variability: setting aside the act of formal RS analysis did not mean setting aside

---

1   Thanks to Gregory Simon for the conversations that helped me to find this point, and for articulating it as such.

the interpretive effects these engagements had on me. An attunement to landscape complexity also shaped my eventual awareness of shared reductionisms and landscape simplifications within my approach to ethnography.

Given my preset research interests in cacao and forests, for example, I was commonly directed to speak to village leaders or those individuals who were seen as knowing the most about cacao agriculture and histories of forest use, nearly all of whom were men. I directed myself similarly given an interest in cacao expansion and in the ways intensification initiatives might be altering how people related to cacao and forests. Over the first months of ethnographic work, and outside more spontaneous encounters, most of my questions and observations thus focused on histories of cacao adoption, field management, and programmatic engagement. My attempts to pose such questions to women reinforced this partiality of vision; many women laughed or became shy, self-identifying as unemployed housewives and mothers.

Landscape observations, however, raised other questions. I wondered about where people went or what they did as cacao fields lay fallow, or about how overbank floods, vegetable cultivation, and forest gardens had also shaped people's engagements with cacao. Observing the forest fruit harvests and peppercorn or patchouli seedling nurseries controlled by women, I also began to wonder how gendered norms and interests contributed to emergent field transitions amid high rates of male out-migration. Staying with these questions rather than abstracting them away ultimately taught me how my fixation on cacao-forest relations within ethnographic work mirrored a fixation on tree cover within RS analyses: both elevated one silenced history of land while reinforcing other silences, parroting a long-standing colonial, state, and corporate focus on the most 'productive' and 'valuable' lands (and labourers).

My initial engagements with RS and ethnography, for instance, both marginalised the significance of floodplain formations or female-dominated cultures of agricultural production and reproduction. In doing so, they reproduced a recurrent politics of erasure, one that had long figured into efforts to expropriate and re-allocate people's land for allegedly more modern, efficient, or intensive modes of protection or production—whether under the Dutch, who designated the wetlands 'muck and more muck' in the process of draining them for resettlement schemes, or under the Indonesian government, who deemed swidden gardens 'unused' and 'under-utilised' and thus property of the state. My focus on (male-dominated) agricultural practices and commodity fields also invisibilised women's labour, echoing the focus of cacao programming on all-male farmer groups.

If staying with the first research tension made me more aware of how both my ethnographic orientations and land cover classifications were political and cultural artefacts (Robbins and Maddock 2000), staying with the second attuned me more fully to how resulting knowledge claims were also methodological artefacts. For example, staying with the tension of lands RS marked in terms of continuous tree cover and ethnography suggested were comprised of fallowed forests, fishing sites, and vegetable gardens taught me this wasn't just because distinct methods distinctly illuminate land's material or ideological lie. Each of these material realities was valid to a degree, depending on the subjected and situated practices of observation engaged. Pursuing multiple ways of visualising the lands in question, by summoning distinctly resolved satellite images from different times, or by conducting repeat ethnographic visits and interviews, helped me to see this.

Among other things, I learned that the RS techniques I had engaged to capture dispersed and long-term patterns of tree cover gain had also systematically obscured smaller-scale land uses from rainy season months. Multi-date image compositing, for instance, derives a median pixel value from all cloud-free pixels within a stack of images. This produces a single wall-to-wall image for the period from which imagery are drawn, enabling spatially extensive analyses. Resampling mosaicked images to the coarse spatial resolution of 1972 imagery (60 m), in turn, renders distinctly resolved sensor data from different time periods comparable, enabling multi-decadal analyses. Nonetheless, these techniques systematically excluded rainy season pixels (due to cloud clover), while obscuring many land openings smaller than about one acre (roughly the size of one 60 m x 60 m pixel). This limited the ability of RS analyses to capture land opening or inundation within montane and swamp forests during rainy season months, when they became small-scale vegetable gardens or fishing sites.

It is in these senses that RS and ethnography not only complemented or confounded one another but reflected and revealed one another by revealing the knowledge partialities and politics structuring each. An awareness of such cross-cutting analytical limitations enabled me to better address cacao-forest questions, which required looking beyond direct cacao-forest relations. It also reminded me that "we ourselves are a part of the mystery that we are trying to solve" (Planck 1932: 217), inextricably enmired in both the conversations and the contradictions. The lesson I take from this is that if integrative research depends on remaining attuned to the potential of process, it also depends on remaining reflexively engaged with how our positionality infuses all elements of this process and its potential. Staying with apparent research tensions, in my experience, fostered such reflexivity.

Embracing the continuities between RS and ethnography

Many conversations on integrating geospatial and ethnographic methods begin with a view to their distinctions, analytically, epistemologically, politically, or otherwise. Such an orientation is valuable in understanding how RS and ethnographic methods can be staggered or stacked to address a given question, how their mutual deployment can expand a given puzzle or problem, or why RS and ethnographic findings might fail to converge. Distinctions between RS and ethnographic methods can also be a barrier to conceiving, justifying, funding, and publishing integrative research, arguably more so as the growing volume, velocity, and variety of RS data seems to push RS science further from an ethnographic insistence on the particular.

But what if the differences between RS and ethnography are not as immutable as it can seem? For example, just because RS is often deployed within a logically positivist position that sees science as an arbiter of capital 'T' truths, RS can also be engaged and embraced as an act of situated and subjective interpretation (Braun 2021). Similarly, even though ethnography is most often engaged within a constructivist frame that emphasises the social relations veiling and mediating truth claims (see Sayre, Chapter 34), doesn't ethnography also have something to say about land's multiple materialities and the processes that beget these materialities at scale? Engaging an epistemological stance of critical realism—one that sees science as a means of producing and revealing situated truths and knowledges—can highlight potential continuities such as these (cf. Lave et al. 2018), avoiding the trap of "setting up distinctions... on the basis of methods and data sources," and "conflating method and epistemology" (Silvey and Lawson 1999: 122)

This section focuses on three attempts to reimagine how I engage RS and ethnography that embrace their shared interpretive strengths and orientations from within a critical realist frame. Each of these attempts emerged in response to the dynamics of place, process, and positionality revealed by ongoing research conversations and contradictions in Sulawesi. Each also ostensibly engages with RS and ethnography independently. I offer them up as reflections on integration, however, because they are attempts to embed integrative instincts into either method from the outset—whether by engaging RS with an ethnographic attunement to the specific, or by engaging ethnography to map the scale-making relations that beget land's changing materiality.

One of my attempts to engage RS more ethnographically in recent years began from an appreciation of the multiple situated and subjective analytical decisions within earlier approaches to land cover classification that shaped what such classifications visibilised and invisibilised. I next asked how such decisions could be altered to make sensibilities palpable within ethnographic work more apparent within resulting land cover classifications. Guided by this process, key changes I have made include reducing the geographical extent of analysis, expanding the set of land cover categories, and engaging in ethnographic observations to ensure fidelity to place-based distinctions in how land is used, localised precipitation regimes, and corresponding distinctions in land's seasonality. Other changes have involved mining the open-access Landsat catalogue not only for imagery but as a time series that reveals how vegetative cover changes throughout a given year (a measure of seasonality) and the responsiveness of vegetation to precipitation at different lags (which can help differentiate, for example, inundated swamps from vegetable fields) (Dean 2021; Kelley 2018).

Resulting land cover classifications do not beget a more accurate, objective, or valid means of mapping landscapes in Sulawesi. Yet they enhance the multiplicity of situated landscape truths captured through RS analyses. Importantly, they also respond to the interests of local constituents in making their historical uses (and claims) over wetland vegetable gardens, swamps, and agroforests more visible as state, corporate, and civil society discourses again position these lands as 'free' for protection or production amid corporate palm and mining expansion.

I have also attempted to engage RS more ethnographically within recent work on the changing frequency and severity of floods, a theme that repeatedly emerged during subsequent ethnographic visits. Were I to situate RS and the biophysical sciences as an inherently more valid means of knowing nature than subjective ethnographic observation, I might have centred RS methods in my approach to studying floods by using Sentinel-C radar data and NASA's MERRA-2 model to correlate precipitation thresholds with recent flood events, engaging in topographic-based hydrologic modelling to delineate the contributing catchment area, and using a derived cumulative catchment precipitation overbank flood equation to infer historical flood events. Doing so, however, would have meant assuming away multiple 'variables' critical in understanding how floods emerge and what they mean, including dynamic, place-based, and relational constructs of flooding, changes in the river and river system, and the situated and complex causality of flooding in any single location (e.g., Pattison and Lane 2012).

I have ultimately drawn upon Sentinel-C radar data (2014–) to map recent flood events and Landsat data to map changes in the river's morphology and in floodplain vegetation. Yet I engage such data as one among many partial means of 'knowing' floods. I also draw on place-based conceptions and histories of flooding from focus group discussions (see Longhurst and Johnston, Chapter 27); archival accounts of prior flooding extremes (see Cope, Chapter 22); colonial and state maps of watershed land covers and property relations; and secondary data on river management and demographic change. Within such an approach, I cannot make quantitative claims about how much more severe or frequent floods have become at regional scales, or quantify the relationship between precipitation and flood events. Nonetheless, embedding RS data within an ethnographically inspired 'thick description' of flooding—and drawing on the RS grid to spatially interrelate historical datasets with spatial attributes—illuminates how racialised settlement schemes, river channelisation initiatives, and agricultural 'improvement' schemes work alongside one another to render anomalous rainfall events disastrous.

The approaches I describe may not be the grounds for further advancing the data, computational, or algorithmic revolutions that orient much of RS science today. Yet they leverage such advances to develop methodological scripts that can help puncture overly dominant regimes of visibility within RS and enable us to do undone science guided by place-based relations and concerns. Shifts in how I engage RS also reflect my attempts to engage the politics of resulting knowledge claims more reflexively. For instance, the RS approach to mapping changing flood regimes I describe above would have implicitly naturalised floods as a product of rainfall. Engaging RS data more ethnographically instead highlights that rainfall extremes only beget flooding disasters where they meet histories of extractivism, expropriation, and erasure. It also reveals that sometimes floods are not a product of rains, they are a product of poorly constructed roads. Such dynamics, importantly, are more amenable to reform than are relatively locked in climatic changes.

A final approach I am experimenting with in Sulawesi involves more explicitly spatialising my ethnographic practice. This approach reflects an attempt to leverage the strengths of ethnography in capturing the path-dependent historical processes and harder-to-detect forms of resistance, collective agency, migration, or kinship that may leave no spectral trace (e.g., because they take place in the understory or are hidden intentionally) but are nonetheless materially meaningful. Different approaches I have tried have included walking oral histories and land genealogies, mobile participant observations, and approaches to analysing ethnographic or survey findings that embed spatial data into the analysis. One approach, for instance, has involved accompanying women as they engage in everyday mobilities, whether to gather medicinal plants at river's edge, to graze livestock within agroforests, or to visit local markets, nearby work sites, and ATMs where remittances and reverse remittances are collected or sent.

These practices recursively inform RS analyses by orienting my attention to those lands most meaningful to people's daily lives. They also open analytical insight. I have learned, for instance, that the relative invisibility of understory cultivations is not only about evading state detection or the limits of optical RS data. It can also reflect the intra-household and village social relations that lead women to invisibilise these engagements within their everyday lives. For example, where cattle grazed in the forest are hidden from neighbours to downplay one's relative wealth, which might beget unwelcome requests for loans, or where forest cultivations are minimised so that husbands don't understand how much income their wives control. I have also learned how the blurriness between agricultural and forested lands at any given moment may reflect embodied perceptions and experiences of safety. Agricultural activities in the forest, for instance, may be abandoned for seasons or years where women lack a network of other women with whom to travel to such lands.

While these methodological adaptions are specific to my research orientations in Sulawesi, emerging conversations in multimodal anthropology and critical RS suggest they can travel. Indeed, if RS were not as much an act of subjective and situated interpretation as ethnography, the history of RS science would reflect far fewer attempts to "make the subjective objective" (Braun 2021) within land cover classification processes, whether this has meant developing methods that 'correct' contingencies of climate, clouds, or sensor technology, or fashioning artificially intelligent algorithms that further remove the analyst from the analysis. Similarly, if ethnography were not a means of observing and sensing land, how would it be possible to attend to the power-laden ways land cover classifications reshape the everyday textures of life and land, or to foreground ways of doing and being that force such categorical closures open? Embracing such continuities can seed the basis for reimagined RS and ethnographic methods that engage integration from within, well before distinct datasets or methods are "brought together".

Conclusion

This chapter reflects on how staying with the conversations, contradictions, and continuities between RS and ethnography can enable approaches to engaging RS and ethnography as partners in seeding alternative landscape stories and scripts. It also highlights the value of deep investment in both methodologies as a basis for such efforts.

While the above reflections are specific to my research orientations and the places engaged, they suggest three insights for other integrative scholarship. First, let place and process start the conversation and lead the way in establishing the need for and means of integrating the two techniques. This approach in Sulawesi involved allowing place-based research insights to inform iterative and recursive approaches to integration. Initial ethnography, for instance, guided RS analyses that eventually structured further ethnographic analysis, which in turn, informed how I methodologically reimagined my engagements with RS. Ultimately, these reconfigurations made it possible to address place-based relations and concerns more reflexively and expansively.

Second, stay with the contradictions that emerge rather than explaining them away as a byproduct of assumed differences between RS and ethnography. Though tensions between RS and ethnographic findings can attest to real differences in their strengths and analytical orientations, meditating on such tensions can also reveal what the two approaches share by illuminating the shared knowledge partialities and politics shaped by researchers' positionality. Within such an approach, integration is not only an analytical paradigm or interpretative practice, but can shape how we ask questions, how we do the work, and our ways of being in the world and relating to others.

Finally, embracing the continuities and not only discontinuities between RS and ethnography can enhance methodological and interpretive possibilities. Such an orientation has allowed me to reimagine how I engage RS and ethnography in response to deepening insights into place, process, and positionality. It has also helped me to better realise the multiple strengths of both methods, enabling a (partial) mapping of the multiple regionalities and globalities that cohere in any single place or pixel. This approach, rooted in a critical realist frame, asks how integration can be built in from the get-go, engaging both methods as situated arbiters of situated truths.

In all these ways, this chapter is also an attempt to converse with emerging work in the fields of anti-colonial environmental science, feminist science and technology studies, and Critical Physical Geography, all of which are engaged in efforts to redeploy and reimagine the methods of dominant science within more critically and humanistically infused accounts of nature-society relations. May the conversations continue, and continue to inspire our own.

References cited

Adey, P. 2010. *Aerial Life: Spaces, Mobilities, Affects* (John Wiley and Sons).

Arce-Nazario, J.A. 2016. 'Translating land-use science to a museum exhibit', *Journal of Land Use Science*, 11, pp. 417–428.

Archer, M. 2021. 'Imagining impact in global supply chains: Data-driven sustainability and the production of surveillable space', *Surveillance and Society*, 19, pp. 282–298.

Belgiu, M. and L. Drăguţ. 2016. 'Random forest in remote sensing: A review of applications and future directions', *ISPRS Journal of Photogrammetry and Remote Sensing*, 114, pp. 24–31.

Bennett, M.M., J.K. Chen, L.F. Alvarez León, and C.J. Gleason. 2022. 'The politics of pixels: A review and agenda for critical remote sensing', *Progress in Human Geography*, 46, pp. 729–752.

Bennett, M.M. and H.O. Faxon. 2021. 'Uneven frontiers: Exposing the geopolitics of Myanmar's borderlands with critical remote sensing', *ISPRS Journal of Photogrammetry and Remote Sensing*, 13, p. 1158.

Biermann, C. and Gibbes, C., Chapter 4, this volume. 'Mixed methods in tension: lessons for and from the research process'.

Braun, A.C., this volume. '(Critical) Satellite remote sensing'.

Braun, A.C. 2021. 'More accurate less meaningful? A critical physical geographer's reflection on interpreting remote sensing land-use analyses', *Progress in Physical Geography: Earth and Environment*, 45, pp. 706–735.

Bryan, J. 2021. 'Mapping resources: Mapping as method for critical resource geographies', in *The Routledge Handbook of Critical Resource Geography*, ed. by M. Himley, E. Havice, and G. Valdivia (Routledge), pp. 441–452.

Bunge, W. 1979. 'Perspective on theoretical geography', *Annals of the American Association of Geographers*, 69, pp. 169–174.

Clough, Y., J. Barkmann, J. Juhrbandt, M. Kessler, T.C. Wanger, A. Anshary, D. Buchori, D. Cicuzza, K. Darras, D.D. Putra, et al. 2011. 'Combining high biodiversity with high yields in tropical agroforests', *Proceedings of the National Academy of Sciences of the United States of America*, 108, pp. 8311–8316.

Clough, Y., H. Faust, and T. Tscharntke. 2009. 'Cacao boom and bust: sustainability of agroforests and opportunities for biodiversity conservation', *Conservation Letters*, 2, pp. 197–205.

Dean, J. 2021. *Effects of State Enclosures and Industrial Concessions on Land Cover Change in Indonesia* (University of Hawai'i at Manoa).

Elden, S. 2013. 'Secure the volume: Vertical geopolitics and the depth of power', *Political Geography*, 34, pp. 35–51.

Elwood, S. 2006. 'Critical issues in participatory GIS: Deconstructions, reconstructions, and new research directions', *Transactions in GIS*, 10, pp. 693–708.

Erasmi, S., A. Twele, M. Ardiansyah, A. Malik, and M. Kappas. 2004. 'Mapping deforestation and land cover conversion at the rainforest margin in Central Sulawesi, Indonesia', *EARSeL EProceedings*, 3, pp. 388–397.

Fairhead, J. and M. Leach. 1996. *Misreading the African Landscape: Society and Ecology in a Forest-Savanna Mosaic* (Cambridge University Press).

Ferring, D. and H. Hausermann. 2019. 'The political ecology of landscape change, malaria, and cumulative vulnerability in central Ghana's gold mining country', *Annals of the American Association of Geographers*, 109, pp. 1074–1091.

Fowler, J.M. and J. Gray. 1988. *Rangeland Economics in the Arid West* (Department of Agricultural Economics and Agricultural Business).

Fujikane, C. 2021. *Mapping Abundance for a Planetary Future: Kanaka Maoli and Critical Settler Cartographies in Hawai'i* (Duke University Press).

Gleason, C.J. and A.N. Hamdan. 2017. 'Crossing the (watershed) divide: satellite data and the changing politics of international river basins', *The Geographical Journal*, 183, pp. 2–15.

Goldstein, J. and E. Nost. 2022. *The Nature of Data: Infrastructures, Environments, Politics* (University of Nebraska Press).

Gorelick, N., M. Hancher, M. Dixon, S. Ilyushchenko, D. Thau, and R. Moore. 2017. 'Google Earth Engine: Planetary-scale geospatial analysis for everyone', *Remote Sensing of Environment*, 202, pp. 18–27.

Graham, S. 2004. 'Vertical geopolitics: Baghdad and after', *Antipode*, 36, pp. 12–23.

Hansen, M.C. and T.R. Loveland. 2012. 'A review of large area monitoring of land cover change using Landsat data', *Remote Sensing of Environment*, 122, pp. 66–74.

Harvey, D. and D. Haraway. 1995. 'Nature, politics, and possibilities: a debate and discussion with David Harvey and Donna Haraway', *Environment and Planning D: Society and Space*, 13, pp. 507–527.

Havice, E., L. Campbell, and A. Boustany. 2022. 'New data technologies and the politics of scale in environmental management: Tracking Atlantic Bluefin tuna', *Annals of the American Association of Geographers*, 112, pp. 1–21.

Hind, S. and S. Lammes. 2016. 'Digital mapping as double-tap: Cartographic modes, calculations and failures', *Global Discourse*, 6, pp. 79–97.

Johnston, R.J. 1986. 'Four fixations and the quest for unity in geography', *Transactions of the Institute of British Geographers*, 11.4, pp. 449–453.

Kelley, L.C. 2018. 'The politics of uneven smallholder cacao expansion: A critical physical geography of agricultural transformation in Southeast Sulawesi, Indonesia', *Geoforum*, 97, pp. 22–34.

Kelley, L.C., S.G. Evans, and M.D. Potts. 2017. 'Richer histories for more relevant policies: 42 years of tree cover loss and gain in Southeast Sulawesi, Indonesia', *Global Change Biology*, 23, pp. 830–839.

Kitchin, R., J. Gleeson, and M. Dodge. 2013. 'Unfolding mapping practices: a new epistemology for cartography', *Transactions of the Institute of British Geographers*, 38, pp. 480–496.

Kwan, M.-P. 2016. 'Algorithmic geographies: Big data, algorithmic uncertainty, and the production of geographic knowledge', *Annals of the American Association of Geographers*, 106, pp. 274–282.

Kwan, M.-P. 2002. 'Feminist visualization: Re-envisioning GIS as a method in feminist geographic research', *Annals of the American Association of Geographers*, 92, pp. 645–661.

Lane, S.N. 2017. 'Slow science, the geographical expedition, and critical physical geography', *Canadian Geographies / Géographies canadiennes*, 61, pp. 84–101.

Lave, R., C. Biermann, S.N. Lane. 2018. 'Introducing critical physical geography', in *The Palgrave Handbook of Critical Physical Geography*, ed. by R. Lave, C. Biermann, and S.N. Lane. (Springer), pp. 3–21.

Lave, R., M.W. Wilson, E.S. Barron, C. Biermann, M.A. Carey, C.S. Duvall, L. Johnson, K.M. Lane, N. McClintock, D. Munroe, et al. 2014. 'Intervention: Critical physical geography', *Canadian Geographies / Géographies canadiennes*, 58, pp. 1–10.

Liboiron, M. 2021. *Pollution is Colonialism* (Duke University Press).

Lukas, M.C. 2014. 'Eroding battlefields: Land degradation in Java reconsidered', *Geoforum*, 56, pp. 87–100.

Nagel, G.W., E.M.L. de Moraes Novo, and M. Kampel. 2020. 'Nanosatellites applied to optical Earth observation: a review', *Revista Ambiente e Água*, 15.

Nost, E. and J.E. Goldstein. 2022. 'A political ecology of data', *Environment and Planning E: Nature and Space*, 5, pp. 3–17.

Öberg, G., L. Fortmann, and T. Gray. 2013. ‹Is interdisciplinary research a mashup?›, *IRES Working Paper Series*, 2013.2.

Ogden, L.A. 2011. *Swamplife: People, Gators, and Mangroves Entangled in the Everglades* (University of Minnesota Press).

19. Engaging remote sensing and ethnography 401

Pattison, I. and S.N. Lane. 2012. 'The link between land-use management and fluvial flood risk: a chaotic conception?', *Progress in Physical Geography*, 36, pp. 72–92.

Peluso, N.L. and P. Vandergeest. 2001. 'Genealogies of the political forest and customary rights in Indonesia, Malaysia, and Thailand', *J. Asian Studies*, 60, pp. 761–812.

Planet | Homepage. n.d. *Planet*, https://www.planet.com/

Robbins, P. 2015. 'The trickster science', in *The Routledge Handbook of Political Ecology*, ed. by P. Robbins (Routledge), pp. 89–101.

Robbins, P. and T. Maddock. 2000. 'Interrogating land cover categories: metaphor and method in remote sensing', *Cartography and Geographic Information Science*, 27, pp. 295–309.

Rothe, D. 2017. 'Seeing like a satellite: Remote sensing and the ontological politics of environmental security', *Security Dialogue*, 48, pp. 334–353.

Ruf, F., G. Schroth, et al. 2004. 'Chocolate forests and monocultures: a historical review of cocoa growing and its conflicting role in tropical deforestation and forest conservation', in *Agroforestry and Biodiversity Conservation in Tropical Landscapes*, ed. by G. Schroth, G.A.B. Fonseca, C.A. Harvey, C. Gascon, H. Vasconcelos, and A.M.N. Izac (Island Press), pp. 107–134.

Ruf, F. and C. Yoddang. 1999. 'The impact of the economic crisis on Indonesia's cocoa sector', *ACIAR Indonesia Research Project*.

Sayre, N.F. 2004. 'The need for qualitative research to understand ranch management', *Journal of Range Management*, 57, pp. 668–674.

Schuurman, N. and G. Pratt. 2002. 'Care of the subject: Feminism and critiques of GIS', *Gender, Place and Culture: A Journal of Feminist Geography*, 9, pp. 291–299.

Silvey, R. and V. Lawson. 1999. 'Placing the Migrant', *Annals of the Association of American Geographers*, 89, pp. 121–132.

Slater, C. 2003. *Fire in El Dorado, or Images of Tropical Nature and their Practical Effects* (Duke University Press).

Steffan-Dewenter, I., M. Kessler, J. Barkmann, M.M. Bos, D. Buchori, S. Erasmi, H. Faust, G. Gerold, K. Glenk, S.R. Gradstein, et al. 2007. 'Tradeoffs between income, biodiversity, and ecosystem functioning during tropical rainforest conversion and agroforestry intensification', *Proceedings of the National Academy of Sciences of the United States of America*, 104, pp. 4973–4978.

TallBear, K. 2014. 'Standing with and speaking as faith: a feminist-indigenous approach to inquiry', *Journal of Research Practice*, 10, p. N17.

Turner, M.D. 2003. 'Methodological reflections on the use of remote sensing and geographic information science in human ecological research', *Human Ecology*, 31, pp. 255–279.

Woodcock, C.E., R. Allen, M. Anderson, A. Belward, R. Bindschadler, W. Cohen, F. Gao, S.N. Goward, D. Helder, E. Helmer, et al. 2008. 'Free access to Landsat imagery', *Science*, 320, p. 1011.

Yang, L., J. Driscol, S. Sarigai, Q. Wu, H. Chen, and C.D. Lippitt. 2022. 'Google Earth engine and artificial intelligence (AI): a comprehensive review', *Remote Sensing*, 14.

Zhu, A. and N.L. Peluso. 2021. 'From gold to rosewood: Agrarian change, high-value resources, and the flexible frontier-makers of the twenty-first century', in *The Routledge Handbook of Critical Resource Geography*, ed. by M. Himley, E. Havice, G. Valdivia (Routledge), pp. 345–357.

Zhu, Z., M.A. Wulder, D.P. Roy, C.E. Woodcock, M.C. Hansen, V.C. Radeloff, S.P. Healey, C. Schaaf, P. Hostert, P. Strobl, J.-F. Pekel, L. Lymburner, N. Pahlevan, and T.A. Scambos. 2019. 'Benefits of the free and open Landsat data policy', *Remote Sensing of Environment*, 224, pp. 382–385.

20. Mixing geoarchaeology, geohistory and ethnology to reconstruct landscape changes on the longue durée

Ninon Blond

What makes it difficult is that research is immersion in the unknown. We just don't know what we're doing. We can't be sure whether we're asking the right question or doing the right experiment until we get the answer or the result. (Schwartz 2008)

Introduction

The study of environments and socio-environmental interactions requires researchers to consider a variety of spatial and temporal scales, and to mobilise diverse but complementary methods that may require very different skills and knowledge.

My research aims to reconstruct socio-environmental interactions and changes in response to severe environmental constraints over the Holocene. The approach is largely inter- and trans-disciplinary (Fig. 20.1) and based on a regressive approach, starting from the contemporary situation and working backwards to earlier events and processes. The aim is thus to identify what in the current landscape is the result of recent processes, what can be linked to older processes, what is the result of changes that are still underway, and what can be linked to past and completed changes.

This requires a range of methods that can speak both to ancient times, when societies left few traces of their activities, their knowledge of the environment or their thought systems, and in particular no (or few) written records or direct testimonies of environmental phenomena; and also to more recent times, when there is a great deal of documentation,

   https://doi.org/10.11647/OBP.0418.20

but for which dating techniques used for ancient periods cannot be applied. It is therefore necessary to combine different methods.

The data I work with has varying granularity, precision, and resolution, and is gathered from complementary but divergent sources ranging from optically stimulated luminescence (OSL) to radiocarbon dating (see King and Abbott, Chapter 28) to travellers' accounts and interviews with contemporary populations. To analyse these diverse data, I use both biophysical (geomorphology, palaeoenvironments, even palaeoclimates) and social (geohistory, environmental history, ethnology, Human Geography, etc.) science and quantitative and qualitative methods (Bergman 2008).

How can such diverse data be assembled and harmonised to reconstruct a coherent environmental history of a particular study region and understand the environmental and landscape changes it has undergone? What is the scientific status and validity of the history reconstructed from such disparate data produced by applying very different methods?

I argue that the scientific value of my work stems from its holistic and systemic approach, which seeks to consider all aspects of an issue, rigorously applying each of the methodological principles involved. The aim is not so much to exhaust the question as to seek the most exhaustive and complete answer, leaving as few grey areas as possible.

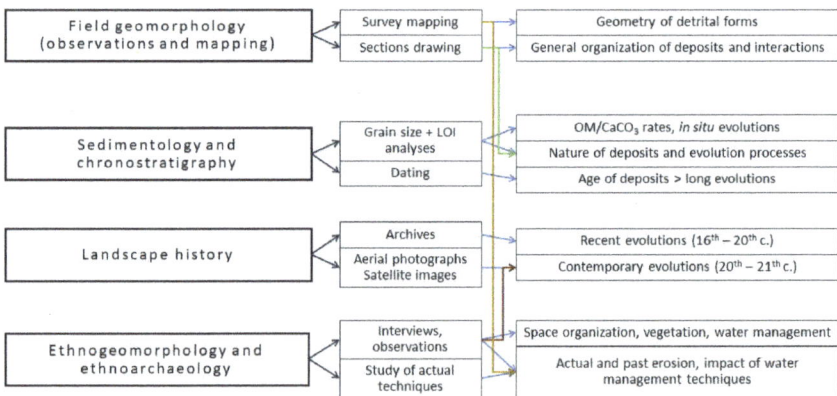

Fig. 20.1 Ninon Blond, 2023, Systemic mixed-methods methodology as applied in the cases studied here

Theoretical and methodological framework

My approach can be described as mixed methods: " the combination of quantitative and qualitative methods in the same research" (Aguilera and Chevalier 2021a; Bergman 2008). In this sense, it is rather commonplace in the social sciences, "because they [the mixed methods] have in fact been at the heart of social science practice from the outset, [...] i.e. mobilising a diversity of methods to consolidate an explanation and strengthen the results" (Aguilera and Chevalier 2021a). Despite this apparent banality, there is still very little mixed-methodological literature, as the authors point out (Aguilera and Chevalier 2021a). In environment studies, the combination is slightly different, in that the quantitative methods, for example, are not the same as those used in social sciences (sedimentological analyses versus questionnaire-based surveys (see Winata and McLafferty, Chapter 43)). The mixed methods presented here are then different from the "traditional" mixed methods used in social sciences.

The use of a mixed-methods approach is a response to the shortcomings of the techniques used individually. The aim is to use complementary methods to shed light on grey areas that remain after the problem has been addressed using a single technique, and to answer new questions that emerge from these initial analyses. More broadly, the goal of mixed-methods environmental research is to provide an overall, systemic response to an issue through the interactions between methods and the results they provide. The strategy is therefore one of enrichment, but also of confirmation (Aguilera and Chevalier 2021b; see also Biermann and Gibbes, Chapter 4).

Here, the quantitative approach ("data analysis methods based essentially on the statistical processing of a large number of data collected as part of the empirical testing of the research problem") (Coman et al. 2022, cited by Aguilera and Chevalier 2021b) is that of processing and statistical representation such as in geomorphological sampling campaigns involving sedimentological data and their analysis (see Lane, Chapter 42). Qualitative methods can be defined as "analysis methods based on the in-depth processing of a small number of data" (Lamont and White, 2005) collected "using ethnographic methods, semi-structured interviews, archival research or the collection of reports" (Aguilera and Chevalier 2021b). In my work, they relate historical approaches in the geosciences, labelled geo-history, and ethnological approaches (Fig. 20.1).

The geomorphological approach is based on observation and mapping in the field prior to the selection of sites of interest from which sediment samples are taken. The sediments are then analysed in the laboratory (carbon-14 dating, laser and sieve granulometry, organic matter and calcium carbonate content). The data from these analyses is statistically processed, enabling the sediment samples to be characterised using indices linked to the statistical ratios of the various elements or the shape of the particle size distribution curves (central values, quantiles, asymmetry and flattening of the curve; see Chapters 25 and 41). Sedimentological characterisation, coupled with the dating of stratigraphic phases (King and Abbott, Chapter 28), enables us to put forward hypotheses to characterise ancient landscapes and environments (for methodological details, see Blond et al. 2018; Blond et al. 2021).

The work of geo-history is based on a wide variety of materials (Jacob-Rousseau 2009) that are apprehended diachronically (i.e., by considering facts as they evolve through time). Most of the evidence consists of old aerial photographs, old ground photographs, accounts and notebooks of travellers and scientific or military expeditions, engravings, old maps and, more rarely, postcards. The treatment of archival materials varies according to their nature, their degree of precision, and the scale at which they are used. General information on the landscape, fauna, flora, and agriculture can be collected from texts and ground photographs, while the digitisation of land use on old aerial photographs and maps makes it possible to quantify evolution through time, both in terms of cultivated areas and the spread of erosive phenomena or the regression of plant cover (for a detailed methodological description, see in particular Blond et al. 2018; Blond et al. 2019).

Finally, the ethnological approach is based on the observation of practices (e.g., farming), but above all on semi-structured interviews (see Johnston and Longhurst, Chapter 32), the main aim of which is to gather first-hand information on local practices. The goal is to integrate Indigenous and local knowledge gathered through observations and interviews in the geomorphological analyses of landscape formation. It is then possible to document contemporary practices, their changes and their impact on the environment, but also to draw a parallel with ancient practices in order to explain current landscapes on the model of ethnoarchaeology (Arthur 2003; Aurenche 2012). Ethnoarchaeology is based on the study of current situations (through observation and/or interviews) in order to provide keys to interpreting archaeological situations through cultural analogy over time. I define ethnogeomorphology on the model of ethnoarchaeology, as a methodology that analyses the relationships maintained by ancient populations with their environments, and their impact on these environments, based on the collection of environmental knowledge and the observation and questioning of the practices of current local populations. Thus, using a regressive approach, it provides keys to interpreting the landscape vestiges that can be observed even if they are highly degraded, and makes it possible to put forward hypotheses on landscape evolution that are not thoroughly documented through archaeological remains. My definition of ethnogeomorphology therefore differs from that of Wilcock et al. (2013) and assumes an important diachronic dimension, based on cultural analogy but also on geological actualism. The landscape remains are indeed the only 'witnesses' that can document ancient practices (with a few exceptions, where, for example, prehistoric engravings bear witness to ancient agricultural practices (see Finneran 2007), or where hieroglyphic texts bear witness to the organisation of ancient quarries (see Gourdon et al. 2022)).

In the sections that follow, I illustrate two broad approaches to data collection and analysis that require mobilisation and combination of a suite of methods and techniques depending on opportunities, needs and constraints. The methodological considerations are based on a variety of fieldwork (in Ethiopia, Egypt, Greece, Sudan, Oman), where scientific and logistical issues diverge (Fig. 20.2).

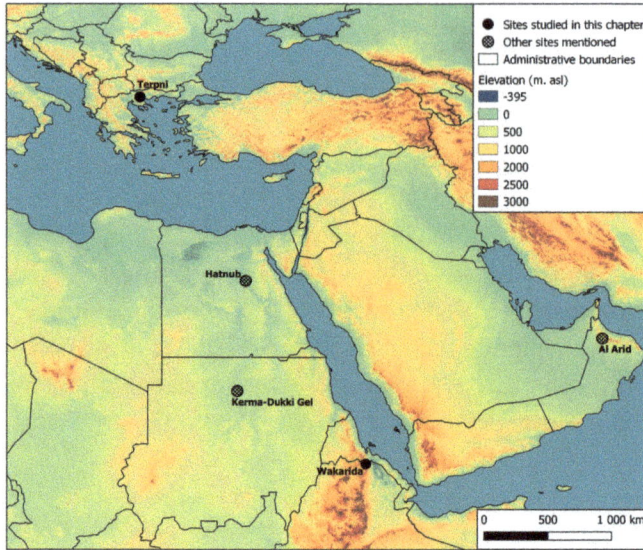

Fig. 20.2 Ninon Blond, 2024. Localisation of sites studied in this paper. After USGS
GTOPO30

Case studies

Ethiopia: implementation of mixed methods and large-scale testing

In Ethiopia, at the Wakarida site in Eastern Tigray (Fig. 20.2), I undertook geoarchaeological work to analyse environmental and landscape changes around the archaeological site during the Holocene and to answer a more specific question of interest to the archaeological team:[1] the age of the cultivation terraces. The archaeologists' initial hypothesis was that the region had been occupied during the pre-Axumite and Axumite periods (1st millennium BC), and that the cultivated terraces currently visible in the landscape were the legacy of this period, bearing witness to their cultural and trade connections with South Arabian populations (Benoist et al. 2020; 2021; 2022; Breton 2015; Gajda et al. 2017).

Initially, the systematic study of stratigraphic sections enabled the reconstruction of an environmental and landscape history from the 7th

---

1   Mission Archéologique Française dans le Tigray Oriental (MAFTOr), directed by A. Benoist, CNRS/Archéorient and J. Schiettecatte, CNRS/Orient Méditerranée.

millennium BC to the 17th century AD (Fig. 20.3). This showed that until the 5th to 4th millennia BC, deposits were formed solely as a result of biophysical processes, without human intervention, in a climate that was wetter and with more regular rainfall than is currently the case (Blond et al. 2018), conditions that can be linked to the African Humid Period (Fig. 20.3, a and b). From the 5th to 4th millennia BC, this humid optimum came to an end, which had consequences for the environment: sedimentary deposits showed a transition towards more irregular processes (erosional events, higher energy depositional events), with less abundant but more concentrated rainfall. While it is possible that these processes were disrupted by societies, as prehistoric sites exist in Tigray (Finneran 2007), these societies do not seem to have left any trace of their presence in sedimentary deposits before the 1st millennium BC (Fig. 20.3c and d). Analysis of the dating (and its perturbations) and of the sedimentary composition of the samples shows an increasing imprint of populations (Fig. 20.3 f and g), particularly between the 14th and 17th centuries AD (Blond et al. 2021). However, chronostratigraphic analysis remains silent on a finer scale and for more recent periods: the radiocarbon dating method has margins of error that can reach several hundred years, which is not compatible with a precise analysis of recent periods (see King and Abbott, Chapter 28).

To document more recent changes in the landscape, I turned to archival evidence: travellers' accounts, expeditions, and conquests from the 16th century onwards, topographical maps from the 17th to the early 20th centuries, engravings from the 19th century, ground photographs from the early 20th century, and aerial photographs from the second half of the 20th century. Descriptions of journeys or expeditions revealed that the eastern edges of Tigray had not been recognised before the beginning of the 20th century (or at least there was no written or cartographic evidence of this recognition). It was therefore not possible to use a quantitative approach such as repeated photography (Frankl et al. 2011). Texts and iconographic documents did, however, provide a qualitative assessment of recent changes in the fauna, with mentions of elephants, lions, rhinoceroses, hyenas, jackals, giraffes and, more generally, game animals that were abundant in the 19[t]h century but have now disappeared from the region (Combes and Tamisier 1838b; Girard 1873; Lejean 1872). On the other hand, the vegetation, while described as abundant, dense and luxuriant by some authors (Combes and Tamisier 1838a; Rüppel 1840; Salt 1816), seems not to have been so idyllic for other authors, such as Lejean (1872), who notes the distance between Bruce's accounts and the landscapes he saw. It is possible that the first explorers idealised the abundance of vegetation they saw.

Combined with sedimentology, this analysis of travellers' accounts confirms that environmental degradation of human origin was already underway in the 17th–19th centuries. However, it does not resolve the question of the origin and age of the terraces. They are hardly ever mentioned by travellers in Tigray until the 1960s (Henze 2001: 64). A diachronic comparison of aerial photographs of the study region between 1965 and 1994 (Blond et al. 2018; Fig. 11) highlights the recent transformation of valley bottoms through the construction of terraces, which are gradually taking over areas further and further away from urban centres, in the upstream parts of the catchment areas.

Why such transformations? Why at this time and not earlier? Could the current terraces be a reworking of older structures, or could the current development be explained by a particular context? Neither sedimentology, nor travellers' accounts, nor aerial photographs could provide answers to these questions. We thus turned to semi-structured interviews with locals to gather their accounts of contemporary and sub-contemporary environmental and landscape changes in the region (see Johnston and Longhurst, Chapter 32).

These interviews enabled me to collect a wealth of data on various themes. The questions were asked in English and then translated into Tigrinya or Amharic; the answers given in Tigrinya or Amharic were translated into English. We learned that the construction of cultivation terraces was a response to growing pressure on food resources in areas where the population had recently increased as a result of political decisions (resettlement, encouraging semi-nomadic populations to settle, measures aimed at increasing the population in this marginal area and above all at reducing the density of central areas affected by major famine episodes). This demographic growth was accompanied by pressure on food resources (the agrarian reforms of the Derg[2] and the period that followed greatly reduced the area of arable land allocated

---

2   Provisional Military Administrative Council (PMAC).

to each person in a movement of nationalisation and redistribution), leading people to do everything they could to conserve water and soil in order to earn a living, which included the construction of terraces at the valley bottoms: the soil is held in place during floods, and as the water flow is slowed down, it can better infiltrate the plots, as part of rain-fed (and therefore non-irrigated) agriculture.

Among the farmers questioned, 14% stated that when they set up, their plots had walls identical to those they have today, while 59% said their plots did not. Finally, 27% said plots had walls, but they were smaller than they are today. When asked why they built walls, 83% of respondents said that it was to conserve water, and 73% to combat erosion. These terraces were also distinguished by the local people from the anti-erosion terraces built on the slopes as part of a government programme to combat erosion, called Food for Work, which aims to exchange the work of building and maintaining these low walls for basic foodstuffs (oil, cereals).

We could not have answered our research questions without this mixed-methods, socio-environmental approach, which allowed us to link transformations in the local landscape with more global contexts. We can thus trace the history of these landscapes from the middle of the Holocene to the present day (Fig. 20.3). The sediments were deposited in a wetter climate during the last phase of the African Humid Period. From the 5th to 4th millennia BC, the gradual aridification of the climate led to the remobilisation of sedimentary stocks, which accelerated with the gradual growth of human impacts in the region, particularly around the 14th to 17th centuries AD. These human impacts and their effects on the landscape were accentuated by political decisions during the 20th century, which encouraged or even forced people to settle permanently in the study area and to develop small plots of farmland. The combination of these political injunctions, developmental incentives for food self-sufficiency and environmental constraints linked to bare, degraded soils and yields that are sometimes uncertain due to the lack of irrigation, have led people to develop ever more densely the smallest plot of land potentially available for agriculture, moving up the valleys to the upstream ends where today there are very small plots at the head of the basin. Reconstructing this entire landscape trajectory would certainly not have been possible, or at least not with such precision, without an integrated, mixed, and transdisciplinary approach.

I believe one of the major contributions of mixed methods here is the fact that it allowed me to tease out the effects of long-term environmental change from long-term societal change to reach the conclusion that the major landscape transformation linked to terracing is a recent one. Without combining biophysical, socio-economic, and political systems, I might not have seen the complexity of the question and have made mistakes in interpreting landscapes evolutions.

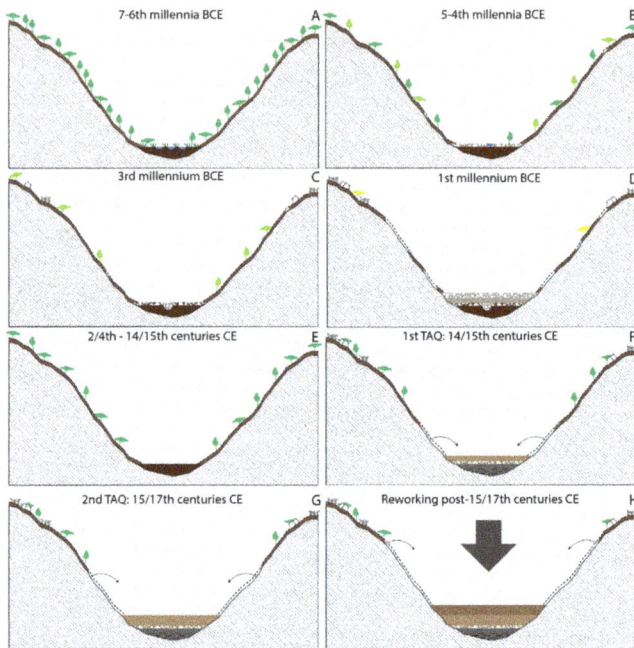

Fig. 20.3 Ninon Blond, 2023. Reconstruction of the evolution of the region's landscapes using sedimentological analyses carried out on three sections: MWb (A and B), KEb (C and D), AMb (E to H). At MWb, the oldest sediments allow us to identify a wetter period (A), between the 7th–6th and 5th–4th millennia, during which the valley floor filled up as a result of run-off and the materials it carried (B); at KEb, we can see that the end of the African Humid Period was marked by a gradual aridification of the climate, leading to a scarcity of vegetation on the slopes (C), and therefore to the erosion of soil on the slopes, which was gradually deposited in the valley floor (D); At AMb, the configuration of the sediments and the organisation of the dates in the sections reveal two main phases of sediment disturbance (F and G), linked to anthropogenic pressure and the continuation of climatic aridification. The impact of human activity was particularly noticeable around the 15th–17th centuries (H)

Greece, adaptation of an iterative approach and complementary methodological elements

In Greece, the site of Terpni, in the lower Strymon valley (Fig. 20.2), was occupied as early as the 6th century BC, although most of the occupation took place during the Hellenistic and Imperial periods. As part of the VitiOrient project[3] and the archaeological mission to Terpni,[4] we studied the vineyards in order to establish a reference system of soils that are currently under vine or have been in the recent past. Through the physico-chemical characteristics of the soils and sediments studied, we hope to enable comparison with archaeological remains and to investigate the possible presence of ancient agricultural soils. More broadly, our goal is to identify an ancient wine-growing terroir[5] in relation to human occupation and pressing activities (Malamidou et al. 2023).

To do this, we implemented a methodology that was modified gradually as new data were acquired—hence the qualification of the methodology as "iterative" (see Chapters 2 and 3). We began with an interview campaign that asked current farmers and winegrowers about the history of their plots (Fig. 20.4a) in order to identify various types: plots currently under vines; plots formerly under vines but now under another crop (olive trees); non-vine-growing fields near the Terpni site, etc. These interviews, combined with observations made on site and on geological maps, led to the mapping of potential sites for core sampling (Fig. 20.4b) and influenced the choice of coring sites.

The interviews revealed that before the 1980s, farms were small and cotton was grown on the plains, while tobacco, vines, watermelons and cereals occupied the foothills. Today maize, cotton, and sunflowers dominate on the plains, while vines and tobacco are declining in the foothills, mainly in favour of olive trees and, to a lesser extent, almonds. Plots are still small in the foothills, while the valley reclaimed from the Strymon is made up of large plots resulting from land consolidation.

---

3   Vigne et vin en Égypte, en Grèce et en Asie Mineure au Ier millénaire avant J.C. Échanges biotechnologiques et transferts techniques en Méditerranée orientale, (resp. C. Pagnoux, MNHN, et B. Redon, CNRS/HiSOMA).

4   Dir. L. Sève, EFA et D. Malamidou, éphorie des antiquités de Serrès.

5   Area of land with certain homogeneous characteristics at this scale, suitable for producing certain agricultural products and giving them a special character.

Fig. 20.4 TER004, surroundings of Terpni archaeological site: using mixed methods to understand and document landscape and environmental changes. (A) Ninon Blond, 2022, Interviews with Vaïos Tasioudis, with the help of Clémence Pagnoux, to document changes in the occupation of his brother's plot (TER004); (B) C. Pagnoux, 2023, CC BY, Core drilling in plot TER004 with Vaïos Tasioudis, Guillaume Bidaut and Laetitia Balaresque; (C) Royal Engineers, British Salonika Force 1917TER004 and its surroundings on the 1:20,000 map of 1917 (Kopaci sheet). Note the indication of "mulberry grove". The cartographic feature corresponds to orchards; (D) Hellenic Army Geographic Service 1945, TER004 and surrounding area on a 1945 aerial photograph. Note the scattered trees and the predominance of fields; (E) ESRI satellite 2023, TER004 and surrounding area on a 2023 satellite image. Many of the fields are still visible, but there are more trees (mostly olive trees, the planting of which was encouraged by national and European governments, we were told during our interviews)

At the same time, we used archival data sources (Cope, Chapter 22) (1:20,000 topographic maps from 1917; plans from the 1920s–1930s; 1:42,000 aerial photographs from 1945; Corona satellite photographs from the second half of the 20th century; contemporary satellite images) with these geomorphological interrogations in mind. This approach made it possible to characterise the evolution of the hydrological network in the Strymon valley before and since its rectification following from construction of the Kerkini Dam. This work, carried out with Line Haza as part of a tutorial project (Ecole Normale Supérieure de Lyon), also helped identify future sites for core drilling in the valley floor as part of a wider project to characterise environmental changes in the lower Strymon valley (Fig. 20.4c, d, e).

It is therefore possible to approach the landscape and socio-economic changes in the region by combining the study of photographic and cartographic archives, interviews and sediment samples studied by geoarchaeology and sedimentology. The formalisation of this somewhat unusual methodology also highlighted the incremental and iterative aspect of this mixed method approach, with the interviews progressively modifying the direction of the sedimentological research (the interviews were used to identify areas where coring could be carried out, see Fig. 20.4a, b), the geomorphological objective guiding the archival research and the content of the interviews (the archives document changes in the landscape and cultivation, confirmed by the interviews, see Fig. 20.4c, d, e), and so on. In this way, interviews allow us to characterise in advance the areas where we could work and to formulate hypotheses about the sediments we might find.

It is thus possible to reconstruct ancient agricultural landscapes and to make the link between viticulture and changes in hydrology. This allowed us to fill a geoarchaeological gap; in this case, whilst the palaeoenvironments of the Strymon valley are well known (Glais 2017), bringing in archaeological information allows us to go beyond this palaeo understanding to make broader links to the historical societies associated with them. These links can deepen our knowledge of past environments based on archaeological data, while also using (palaeo) environmental data to better understand and explain the sedimentary accumulations found during the excavations.

Discussion: Benefits and limitations of this approach

One of the most important contributions of the work described above is the triangulation between qualitative and quantitative methods, and between biophysical and social science methods (see Biermann and Gibbes, Chapter 4). I feel it is more appropriate to adapt my research design to the particular case than to privilege a particular type of method. I thus chose methodological and technical options according to my research objectives rather than relying on one method to be applied everywhere and in every circumstance (see Lane and Lave, Chapter 2). I generally try to enter a field with a fresh eye and as few preconceptions as possible to try to see what adjustments I can make to adapt to what is available and let the field speak for itself. I adapt my methodology to the field site, rather than trying to adapt the field site to my methodology.

Perhaps this is also why mixed methods are so important in my work: they give flexibility and a toolbox from which to draw research approaches best suited to the space and the moment.

As the examples above show, the use of a multi-method approach that combines quantitative and qualitative social and biophysical approaches enables a systemic, diachronic, and integrated approach to understanding socio-ecological systems. This integrated approach made it possible to trace pre-anthropogenic environmental changes and to highlight the periods when actions began to disrupt the Tigray sedimentary system; to identify phases when the impact of human activity was particularly noticeable; and to examine the influence of political and economic decisions, demographics, culture, and agricultural practices on landscape changes and the disruption of sedimentary cascades. This type of mixed-methods research makes up for the shortcomings of a single technique or method when used on its own (e.g., the limited accuracy of radiocarbon dating for recent periods, lack of information in geoarchaeological or archaeological studies on the tools and equipment used in agriculture or quarrying, and lack of precise data on long-term environmental changes in interviews). The synergy between these different data results in a complete, synthetic and systemic reconstruction that enables the research hypotheses to be tested as fully as possible with a relatively moderate deployment of resources (see Biermann and Gibbes, Chapter 4). This has shown encouraging results at other field sites that it was not possible to present in detail here: in Egypt, on the quarry site of Hatnub (Gourdon et al. 2022); in Sudan, about earthen construction on the site of Kerma Dukki-Gel; and in Oman, on ancient hydraulic structures (*aflaj*) on the site of Al Arid (see Fig. 20.2 for location).

One of the central contributions of this work is that it takes a truly interdisciplinary approach in that it does not simply rely on entrusting various tasks to specialists in these different methods and then assembling the results, as we see, for example, in many palaeoenvironmental studies, but rather on integrating these different methods from the outset to develop hypotheses, choose sites and anticipate modes of interpretation. As a result, the researcher here is something of a specialist in the three approaches and their discussion, which facilitates a holistic and global understanding of the issue, rather than a sector-based approach. This seems to me to be one of the advantages of mixed methods: one touches on all that a particular research question or place needs (qualitative, quantitative, social sciences, environmental sciences), but all the levers are in the hands of a single operator, which facilitates an overall vision and a global interpretation of the results.

In this context, the use of ethnogeomorphology, in a version that differs from that of Wilcock et al. (2013), stems from similar dynamics. We would define it as a methodology that, coupled with thorough geomorphological analysis, targets the understanding of the relationships maintained by ancient populations with their environments and their impact on these environments based on the environmental knowledge and agricultural practices of current local populations. Ethnogeomorphology examines current populations' practices and perceptions of the landscape and their collective memory of past morphogenic events (floods, landslides, erosion) through interviews. Thus, using a regressive approach[6] provides keys to read the landscape evolution in the actual state of present landscapes, even when those traces have been degraded (e.g., by erosion), and makes it possible to put forward hypotheses on past agricultural uses, which cannot be thoroughly documented through archaeological remains.

Overall, the field of environmental and/or archaeological studies could benefit from the contributions of mixing geoarchaeology, geomorphology, sedimentology, geohistory, and ethnogeomorphology. Starting from concrete objects (terraces, quarries, abandoned settlements or holy sites) and/or practices (agriculture, mining) provides immediate and easily understood support for intellectual exchanges among disciplines. Agriculture, for example, seems to be a relevant theme, as practices have major impacts on the environment and land uses. Its—current or former—central position in many societies also makes it suitable for overlapping studies. However, other areas relating to resource exploitation can also be considered (e.g., mineral or forest resources) as long as it is possible to have access to soil data, to precise archives on the area and to collective and individual memory on current

---

6   In history, for example, the regressive method is used when an object is approached by comparing several states at different times, starting with the most recent and working backwards. The basic premise of this approach is that the most recent phase contains vestiges of earlier phases that can be deciphered to reconstruct these earlier phases. It's "reading history backwards", to use Marc Bloch (1931)'s expression.

and/or sub-actual practices and landscapes. This implies a continuity, at least of occupation if not of practice, in order to guarantee the transmission of memory. However, not all topics are equally amenable to this method. It is therefore important to think carefully beforehand about the research issues and the precise contribution expected from each method.

The question of temporality must also be raised. Some societies have oral traditions that are passed from generation to generation, making it possible to gather information on ancient states of the environment or practices. If an area has been occupied for a long time, local archives may exist (like in Serres (Greece), near the site of Terpni, where there are maps and drawings of the construction of the Kerkini Dam, and the memory of Bulgarian incursions in the area has been preserved, even regarding ancient place names that have now disappeared). On the other hand, in recently settled areas, this mixed-methods approach might be hampered by the lack of rich documentation from archives and/or oral memory.

It must also be noted that the preparation and use of questionnaires or other kinds of interviews require significant bibliographical work as well as prior knowledge of the area, its problems and specificities. Ethnogeomorphological interview guides cannot be transposed indiscriminately from one territory to another. As in any ethnological or anthropological study, it is important to have local connections in order to be introduced to the community and to break down barriers (see Sayre, Chapter 34). Similarly, it is important to have an overview of the archival data that might be available. Some areas are very well documented (e.g., because they were a central place during war) and have been covered by maps and aerial photographs while other areas have long remained marginal and have not been well documented.

There are some important limitations to the qualitative and quantitative, social and biophysical mixed-methods approach I have described here. First, it is a complex challenge, in that it requires the use of different methods, training, the ability to manipulate and interpret the results of a wide range of research, knowledge of the biases involved, time, skills, and a significant initial investment. The researcher may be a bit of a 'one-person band', which is challenging when there are so many methods where expertise is required. I learned new methods when I felt the need to use them. Then, by using them again and again, I was able to deepen my knowledge of the biases and limitations through reading and interdisciplinary exchanges with other researchers.

Furthermore, this approach brings together data from different sources, which are not of the same granularity or precision, and which have been acquired by different means. This means going beyond the usual considerations of the relationship between quantitative and qualitative data and between biophysical and social sciences: the qualitative approach does not simply serve to illustrate the quantitative data, and the biophysical sciences are not, in essence, more precise or more scientific than the social sciences. On the contrary, it is the development of a convergent set of clues using these different methods that makes it possible to construct solid, realistic and viable working hypotheses (see Biermann and Gibbes, Chapter 4) through the framework of falsification: here, I've been able to prove what's false—e.g., Tigray terraces were not ancient—rather than demonstrate something that is true for sure. Thanks to that method, interviews can be used to prepare the ground for a coring campaign, or archives can complement sedimentological and pedological analysis.

Finally, it should be emphasised that each of these methods has its biases, which can also be apparent in a mixed approach. For example, as in any social science approach, it is important to reflect on the possible impacts of social structures of dominance (Palsky 2013) based on race, training (the researcher or expert vs. the layman), or gender (the first translator working with me in Ethiopia stated, with regard to a potential interviewee: 'she's a woman, she probably won't know', referring to the fact that agricultural work is mainly a male affair).

The presence of a translator makes it possible to have an interlocutor whose presence may seem more legitimate in the eyes of the population: beyond the knowledge of the language, it is the knowledge of the local and regional context and of the proximal problems of the populations that can promote exchanges. But the to-and-fro between French, English, and the local languages (Tigrinya, Amharic, Egyptian, Greek, Omani, Sudanese, and Nubian) can be responsible for loss of information, errors in interpretation on the part of all the actors involved (interviewee, translator, researcher) or personal biases in the collection of the answers (e.g., one translator in Ethiopia ended up trying to find what he considered the right answer).

As in any study based on interviews, a 'positivity bias' may also exist (a tendency to answer positively rather than negatively), as well as a 'social desirability bias' (a desire to give a good image of oneself) (Berthier 2023). In Ethiopia, respondents tended to highlight the local origin of terraces and the absence of state intervention. This could indicate a desire to set local history against the history of the federal state through a founding myth around the terrace object. The knowledge produced in interviews "is nothing more than plausible approximations" (Olivier de Sardan 2004).

Texts and graphic or cartographic representations of archival documents are also situated products of the context in which they were produced: travellers' accounts may reflect their authors' view of the areas they travelled through (as in the idyllic descriptions of landscapes reported by 19th-century explorers such as Lejean (1872) or Girard (1873)) or the populations they encountered, who may be considered, in a context marked by colonialism, as indolent (Andree 1869: 235) or uncivilised. Engravings could reflect these ideas. Maps, as tools of power[7] (Harley 1989; Hirt 2009; Hirt and Lerch 2013), can reflect territorial conflicts or military issues (many of the maps of the region studied in Greece come from an international production focused on the front lines during the Balkan War).

Finally, even though sedimentological analysis techniques are based on well-established scientific protocols, they too have biases. The multiplication of treatments applied to samples, as well as the repetition of certain manipulations, entail the risk of falsifying results (pollution, loss of material during manipulation, approximations linked to instrumentation). Further, sedimentological analysis represents only part of the sample, which is itself only a fraction of the stratigraphic unit sampled. As the sample is assumed to be representative of the stratigraphic unit, the results obtained for the first are extended to the second. Nevertheless, the sample may only represent an extremely localised event. Finally, radiocarbon dating can be biased by the sampling itself, but also by the remobilisation of earlier charcoal during more recent episodes (Blond et al. 2021) and the "old wood effect" (Stouvenot et al. 2013).

---

7   "Geography is first and foremost a way of waging war", to quote the provocative title of a book by Y. Lacoste (1976).

Conclusion

This chapter seeks to show that mixed methods can make a significant contribution to scientific knowledge of diachronic environmental and landscape changes. The combination of techniques with different epistemological foundations and instruments, far from hampering understanding, actually makes it possible to address complementary aspects of the same question and shed light on grey areas. This integrated approach is built up over the course of encounters and the results provided by the methods previously used, in a mode of constant adjustment or accommodation. However, the possibility of using mixed methods should be factored in from the outset of a project to prepare the hypotheses, the fieldwork and the practicalities of implementing certain techniques, which may, for example, require permits. Above all, the idea is not to restrict oneself a priori to a single type of method, but to remain open to the possibility of inventing new methodological combinations that correspond to the research question, the research strategy and the constraints of the field.

While this use of mixed methods is now an integral part of my research strategy and I have no hesitation in resorting to it if the problem or the field requires it, it should not be considered a 'miracle' solution that can be adapted to any subject, any situation or any object. This approach requires some conditions to be met to ensure its feasibility, and it does not resolve the biases associated with each of the methodologies employed. Similarly, assembling data of different types and precision can be a real challenge for research. This is what makes this approach so difficult, but also so interesting and exciting.

References cited

Aguilera, T. and T. Chevalier. 2021a. 'Les méthodes mixtes : vers une méthodologie 3.0 ?', *Revue française de science politique*, 71.3, pp. 361–363. https://doi.org/10.3917/rfsp.713.0361

Aguilera, T. and T. Chevalier. 2021b. 'Les méthodes mixtes pour la science politique. Apports, limites et propositions de stratégies de recherche', *Revue française de science politique*, 71.3, pp. 365–389. https://doi.org/10.3917/rfsp.713.0365

Andree, R. 1869. *Abessinien, das Alpenland unter den Tropen und seine Grenzländer* (Otto Spamer).

Arthur J.W. 2003. 'Ethnoarchaeology, pottery, and technology: Bridging ethnographic and archaeological approaches', *Reviews in Anthropology*, 32.4, pp. 359–378. https://doi.org/10.1080/00988150390250881

Aurenche, O. 2012. *Vous avez dit ethnoarchéologue ? Choix d'articles (1972–2007)* (Jean Pouilloux).

Benoist, A., I. Gajda, S. Matthews, J. Schiettecatte, N. Blond, S. Büchner, and P. Wolf. 2020. 'On the nature of South Arabian influences in Ethiopia during the late first millennium BC: late pre-Aksumite settlement on the margins of the eastern Tigray plateau', *Proceedings of the Seminar for Arabian Studies: Papers from the fifty-third meeting of the Seminar for Arabian Studies held at the University of Leiden from Thursday 11th to Saturday 13th July 2019 50*, pp. 19–36.

Benoist, A., I. Gajda, J. Schiettecatte, N. Blond, and S. Antonini. 2021. 'What was the South Arabian impact on the development of Ethiopian margins in antiquity? Evolution of settlement patterns in the Wakarida region from pre-Aksumite to late Aksumite periods', in *South Arabian Long-Distance Trade in Antiquity: "Out of Arabia"*, ed. by G. Hatke and R. Ruzicka (Cambridge Scholars Publishing), pp. 111–153.

Benoist, A., I. Gajda, J. Schiettecatte, N. Blond, O. Barge, D. Capra, E. Régagnon, and E. Vila. 2022. 'Emprises et déprises agricoles aux marges du Tigray oriental : Les régions de Wolwalo et Wakarida de la période pré-aksumite à la fin de la période aksumite', in *Networked Spaces : The Spatiality of Networks in the Red Sea and Western Indian Ocean*, ed. by C. Durand, J. Marchand, B. Redon, and P. Schneider (Open Edition Books), pp. 531–560. http://books.openedition.org/momeditions/16496

Bergman, M.M. 2008. *Advances in Mixed Methods Research: Theories and applications* (Sage Publications).

Berthier, N. 2023. *Les techniques d'enquête en sciences sociales—Méthodes et exercices corrigés* (Armand Colin).

Biermann, C. and Gibbes, C., Chapter 4, this volume. 'Mixed methods in tension: lessons for and from the research process'.

Bloch, M. 1931. *Les caractères originaux de l'histoire rurale française* (Armand Colin).

Blond, N. et al. 2018. 'Terrasses alluviales et terrasses agricoles. Première approche des comblements sédimentaires et de leurs aménagements agricoles depuis 5000 av. n. è. à Wakarida (Éthiopie)', *Géomorphologie : Relief, Processus, Environnement*, 24.3, pp. 277–300. https://doi.org/10.4000/geomorphologie.12258

Blond, N. et al. 2019. 'Étude de l'évolution du ravinement dans les jessour du Sud Tunisien grâce aux images aériennes', *Cybergeo*, 10.4. https://doi.org/10.4000/cybergeo.32495

Blond, N. et al. 2021. 'From section to landscape(s): reconstructions of environmental and landscape changes for the past 8000 years around the site of Wakarida (Ethiopia) using chronostratigraphy', *BSGF—Earth Sciences Bulletin*, 192.1, p. 53. https://doi.org/10.1051/bsgf/2021041

Breton, J.-F. 2015. *Les bâtisseurs sur les deux rives de la Mer Rouge* (Centre français des études éthiopiennes, Addis-Abeba, Éthiopie).

Combes, E. and M. Tamisier. 1838a. *Voyage en Abyssinie, dans le pays des Galla, de Choa et d'Ifat : précédé d'une excursion dans l'Arabie-Heureuse, et accompagnée d'une carte de ces diverses contrées, 1835–1837.* Vol. 1 (Louis Desessart).

Combes, E. and M. Tamisier. 1838b. *Voyage en Abyssinie, dans le pays des Galla, de Choa et d'Ifat : précédé d'une excursion dans l'Arabie-Heureuse, et accompagnée d'une carte de ces diverses contrées, 1835–1837.* Vol. 2 (Louis Desessart).

Cope, M., Chapter 22, this volume. 'Archival methods'.

Finneran, N. 2007. *The Archaeology of Ethiopia* (Routledge).

Frankl, A. et al. 2011. 'Linking long-term gully and river channel dynamics to environmental change using repeat photography (Northern Ethiopia)', *Geomorphology*, 129.3–4, pp. 238–251. https://doi.org/10.1016/j.geomorph.2011.02.018

Gajda, I., A. Benoist, J. Charbonnier, S. Antonini, X. Peixoto, C. Verdellet, V. Bernard, O. Barge, E. Régagnon, and Y Callot. 2017. 'Wakarida, un site aksumite à l'est du Tigray : fouilles et prospections 2011-2014', *Dossier : Women, Gender and Religions in Ethiopia*, 30, pp. 175–222.

Girard, A.-R.-A. 1873. *Souvenirs d'un voyage en Abyssinie (1868–1869)* (Ebner).

Glais, A. 2017. *Interactions sociétés-environnement en Macédoine orientale (Grèce du Nord) depuis le début de l'holocène : approche multiscalaire et paléoenvironnementale* (Université de Caen, Normandie).

Gourdon, Y. et al. 2022. 'Hatnoub (2021)', *Bulletin archéologique des Écoles françaises à l'étranger. https://journals.openedition.org/baefe/6305#tocto1n5*

Harley, J.B. 1989. 'Deconstructing the map. Cartographica', *The International Journal for Geographic Information and Geovisualization*, 26.2, pp. 1–20.

Hirt, I. 2009. 'Cartographies autochtones. Éléments pour une analyse critique', *L'Espace géographique*, 38.2, pp. 171–186. https://doi.org/10.3917/eg.382.0171.

Hirt, I. and L. Lerch. 2013. 'Cartographier les territorialités indigènes dans les Andes boliviennes : enjeux politiques, défis méthodologiques', *Cybergeo: European Journal of Geography*, 638. https://doi.org/10.4000/cybergeo.25843

Jacob-Rousseau, N. 2009. 'Géohistoire/géo-histoire : quelles méthodes pour quel récit ?', *Géocarrefour*, 84.4, pp. 211–216. https://doi.org/10.4000/geocarrefour.7598

King, G.E. and Abbott, P.M., Chapter 28, this volume. 'Geochronological methods'.

Lacoste, Y. 1976. *La géographie, ça sert, d'abord, à faire la guerre, La Découverte* (Maspero).

Lane, S.N. and Lave, R., Chapter 2, this volume. 'Introduction to building the research "kitchen"'.

Lane, S.N. and Lave, R., Chapter 3, this volume. 'Frames, disciplines and mixing methods in environmental research'.

Lane, S.N., Chapter 25, this volume. 'Descriptive statistics'.

Lane, S.N., Chapter 42, this volume. 'Statistical inference'.

Lamont, M. and P. White. 2005. 'Workshop on interdisciplinary standards for systematic qualitative research. Cultural anthropology, law and social science, political science, and sociology programs', *Proceedings of Workshop on Interdisciplinary Standards for Systematic Qualitative Research*, p. 180.

Lejean, G.M. 1872. *Voyage en Abyssinie éxécuté de 1862 à 1864* (Hachette et Cie).

Longhurst, R. and Johnston, L., Chapter 27, this volume. 'Focus groups'.

Johnston, L., and Longhurst, R., Chapter 32, this volume. 'Interviews: Structured, semi-structured and open-ended'.

Malamidou, D., et al. 2023. 'Terpni, Campagne de terrain 2022', *Bulletin Archéologique des Écoles Françaises à l'Étranger.* https://doi.org/10.4000/baefe.9759

Olivier de Sardan, J.-P. 2004. 'La rigueur du qualitatif. L'anthropologie comme science empirique', *Espaces Temps*, 84, pp. 38–50. https://doi.org/10.3406/espat.2004.4237

Palsky, G. 2013. 'Cartographie participative, cartographie indisciplinée', *L'Information géographique*, 77.4, pp. 10–25. https://doi.org/10.3917/lig.774.0010.

Rüppell, E. 1840. *Reise in Abyssinien* (Siegmund Schmerber).

Salt, H. 1816. *Voyage en Abyssinie* (Magimel).

Sayre, N.F., Chapter 34, this volume. 'Participant observation and ethnography'.

Sauvayre, R. 2013. *Les méthodes de l'entretien en sciences sociales* (Dunod).

Schwartz, M.A. 2008. 'The importance of stupidity in scientific research', *Journal of Cell Science*, 121.11, pp. 1771–1771. https://doi.org/10.1242/jcs.033340

Stouvenot, C., J. Beauchêne, D. Bonnissent, and C. Oberlin. 2013. 'Datations radiocarbone et le "problème vieux bois" dans l'arc antillais : état de la question', *25to Congreso internacional de arqueología del Caribe—25th International Congress for Caribbean Archeology—25e Congrés International de l'Archéologie de La Caraïbe*, pp. 459–494.

SECTION 3:
INTRODUCTION TO THE LIST OF INGREDIENTS

21. Introduction to the list of ingredients

Rebecca Lave and Stuart N. Lane

Now that you have seen compelling "recipes" for mixing biophysical and social methods in environmental research, it is time to compile your own methodological "shopping list". The chapters that follow describe a subset of possible "ingredients", from modelling to statistical analysis to archival research, focusing only on those methods you have already seen in practice in the chapters in Section 2. A quick look at the Table of Contents for this Section will make it clear that many more social than biophysical methods are covered. This is because biophysical methods tend to have narrower, more specific purposes than those that come from the social sciences or the humanities. They also tend to change more quickly. Thus, the chapters below address a selection of broadly used approaches common in the biophysical sciences (e.g., statistical analysis and remote sensing) rather than a large number of very specific approaches.

Please keep in mind that each of the methods addressed below is a single ingredient. To make something edible it is the recipe that matters, defining which ingredients are combined and how this is done. Thus, it is important not to consider any one method in isolation but to think about how methods compare and contrast, and how they may be combined (and under what circumstances). It is also important to remember that some methods cannot be combined but can still be juxtaposed (see Biermann and Gibbes, Chapter 4; Kelley, Chapter 19).

The chapters in this Section do not provide comprehensive instructions for using a particular approach. Instead, the goal is to provide a succinct overview of each method's strengths, weaknesses, and compatibility with other methods. The chapters also address ethical issues associated with each method and provide a few key references (open access where

 https://doi.org/10.11647/OBP.0418.21

possible) for learning more about how to use it. Our hope is that this will allow you to consider a wide range of methods before doing the time-intensive work of learning to use a few methods well. We encourage you to pay particular attention to ethical issues; some proponents of certain methods overlook important ethical questions.

How to choose among the various options? We recommend that you start by attending carefully to the site or object you wish to study (e.g., changing land use in a situation of ongoing colonialism or researchers' impacts on the Antarctic Dry Valleys). What particular combination of biophysical and social forces is shaping your field site? What does the site tell you if you listen to it, watch it, etc.? Answering these questions will likely require an initial round of fieldwork, since our preconceptions about the forces at work may be incomplete or even incorrect (e.g., Kelley, Chapter 19; Lebek and Krueger, Chapter 12).

Once you have an initial understanding of the eco-social forces shaping your field site, use the chapters below to get a sense of the range of options for examining them. Think about which approaches will allow you to "listen" to what your site is saying. This is likely to yield a long list of potential ingredients, thus the final step is narrowing down your shopping list. A good place to start is the compatibility of the particular biophysical and social approaches among your initial set of potential methods. We think it is possible to combine almost any set of methods, but some are certainly easier to combine than others. Even those that can't be combined may provide fascinating insights when used in juxtaposition (Biermann and Gibbes, Chapter 4; Kelley, Chapter 19).

Then think carefully about what each potential approach would require of you. Exploring new methods can expand your capacities in exciting ways, but some approaches may be out of reach. We recommend that you consider a range of factors, from personal preferences to logistical feasibility to ethics. First, take your own preferences seriously. If you suffer from anxiety around numbers and data analysis, statistics may not be for you. Similarly, if you are deeply uncomfortable talking with people one-on-one, focus groups might suit you better. Ideally, your research should be deeply enjoyable, even fun. If you discover you hate a particular method or field site, we recommend choosing a different path.

Logistics also have a major impact on what methods you choose. Some methods are far more expensive than others, so it is important to be realistic about your budget. Similarly, available expertise is surprisingly important. While it is possible to learn new methods just from the published literature, hands-on training is much easier in our experience. Thus, when choosing among potential methods, it is worth considering whether there is training or experts you could consult at your institution or those nearby.

A final, crucial concern is the ethics of your proposed methods. Among the key ethical commitments of any environmental research project should be minimising harm to landscapes, humans, and extra-human nature (see Meadow et al., Chapter 5). To minimise harm to the people and places you study, it is important to give careful thought to the methods you employ, the GHG emissions associated with your research, etc. (see Lane, Chapter 8). It is also critical not to proceed without consent from local Indigenous populations and land managers, who must be allowed to refuse research projects of their communities and territories (Liboiron 2021). Lane (Chapter 8) provides an approach for thinking through the impacts of various approaches to answering your research questions.

It also crucial to keep yourself and any other members of your research team safe. While research should be as inclusive as possible, race, gender, and ability have substantive impacts on fieldwork, whether solo or in a team (Miesen and Gevers, Chapter 9). In the US, for example, scholars of colour report hostile reactions from landowners and law enforcement in some rural areas (Anadu et al. 2020), while female scholars report disturbingly high levels of sexual abuse and harassment when working in groups in the field (e.g., Clancy et al. 2014). There are also important safety questions to consider at any field site and Miesen (Chapter 10) proposes a means for assessing risks associated with fieldwork. It is very important to take these considerations seriously.

References cited

Anadu, J., H. Ali, and C. Jackson. 2020. 'Ten steps to protect BIPOC scholars in the field', *Eos*, http://eos.org/opinions/ten-steps-to-protect-bipoc-scholars-in-the-field

Biermann, C. and Gibbes, C., Chapter 4, this volume. 'Mixed methods in tension: lessons for and from the research process'.

Clancy, K.B.H., R.G. Nelson, J.N. Rutherford, and K. Hinde. 2014. 'Survey of Academic Field Experiences (SAFE): Trainees report harassment and assault', *PLOS ONE*, 9.7, p. e102172. https://doi.org/10.1371/journal.pone.0102172

Kelley, L., Chapter 19, this volume. 'Engaging remote sensing and ethnography to seed alternative landscape stories and scripts'.

Lane, S.N., Chapter 8, this volume. 'The environmental impacts of fieldwork: making an environmental impact statement'.

Lebek, K. and Krueger, T., Chapter 12, this volume. 'On the dialogue between ethnographic field work and statistical modelling'.

Liboiron, M. 2021. 'Decolonizing geoscience requires more than equity and inclusion', *Nature Geoscience*, 14.12, pp. 876–77. https://doi.org/10.1038/s41561-021-00861-7

Miesen, F., Chapter 10, this volume. 'Fieldwork safety planning and risk management'.

22. Archival methods

Meghan Cope

Definition

Archival methods involve using records of the past to understand historical dimensions of social relationships, economic processes, the production of spaces/places, migration, the physical environment, and other geographic topics of interest. Records have traditionally included text-based documents, maps, photos, oral histories, material objects, numerical data, and audio and video sources. Recent developments have considered landscapes, 'the city', buildings, and even 'the body', as an archive, which illuminate the ways that *archive* could refer to many forms of accumulated knowledge stored in a real or virtual repository.

The basics of archival methods

Historical geographers (e.g., Roche 2021; Nicholson 2023) identify two initial steps every researcher is advised to take: 1) start by reading everything available in the relevant secondary literature to familiarise oneself with what others have found in the historical records about your topic, place, or time period; and 2) once a particular repository or collection has been identified, make contact with the archivist and meet with them to learn as much as possible about the sources. Archivists and collection managers are invaluable resources for guiding researchers through territory they have spent their careers getting to know.

Importantly, research questions are significantly constrained and influenced by available materials and data. Historical projects thus always require a degree of flexibility, especially at the beginning: if there are no records, objects, or accounts of the topic of interest, it may be

 https://doi.org/10.11647/OBP.0418.22

necessary to change the research questions to accommodate what *is* in the archive, particularly for students on a tight timeline. Alternatively, thinking broadly about what might constitute evidence may lead to exploring more creative and expansive sources that are typically missing from the 'official' record, such as family histories, artifacts, artwork, and even clues from the physical environment.

Archival methods in depth

Archives present rich opportunities to understand more about a topic, a person, or an event, but also involve important decisions for the researcher. Embarking on historical research may involve a relatively targeted search for specific information, or open-ended browsing of a set of materials without necessarily having a clear sense of what one is looking for. Both of these entry points have value. Targeted searches are important for corroborating other sources, for deepening the knowledge one has about a topic or place, and for discovering information from a different angle or time period. Browsing enables the researcher to think broadly about the context, learn about related topics, and encounter unexpected sources and information. However, browsing can be engrossing and all-consuming, occasionally resulting in the mysterious experience of realising four hours have gone by without notice. Inevitably, archival research involves both targeted searches and browsing in order to fill out the twists and turns of history.

Although many textual records are now digitised and widely available online (and this number is growing fast), millions more sit in their original material form awaiting discovery through in-person visits. Sometimes those materials are held in climate-controlled rooms, are carefully maintained based on provenance and 'original order', indexed by keywords, and have 'finding aids' and helpful staff available for consultation. Other times, especially if the records are not regarded as important because their creators were not powerful elites, they are jumbled in an attic, haphazardly shoved in a drawer, or accumulate in the obscurity of a local historical society that has limited funds for accession and preservation. Those responsible for such informal archives may be unaware of what exists or how to find information in the records. Thus, researchers need to be cognizant of the possibilities of alternative primary data sources and sites and to be open to opportunistic discoveries of 'fragments' (Mills 2013) of people's lives or biophysical information that lead to new insights.

Because the sources of data are so varied in archival research, the methods of data collection and analysis also differ depending on the project, the research questions, the time available, and the scope of the research. Some quantitative applications for archives include statistically analysing and/or mapping numerical data such as residential census data, government spending, institutional and corporate records, or labour and employment reports. In biophysical terms, it could be digitising maps of past river channels, aligning spatially referenced tree-ring data with historical human events, or charting the impact of the Medieval Climate Anomaly (MCA) on agricultural yields. These techniques are particularly helpful for setting the broader context of a phenomenon of interest, assessing its geographic extent, and identifying patterns in large-scale processes. The data that result may be comparable with contemporary datasets allowing for quantitative understanding of change through time.

On the qualitative side, archival methods are eclectic in both technique and topic. Examples include using thematic coding to untangle lesbian and queer spaces of late 20th-century New York City (Gieseking 2016); narrative analysis of Black oral histories (see Chakov et al., Chapter 33) to understand the meanings and uses of an unpaved road in South Carolina (Scott 2019); and discourse analysis of a set of texts—or even street signs—to understand the ways a particular issue was 'framed' as common sense but held much more significant meanings (Alderman 2002).

The critical construction of 'historical GIS' (Knowles and Hillier 2008) enlivens the spatial dimensions of archives, which might otherwise be seen solely through the lens of time, and has sparked a fresh era of mixed-methods historical-geographical inquiry. Mixed-methods approaches using archives could include the use of qualitative GIS to analyse women's representation in Texas historical markers (Ansah et al. 2024), or they could blend social and physical data sources, as Mourey and Ravanel (2017) used in their assessment of mountaineering routes in the context of glacial shrinkage through a "collection and analysis of maps, climbing guidebooks and photographs; semi-structured interviews;

and analysis of high-resolution digital terrain models obtained through terrestrial laser scanning".

Why are archival methods important?

Archives enable a much deeper understanding of our past, but also of our present and future. We can see this aspect of archives through physical records such as ice cores, tree-rings, sediment profiles, and historical maps of river courses, which indicate past climates and human impacts, and suggest important insights for the present and future. Archives' quality of informing the present and future is also true for critical research that seeks to find and share the experiences of people who are typically hidden from official records, such as children, enslaved people, poor people, women, racially and ethnically marginalised people, disabled people, and other social groups who were (or are) seen as inferior, unworthy of attention, or otherwise inconsequential. In this way, recovering the stories, cultural practices, experiences, and ideas of subordinate groups becomes a radical act, a recovery of the pasts of those who were otherwise silenced, and a refusal to accept one-dimensional historical accounts from a dominant group.

In a further extension of this, some of the most creative work in mixed-methods approaches using archives combines the physical and social records of a place to generate new insights, but also generates opportunities for citizen engagement and education. For example, McDonagh et al. (2023) used 750 years of flood data in Hull, UK, derived from sources ranging from climate data and historical maps of rivers and estuaries to centuries-old civil complaints about people's losses due to floods, in order to create an interactive, community-based, geo-humanities project called *Risky Cities*. In one of their resulting publications, they noted that the project's approach

> combines archival research into histories of living with water and flooding in Hull and the surrounding areas, with participatory and creative methods in order to co-create 'learning histories' that are used to engage people in climate and flood resilience actions. In doing so, it also builds on the work of historical geographers who have conducted archival research alongside community work, the gathering of oral histories and other participatory methodologies. (McDonagh et al. 2023: 92)

The *Risky Cities* project also serves as an important example of the need for both skepticism and creativity when working with archives. There were no stream gauges from the 14th century for McDonagh and her collaborators to use to establish hydrological patterns, extent of tides, or magnitude of storm surges. Instead, they had to be innovative in tracing historical property and municipal records in which land-holders and residents submitted complaints about villages flooding, saltwater incursion, and agricultural losses; they found historical maps of dikes and aqueducts being built or destroyed; and they considered local stories of pre-20th-century floods.

Relationship of archival methods with other methods

Historical data are relevant for the full range of geographic inquiry. From gleaning past weather and climate data in farmers' journals (Dupigny-Giroux 2009) to finding nationalist rhetoric in the blueprints of playgrounds (Gagen 2004), and from "interpreting the politics of venereal disease" in mid-20th-century Seattle (Brown 2009: 1) to "the radical potential of [marginalised communities'] counter-narratives in seeing the city itself as an archive" (Burgum 2022: 504), archival methods are readily taken up across the discipline. Indeed, the 'cultural turn' in geography has sparked a renewed interest in historical meanings, practices, and discourses. Further, feminist, queer, and anti-racist critical approaches have rejected the assumed 'authority' of archives as an official repository of power and its documents while exploring broader interpretations of "anything that can hold and spark memory" (Berry 2021: 3). Of course, geographers are particularly attuned to the historical data produced by and stored in the environment that stretch back millennia; sources such as lake sediment cores, ice cores, tree-ring data, and even shrubs (Rayback et al. 2012) provide essential archives for reconstructing past environmental conditions. Thus, archival methods, taken in their broadest terms, can be combined with most geographical research.

Ethical issues and archival methods

The ethical and moral dilemmas of archival work hinge on similar concerns to other methods, such as privacy/confidentiality, avoidance of harm, and respect for persons. However, one of the principles of ethical research—informed consent—is much trickier for archival materials because they were usually preserved without the consent, or even knowledge, of their original authors or creators. This is further complicated by a general lack of guidelines for use and representation of archival materials beyond permission of the current owner; this is mirrored by most institutional review boards' exemption of historical research from ethical scholarship reviews. Thus, when research touches on sensitive issues (such as Moore's (2010) investigation of abortion practices from a time when they were illegal in the UK), reveals past criminal acts, or even documents racist and sexist playground song lyrics that have the contemporary potential to be hurtful (Mills 2017), scholars are faced with dilemmas of how to proceed. As Crawford (2024: 1) suggests in her examination of emancipatory approaches to historical geographies of disability, "the explorative nature of archival research necessitates that ethics should be an iterative undertaking, with archival sources having the potential to shape both the content and conduct of the research".

At another level is the question of how historical evidence can re-traumatise members of a group who have suffered violence, oppression, and loss, as Simmons (2021) documents among Black mothers recalling their own and their foremothers' infant losses. Similarly, Hartman (2008: 4) ponders the archive and wonders, "How does one revisit the scene of subjection without replicating the grammar of violence?". A rough general practice for ethical archival research, then, would be to mask identities when possible, contextualise quotes and actions within the structures of oppression operating at a given time and place, and avoid harming descendants.

Issues to be aware of in using archival methods

A foundational principle of archival research is recognising that records of the past are inherently partial, fragmented, and should be treated with skepticism. On the Physical Geography side, ancient humans' forest management practices spur caution in tree-ring analysis (Skiadaresis et al. 2021), whilst in Human Geography, we note that archives are skewed toward social groups who were/are in power, and whose actions and records were more likely to be considered worthy of preservation. For instance, property records of male landowners in 18th-century America are far more likely to have been saved than women's diaries of the same time period; texts written by adults are more likely to be preserved than children's drawings; and a lavish plantation home owned by enslavers is more likely to remain intact than modest shelters of enslaved workers. While there are methods for exploring the past life-worlds of social groups who did/do not hold power, they often require more creative searching and reading 'against the grain' (Duncan 2001); in an example from my own work using records from a children's home from the early 20th century, I was able to discern a lot about children's daily lives by scrutinising the logbooks kept by the home's Matron and meeting minutes of its board of directors (Cope 2024). More broadly, by attending to the hierarchies and mechanisms of power while centring the actions, voices, and experiences of subordinate or marginalised groups, researchers can confront past geographies with a critical eye.

Commented further reading

The annual journal, *Historical Geography*, and the quarterly *Journal of Historical Geography* are excellent starting points for examples of using archival materials and methods in diverse ways.

For general introductions, chapters in methods handbooks can be useful for the beginning researcher; these include Roche (2021); Nicholson (2023); and Lorimer (2009).

For more on uses of unconventional sources, see DeLyser, Sheehan, and Curtis (2004) regarding buying materials on eBay; DeSilvey (2007) on excavating an abandoned homestead; and Cowen (2020) on "following the infrastructure" of cities as "archives of Indigenous dispossession and genocide".

References cited

Alderman, D.H. 2002. 'Street names as memorial arenas: the reputational politics of commemorating Martin Luther King Jr. in a Georgia county', *Historical Geography*, 30.1, pp. 99–120. https://nebraskapressjournals.unl.edu/07hg30-alderman/

Ansah, H., Y. Lu, and Y. Choi. 2024. 'Recognizing Texas women in time and space: A qualitative GIS inquiry into historical markers', *Geographical Review*, pp. 1–20. https://doi.org/10.1080/00167428.2024.2319252

Berry, D. 2021. 'The house archives built', *up//root*, 22. https://www.uproot.space/features/the-house-archives-built

Chakov, A., Chang, T., Covey, H., Dickson, T., Goggins, S., Harris, N., Purna, S., Widell, S. and Druschke, C.G., Chapter 33, this volume. 'Oral history'.

Cope, M. 2024. '"Fixing" destitute children: the relational geography of an early twentieth century children's home through its archives', *Area*, 56.1. https://doi.org/10.1111/area.12882

Cowen, D. 2020. 'Following the infrastructures of empire: Notes on cities, settler colonialism, and method', *Urban Geography*, 41.4, pp. 469–486. https://doi.org/10.1080/02723638.2019.1677990

Crawford, L. 2024. 'Emancipatory archival methods: Exploring the historical geographies of disability', *Area*, 56.1. https://doi.org/10.1111/area.12844

DeLyser, D., R. Sheehan, and A. Curtis. 2004. 'eBay and research in historical geography', *Journal of Historical Geography*, 30.4, pp. 764–782. https://doi.org/10.1016/j.jhg.2005.01.001

DeSilvey, C. 2007. 'Art and archive: memory-work on a Montana homestead', *Journal of Historical Geography*, 33.4, pp. 878–900. https://doi.org/10.1016/j.jhg.2006.10.020

Duncan, J. 2001. 'Notes on emancipatory collaborative historical research', *Historical Geography*, 29, pp. 65–67.

Dupigny-Giroux, L.A. 2009. 'Backward seasons, droughts and other bioclimatic indicators of variability', in *Historical Climate Variability and Impacts in North America*, ed. by L.A. Dupigny-Giroux and C. Mock (Springer).10.1007/978-90-481-2828-0_14

Gagen, E.A. 2004. 'Making America flesh: Physicality and nationhood in early twentieth-century physical education reform', *Cultural Geographies*, 11.4, pp. 417–442. https://doi.org/10.1191/1474474004eu321oa

Gieseking, J.J. 2016. 'Dyked New York: the space between geographical imagination and materialization of lesbian–queer bars and neighbourhoods', in *The Routledge Research Companion to Geographies of Sex and Sexualities*, ed. by G. Brown and K. Browne (Routledge), pp. 29–36.

Hartman, S. 2008. 'Venus in two acts', *Small Axe: A Caribbean Journal of Criticism*, 12.2, pp. 1–14. https://doi.org/10.1215/-12-2-1

Knowles, A.K. and A. Hillier. 2008. *Placing History: How Maps, Spatial Data, and GIS Are Changing Historical Scholarship* (ESRI, Inc).

Lorimer, H. 2009. '"Caught in the nick of time: archives and fieldwork"', in *The Sage Handbook of Qualitative Geography*, ed. by S. Aitken, M. Crang, D. DeLyser, S. Herbert, and L. McDowell (Sage Publications), pp. 248 – 272.

McDonagh, B., E. Brookes, K. Smith, H. Worthen, T.J. Coulthard, G. Hughes, and J. Chamberlain. 2023. 'Learning histories, participatory methods and creative engagement for climate resilience', *Journal of Historical Geography*, 82, pp. 91–97. https://doi.org/10.1016/j.jhg.2023.09.002

Mills, S. 2013. 'Cultural–historical geographies of the archive: Fragments, objects and ghosts', *Geography Compass*, 7, pp. 701–713. https://doi.org/10.1111/gec3.12071

Mills, S. 2017. 'Voice: sonic geographies of childhood', *Children's Geographies*, 15.6, pp. 664–677. https://doi.org/10.1080/14733285.2017.1287879

Mourey, J. and L. Ravanel. 2017. 'Evolution of access routes to high mountain refuges of the Mer de Glace basin (Mont Blanc Massif, France)', *Journal of Alpine Research | Revue de géographie alpine*, 105. https://doi.org/10.4000/rga.3790

Nicholson, J.F. 2023. 'Historical and archival research', in *Key Methods in Geography*, ed. by N. Clifford, M. Cope, and T. Gillespie (Sage Publications), pp. 135–152.

Rayback, S.A., A. Lini, and D.L. Berg. 2012. 'The dendro-climatological potential of an alpine shrub, Cassiope mertensiana, from Mount Rainier, WA, USA', *Geografiska Annaler: Series A, Physical Geography*, 94, pp. 413–427. https://doi.org/10.1111/j.1468-0459.2012.00463.x

Roche, M. 2021. 'From dusty to digital: Archival research', in *Qualitative Research Methods in Human Geography*, ed. by I. Hay and M. Cope (Oxford University Press).

Skiadaresis, G., B. Muigg, and W. Tegel. 2021. 'Historical forest management practices influence tree-ring based climate reconstructions', *Frontiers in Ecology and Evolution*, 9. https://doi.org/10.3389/fevo.2021.727651

Scott, D. 2019 'Oral history and emplacement in "nowhere at all": the role of personal and family narratives in rural Black community-building', *Social and Cultural Geography*, 20:8. https://doi.org/10.1080/14649365.2017.1413205

Simmons, L.M. 2021. 'Black feminist theories of motherhood and generation: histories of Black infant and child loss in the United States', *Signs: Journal of Women in Culture and Society*, 46.2, pp. 311–335. https://doi.org/10.1086/710805

23. Arts-based environmental research

Mrill Ingram

Definition of arts-based environmental research

Arts-based methods provide creative avenues for researchers and research participants to gather environmental data and to communicate ideas, experiences, and feelings or emotions via a range of media and genres. By engaging with sensory, embodied, and affective environmental experiences and perceptions, art-based research can be used independently or in concert with other environmental research methods to expand ways of knowing. Rooted in the creative arts, arts-based research frameworks are adaptive, anticipating imaginative, emergent, and unpredictable results rather than a clear confirmation or rejection of a hypothesis (Leavy 2023). Some arts-based research focuses on social engagement and transformation.

The basics of arts-based methods and research

There are at least three common paths for engaging with the diverse methodological and theoretical orientations within arts-based research. First, arts-based environmental methods offer creative tools to expand and diversify environmental data generation methods and analysis. For example, researchers might use photography, storytelling, collage-making, or drawing to investigate how people experience their environments and to expand the ways in which environmental perceptions can be expressed. Second, the literary, performative, or visual arts can be used to communicate research results by producing a novel, film, or dance performance in place of a paper or thesis. Such alternative research products can convey the emotionality, complexity,

   https://doi.org/10.11647/OBP.0418.23

or contingency frequently encountered in environmental research but not easily captured in a text-based product. Third, arts-based research can be pursued within a transformational framework in which the creative power of the arts is engaged to generate new knowledge to address problems and aid in social transformation. This "performative" emphasis shares goals with community-based and participatory research traditions (see Mokos, Chapter 36).

Arts-based research in depth

In environmental research, arts-based methods may offer data gathering approaches sensitive to diverse contexts, personalities, abilities, cultures, and even species. How a particular element of the environment is known and experienced can be explored visually, aurally, orally, and performatively, as well as via conventional methods such as interviews (Longhurst and Johnston, Chapter 27) or survey research (Winata and McLafferty, Chapter 43). Creative methods can provide additional modes of communication between a researcher and research subjects of different cultures, languages, or abilities.

Specific arts-based techniques include the creation and performance of music and poetry, including the spoken word; dance or other forms of body movement; storytelling and autoethnography; drawing, painting, comics, and collage, as well as film, video, and other visual arts; and installations combining any of these approaches. Photo elicitation, video ethnography, writing fiction, and creating shared performances are all strategies by which environmental researchers can expand their range of data gathering and communication.

Creative techniques can aid in the exploration of, and communication about, non-human environments, such as creating a musical or dance performance integrating sounds or movements generated by animals or insects. Hamilton and Taylor look to arts-based creative methods in ethnography to "bring animals in ... and reveal their significance as social actors both in relation to humans and in their own right" (2017: 89).

While arts-based methods are often deployed within epistemological norms of biophysical and social science environmental research, they can also be embraced as tools to support more reflexive socially engaged research (Bourriaud 2002: Leavy 2017: Seppälä et al. 2021). Socially engaged arts research has an explicit goal of harnessing emotional experience and expression for the purpose of social transformation; for example, Suzanne Lacy's "new genre" public art (1994). In "The Oakland Projects (1991–2001)" Lacy collaborated with young people in Oakland, California to create a series of installations, performances, and policy efforts to address conflict with police and other issues.

Why arts-based methods are important

The tremendously diverse range of creative arts-based methods greatly expands traditional environmental data gathering tools and offers new techniques for capturing otherwise elusive environmental experiences (e.g., Fujikane 2021; Wylie 2017). Arts-based techniques offer ways to explore the meaning of different landscapes to people, and to capture sensory experiences and memories as important forces that shape environmental behaviour, institutions, and policy. By placing value on ways of environmental knowing other than that produced by conventional environmental research methods, arts-based research can provide tools for researchers working to honour disenfranchised perspectives on the environment, such as those of Indigenous peoples, non-human beings, and others.

Another benefit of arts-based approaches is the diversification of ways to interpret and communicate findings. More creative approaches to communication can be helpful in connecting to interdisciplinary audiences as well as to groups separated by age, language, and cultural differences. Arts-based methods expand definitions of valid environmental data beyond quantitative measures to reflect the diversity of environmental experiences.

Finally, tensions or ambiguities in how people might define their own environments may be better reflected in a collage or performance piece rather than a numerical rating such as a Likert scale-based survey (see Winata and McLafferty, Chapter 43).

Relationship of arts-based research with other methods

Arts-based inquiry can be used in concert with many other environmental research methods (e.g., Gleiniger et al. 2010; Harrison et al. 2016). For example, Wylie (2017) used photographic paper's reactivity to hydrogen sulfide to visualise environmental impacts she was also measuring quantitatively. Arce-Nazario (Chapter 17) used remote sensing images in interactive museum exhibits as a way to engage local communities in land use change.

Arts-based inquiry can also be an integral part of participatory research (see Mokos, Chapter 36) and this may have the goal of action, empowering disenfranchised groups, aiding them in communicating challenges, and developing tools to solve the environmental problems they identify as most urgent (Jackson 2011).

Ethical issues in using arts-based research

The ethics of engaging arts-based methods should be critically examined. Evoking strong emotion or feelings of empathy is often an explicit goal of art, but this can seriously damage relationships among and between researchers and research subjects if researchers do not responsibly hold space for generated emotions and information (especially when they are unanticipated). A strong ethical commitment is key to avoiding manipulation and respecting participants' time and emotional sovereignty.

Arts-based research is often oriented to creating a multiplicity of pathways, or "answers" in order to explore and to understand a research question, rather than the establishment of a hierarchy of explanatory influences. Researchers will benefit from being clear with themselves and their collaborators on the purpose of engaging in arts-based research, and how its application achieves research goals.

Issues to be aware of in using arts-based research

Biophysical and social science engagement with arts-based research methods can range from a more circumscribed adoption of specific data collection methods to a broader exploration of research traditions in the arts that deeply informs the objectives and process of a research project.

The creative arts and environmental sciences share many research concepts including place, space, site, bodies, landscapes, objects, and ways of knowing. The arts engage with these concepts differently, however, and often for different ends. Biophysical and social scientists need some understanding of these histories and contexts to engage with arts-based approaches.

Although environmental science research engagement with the arts is expanding, many traditional scientists find it unfamiliar and thus difficult to evaluate. In concert with available tools to help assess arts-based projects, researchers should be prepared to defend creative practice as necessary and adequately rigorous research with great value to understanding environments (Hawkins 2013).

As explained above, environmental science researchers should carefully consider the ethics of arts-based methods.

Sometimes the embrace of arts-based methods includes collaborating with artists. In these cases, it is critical for all involved to be clear on goals and commitments to avoid exploiting artists, whose skills are too often undervalued as "window dressing" or as communication tools rather than as legitimate avenues for producing valuable knowledge.

Suggested further reading

Bourriaud, N. 2002. *Relational Aesthetics* (Presses du réel).

Fujikane, C. 2021. *Mapping Abundance for a Planetary Future: Kanaka Maoli and Critical Settler Cartographies in Hawai'i* (Duke University Press). https://doi.org/10.1215/9781478021247

Gleiniger, A., A. Hilbeck, and J. Scott. 2011. *Transdiscourse: Mediated Environments* (Springer). https://doi.org/10.1007/978-3-7091-0288-6

Hamilton, L., and N. Taylor. 2017. 'Visual methods', in *Ethnography After Humanism: Power, Politics and Method in Multi-Species Research*, ed. by L. Hamilton and N. Taylor (Palgrave Macmillan), pp. 89–109. https://doi.org/10.1057/978-1-137-53933-5_5

Harrison, H.M., A. Douglas, N. Harrison, C. Fremantle, W.L. Fox, E. Heartney, R. Malina, P. Mankiewicz, D. Sagan, and A.W. Spirn. 2016. *The Time of the Force Majeure: After 45 Years, Counterforce is on the Horizon* (Prestel). SBN-13 978-3791355498

Hawkins, H. 2013. 'Geography and art. An expanding field: Site, the body and practice', *Progress in Human Geography*, 37.1, pp. 52–71. https://doi.org/10.1177/0309132512442865

Jackson, S. 2011. *Social Works: Performing Art, Supporting Publics* (Routledge). https://doi-org.ezproxy.library.wisc.edu/10.4324/9780203852897

Lacy, S. 1994. *Mapping the Terrain: New Genre Public Art* (Bay Press).

Leavy, P. 2023. *Research Design: Quantitative, Qualitative, Mixed Methods, Arts-Based, and Community-Based Participatory Research Approaches* (Guilford Press).

Leavy, P. 2017. *Handbook of Arts-Based Research* (Guilford Press).

Longhurst, R. and Johnston, L., Chapter 27, this volume. 'Focus groups'.

Mokos, J., Chapter 36, this volume. 'Participatory methods'.

Pearson, H.E. 2017. 'Moving towards the trialectics of space, disability, and intersectionality: Intersecting spatiality and arts-based visual methodologies', *Knowledge Cultures*, 5.5, pp. 43–68. https://doi.org/10.22381/kc5520174

Seppälä, T., M. Sarantou, and S. Miettinen. 2021. *Arts-Based Methods for Decolonising Participatory Research* (Taylor and Francis). https://doi.org/10.4324/9781003053408

Springgay, S., R. Irwin, and S. Kind. 2005. 'A/r/tography as living inquiry through art and text', *Qualitative Inquiry*, 11.6, pp. 897–912. https://doi.org/10.1177/1077800405280696

Wang, C., and M. Burns. 1997. 'Photovoice: Concept, methodology, and use for participatory needs assessment', *Health, Education, and Behaviour*, 24.3, pp. 369–387. https://doi.org/10.1177/109019819702400309

Winata, F. and McLafferty, S., Chapter 43, this volume. 'Survey and questionnaire methods'.

Wylie, S., E. Wilder, L. Vera, D. Thomas, and M. McLaughlin. 2017. 'Materializing exposure: Developing an indexical method to visualize health hazards related to fossil fuel extraction', *Engaging Science, Technology, and Society*, 3, p. 426. https://doi.org/10.17351/ests2017.123

24. Case studies

Stuart N. Lane

Definition

A case study is a type of research developed around an in-depth focus upon a geographically constrained environment. It is not a method in and of itself; instead, case studies involve bringing together those methods needed to answer a particular set of research questions. Importantly, case study research focuses on revealing the causes of phenomena rather than documenting their prevalence. Case studies are common in both biophysical and social analysis.

The basics of a case study

A case study involves the "detailed examination of an event (or series of related events) which the analyst believes exhibits (or exhibit) the operation of some identified general theoretical principle" (Mitchell 1983: 192). In-depth analysis of a case enables the researcher to identify general underlying processes, factors and mechanisms that contribute to the characteristics of that and other cases. It may also reveal the local (in space and time) conditions that shape which of these factors matter and how they work. Case studies are always contingent, and a particular skill in case study research is separating the general from the local.

Case studies in depth

The easiest way to understand the strengths of case studies is to compare them with classical statistical inference. In the latter, generalisation is achieved by searching for the patterns revealed by studying many

  https://doi.org/10.11647/OBP.0418.24

individually sampled events. Sampling in an efficient way allows inferences to be made about the population from which the sample is taken. In case study research, a much smaller number of individual samples (or even a single sample) may be studied in-depth. Inference is then drawn from the observed behaviour of only one or a small number of cases rather than from the observed pattern that comes from generalising a very large number of cases. For this reason, statistical inference is often described as large-N research and case studies as small-N research (Richards 1996).

Because of the small sample size, choosing the case study is critical. The goal is to find a case that (1) likely represents the phenomena of interest to the research project, and (2) can be studied in a way that reveals the underlying properties of that phenomena. Choice of a case study should be based on careful and detailed consideration of how its characteristics manifest the underlying and general properties the researcher hopes to study, and of how the researcher can tease out these general properties from local (sometimes called "contingent") conditions. Presentation of case study research must include a careful and detailed description of those characteristics.

Case study research is intensive, commonly involving the simultaneous study of a suite of phenomena in a particular place through time. This is partly by necessity as identifying how things interact, how general properties are shaped by local or contingent conditions, and simply being present when something happens all require a significant amount of time.

Case study research may be multivariate in nature (measuring many phenomena) even if it is not clear whether a particular measurement will be needed. It may also be strongly emic, where the case study itself informs what should be studied and how, and there is a strong "insider's" perspective. This is distinct from large-N research, which is commonly etic, the researcher acting as an observer separated from what they are researching. As compared with large-N research, where the questions to be asked (and thus the findings) are bounded in advance, there is commonly an open-endedness in case-study research; what the case was thought to be representative of may change as the research progresses.

Every case study is unique. Given the importance of contingence, it is rare for a case study to be reproducible in a classical experimental sense. Instead, its reproducibility lies in what processes, factors, and mechanisms can be generalised from the case study (Richards 1996). This makes case study approaches especially relevant for environmental research that involves fieldwork where the local characteristics of the field matter (and change in time).

Case studies can be distinguished from other methods in the kinds of closure used to make the research tractable. In laboratory studies, a research question is commonly made answerable by designing an experiment that holds other variables constant, and that is a form of closure. In surveys, research is made tractable through choices of what variables to measure, how to sample and so on, another form of closure. In a case-study approach, closure is associated with the choice of which general processes, factors, or mechanisms to study, where and when (Lane 2001).

Why are case studies important?

Case studies are important because they solve one of the fundamental problems of more extensive research methods: the number of times something happens does not necessarily tell you why it happens. This is best illustrated with an example. Consider the classical relationship of Leopold and Wolman (1957) in which they identified a statistical threshold between meandering and braiding rivers that could be defined by a river's valley slope (s) and mean bankfull discharge (Q, in cubic feet per second); braiding tends to occur if

$$s > 0.06 \ Q^{-0.44}$$

This is a statistical generalisation; a useful one as it allows a broad-brush identification of what kind of channel pattern might be expected for different combinations of valley slope and river discharge. It reflects the observation that braided rivers can occur both in mountain environments, such as in front of Alpine glaciers (low Q, higher s), and in river deltas (higher Q, low s). However, it does not explain why. Not does it describe in its own right the conditions required for a river to follow this statistical generalisation (in general, the rivers must be self-formed and alluvial).

In the history of river channel pattern research it was possible to mobilise theory to explain why this generalisation exists. Ferguson (1987) showed mathematically that the underlying physical mechanism that leads to the relation relates to a threshold of stream power, with high stream power being required for a river to braid, which can be achieved via high values of either s or of Q. However, Hickin's (1984) account of the history of research into relations between vegetation and fluvial geomorphology shows how case studies of the relationship between vegetation encroachment and channel pattern in the 1960s and 1970s suggested a crucial role for vegetation, likely lost in the scatter in the Leopold and Wolman (1957) relation.[1] This role could not be determined from theory. Subsequent research, much of it based on individual case studies, revealed why vegetation influenced channel pattern: the effects of vegetation on resistance to flow and river bank strength, its role as nucleus for bar sedimentation, the importance of construction and breaching of jams formed by woody debris, and the importance of vegetation for concave-bank bench deposition (Hickin 1984).

This example emphasises that case studies may serve one or both of two functions. First, they can reveal the reasons behind the patterns or statistical generalisations observed in large-N research. Second, case studies may often be motivated by their appearance as an exception to some dominant generalisation or accepted theory. Case studies commonly have a geographical and temporal grounding making them, in many senses, "truth spots", providing an unsettling role in challenging dominant forms of explanation and revealing new questions, hypotheses and explanations.

Relationship of case studies to other methods

A case study is not a method. Instead, it requires researchers to identify a set of methods appropriate for answering their research questions. The process of method selection should be reflexive, requiring the researcher to let the case study frame how it should be studied (that is to be emic; see Chapters 2 and 3). "Listening" to what the study-site can "say" may require adding new methods after the research begins, which can be

---

1   It should be noted that Leopold and Wolman (1957) recognised the potential role of vegetation.

logistically complicated if researchers need to learn new approaches in the middle of a project. This methodological responsiveness is central to the success of the case study approach, though, because of its potential to reveal new understandings not apparent within previously used methodological frameworks.

A second strength and complexity of case study research is that its findings may need to remain provisional until they can be compared with other case studies or other research approaches (e.g., extensive statistical data analysis). An example of this is provided in Egli et al. (2021) who studied collapse features in retreating Alpine glaciers. An intensive study of one glacier suggested that collapse was driven by the development and migration of a meandering river channel under the ice at the ice margin. Their case study revealed this previously unexplored mechanism through a novel combination of methods (ground penetrating radar, drone-based photogrammetry) and intensive study of a collapse feature as it developed. However, such a result could be limited, the result of local causes particular to that glacier rather than generalisable processes. In order to assess the wider relevance of this mechanism an extensive statistical study was undertaken that (1) used the detailed understanding that came from that mechanism to identify more readily measurable variables that described when collapse features could form due to subglacial river channels; (2) identified other glaciers that have both a history of the same kind of collapse feature, and otherwise; and (3) tested the extent to which those glaciers showing collapse had variable properties that indicated the wider mechanism. This mixing of methods is often integral both within case study research and in seeking to draw out wider conclusions.

Ethical issues and case studies

The ethical issues case studies raise are not specific to them but instead associated with the particular methods used. First, because of the intensive nature of a case study, ethical issues associated with participatory research (see Mokos, Chapter 36) may become important. Permission, documentation of research, and authorisation to use results may be critical. Where case study research is geographically delimited there will be a growing likelihood that the research could impact local policy development, communities and so on. This requires a strong sensitivity to the impacts of research, especially unintended ones (see Meadow et al., Chapter 5). Second, the intensive nature of case study research, notably the extended presence of researchers in a geographical location, may lead to impacts on the environment, upon local people, etc. (see Lane, Chapter 8). In the Egli et al. (2021) study above, for instance, the isolated nature of the research meant it was necessary to camp in an environmentally protected zone with the presence of people in an area that is special to passing guides and their tourists precisely because of its wilderness nature.

Issues to be aware of in using case studies

The most difficult part of any case study is choosing a case that balances practical issues, such as access, with the scientific questions the researcher wishes to answer. The first challenge is to balance the general and the local, choosing a case in which it is possible to separate the intricacies and specifics of a particular place from the general processes, factors and mechanisms that transcend it. This generalisation can take a number of forms and may be methodological or theoretical, but must produce ideas, hypotheses, or theories that then are subject to ongoing testing, refinement and development. A failure to attempt this risks a study of detail with little broader relevance.

The second challenge of case study research stems from one of its strengths: the intensity of the relationship between the researcher and what is being researched. Case study research is deeply immersive, leading to personal investment and connections that can make it harder to identify other perspectives and explanations than in approaches where there is more separation between the researcher and the researched. This is reflected in debates in anthropology over the challenges of undertaking truly emic research (Mostowlansky and Rota 2020). Case studies require very careful documentation and reporting of those connections to help the researcher to identify what can be generalised from the case, and to guide other scholars who wish to use their findings to support other research projects. They also require a self-reflexivity that can be challenging to adopt when the issues revealed by the research are strongly emotive.

Suggested further reading

Richards, K.S. 1996. 'Samples and cases: Generalisation and explanation in geomorphology', in *The Scientific Nature of Geomorphology: Proceedings of the 27th Binghamton Symposium in Geomorphology held 27–29 September 1996*, ed. by B.L. Rhoads and C.E. Thorn (John Wiley and Sons), pp. 171–90.

Sayer, A. 2010. *Method in Social Science* (Routledge). https://doi.org/10.4324/9780203850374

Yin, R.K. 2017. *Case Study Research: Design and Methods, 6th Edition* (Sage Publications).

References cited

Egli, O., B. Belotti, B. Ouvry, J. Irving, and S.N. Lane. 2021. 'Subglacial channels, climate warming, and increasing frequency of Alpine glacier snout collapse', *Geophysical Research Letters*, 48. https://doi.org/10.1029/2021gl096031

Ferguson, R.I. 1987. 'Hydraulic and sedimentary controls of channel pattern', in *River Channels: Environment and Process*, ed. by K.S. Richards (Blackwell), pp. 129–58.

Hickin, E.J. 1984. 'Vegetation and river channel dynamics', *The Canadian Geographer/ Le Géographe canadien*, 28, pp. 111–26. https://doi.org/10.1111/j.1541-0064.1984.tb00779.x

Lane, S.N. 2001. 'Constructive comments on D. Massey Space-time, "science" and the relationship between physical geography and human geography', *Transactions of the Institute of British Geographers*, 26, pp. 243–56. https://doi.org/10.1111/1475-5661.00018

Lane, S.N. and Lave, R., Chapter 2, this volume. 'Introduction to building the research "kitchen"'.

Lane, S.N. and Lave, R., Chapter 3, this volume. 'Frames, disciplines and mixing methods in environmental research'.

Lane, S.N., Chapter 8, this volume. 'The environmental impacts of fieldwork: making an environmental impact statement'.

Leopold, L. and M.G. Wolman. 1957. 'River channel patterns: braided, meandering and straight', *United State Geological Survey Professional Paper*, 282-B.

Mitchell, J.C. 1983. 'Case and situation analysis', *The Sociological Review*, 31, pp. 187–211. https://doi.org/10.1111/j.1467-954x.1983.tb00387.x

Mokos, J., Chapter 36, this volume. 'Participatory methods'.

Mostowlansky, T. and A. Rota. 2020. 'Emic and etic', in *The Open Encyclopedia of Anthropology*, ed. by F. Stein (Cambridge University Press).

Richards, K.S. 1996. 'Samples and cases: Generalisation and explanation in geomorphology', in *The Scientific Nature of Geomorphology: Proceedings of the 27th Binghamton Symposium in Geomorphology Held 27–29 September 1996*, ed. by B.L. Rhoads and C.E. Thorn (John Wiley and Sons), pp. 171–90.

25. Descriptive statistics

Stuart N. Lane

Definition

Descriptive statistics are quantitative measures that summarise key data collected in a study, such as their number and distribution.

The basics of descriptive statistics

Descriptive statistics are often an initial step in data analysis and have two broad roles: (1) to provide summary information that characterises the data that have been acquired; and (2) to inform subsequent analysis (e.g., attempts to predict based on a sample or draw inferences about relationships among variables). For instance, if the aim of the study is to infer characteristics of a population (e.g., the effects of gender on salaries) from the data (e.g., a sample of the population), then descriptive statistics can help determine the kind of inferential statistics (Lane, Chapter 42) that may be legitimately used for further analysis (e.g., if the salaries follow a normal distribution, or are Gaussian, testing for the effects of gender on salary may use certain kinds of statistical tests).

Descriptive statistics in depth

Four basic kinds of data—nominal, ordinal, interval, and ratio—determine the kinds of descriptive statistics that may be used.

Nominal data are based upon categories where there is no a priori relationship (e.g., a ranking) between class values. Gender is a good example. Traditionally this had two categories, female and male; now it is more common to have three: female, male, and non-binary. These

    https://doi.org/10.11647/OBP.0418.25

categories can't themselves be related to one another (e.g., we can't say that "non-binary" is "better" than "male"). Descriptive statistics about nominal data are commonly presented in absolute terms (e.g., the number of people identifying themselves as female, male, or non-binary) or in relative terms by percentages (e.g., the number of people in each gender category as a percentage of the total number of people in the survey).

There is limited additional description that can be done with nominal data for an individual variable. However, when there are two or more variables (with any number of categories), there are some very important descriptions of the data that become possible by associating membership of categories and variables. This is called cross-classification or contingency analysis. It is illustrated in Table 25.1 for two nominal variables; (1) gender, and (2) kind of enrolment in an environmental science undergraduate programme. Contingency analysis allows analysis of pattern in data. A specific property of a contingency table is that the expected membership of category combinations can be determined by probability. In Table 25.1 for instance, if 77 out of 162 students were female and 84 students out of 162 were in a biophysical science programme we would expect the proportion of students who are *both* female *and* in a natural science programme to be (77/162)*(84/162), or 0.246. With 162 students this would be 39.9 students. Deviations between measured and expected then start to hint at association or bias. Initial reading of Table 25.1 suggests that female students are more biased towards the biophysical science programme. However, the meaningfulness of these differences depends on sample size. To actually claim that this bias is indeed a characteristic of the population we need to move to inferential statistics (in this case using a Chi-square analysis), see Lane (Chapter 42).

Table 25.1. A contingency table based upon nominal data. Measured values are given in bold, expected values in brackets.

	Biophysical Science programme	Social Science programme	Total (by gender)
Female	**45** (39.9)	**32** (37.1)	77
Male	**32** (36.3)	**38** (33.7)	70
Non-binary	**7** (7.8)	**8** (7.2)	15
Total (by programme)	84	78	162

Ordinal data are also based upon categories but where there is some known relationship between them. For instance, with the broad categories of children and adolescents, working age, and retired, we also know something about the relative ages of each category. There are more kinds of descriptive statistics available for ordinal data, such as median and mode, but the meaningfulness of these descriptions is dependent on the nature of the categories. For instance, the number of years in the three broad categories above (assuming retirement at 65) are likely 0–18, 18–65, and >65. However, these are not equal in their width. It is likely that the median membership is for the category 18–65 but we cannot say more than that about where the median lies within this category (unless we still have the raw data).

Interval data are relative numbers that are based upon quantities expressed on a continuous measurement scale. They are relative in that they describe how one value relates to another but are not absolute because they are set on a scale where the value 0 is largely arbitrary. The classic example of this is temperature measured on the Fahrenheit or Celsius scales. These scales both have zero values that have some meaning (in Fahrenheit, 0° indicates the temperature at which saltwater with a certain concentration freezes; in Celsius it is the temperature at which pure water freezes). But both are arbitrary in the sense that they are defined with respect to some other kind of property (exactly what is being frozen). As interval data can be added and subtracted, they can be described in many more ways than nominal or ordinal data. However, properties of the data determine which descriptive statistics are meaningful. Some kinds of data can be used generally (percentiles, quartiles, the inter-quartile range, the median). Other descriptive statistics (e.g., mean, standard deviation, skewness, kurtosis) may only be valid if the data themselves fulfil certain properties (e.g., a normal distribution). The mean is a good example. It is possible to calculate the mean of any interval dataset, but if the dataset contains many extreme values or a very skewed distribution, the mean may be a very misleading description of the data. For this reason, it is good practice to look at the distribution of interval data before choosing what kind of descriptive statistic is most appropriate.

Ratio data differ from interval data in that they have a pre-defined and logical zero point. Meaningful physical quantities are commonly measured on ratio scales (time, mass, volume, distance, age). Ratio data can be described using the same statistics as interval data (see above). They have one important additional property which is that multiplying and/or dividing numbers on a ratio scale is possible. In some cases, this is needed to allow derivation of additional and important quantities such as density (i.e., mass per unit volume). All ratio data have units and when using them for description those units should be systematically presented.

One final observation is important. It is possible to move from ratio/interval scales to ordinal scales by ranking data, and from ordinal scales to nominal scales by simple classification. This downwards movement is straightforward, but it is not possible to go in the opposite direction.

Why are descriptive statistics important?

Descriptive statistics are critical for allowing a reader unfamiliar with a research project to grasp the focus of the research project and how it compares with other projects and datasets. It is common to include summary tables with descriptive statistics in the methodology section of a paper to justify methodological choices. Descriptive statistics may also be an important part of a results section when they describe findings.

Relationship of descriptive statistics to other methods

Descriptive statistics are a crucial part of any project that acquires quantitative data, whether biophysical or social (or both). Descriptive statistics also set the stage for inferential statistics (Lane, Chapter 42) by identifying what kind of inferential approaches may be most appropriate.

Ethical issues and descriptive statistics

The data that feed into descriptive statistics should not raise ethical issues if ethical safeguards have been correctly incorporated into the research process (e.g., in a questionnaire survey, or in applying remote sensing). However, in describing such data, new ethical issues can result if a descriptive statistic is used to attribute a characteristic to a particular unit (e.g., a spatial unit, or to a kind of person). This can lead to basic challenges (see the discussion of ecological fallacy below) but it becomes an ethical issue because of the perceptions and understandings that are then created by that attribution.

Additional ethical issues may arise when description goes beyond the use to which it was intended. A good example of this is downscaling or small area estimation. Census organisations often geographically aggregate publicly available Census data to avoid a level of granularity that might start to breach confidentiality. However, researchers have developed tools that can combine Census data with other datasets to disaggregate down to much smaller areas, even households or individuals (Mervis 2019). This may breach stated confidentiality policies during Census data collection.

Issues to be aware of in using descriptive statistics

As with all statistics, descriptive statistics must be treated and interpreted with caution as how data are presented may have a lasting effect on how something is interpreted. As an example, one of the classical errors made in statistics is the ecological fallacy. Switzerland, for instance, has the highest mean global wealth per capita by country in the world in 2023.[1] It is indeed a "rich" country in terms of wealth. The same report shows that in terms of equality in income distribution it is 68th out of 164 countries. Indeed, Switzerland has substantial poverty, with 8.7% of the Swiss population living below the Swiss poverty line in 2021.[2] It may be justifiable to use a mean to describe a dataset, such as global wealth per capita, but an ecological fallacy arrives when that mean value is ascribed to all members of that dataset used to calculate the mean. It would be an ecological fallacy to conclude that all individual inhabitants in Switzerland are rich.

A very specific kind of ecological fallacy known as the modifiable areal unit problem (Openshaw 1983) relates to how the choice of a spatial unit influences the statistics used to describe that unit. If you divide space up in different ways, you get different statistics. This

---

1   https://www.ubs.com/global/en/family-office-uhnw/reports/global-wealth-report-2023.html
2   https://www.bfs.admin.ch/bfs/en/home/statistics/economic-social-situation-population/economic-and-social-situation-of-the-population/poverty-deprivation/poverty.html

problem is not simply an arbitrary statistical one but may be used to deliver political or other outcomes. This occurs in political districting, where the boundaries between jurisdictions in a majority-voting system are redrawn to give advantage to a particular socio-economic or political group (also known as gerrymandering). For instance, in the 2018 House of Representatives elections in the state of Wisconsin, the state voted Democrat (53.2% of votes) but the Republicans took 62.5% of seats available in the state. This imbalance is largely attributed to a congressional redistricting plan dating from 2011. Thus, descriptive statistics should always be approached with caution, perhaps even skepticism, to avoid making basic ecological fallacies.

Suggested further reading

The following is a very good entry point into different basic statistical concepts with links to where to find more detail.

Everitt, B.S. and A. Skrondal. 2010. *The Cambridge Dictionary of Statistics* (Cambridge University Press), p. 480. https://doi.org/10.1017/cbo9780511779633

References cited

Lane, S.N., Chapter 42, this volume. 'Statistical inference'.

Mervis, K. 2019. 'Can a set of equations keep U.S. census data private?', *Science*, 366, https://www.doi.org/10.1126/science.aaw5470

Openshaw, S. 1983. 'The modifiable areal unit problem', *Concepts and Techniques in Modern Geography*, 38, p. 41.

26. Environmental modelling

Tobias Krueger

Definition

Environmental modelling represents environmental processes using mathematical equations and eventually computer programmes. These models allow virtual experiments on a system to gain understanding of its dynamics and to predict system behaviour in time and space under different driving forces.

The basics of environmental modelling

Environmental modelling is a common research tool in environmental science, geography, ecology, and engineering, among other disciplines, which builds on fields such as physics, chemistry, mathematics, and computer science. Environmental models are also frequently used to inform policies and interventions. Models are commonly of subsystems of the Earth system such as climate, land, water or ecosystems, but these subsystems may be coupled together. An important motivation is to model environmental change, including impacts of human societies on environmental systems. When models describe the two-way interactions between environmental systems and human societies they may be referred to as social-ecological, human-environment, or coupled human and natural models. Underlying environmental modelling is thus a separation of humans and environment, even if the two realms are interacting in a coupled treatment.

Fundamental to environmental modelling is the notion of an external reality that can be abstracted in a model. Environmental modellers generally accept (however implicitly) that this abstraction is necessarily

 https://doi.org/10.11647/OBP.0418.26

a simplification of that reality. The practice of modelling then typically revolves around the level of simplification that may be adequate for a particular application (see below). To this end, some researchers might develop a model from scratch, while others might apply a pre-existing model. Both might be called "modelling".

Environmental modelling owes much to systems theory from which it has inherited some basic definitions. The 'system boundary' demarcates what is in the model and what is out, e.g., a hydrologically defined river basin. 'Variables' are time-variant states of the system or external forcings, e.g., groundwater levels or precipitation falling on the basin, respectively. 'Processes' relate variables to each other through mathematical equations, e.g., infiltration. These equations contain 'parameters' which are properties of the system that are sometimes assumed to be time-invariant, e.g., hydraulic conductivity of soil (though time-varying parameters are a way to represent system change). External drivers are often called 'boundary conditions', while the states of variables when model simulation begins are called 'initial conditions'.

Environmental modelling in depth

Researchers developing a model from scratch typically go through a sequence of perceptual, formal, and procedural modelling (Beven 2009). This sequence idealises how a model progressively abstracts from reality. The perceptual model is a rather qualitative understanding of the system, existing theory, hypotheses and intended purposes of the model. The term "perception" is misleading here because this understanding is not purely cognitive but involves disciplinary norms, habits and the political-economic conditions of the research (Krueger et al. 2012; Rusca and Mazzoleni, Chapter 14). These factors also condition the subsequent modelling steps during which technical factors may also enter (what is computationally feasible to model at an acceptable cost). The formal model translates the qualitative understanding into mathematical equations at which point some detail gets lost because not everything can be represented mathematically. The procedural model encodes the equations in the form of computer code which requires further simplifications to make the equations solvable. Spatial and temporal discretisation of the system are part of this step.

Modelling practice involves more iterations than the linear sequence suggests. The modeller must decide on the spatial, temporal and process complexity of their model in light of its intended purpose, theory and the data available to parameterise, drive and evaluate the model. Again, these decisions are not only made in relation to the system to be modelled and what the modeller knows about it (epistemic decisions) but involve norms, habits, political-economic and technical factors (however implicitly). A scale of process complexity can be constructed between statistical models at one end of the spectrum and mechanistic models at the other, with so called conceptual models in between. (Note that the term conceptual model is also often used for the perceptual modelling stage.) Statistical models are functional relationships fitted to empirical data with little (but often some) consideration of the underlying mechanisms. The parameters of statistical models cannot be measured independently. Mechanistic models aim to represent all processes by small-scale physics, with parameters that can theoretically be measured, but practically not at the scale of model discretisation. The model scale and the measurement scale are said to be incommensurable, and parameters are therefore "effective parameters" with only a loose correspondence to theory and thus require calibration against observations of model output (Beven 2009). This realisation has spawned the development of conceptual models of intermediate complexity, which use simplified physics with few parameters that are computationally not too demanding to calibrate.

A procedural model needs to be parameterised for a given application, i.e., values for the parameters that represent the system under study need to be provided. When developing a model from scratch, this step often involves iterations with perceptual, formal, and procedural modelling. Those applying a pre-existing model enter modelling here. Values for some parameters may be inferred from field or laboratory measurements, though not without (implicit) scaling relationships that translate between the measurement and the modelling scale. In general, however, parameters require some level of 'calibration' by changing them, so the model predicts independent observations. Today, calibration is commonly done using automated search algorithms, often with a probabilistic component to account for uncertainties (or some other uncertainty model). In principle, this is a regression problem familiar from statistics, only that the model can be much more complex.

After calibration, it is good practice to 'validate' a model by comparing model predictions to independent observations not used in calibration, without further parameter value changes. Calibration and validation require the choice of objective functions or error models that encode what is an acceptable fit between model predictions and observations, which again is a matter of disciplinary norms, habits and the intended purpose of the model. Validation in particular leads into uncertainty and sensitivity analysis of models (see below).

Given the spectrum of process complexity, 'model parsimony' has been a guiding principle in environmental modelling, i.e., keeping the model as simple as possible (so it can be parameterised, driven and evaluated with the available data and computational resources) but as complex as necessary for the intended purpose (so it can answer the research or policy questions at hand). Ultimately, a model should meet user needs and intended purposes, have acceptable levels of uncertainty and trust, and be developed or used given the data, time, funding, and other resources available (Hamilton et al. 2022).

Why is environmental modelling important?

Environmental modelling can formalise process understanding, make predictions, and inform policies and interventions. These three aims may go together but they do not need to. Predictions useful for policy and intervention may result from statistical models with little formalised understanding of underlying mechanisms, especially over short-term horizons (forecasting). This approach has become ever more prominent with the advent of machine learning.

The formalisation of process understanding is related to the ideas of reconciling theory with observation, hypothesis testing and eventually arriving at a general representation of environmental processes. The process of modelling itself may be a useful integration tool in inter- and transdisciplinary projects (see below). Predictions may be made for the future but also for the past or for places where no observations exist (or where they do exist when the aim is to evaluate a model). Most environmental models are 'stationary' in that the model structure and parameter values are time-invariant, while the system represented is not. Hence, future prediction depends on the time horizon over which stationarity holds in model structures or parameter values. Some models adapt model structures and parameter values over time.

Relationship of environmental modelling with other methods

Some models covered in this book can be considered subcategories of environmental models, e.g., hydraulic models (Lane, Chapter 30) or hydrological models (Melsen, Chapter 31). Other methods provide data that may be used to parameterise, drive, or evaluate models. Quantitative data from questionnaires (Johnston and Longhurst, Chapter 32), surveys (Winata and McLafferty, Chapter 43) or remote sensing (Braun, Chapter 39) may thereby enter models directly. Qualitative data from participant observation (Sayre, Chapter 34), document analysis or interviews may be translated into quantitative data (though this is contested) or inform model development at the perceptual stage. Some environmental models are statistical models (see above) and statistical methods (Lane, Chapter 42) certainly come in during parameter calibration, validation, and uncertainty analysis. In inter- and transdisciplinary projects, modelling can have a processual role in providing a focal point for the integration or negotiation of different disciplinary and non-academic knowledges. This function connects to participatory modelling (Landström, Chapter 35) and participatory research (Mokos, Chapter 36) more broadly. One needs to be mindful, however, that using a model as the focal point for collaboration runs the risk of subordinating research methods that are more qualitative or interpretive.

Ethical issues and environmental modelling

In addition to ethical issues concerning the data that go into models, environmental models amplify ethical concerns regarding the generation of environmental information at increasingly fine resolution. In environmental pollution research, for example, models have become so fine scaled that they make individual businesses, communities or people visible (Lane, Chapter 30) and therefore implicate them in environmental pollution, for better or worse, which may be called "surveillant science" (Lane et al. 2006).

Another ethical issue is the choices and simplifications encoded in models (see above), through which models afford certain discourses and interventions and not others and thereby create realities that affect people differently. The notion of models as neutral and merely representing a pre-existing reality obscures this political force of models (Krueger and Alba 2022).

Issues to be aware of in using environmental modelling

Model commissioners, developers and users should be clear about the intended purpose of the model and select a modelling framework, model or suite of models accordingly. They should end up with models that can answer the research or policy questions at hand. Given that models are simplifications and prone to uncertainties, it is important to apply sensitivity and uncertainty analysis (Beven 2009; Saltelli et al. 2008). Even so, a model might fit the observations well for the wrong (mechanistic) reasons, and several models might fit the observations equally well ('equifinality'). It is thus paramount to document and communicate model assumptions and limitations transparently. Lastly, we must not forget that a model is never neutral but always reproduces particular disciplinary norms, habits, and political, economic, and technical contexts.

Suggested further reading

Beven, K.J. 2009. *Environmental Modelling: An Uncertain Future?* (https://doi.org/10.1201/9781482288575). Gives an introduction to model development, modelling philosophy and uncertainty analysis.

Hamilton, S.H., C.A. Pollino, D.S. Stratford, B. Fu, and A.J. Jakeman. 2022. 'Fit-for-purpose environmental modeling: Targeting the intersection of usability, reliability and feasibility', *Environmental Modelling and Software* 148: 105278 https://doi.org/10.1016/j.envsoft.2021.105278. Formalises the concept of a model fit for purpose along the dimensions: usefulness, reliability and feasibility; and lays out modelling guidelines.

Wainwright, J. and M. Mulligan. 2013. *Environmental Modelling: Finding Simplicity in Complexity* (https://doi.org/10.1002/9781118351475). Provides introductions to many contemporary kinds of environmental models.

References cited

Braun, A., Chapter 39, this volume. '(Critical) Satellite remote sensing'.

Krueger, T., T. Page, K. Hubacek, L. Smith, and K. Hiscock. 2012. 'The role of expert opinion in environmental modelling', *Environmental Modelling and Software*, 36, pp. 4–18. . https://doi.org/10.1016/j.envsoft.2012.01.011

Krueger, T. and R. Alba. 2022. 'Ontological and epistemological commitments in interdisciplinary water research: Uncertainty as an entry point for reflexion', *Frontiers in Water*, 4. https://doi.org/10.3389/frwa.2022.1038322

Landström, C., Chapter 35, this volume. 'Participatory modelling'.

Lane, S.N., Chapter 30, this volume. 'Hydraulic modelling'.

Lane, S.N., Chapter 42, this volume. 'Statistical inference'.

Lane, S.N., C.J. Brookes, A.L. Heathwaite, and S. Reaney. 2006. 'Surveillant science: Challenges for the management of rural environments emerging from the new generation diffuse pollution models', *Journal of Agricultural Economics*, 57.2, pp. 239–257. https://doi.org/10.1111/j.1477-9552.2006.00050.x

Johnston, L., and Longhurst, R., Chapter 32, this volume. 'Interviews: Structured, semi-structured and open-ended'.

Melsen, L., Chapter 31, this volume. 'Hydrological modelling'.

Mokos, J., Chapter 36, this volume. 'Participatory methods'.

Rusca, M. and Mazzoleni, M., Chapter 14, this volume. 'The interface between hydrological modelling and political ecology'.

Saltelli, A., M. Ratto, T. Andres, F. Campolongo, J. Cariboni, D. Gatelli, M. Saisana, and S. Tarantola. 2008. *Global Sensitivity Analysis: The Primer* (Wiley-Blackwell). https://doi.org/10.1002/9780470725184

Sayre, N.F., Chapter 34, this volume. 'Participant observation and ethnography'.

Winata, F. and McLafferty, S., Chapter 43, this volume. 'Survey and questionnaire methods'.

27. Focus groups

Robyn Longhurst and Lynda Johnston

Definition

Focus groups (sometimes referred to as focus group interviews or focus group discussions) are groups of people, usually between 6 and 12, who meet informally to discuss a specific topic (i.e., focus) that has been determined by the researcher. The researcher facilitates the discussion, keeping the group on topic or 'focused' but is otherwise non-directive. This empowers participants to examine the subject from as many angles as they wish.

The basics of focus groups

The principle behind focus groups is that talking with people is an excellent way of gathering information. Sometimes in everyday discussions or conversations, however, we tend to speak too quickly, do not listen carefully, and/or talk over others. Focus groups are about talking with people but in ways that are more structured. At the same time, they seek to avoid the "over-structuring" typical of questionnaire surveys where the responses elicited are almost entirely pre-defined by the questions asked. Historically, focus groups tended to be conducted in-person but over the past decade—and especially since the COVID-19 pandemic—researchers have begun conducting them online using mobile phones, tablets, or computers with audio-visual interfaces.

 https://doi.org/10.11647/OBP.0418.27

Focus groups in depth

Focus groups usually last between one and two hours. Their most defining feature is the interaction that happens between group members, unlike in individual interviews where the interaction happens between the interviewee and the researcher. With focus groups, again as opposed to individual interviews, it is possible to gather the opinions of many people within a short space of time. Sometimes researchers carry out just one focus group, other times a series.

Focus groups tend to work best when the researcher brings together a reasonably homogeneous group, so they are relaxed, talking, and even laughing (see Browne 2016) with each other. This could be a group of friends, a sports team, colleagues, or individuals who have something else in common.

Because focus groups are not just casual conversations, each requires its own careful preparation. The first step in preparing is for the researcher to create a list of themes or questions to ask. Some questions are designed to draw out 'factual' information, others are designed to draw out more subjective information. Combining different types of questions can be an effective strategy. It is a good idea for the researcher to start with a question that participants will likely feel relaxed answering. The aim is to generate easy and free-flowing conversation and to allow the participants to explore questions and topics that are related to the focus but that perhaps the interviewer had not envisioned when preparing the focus group. It is important, however, to check back at the end to make sure all the planned questions or topics have at some point been addressed.

The second step in preparing for a focus group is to think of an activity that gets participants engaged—or focused—on the topic at hand. For example, participants might be asked to draw a picture, put items on a list in order of priority, taste a food, or react to a photograph. This focuses the group but also helps them to 'warm up' to the conversation. It can also serve as a "levelling device" encouraging a shift away from prior conceptions by those in the group and towards the focus at hand.

The third step is selecting participants. Participants tend to be selected on account of their experience related to the topic, and/or whether they have something in common, not on whether they constitute a random or representative sample (as with quantitative methods). Objectivity is not the aim; instead, the goal is to understand how people make sense of their own and others' experience.

A fourth consideration, once participants have been selected, is how they will be recruited. They could be recruited by advertising in local newspapers or on radio stations, or by posting a message on social media (such as Facebook, Instagram, and TikTok), an approach more commonly used today. It is also possible to recruit on-site, such as at sports clubs, cultural centres, health clinics, churches, and community facilities. Recruitment can be challenging, however, as where recruitment takes place may influence how the group develops and the nature of the discussions that result.

A fifth step is deciding where to conduct the focus-group including whether it would be best held in-person or online. If possible, the setting should be relatively neutral, that is, a place where everyone feels comfortable. If place is integral to the topic being discussed, it may make sense to hold the focus group in that place. If the focus group is to be conducted online then a software package needs to be chosen (e.g., Zoom, Microsoft Teams, WhatsApp). It is imperative to test the technology ahead of the meeting especially if it is a large focus group. It is also important to be aware that how people interact online may not be the same as how they interact in person and so this choice may influence the outcomes of the group.

Finally, it is important to decide how to take notes or audio/video record the discussion. Recording enables the facilitator to concentrate on the interaction taking place instead of feeling pressure to write or type the participants' words. If it is an online focus group, inbuilt software (e.g., Sound Recorder on Windows; QuickTime on Apple) or commercial applications can be used to record. If it is an in-person group there is now recording functionality built into most mobile phones as well as apps available to record and transcribe conversations (e.g., Otter. ai). This transcript can then be analysed using a range of techniques (e.g., reading, coding, generating themes, comparing transcripts). Specific agreement of all participants and an accepted statement of what the recording will be used for are normally both required if recording.

Why are focus groups important?

Focus groups can help researchers collect a wide range of opinions and experiences. They do not offer a route to discovering 'the truth' but they do offer a way of understanding what people think and do in their own words.

Focus groups are often recommended to researchers who want to get to know a new field including how to frame questions, what is the appropriate terminology to use, and what is the breadth of opinion on topic. In cases where there is no or little research on a topic, focus groups can establish some of the parameters of the project before using other methods.

Relationship of focus groups with other methods

Focus groups can be used as a stand-alone method or with a suite of other methods. Most commonly they are used in combination with other methods, especially qualitative methods such as semi-structured interviews (Johnston and Longhurst, Chapter 32). They may also help inform the focus of more traditional questionnaire-based surveys (Winata and McLafferty, Chapter 43) that cover a larger population. Focus groups tend not to be used as often as individual interviews, although they have become more popular with social scientists over the last few decades.

Ethical issues and focus groups

The first issue to be aware of in using focus groups is confidentiality. Participants need to be assured that all the information they provide will remain secure. In addition, researchers should note that it is not only themselves or a facilitator who will be privy to information but also members of the focus group. Therefore, participants need to be asked to treat discussions as confidential. This cannot be guaranteed, though, and so participants need to be advised to disclose only those things they feel comfortable being repeated outside the group. Other methods such as individual interviews might be better for controversial or sensitive topics.

The second ethical issue, and one that goes hand in hand with confidentiality, is anonymity. Anonymity can only be guaranteed if confidentiality is maintained, that is, if individuals do not become identifiable. Therefore, participants ought to remain anonymous unless they specifically request that the researcher use their name. In addition, they have the right to withdraw from the research at any time without having to explain to the researcher or other group members.

A third ethical issue is that participants, ideally, ought to be provided with a summary of the research results at the end of the project. This could be a hard or an e-copy, preferably written or presented in a way that participants can understand (which may require preparing a summary specifically for them).

A final issue is that participants may express discriminatory or offensive views during the discussion. It could be argued that researchers ought to listen and not judge but the difficulty with this is that being non-judgmental can reproduce and even give legitimacy to a participant's comments. Researchers need to carefully consider how best to deal with such situations.

Issues to be aware of in using focus groups

Without careful preparation several issues may arise in using focus groups. First, a common issue is that the conversation does not flow easily because group members are not comfortable with each other and/or the place that has been chosen (real or online) does not help participants to feel relaxed. A second issue is the conversation can flow rather too easily to a point where the researcher struggles to keep the group on topic. Failing to record the discussion properly or at all is yet another issue. Getting to the end of a rich discussion only to find that the recording device has not been activated or the discussion has not been recorded clearly can be devastating. All equipment needs to be tested ahead. Finally, to reiterate, there is a web of ethical issues and power relations that need to be considered when using all qualitative methods but perhaps especially for focus groups because managing a group of people rather than just single individuals (as in the case in individual interviews) can be challenging.

None of these issues, however, should be seen as deterrents because the group dynamic produced by focus groups can provide researchers from a range of disciplines with deep insights into understanding people and places.

Suggested further reading

Unfortunately, open-access resources on focus groups are difficult to find but an excellent 'closed' source is: Cameron, J. 2021. 'Focusing on the focus group', in *Qualitative Research Methods in Human Geography*, ed. by I. Hay and M. Cope (Oxford University Press), pp. 200–21.

On both focus groups and interviews see: Longhurst, R. and L. Johnston. 2023. 'Semi-structured interviews and focus groups', in *Key Methods in Geography*, ed. by N. Clifford, M. Cope, and T. Gillespie (Sage Publications), pp. 168–183.

References cited

Browne, A.L. 2016. 'Can people talk together about their practices? Focus groups, humour and the sensitive dynamics of everyday life', *Area*, 48.2, pp. 198–205.

Johnston, L. and Longhurst, R., Chapter 32, this volume. 'Interviews: Structured, semi-structured and open-ended'.

Winata, F. and McLafferty, S., Chapter 43, this volume. 'Survey and questionnaire methods'.

28. Geochronological Methods

Georgina E. King and Peter M. Abbott

Definition

Geochronology provides information on the timing, rates and frequency of environmental processes and events that occur over timescales that exceed laboratory measurements and/or human observation.

The basics of geochronology

A wide variety of geochronological methods exist and these can be applied over a range of timescales and in different environmental settings. The methods can be broadly categorised into direct and indirect dating techniques. Direct dating methods yield an age or a date for when a change or event occurred, or a process started. Indirect dating techniques provide marker horizons that can be correlated between different environments; while a marker horizon may not yield a specific age, it has the same age at all sites, providing a "younger than" or "older than" boundary.

Geochronology in depth

Geochronological methods can be categorised depending on whether they provide direct or indirect age information, their timescale of applicability, the physical process on which they are based, and the environmental setting in which they can be used. Different geochronological methods are based on different time-dependent physical processes. The techniques with perhaps the broadest applicability range are radiometric methods that measure the ratio between unstable (radioactive) parent and daughter isotopes. These techniques can be applied to organic matter (e.g., radiocarbon dating, 14C), minerals (e.g., U-Pb dating of zircon grains) or sediments (e.g., 210Pb dating). As the decay rate of radiogenic isotopes is known, it is possible to obtain very precise age estimates by measuring the accumulation of daughter products, providing the initial concentration of parent isotopes can be estimated. As decay rates vary between different systems, isotope geochemistry can be used to date processes on the order of tens (e.g., 137Cs), hundreds (e.g., 210Pb), thousands (e.g., 10Be cosmogenic nuclide dating), millions (e.g., U-series), or even billions of years (e.g., U-Pb dating).

Other direct geochronological methods rely on the accumulation of electrons within minerals in response to natural ionising radiation (e.g., luminescence dating), annual physiological changes within organisms, or the recording of seasonal variations in sedimentary archives. Dendrochronology exploits the production of annual growth rings within the trunks of trees, which can be counted and correlated between living and fossilised specimens. This dating method is widely used in environmental research to understand the timing and magnitude of environmental changes as ring width and wood density/anatomy are highly sensitive to climatic, environmental, and geomorphological changes. Furthermore, as dendrochronology comprises annual layer counting, the dates obtained can be very precise. Sclerochronology is a similar technique that exploits annually deposited layers in marine organisms, such as bivalves (e.g., *Arctica islandica*). The total growth of some lichen species can also be used to determine the age of landforms like moraines over the past few hundred years (lichenometry).

The isotopic and chemical composition of snow deposited over ice sheets and glaciers can vary seasonally. After ice cores have been extracted these annual variations can be identified through high-resolution measurements and counted to define high-precision chronologies. These chronologies can be tested and anchored with markers of a known age, such as volcanic eruptions. Sediments in some lake sequences can show seasonal variations in the properties of the particles deposited over the course of a year, with a coarse and a fine particle layer deposited. These pairs of layers are termed varves and, akin to ice core layers, can be counted to derive an annually resolved varve chronology.

In addition to direct geochronological methods, indirect geochronological methods are exceptionally valuable for reconstructing the timing and rates of changes. Tephrochronology involves the isolation and characterisation of volcanic ash particles (tephra) that are deposited in a diversity of environmental settings from ice cores to peat bogs and lake sediments. Following a volcanic eruption tephra is deposited rapidly (over days to weeks) and can be fingerprinted and correlated between sites using their major and trace element chemical composition. Thus, tephrochronology provides time-stratigraphic markers that can be traced between different sedimentological records (e.g., marine cores and terrestrial records). If the tephra shards from a particular eruption can be dated using a radiometric method (e.g., K-Ar, Ar-Ar dating), the method becomes a direct, rather than indirect dating technique, providing further geochronological constraint.

Other examples of indirect dating methods include Schmidt hammer analyses, which provide an indication of the relative degree of weathering of bedrock based on its hardness and compressive strength, which is related to exposure duration. When sediments are deposited, Fe-bearing minerals are oriented in a particular direction that is linked to the magnetic field of the Earth. Palaeomagnetism exploits these past variations in the magnetic field to provide a framework against which different sedimentary deposits (e.g., loess deposits, or marine sediments) can be contrasted using notable globally synchronous transitions, e.g., the Brunhes-Matuyama boundary, some of which have precise independent age estimates.

Pollen from terrestrial records, oxygen isotope records from ice cores, or foraminifera in marine sequences provide extensive records of past temperature and environmental changes. Whilst these records do not comprise geochronological techniques in themselves, they allow correlations to be made, based on "wiggle-matching" of climatic or environmental changes between sequences e.g., in the marine environment or between marine and terrestrial records. This can provide a first indication of chronology where independent ages are difficult to obtain or are unavailable but does prohibit the assessment of climatic phasing between the archives. Finally, biostratigraphy, which tracks the emergence and extinction of different species, can also be used to make inferences about the relative timing not only of speciation but also of major tectonic, geomorphic, and climatic events.

Why is geochronology important?

Without knowing the timing of the formation of a landform or a climatic event, or the date of onset of a particular process, it is impossible to truly understand that process. Only when the rate of change or frequency of events can be established can we begin to understand environmental change. Time is always constrained in laboratory experiments; geochronological methods allow us to apply the same principle to the natural environment, albeit with greater uncertainty.

Relationship of geochronology with other methods

Ages, dates, or rates of change are of limited use in isolation; thus, geochronology is commonly used together with other techniques such as geomorphological mapping or geochemical analyses to understand how a particular environment changed and when that change occurred. Ages allow researchers to link the formation of landforms (e.g., a fluvial terrace) with regional environmental records, providing insights into the processes that lead to the landform's formation. Geochronology can also be used to determine the frequency of events, such as the recurrence interval of floods, earthquakes, or landslides, and so in addition to geomorphological, tectonic or climatic data, geochronological data are often combined with statistical analyses. Beyond the geosciences, geochronological data are key to understanding human evolution and archaeological records, as well as rates of biological speciation.

Ethical issues and geochronology

The expense of geochronological analyses means that laboratories tend to be located within the Global North, and historically project partners from the Global South have been excluded from studies where samples have been taken from their countries. This problem is being reduced via greater awareness of the need to avoid so-called "helicopter science", whereby scientists from the Global North with significant research funding do not include local collaborators, although there remains a significant amount of progress to be made. Other ethical issues revolve around data sovereignty and the right of local communities to keep data private. For example, the location of archaeological sites is often highly sensitive and thus in some instances should be kept confidential. For some research questions, it may be necessary to receive prior approval from local stakeholders to collect samples from culturally significant or sacred sites. A final ethical consideration is that sometimes geochronological methods such as luminescence or radiocarbon dating are used to date antiques to determine their authenticity. In such instances, determining that an antique is not authentic has considerable implications for its value.

Issues to be aware of in using geochronology

Geochronological methods are highly diverse, and their applicability can be restricted to specific time periods, environments, or materials. Thus, it is essential to select the appropriate technique(s) for the research question being posed and for the environmental setting under investigation. All geochronological techniques have challenges and limitations. Ideally multiple techniques should be combined to confirm that the established chronology is correct. All geochronological methods have an analytical uncertainty that must be considered when the data are used to interpret an environmental change, process, or archaeological site. The quality of an age is only as good as the material that has been sampled. Thus, when attempting to date a particular event, process, or landscape, it is essential to consider what is being dated and what the age will likely mean. For example, dating of a glacial moraine will yield the age at which the moraine was last active but will not necessarily yield the age of the moraine's formation, which may be much older if the glacier occupied that limit on multiple occasions. Indirect dating methods are highly powerful for allowing correlations between sequences. However, in the absence of any direct dating control, unless the record is continuous (e.g., the Greenland ice cores) it is not possible to assign a specific age.

Suggested further reading

There are several textbooks that address geochronology. The textbooks by
   Walker, M., 2005. *Quaternary Dating Methods* (John Wiley and Sons) and
   R. Bradley., 2015. *Paleoclimatology: Reconstructing Climates of the Quaternary*
   (Academic Press) both provide an excellent overview of many different
   types of geochronological technique, focusing on methods that are
   applicable over the past 2 Ma.

The textbooks by A.P. Dickin, 2005. *Radiogenic Isotope Geology* (Cambridge University Press) and by G. Faure and T.M. Mensing, 2004. *Isotopes: Principles and Applications* (Wiley) provide detailed information on different radiometric techniques.

29. Historical ecology

Diana K. Davis

Definition

Historical ecology is an increasingly recognised framework for analysing and interpreting environmental and landscape change over time. It incorporates a diverse collection of interdisciplinary methods and is utilised primarily by those in the disciplines of ecology, anthropology (and archaeology), geography, and history.

The basics of historical ecology

The toolkit of historical ecology is mixed, interdisciplinary, diverse, and eclectic (see Walters et al., Chapter 13). Depending on the research question and the researcher, it may involve some or most of the following individual methods: biological and physical data collection such as fossil pollen analysis, sediment records (see Blond, Chapter 20; King and Abbott, Chapter 28), tree-ring or charcoal analyses, isotopes of oxygen, remote sensing data (see Braun, Chapter 39), and floral/faunal data (see Walters et al., Chapter 13) including archaeozoological data, species lists or inventories in combination with social, humanistic data such as historical archival records (see Cope, Chapter 22), oral histories (see Chakov et al., Chapter 33), maps, photography, interviews (see Johnston and Longhurst, Chapter 32), agriculture and forestry manuals, and various written documents including original travel and exploration accounts (see Doel, Chapter 44), among others.

https://doi.org/10.11647/OBP.0418.29

Historical ecology in depth

Despite its pre-twentieth-century origins, which are usually traced to early geographers including Alexander von Humboldt, Elisée Reclus and Paul Vidal de la Blache, the first use of the term "historical ecology" appears to have been in the 1940s by ecologists interested in comparing contemporary plant distributions with earlier distributions. Since then, the term has been invoked primarily by ecologists, environmental anthropologists, and other researchers interested in the applied work of ecological restoration and historical and environmental conservation projects.

The mixed-methods approach used in historical ecology is also widely utilised by other disciplines and subdisciplines, most prominently by historical political ecologists, historical geographers, and critical physical geographers, as well as by environmental historians and sometimes by environmental sociologists. Geographers and historians tend to incorporate more extensive archival research, especially drawing on colonial/imperial archives and colonial-era publications to illuminate and excavate constructed narratives of environmental change. At its best, such research triangulates (see Biermann and Gibbes, Chapter 4; Blond, Chapter 20) among several data sets (e.g., archival, published written, palaeoecological, longitudinal photography, biophysical, etc.) to arrive at the most likely and accurate representation of environmental change over time. It is particularly important to try to assess archival/written descriptions of the environment and ecological change against what the most recent biophysical record reveals in order to uncover potential bias in conventional or scientific understandings.

Why is historical ecology important?

For several decades research termed "historical ecology" was used primarily to construct baselines for applied work in environmental conservation and in restoration ecology along with some use in historical preservation work. Over the last decade or so some practitioners and scholars who utilise the term "historical ecology" have begun to take into consideration Indigenous knowledges and practices, to recognise the false dichotomy of the nature/human divide, and to consider the problematic narratives and related policies developed by various imperial powers. The Indigenous use of fire to manage the land, for example, has begun to be taken more seriously in historical ecology, especially in terms of understanding baselines for future management.

Research by geographers and environmental historians, on the other hand, which utilises much of the historical ecology approach without often invoking the term, has been analysing unequal power relations in the construction of environmental narratives since at least the early 1990s. This research has long pointed out that false or misleading discourses of environmental change have been deployed by those in power (imperial or otherwise) to appropriate land and other resources from a wide variety of Indigenous groups around the globe. Such erroneous imperial/colonial knowledge continues to misdirect policy, and even restoration efforts, in the form of maps, data sets, and scientific understandings that are sometimes used unquestioningly.

In addition to the important work of ecological restoration of damaged environments, then, historical ecology is also important in terms of equity, social justice and growing efforts at decolonisation and reparations.

Relationship of historical ecology with other methods

Historical ecology does not have its own well-defined method(s) but rather "borrows" or mixes a wide variety of methods from other disciplines ranging from archaeology and ecology to geography and history.

Ethical and other issues to be aware of in using historical ecology

Questionable narratives of landscape change over time have long been used by the powerful to disenfranchise the less powerful and marginalised, especially during the colonial period, and since. Researchers need to be mindful to triangulate all available sources (social and biophysical) in order to construct the most accurate history of landscape/environmental change possible, and thus to ensure the most equitable and socially just outcomes possible. Not doing so can lead to land-grabbing, exclusion, marginalisation, and other inequities.

Doing so can support decolonisation, emancipatory policies, and outcomes that are more socially and environmentally just.

The results of historical ecology and related research are frequently used to inform "baselines" of what is determined to be a "normal and healthy" environment and to guide future policies. Bias in the data or analysis can lead to biased and inequitable ideas of baselines that can then skew environmental conservation and restoration projects towards inequitable, unrealistic or ecologically inappropriate outcomes. The current "Trillion Tree Campaign" is a case in point because this programme advocates planting trees even in naturally relatively tree-less ecosystems (often claimed to be deforested without proper evidence). Planting lots of trees in such regions, resulting in high afforestation rates, is causing ecological harm, such as lowered water tables, in addition to expropriating marginalised peoples from their land and resources. Choosing a baseline is ultimately political and all stakeholders should be involved.

Due to the necessity of creating or utilising multiple bodies of data/ research, the amount of time, work, money, and energy needed for top-quality historical ecology research is substantial and should be factored into any research plan that will incorporate this approach. The time needed for archival work, for instance, is often underestimated by those who are new to such research, as is that for palaeoecological research.

Each individual method used in this 'mixed-methods' approach may have its own separate ethical issues that need to be considered, as with oral history (see Chakov et al., Chapter 33). Written archives, while very useful records of those in power, are also often biased in terms of the exclusion of subaltern or Indigenous voices most of the time. Written records of exploration and travel may have similar biases towards the powerful that should be carefully considered. One needs to be mindful of this and try to include Indigenous and subaltern voices and perspectives through different means that might include interviews with elders, oral histories, stories, or rituals, among others in a responsible, non-extractive manner.

The writing of history is always situated in particular socio-political-economic contexts and it is important to keep this in mind so as not to replicate biased Euro-triumphalist or neoliberal histories in the research (see Doel, Chapter 44). Scientific knowledge and data sets are similarly situated and can be biased by, for example, lingering environmental determinism. Thus, biophysical data sets should be assessed using some of the tools of science and technology studies, or other critical approaches.

Suggested further reading

Lave, R., C. Biermann, and S. Lane. 2018. *The Palgrave Handbook of Critical Physical Geography* (Palgrave MacMillan). This edited volume is an excellent source for concrete examples of many of the methods of historical ecology put into action by a wide variety of human-environment researchers.

Davis, D. 2007. *Resurrecting the Granary of Rome: Environmental History and French Colonial Expansion in North Africa* (Ohio University Press). This book provides a concrete example of several of these methods to analyse a colonial environmental narrative (desertification) from its origins to its use in expropriating Indigenous groups during the period of colonisation in North Africa. It includes the use of palaeoecological data (like fossil pollen), archival, and contemporary ecological data to reveal the many problems and inaccuracies in the colonial environmental history and shows how it remains influential today in some policy circles despite its serious flaws.

Leach, M. and R. Mearns. 1996. *The Lie of the Land: Challenging the Received Wisdom on the African Environment* (Heinemann). This highly influential book contains many chapters with excellent examples of many of the methods used in historical ecology and related fields all of which incorporate serious consideration of power, equity, and social justice.

References cited

Beller, E., L. McClenachan, A. Trant, E.W. Sanderson, J. Rhemtulla, A. Guerrini, et al. 2017. 'Toward principles of historical ecology', *American Journal of Botany*, 104.5, pp. 645–648. https://doi.org/10.1016/j.gecco.2019.e00836

Biermann, C. and Gibbes, C., Chapter 4, this volume. 'Mixed methods in tension: lessons for and from the research process'.

Blond, N., Chapter 20, this volume. 'Mixing geoarchaeology, geohistory and ethnology to reconstruct landscape changes on the longue durée'.

Braun, A., Chapter 39, this volume. '(Critical) Satellite remote sensing'.

Chakov, A., Chang, T., Covey, H., Dickson, T., Goggins, S., Harris, N., Purna, S., Widell, S., and Druschke, C.G., Chapter 33, this volume. 'Oral history'.

Cope, M., Chapter 22, this volume. 'Archival methods'.

Crumley, C. 2014. 'What is historical ecology', *The Hercules Project*, http://www.hercules-landscapes.eu/blog.php?what_is_historical_ecologyandid=10

Doel, M., Chapter 44, this volume. 'Textual Analysis'.

King, G.E. and Abbott, P.M., Chapter 28, this volume. 'Geochronological methods'.

Johnston, L. and Longhurst, R., Chapter 32, this volume. 'Interviews: Structured, semi-structured and open-ended'.

Szabó, P. 2015. 'Historical ecology: past, present and future', *Biological Reviews*, 90.4, pp. 997–1014. https://doi.org/10.1111/brv.12141

Walters, G., Hymas, O., Touladjan, S. and Ndong, K., Chapter 13, this volume. 'Revealing the social histories of ancient savannas and intact forests using a historical ecology approach in Central Africa'.

Wolverton, S., R.M. Figueroa, and C.G. Armstrong. 2023. 'Integrating historical ecology and environmental justice', *Journal of Ethnobiology*, 43.1, pp. 57–68. https://doi.org/10.1177/02780771231162196

30. Hydraulic modelling

Stuart N. Lane

Definition

Hydraulic modelling is used to solve the basic equations of conservation of mass and momentum for water bodies such as lakes and rivers. It is a particularly important tool in water resource and risk management, such as for determining flood risk, evaluating river habitat or assessing lake mixing.

The basics of hydraulic modelling

Hydraulic modelling is based upon numerical analysis of a set of governing equations that are designed to conserve momentum and mass. It differs from hydrological modelling (Melsen, Chapter 31) because it is primarily concerned with surface waters (as opposed to groundwater) where the physics of the fluid itself (e.g., turbulence, water surface gradients) needs to be included in the analysis. The specific equations used depend on the assumptions that can be made about the fluid being modelled (e.g., incompressible, constant viscosity) and the scale of the analysis. As more assumptions are made, so the analysis becomes more similar to hydrological modelling.

Increasingly, hydraulic modelling has been built into software systems that may be freely available (such as in the US Army Corps of Engineers Hydrologic Engineering Center's River Analysis System (HEC-RAS) software; or the Swiss Basic Environment Simulation (BASEMENT) system for rivers). The practice of hydraulic modelling is thus more about making sure the software systems are used correctly (e.g., numerical solution; specification of boundary conditions and

https://doi.org/10.11647/OBP.0418.30

parameter values; choice of correct auxiliary relations for the problem at hand; model sensitivity) than it is about writing code to solve sets of equations.

The basic workflow involves (a) choosing an appropriate model; (b) assimilating datasets (e.g., river or lake bathymetry) needed to run the model; (c) setting up an appropriate numerical solution approach (notably the grid or mesh used in numerical solution); (d) specifying boundary conditions (e.g., water inflows and outflows) and parameter values; (e) running the model (e.g., checks on numerical accuracy including mass conservation); (f) calibration of the model using relevant parameters; and (g) independent validation of model predictions. Once a model is sufficiently valid that it can be deemed fit-for-purpose, it may be used to make predictions for conditions (e.g., flood events) that have not yet been observed.

Hydraulic modelling in depth

Hydraulic modelling uses basic physical principles to do hydraulic calculations. Primarily, it assumes that the fluid being described is Newtonian (where the viscous stresses in a fluid are a linear function of the local strain rate) but there are environmental flows that are non-Newtonian e.g., mudslides) and these requires different analysis.

The basic Navier-Stokes equations describe the case for three-dimensional unsteady flows (Ingham and Ma 2005). These may be time-averaged (known as Reynolds-averaging), integrated in the vertical to produce the two-dimensional depth-averaged equations (known as the Shallow Water Equations), and then integrated in the lateral to produce the one-dimensional St. Venant equations. Hydraulic modelling then involves choosing the right form of the equations according to the scale of the analysis (e.g., routing a flood wave through a river basin can likely be represented through the St. Venant equations; but predicting the spatial patterns of flood inundations likely needs the Shallow Water Equations).

Once the appropriate set of equations is chosen, relevant boundary conditions should be supplied or assumed (e.g., do wind friction effects on the surface of the water body need to be represented?). Such questions are a function of the system being modelled and the form of equations used. As the Navier-Stokes equations are simplified so terms may appear that also have to be modelled. For instance, Reynolds- (or time-) averaging the Navier Stokes equations creates the Reynolds stresses. These describe the momentum transported by turbulent fluctuations. In natural flows, they are never zero. Such terms are represented by auxiliary relations. These are equations with varying physical-bases used to parameterise such terms (e.g., specifying an eddy viscosity to parameterise turbulence-related energy losses). The more the Navier-Stokes are reduced in dimensions the more important auxiliary relations become in determining model predictions.

A major issue in applying hydraulic models is numerical solution (Wright 2005). This commonly needs a high level of training and experience if it is to be done properly. Choices regarding numerical solution will have a major impact upon results and their validity. These include specification of the domain upon which computations will be made as well as checking for artificial diffusion and numerical instabilities. Hydraulic modelling often requires more expertise than basic software packages suggest.

Once a solution is obtained, three steps are needed before a model may be deemed fit for purpose (Lane et al. 2005). The first is verification, where it is determined the model has approximated the physical equations correctly. This normally involves basic checks on mass balance (for instance, in modelling a river, whether the amount of water entering the solution domain matches the amount leaving, taking into account changes in storage). It should normally involve some grid or mesh sensitivity testing (for instance, do the basic properties of interest not change significantly if the grid or mesh is made finer?).

The second step is model calibration in which model predictions are optimised against some independent measure of those predictions by modifying parameters in the auxiliary relations. In hydraulic modelling using the Saint Venant equations, for instance, energy losses have to be represented in an auxiliary relation that has a roughness parameter (such as the Manning's (n) parameter). This parameter may have a major impact on model predictions. In calibration, independent observations (e.g., of maximum flood inundation extent) are used to improve model predictions by changing parameter values. The resulting parameter values are "effective" in that they make the model perform (Lane 2012) but their values may be very different to measured ones. There can be very good physical reasoning for this divergence (see for example, Lane, 2014).

The third step is model validation. Data not used in model calibration are used to provide an independent assessment of how the model performs in the absence of further parameterisation. By comparing a model against independent data, its transferability is evaluated and therefore helps the modeller to decide if the model is fit for the purpose of scenario analysis.

Why is hydraulic modelling important?

Hydraulic modelling is important to a wide range of water management domains because it predicts fluid flows that are difficult to measure directly. Hydraulic processes have very strong spatial gradients, requiring high spatial densities for measurement. They change rapidly in time, necessitating re-measurement. Very high or very low flow events in a river and over-turning events in a lake may be rare and so do not appear in measurements. Extreme events can be too difficult to measure. In order to inform preventative measures (e.g., flood protection systems) predictions beyond the range of measurement may be crucial. Hydraulic model outputs may inform other models and analyses. For instance, they can be combined with ecological knowledge to develop coupled eco-hydraulic models for river management (e.g., Leclerc 2005).

Relationship of hydraulic modelling to other methods

Hydraulic modelling shares theoretical and practical issues with environmental modelling more broadly (Krueger, Chapter 26) and lessons and experience with the latter apply in equal measure. Notably, there are strong similarities with hydrological modelling (Melsen, Chapter 31) in terms of model calibration and dependency upon data. Developments in remote sensing (Braun, Chapter 39) in LiDAR and UAV surveys (Kasvi, Chapter 45) have underpinned the move from 1D to 2D and 3D approaches to hydraulic modelling meaning that hydraulic modelling is commonly linked to geomatics and GIS. New methods for obtaining data for validation (e.g., spatial patterns of flow velocity; Legleiter and Kinzel 2024) make model parameterisation easier.

Ethical issues and hydraulic modelling

Hydraulic modelling is a form of surveillance science. Researchers (or authorities) take spatially restricted data (e.g., a river flow history at a gauging station in a floodplain); combine it with remotely sensed data obtained with no authorisation from those to whom it pertains; and then make spatially explicit predictions of, for example, likely exposure to flooding. Resulting flood maps label people and properties at flood risk, feeding into insurance premiums and availability as well as development planning and policy (Porter and Demeritt 2012). In this sense, ethical issues should be a basic question in all hydraulic modelling, but this is rarely if ever the case.

Issues to be aware of in using hydraulic modelling

Guaranteeing that the numerical solution used is robust and reliable is crucial. Many of the data and parameters that hydraulic models require, including those needed for auxiliary relations, can be uncertain. Modellers should thus undertake simulations to quantify how such uncertainty propagates into model predictions. Careful thought needs to be given to the meaning of parameters in auxiliary relations. For example, some 1D and 2D models use friction treatments that are based upon Manning's n. As noted above, Manning's n is an effective parameter and the processes that it is meant to represent differ between 1D and 2D models. Required values may have very little relation with any field measurements. Model parameterisation needs very careful attention. Finally, when applying hydraulic models, there is a class of situations associated with engineering infrastructure that requires special representation (e.g., weirs, culverts). This should not be overlooked.

Suggested further reading

This edited volume addresses many of the issues associated with hydraulic modelling in more detail:

Bates, P.D., S.N. Lane, and R.I. Ferguson. 2005. *Computational Fluid Dynamics: Applications in Environmental Hydraulics* (Wiley-Blackwell). https://doi. org/10.1002/0470015195.ch1

References cited

Braun, A., Chapter 39, this volume. '(Critical) Satellite remote sensing'.

Ingham, D.B. and L. Ma. 2005. 'Fundamental equations for CFD in river flow simulations', in *Computational Fluid Dynamics: Applications in Environmental Hydraulics*, ed. by P.D. Bates, S.N. Lane, and R.I. Ferguson (Wiley-Blackwell), pp. 19–50. https://doi.org/10.1002/0470015195.ch1

Kasvi, E., Chapter 45, this volume. 'Uncrewed airborne vehicle surveys (drones)'.

Krueger, T., Chapter 26, this volume. 'Environmental modelling'.

Landstrom, C., S.J. Whatmore, and S.N. Lane. 2011. 'Virtual engineering: computer simulation modelling for UK flood risk management', *Science Studies*, 24, pp. 3–22. https://doi.org/10.23987/sts.55261

Lane, S.N., R.J. Hardy, R.I. Ferguson, and D.R. Parsons. 2005. 'A framework for model verification and validation of CFD schemes in natural open channel flows', in *Computational Fluid Dynamics: Applications in Environmental Hydraulics*, ed. by P.D. Bates, S.N. Lane, and R.I. Ferguson (Wiley-Blackwell), pp. 429–60. https://doi.org/10.1002/0470015195.ch1

Lane, S.N. 2012. 'Making mathematical models perform in geographical space(s)', in *Handbook of Geographical Knowledge*, ed. by J. Agnew and D. Livingstone (Sage Publications). https://doi.org/10.4135/9781446201091

Lane, S.N. 2014. 'Acting, predicting and intervening in a socio-hydrological world', *Hydrology and Earth System Sciences*, 18, pp. 927–52. https://doi.org/10.5194/hess-18-927-2014

Leclerc, M. 2005. 'Ecohydraulics: a new interdisciplinary frontier for CFD', in *Computational Fluid Dynamics: Applications in Environmental Hydraulics*, ed. by P.D. Bates, Lane, S.N. and R.I. Ferguson (Wiley-Blackwell), pp. 429–60. https://doi.org/10.1002/0470015195.ch1

Legleiter, C. and P. Kinzel. 2024. 'A framework to facilitate development and testing of image-based river velocimetry algorithms', in *Earth Surface Processes and Landforms*, 49, 1361-1382. https://doi.org/10.1002/esp.5772

Melsen, L., Chapter 31, this volume. 'Hydrological modelling'.

Porter, J. and D. Demeritt. 2012. 'Flood-risk management, mapping, and planning: the institutional politics of decision support in England', *Environment and Planning A: Economy and Space*, 44, pp. 2359–2378. https://doi.org/10.1068/a44660

Wright, N.G. 2005. 'Introduction to numerical methods for fluid flow', in *Computational Fluid Dynamics: Applications in Environmental Hydraulics*, ed. by P.D. Bates, S.N. Lane, and R.I. Ferguson (Wiley-Blackwell), pp. 147–68. https://doi.org/10.1002/0470015195.ch1

31. Hydrological modelling

Lieke Melsen

Definition of hydrological modelling

Numerical hydrological computer models can be defined as "a hypothesis of the real world's functioning, codified in quantitative terms" (Savenije 2009). Hydrological models describe (parts of) the hydrological cycle with equations, and are used to simulate and quantify different states (e.g., soil moisture, groundwater level) and fluxes (e.g., discharge, evaporation) of the hydrological cycle. Hydrological models are applied to investigate system functioning and to support practical applications such as flood forecasting.

The basics of hydrological models

In general, hydrological models tend to describe the processes of the hydrological cycle quite explicitly, in contrast to, for example, statistical approaches. Still, there is a wide range of different models available, varying in how explicit processes are represented in the model. The more 'conceptual' models tend to simplify the described system quite a bit, for example by representing the soil as a single simple bucket, while the so-called 'physically-based' or 'physics-informed' models include more explicit representations, such as vegetation growth, rooting depth, different soil layers, and different flow paths through the soil. All these models require 'forcing', meteorological input variables that drive the hydrological cycle, such as precipitation (P) and temperature (T). Generally speaking, the more complex the model is, the more input data (for instance, land-use or soil type data) and forcing (for instance, beyond P and T also vapour pressure deficit or wind speed) it requires, but

https://doi.org/10.11647/OBP.0418.31

also the more output the model creates (for instance, not only discharge values, but also soil moisture at several depths, evapotranspiration). Hydrological models simulate these states and fluxes over time, and can either be spatially lumped (representing a whole catchment as one) or distributed (cutting the area of interest into smaller pieces and run the model for each of these small pieces), allowing the model to represent and investigate spatial heterogeneity.

Hydrological modelling in depth

The process of developing and applying hydrological models is often guided by several distinct steps (Beven 2012; Knoben et al. 2019; Jansen et al. 2021). First, a perceptual, mental, model is developed by the expert. Modelling is often described as an art (Savenije 2009), because it requires imagination, inspiration, and creativity. That particularly applies to defining the perceptual model, where the expert has to formulate and identify the processes relevant to represent in the model. As such, it is already a selective perspective on the world (Krueger et al. 2012). This highly critical step is often informed by hydrological data (Odoni and Lane 2010), which are also subject to partiality and bias (Lane 2011; Comber et al. 2005). Once the perceptual model is defined, the formal or mathematical model is developed, which contains the equations describing the processes. After that, these equations have to be implemented numerically to run on a computer (the procedural model). Finally, the model is often subject to calibration (tuning the parameters to fit some observations), sensitivity analysis (understanding the relative importance of different factors in the model), uncertainty analysis (quantifying the uncertainty in the output of the model), and evaluation (comparison against a benchmark—which can for instance be observations or output from another model). These last steps are often done interactively. Global sensitivity analysis, for example, can be used to identify the most important parameters for calibration (Pianosi et al. 2016), while local sensitivity analysis can contribute to estimate predictive uncertainty.

Procedures have been developed to support model construction. For instance (Gupta et al. 2012) and (Clark et al. 2008; Clark et al. 2015; Gharari et al. 2021) provide methodological frameworks to formulate model structures. There is awareness of the relevance of choosing your numerical scheme wisely, see for instance (Clark and Kavetski 2010; Kavetski and Clark 2010; La Follette et al. 2021). There are ample of papers describing calibration strategies, uncertainty estimation methods (Liu and Gupta 2007), sensitivity analysis (Pianosi et al. 2016) and evaluation strategies (Andr´eassian et al. 2009; Bennett et al. 2013; Best et al. 2015), where the citations mentioned here are just a tiny selection.

However, in practice, the first three steps of developing a model (the perceptual, mathematical, and numerical steps) are often replaced by selecting an already existing hydrological model—an 'off-the-shelf' model. A plethora of hydrological models are available to choose from (Weiler and Beven 2015; Horton et al. 2022), ranging from highly conceptual to highly physics-informed. Theory exists on how to select "the right" model, for example as a trade-off between available data and model complexity (Höge et al. 2018) or based on a fit-for-purpose evaluation (Hamilton et al. 2019). However, in practice, legacy-reasons such as habit (Babel et al. 2019) and experience seem to be main drivers of model selection (Addor and Melsen 2019). Also for the modelling steps after model selection, such as calibration and sensitivity analysis, experience and familiarity appear to be the main drivers for selecting a method (Melsen 2022), which is further reinforced through recruitment strategies for modellers (Melsen 2023). As such, cognitive and motivational biases of the modeller and the environment of the modeller co-shape the model and its results (Hämäläinen 2015; Hämäläinen and Lahtinen 2016).

Besides numerical hydrological models, which are usually meant when the word 'hydrological model' is used, other modelling approaches exist, such as statistical modelling and Machine Learning (ML) applications. Statistical hydrological models can make predictions without trying to explain the underlying mechanisms. This makes them mainly suitable for practical applications. An advantage of statistical models compared to numerical models is that they generally require less calculation time. More advanced statistical models trained on large amounts of data are so-called Machine Learning applications (Nearing et al. 2021). Neural Networks (NN) are frequently explored as an alternative to numerical models (e.g., Kratzert et al. 2018). Relative weights of the nodes in a NN might provide some insights in process relevance. Recently, blended approaches are increasingly explored, where ML applications are mixed with process-based hydrological models (Razavi et al. 2022). Just like numerical models, also statistical and ML-based models are sensitive to partiality and bias.

Why is hydrological modelling important?

Hydrological models are often mentioned to serve two purposes. First, a model can be used to investigate system dynamics and interactions—in that sense, it is a way of testing a hypothesis of how the system functions and behaves, and it can help to identify directions that need further exploration (Clark et al. 2011; Dooge 1986). The second application is that models are used for practical purposes to support decision making. Examples include real-time forecasting for flood and drought preparedness, or long-term strategic planning for climate change impacts (Arnell et al. 2011).

At a more reflective level, hydrological models also serve the purpose of making the inaccessible accessible by quantifying otherwise unmeasurable processes (Lane 2011). Besides, hydrological models can serve as boundary objects: They can facilitate communication between different stakeholders with different perspectives on the world (Lim et al. 2023)—although, as noted by (Lim et al. 2023), in the same fashion models can also be sites of contestation.

Relationship of hydrological modelling with other methods

Hydrological models rely on input data, both for their set-up and for calibration and validation. Many of these data are also (partially) model products, with their own assumptions and uncertainties. Input data for example include land use/land cover maps, soil maps, and meteorological forcing. The Harmonized World Soil Database, for example, is a global product that combines local observations with interpolation techniques to create global coverage. Many land cover products use satellite remote sensing data (Braun, Chapter 39), and thereby assume a relation between wavelengths of light that are absorbed and reflected by different vegetation types. Discharge data, usually used for calibration and validation, are often based on a rating-curve (or a "stage-discharge relation") that links water level to water volume, thereby assuming a certain geometry of the channel. Meteorological input data can be based on local point observations from a meteorological station, but are also often products with spatial coverage, for which a model is used. Climate change impact analysis requires the use of climate model output as input for the hydrological model.

Ethical issues and hydrological modelling

Models are not neutral (Saltelli et al. 2020). For instance, in defining the perceptual model, the expert evaluates which processes are considered relevant enough to include in the model, but different experts will make different choices—also when it comes to selecting an already existing model or to making choices in the configuration of the model (Melsen 2022). The implication is that certain perspectives and values are favoured by the model and its results. An example is that gender related perceptions can be mainstreamed into the model (Packett et al. 2020). Such a biased model can favour certain groups at the expense of others. This is particularly problematic in the context of decision support modelling. The model results have agency and power and interfere with the real world (Lane 2014; Melsen et al. 2018). The model can for instance be used to determine which farmers should lower their water use and therefore can apply for compensation from the government (Sanz et al., 2018), but different models would lead to different results on who would be eligible for the compensation programme.

Issues to be aware of in using hydrological modelling

The single main issue to be aware of when using hydrological models is that models are simplified representations of reality. This makes them inherently prone to uncertainty (Oreskes et al. 1994; Walker et al. 2003; Puy et al. 2022). This surfaces in different ways.

In general, three sources of uncertainty are distinguished in hydrological modelling: data uncertainty, parameter uncertainty and model structural uncertainty. These three sources are linked and dependent in many ways—parameter uncertainty can for example stem from data uncertainty through calibration. When properly accounted for, these sources of uncertainty can lead to a considerable spread in model output.

The sources of uncertainty above only cover technical model aspects, but social and political dynamics also drive hydrological model development and use and as such contribute to different model results. The way a question is asked can for example already create a bias: certain aspects will not be investigated with the model, or will be considered irrelevant and thus marginalized. Such a bias will not be made visible with an uncertainty analysis focused on the technical aspects mentioned above.

Models have agency and power, and should therefore be used cautiously—especially in the face of large uncertainties. Their results should not be over-interpreted and the use of models should come with humility.

Suggested further reading

For the basics of (rainfall-runoff) modelling: Beven, K.J. 2012. 'Rainfall-runoff modelling, the primer', in *Down to Basics: Runoff Processes and the Modelling Process* (John Wiley and Sons). https://doi.org/10.1002/9781119951001

References cited

Addor, N. and L. Melsen. 2019. 'Legacy, rather than adequacy, drives the selection of hydrological models', *Water Resources Research*, 55. https://doi.org/10.1029/2018wr022958

Andréassian, V., C. Perrin, L. Berthet, N. Le Moine, J. Lerat, C. Loumagne, L. Oudin, T. Mathevet, M. Ramos, and A. Valéry. 2009. 'HESS opinions: Crash tests for a standardized evaluation of hydrological models', *Hydrology and Earth System Sciences*, 13, pp. 1757–1764. https://doi.org/10.5194/hess-13-1757-2009

Arnell, N., D. van Vuuren, and M. Isaac. 2011. 'The implications of climate policy for the impacts of climate change on global water resources', *Global Environmental Change*, 21.2, pp. 592–603. https://doi.org/10.1016/j.gloenvcha.2011.01.015

Babel, L., D. Vinck, and D. Karssenberg. 2019. 'Decision-making in model construction: Unveiling habits', *Environmental Modelling & Software*, 120. https://doi.org/10.1016/j.envsoft.2019.07.015

Bennett, N., B. Croke, G. Guariso, J. Guillaume, S. Hamilton, A. Jakeman, S. Marsili-Libelli, L. Newham, J. Norton, C. Perrin, S. Pierce, B. Robson, R. Seppelt, A. Voinov, B. Fath, and V. Andréassin. 2013. 'Characterising performance of environmental models', *Environmental Modelling & Software*, 40, pp. 1–20. https://doi.org/10.1016/j.envsoft.2012.09.011

Best, M., G. Abramowitz, H. Johnson, A. Pitman, G. Balsamo, A. Boone, M. Cuntz, B. Decharme, P. Dirmeyer, J. Dong, M. Ek, Z. Guo, V. Haverd, B. van den Hurk, G. Nearing, B. Pak, C. Peters-Lidard, J. Santanello Jr., L. Stevens, and N. Vuichard. 2015. 'The plumbing of land surface models: Benchmarking model performance', *Journal of Hydrometeorology*, 16, pp. 1425–1442. https://doi.org/10.1175/jhm-d-14-0158.1

Beven, K.J. 2012. 'Rainfall-runoff modelling, the primer', in *Down to Basics: Runoff Processes and the Modelling Process*, by K.J. Beven (John Wiley and Sons). https://doi.org/10.1002/9781119951001

Clark, M. and D. Kavetski. 2010. 'Ancient numerical daemons of conceptual hydrological modeling: 1. Fidelity and efficiency of time stepping schemes', *Water Resources Research*, 46. https://doi.org/10.1029/2009wr008894

Clark, M., H. McMillan, D. Collins, D. Kavetski, and R. Woods. 2011. 'Hydro- logical field data from a modeller's perspective: Part 2: Process-based evaluation of model hypotheses', *Hydrological Processes*, 25, pp. 523–543. https://doi.org/10.1002/hyp.7902

Clark, M., B. Nijssen, J. Lundquist, D. Kavetski, D. Rupp, R. Woods, J. Freer, E. Gutmann, A. Wood, L.D. Brekke, J. Arnold, D. Gochis, and R. Rasmussen. 2015. 'A unified approach for process-based hydrologic modeling: 1. Modeling concept', *Water Resources Research*, 51, pp. 2498–2514. https://doi.org/10.1002/2015wr017198

Clark, M. P., A.G. Slater, D.E. Rupp, R.A. Woods, J.A. Vrugt, H.V. Gupta, T. Wagener, and L.E. Hay. 2008. 'Framework for Understanding Structural Errors (FUSE): a modular framework to diagnose differences between hydrological models', *Water Resources Research*, 44. https://doi.org/10.1029/2007wr006735

Comber, A., P. Fisher, and R. Wadsworth. 2005. 'What is land cover?', *Environment and Planning B: Planning and Design*, 32, pp. 199–209. https://doi.org/10.1068/b31135

Dooge, J. 1986. *Reflections in Hydrology: Science and Practice* (American Geophysical Union).

Gharari, S., H.V. Gupta, M.P. Clark, M. Hrachowitz, F. Fenicia, P. Matgen, and H.H.G. Savenije. 2021. 'Understanding the information content in the hierarchy of model development decisions: Learning from data', *Water Resources Research*, 57.6. https://doi.org/10.1029/2020wr027948

Gupta, H.V., M.P. Clark, J.A.V.G. Abramowitz, and M. Ye. 2012. 'Towards a comprehensive assessment of model structural adequacy', *Water Resources Research*, 48. https://doi.org/10.1029/2011wr011044

Hämäläinen, R. 2015. 'Behavioural issues in environmental modelling—the missing perspective', *Environmental Modelling & Software*, 73, pp. 244–253. https://doi.org/10.1016/j.envsoft.2015.08.019

Hämäläinen, R. and T. Lahtinen. 2016. 'Path dependence in operational research—how the modeling process can influence the results', *Operations Research Perspectives*, 3, pp. 14–20. https://doi.org/10.1016/j.orp.2016.03.001

Hamilton, S., B. Fu, J. Guillaume, S. Pierce, and F. Zare. 2019. 'A framework for characterising and evaluating the effectiveness of environmental modelling', *Environ. Modell. Softw.*, 118, pp. 83–98. https://doi.org/10.1016/j.envsoft.2019.04.008

Höge, M., T. Wöhling, and W. Nowak. 2018. 'A primer for model selection: The decisive role of model complexity', *Water Resources Research*, 54.3, pp. 1688–1715. https://doi.org/10.1002/2017wr021902

Horton, P., B. Schaefli, and M. Kauzlaric. 2022. 'Why do we have so many different hydrological models? a review based on the case of Switzerland', *WIREs Water*, 9.1. https://doi.org/10.1002/wat2.1574

Jansen, K. F., A.J. Teuling, J.R. Craig, M. Dal Molin, W.J.M. Knoben, J. Parajka, M. Vis, and L.A. Melsen. 2021. 'Mimicry of a conceptual hydrological model (HBV): What's in a name?', *Water Resources Research*, 57.5. https://doi.org/10.1029/2020wr029143

Kavetski, D. and M. Clark. 2010. 'Ancient numerical daemons of conceptual hydrological modeling: 2. Impact of time stepping schemes on model analysis and prediction', *Water Resources Research*, 46. https://doi.org/10.1029/2009wr008896

Knoben, W. J. M., J.E. Freer, K.J.A. Fowler, M.C. Peel, and R.A. Woods. 2019. 'Modular assessment of rainfall–runoff models toolbox (MARRMoT) v1.2: an open-source, extendable framework providing implementations of 46 conceptual hydrologic models as continuous state-space formulations', *Geoscientific Model Development*, 12.6, pp. 2463–2480. https://doi.org/10.5194/gmd-12-2463-2019

Kratzert, F., D. Klotz, C. Brenner, K. Schulz, and M. Herrnegger. 2018. 'Rainfall–runoff modelling using long short-term memory (LSTM) networks', *Hydrology and Earth System Sciences*, 22.11, pp. 6005–6022. https://doi.org/10.5194/hess-22-6005-2018

Krueger, T., T. Page, K. Hubacek, L. Smith, and K. Hiscock. 2012. 'The role of expert opinion in environmental modelling', *Environmental Modelling & Software*, 36, pp. 4–18. https://doi.org/10.1016/j.envsoft.2012.01.011

La Follette, P.T., A.J. Teuling, N. Addor, M. Clark, K. Jansen, and L.A. Melsen. 2021. 'Numerical daemons of hydrological models are summoned by extreme precipitation', *Hydrology and Earth System Sciences*, 25.10, pp. 5425–5446. https://doi.org/10.5194/hess-25-5425-2021

Lane, S.N. 2011. 'Making mathematical models perform in geographical space(s),' in *The Sage Handbook of Geographical Knowledge*, ed. by J.A. Agnew and D.N. Livingston (Sage Publications).

Lane, S.N. 2014. 'Acting, predicting and intervening in a socio-hydrological world', *Hydrology and Earth System Sciences*, 18.3, pp. 927–952. https://doi.org/10.5194/hess-18-927-2014

Lim, T., P. Glynn, G. Bitterman, J. Guillaume, J. Little, and D. Webster. 2023. 'Recognizing political influences in participatory socio-ecological systems modeling', *Socio-Environmental Systems Modelling*, 5. https://doi.org/10.18174/sesmo.18509

Liu, Y. and H.V. Gupta. 2007. 'Uncertainty in hydrologic modeling: Towards an integrated data assimilation framework', *Water Resources Research*, 43. https://doi.org/10.1029/2006wr005756

Melsen, L. 2022. 'It takes a village to run a model: the social practices of hydrological modelling', *Water Resources Research*, 58.2. https://doi.org/10.1029/2021wr030600

Melsen, L. 2023. 'The modeling toolkit: how recruitment strategies for modeling positions influence model progress', *Frontiers in Water*, 5. https://doi.org/10.3389/frwa.2023.1149590

Melsen, L., J. Vos, and R. Boelens. 2018. 'What is the role of the model in socio-hydrology? discussion of 'prediction in a socio-hydrological world', *Hydrological Sciences Journal*, 63, pp. 1435–1443. https://doi.org/10.1080/02626667.2018.1499025

Nearing, G. S., F. Kratzert, A.K. Sampson, C.S. Pelissier, D. Klotz, J.M. Frame, C. Prieto, and H.V.Gupta. 2021. 'What role does hydrological science play in the age of machine learning?', *Water Resources Research*, 57.3. https://doi.org/10.1029/2020wr028091

Odoni, N. and S. Lane. 2010. 'Knowledge-theoretic models in hydrology', *Progress in Physical Geography: Earth and Environment*, 34, pp. 151–171. https://doi.org/10.1177/0309133309359893

Oreskes, N., K. Shrader-Frechette, and K. Belitz. 1994. 'Verification, validation, and confirmation of numerical models in the Earth Sciences', *Science*, 263.5147, pp. 641–646. https://doi.org/10.1126/science.263.5147.641

Packett, E., N. Grigg, J. Wu, S. Cuddy, P. Wallbrink, and A. Jakeman. 2020. 'Mainstreaming gender into water management modelling processes', *Environmental Modelling & Software*, 127. https://doi.org/10.1016/j.envsoft.2020.104683

Pianosi, F., K. Beven, J. Freer, J. Hall, J. Rougier, D. Stephanson, and T. Wagener. 2016. 'Sensitivity analysis of environmental models: A systematic review with practical workflow', *Environmental Modelling & Software*, 79, pp. 214–232. https://doi.org/10.1016/j.envsoft.2016.02.008

Puy, A., R. Sheikholeslami, H. Gupta, J. Hall, B. Lankford, S. Lo Piano, J. Meier, F. Pappenberger, A. Porporato, G. Vico, and A. Saltelli. 2022. 'The delusive accuracy of global irrigation water withdrawal estimates', *Nature Communications*, 13.3183. https://doi.org/10.1038/s41467-022-30731-8

Razavi, S., D.M. Hannah, A. Elshorbagy, S. Kumar, L. Marshall, D.P. Solomatine, A. Dezfuli, M. Sadegh, and J. Famiglietti. 2022. 'Coevolution of machine learning and process-based modelling to revolutionize earth and environmental sciences: A perspective', *Hydrological Processes*, 36.6. https://doi.org/10.1002/hyp.14596

Saltelli, A., L. Benini, S. Funtowicz, M. Giampietro, M. Kaiser, E. Reinert, and J.P. van der Sluijs. 2020. 'The technique is never neutral. How methodological choices condition the generation of narratives for sustainability', *Environmental Science and Policy*, 106, pp. 87–98. https://doi.org/10.1016/j.envsci.2020.01.008

Sanz, D., Vos, J., Rambags, F., Hoogesteger, J., Cassiraga, E., & Gómez-Alday, J. J. (2018). The social construction and consequences of groundwater modelling: insight from the Mancha Oriental aquifer, Spain. *International Journal of Water Resources Development*, 35(5), 808–829. https://doi.org/10.1080/07900627.2018.1495619

Savenije, H. 2009. 'HESS opinions: The art of hydrology', *Hydrology and Earth System Sciences*, 13, pp. 157–161. https://doi.org/10.5194/hess-13-157-2009

Walker, W., P. Harremoës, J. Rotmans, J. van der Sluijs, M. van Asselt, P. Janssen, and M. Krayer von Krauss. 2003. 'Defining uncertainty: A conceptual basis for uncertainty management in model-based decision support', *Integrated Assessment*, 4.1, pp. 5–17. https://doi.org/10.1076/iaij.4.1.5.16466

Weiler, M. and K. Beven. 2015. 'Do we need a Community Hydrological Model?', *Water Resources Research*, 51, pp. 7777–7784. https://doi.org/10.1002/2014wr016731

32. Interviews: Structured, semi-structured and open-ended

Lynda Johnston and Robyn Longhurst

Definition

Interviews are verbal exchanges in which one person—the interviewer—attempts to gather information, opinions, and/or beliefs from another person, the interviewee or research participant, by asking questions. Interviewing requires careful planning, including the interviewer preparing a list of predetermined questions and conversation prompts. It is one of the most used qualitative research methods for geographers because it generates rich data.

The basics of interviews

The principle behind interviewing is that talking with people is an excellent way to access information about places, people, and experiences. Information gathered through interviewing varies depending on the interviewee's gender, sexuality, ethnicity, age, and disability. Interviews can be face-to-face or online. Digital interviewing, also called video interviewing (using Zoom, Skype, FaceTime, Messenger, WhatsApp, etc.), is increasingly popular and enables the interviewer to include participants from a wide range of locations, saving time and travel costs.

Interviews in depth

Interviews are an intensive method and are key for examining social and spatial relations. There are three major types of interviews: structured; semi-structured; and unstructured. Structured interviews follow a

 https://doi.org/10.11647/OBP.0418.32

predetermined list of questions, asked in the same way and same order. Semi-structured interviews have some predetermined questions and prompts but maintain flexibility in the way topics are addressed by the interviewee. Unstructured interviews are without predetermined questions so that the interviewee may construct their own accounts of their experiences (an oral history, for example (see Chakov et al., Chapter 33)), in their own words.

Interviews are a dialogue rather than an interrogation. They are conversations with purpose. Interviews are particularly useful for sensitive people-oriented topics as they may prompt a far more in-depth discussion than, say, a survey questionnaire (see Winata and McLafferty, Chapter 43).

In general, interviews tend to be used for four main purposes: to investigate behaviours and motivations; collect a diversity (and consensus) of meanings, opinions, and experiences; be respectful of the people who share their knowledge; and fill gaps in knowledge that other methods are unable to. They may be the only method used for a research project or used in conjunction with other methods.

The first step is to decide who to interview. The aim is not to seek a representative sample, but rather to understand how individuals experience and make sense of their places and lives. Hence the sample must be illustrative and choosing who to interview is often influenced by theories and contexts. For example, Johnston (2019)—in conducting a research project about gender variant geographies and feelings of belonging—interviewed people for whom gender diversity was key to their identities. Diverse gender identities also intersect with other identities such as ethnicity, age, class and so on which may also have a bearing on participants' feelings of belonging.

The second step is to reflect on who you are as an interviewer, and how your own embodied identities shape interactions between yourself and interviewees. This process is about recognising one's positionality and being critically reflexive to address different power relationships that may exist between yourself and the interviewees.

The third step is to decide on who to approach and recruit potential interviewees. This may be via a questionnaire survey to gather initial project information with an option at the end for people to include contact details if they would like to be in a follow up interview. Another way to recruit participants is through 'cold calling'. Electoral rolls, for example, locate where someone is living. 'Cold calling' (a name derived from sales occupations) involves knocking on people's doors and asking if they would like to be interviewed. Approaching strangers for interviews can also be done at events (Johnston and Waitt 2021), parks or other public places. Another way to recruit interviewees is through approaching organisations and institutions, making it clear what type of information you want and who you would like to interview. This is called gaining access via gatekeepers. Gatekeepers may, however, try to select the interviewees for you, which may reduce the diversity of views from within the organisation. Finally, another well used method of recruitment is 'snowballing' where you may recruit one person and they help you find more interviewees. Recruiting, then, gains momentum or 'snowballs' as the recruiter builds up layers of potential interviewees.

The fourth step is to decide where to hold the interviews. Online options may be useful when interviewing people that are living in different parts of your country, or indeed different countries. When meeting online or in person, it's best to avoid busy, noisy and public spaces like open-plan workspaces, cafes, or clubs. Talking with people in their own spaces—their homes, for example—can help create a relaxed conversation where interviewees may even illustrate their discussion with objects, as was the case for Johnston (2019) when interviewing people about the importance of home spaces in relation to gender diversity. One needs to consider personal safety, however, when in a stranger's home.

The fifth step is to prepare questions and conversation prompts for the interview. Each interview will be different, as these are social encounters that unfold around interpersonal and place dynamics. Successful interviewing is likely to occur when the interviewer is fully prepared, understands the research topic, and does not need to rely on a rigid set of questions but rather lets the interviewee discuss the most important issues. It is useful to have themes and questions ready in case the conversation does not flow. Having some key questions may prompt the interviewee to think about their experiences. Questions might be descriptive (asking for information on activities or experiences); structural (focusing on how and when events took place); and thoughtful (exploring meanings, feelings and opinions). Good interviews empower the interviewee to 'tell their story'.

The sixth step is to listen. Active listening means being alert and quick to pick up on ideas and themes, then prompt the interviewee to elaborate through phrases such as 'tell me more about that ...'. The main aim of interviewing is to enable people to speak their own versions of events, in their own words, hence it is important to ask follow-up questions in an encouraging way. An interview tends to last about one to two hours, after which it may be difficult to keep the momentum going.

Finally, recording and transcribing interviews lets the researcher concentrate on the interview—in the moment—without trying to take notes. A recorded interview produces an accurate record of the conversation and can be listened to many times. Transcription is best done as soon as possible after the interview, when the conversation is fresh in your mind. It is best practice to follow up with interviewees, sending them a thank you message and a short summary of the findings.

Why is interviewing important?

Surveys tend to ask a rigid set of questions, whereas an interview allows for the participant to express their experiences, feelings, and opinions. Thus, interviewing is used routinely across the social sciences to understand complexities and differences as well as where there is consensus on issues. Interviews are often conversational and adaptable to interviewees' interests, experiences, and views. It is a method that is sensitive and people-oriented, allowing interviewees to construct their own accounts of their experiences.

Relationship of interviews with other methods

The aim of an interview is not to be representative but to understand how individuals make sense of their lives. Interviews are often used in conjunction with questionnaire surveys (Winata and McLafferty, Chapter 43), whether as part of a recruitment device to secure interviewees, or to gather representational empirical data. Interviews can be used side-by-side with many methods.

Ethical issues and interviews

With the rise of video call interviewing a powerful record may be generated. Interviewers need to be explicit about gaining consent to record a video interview and provide details on how the digital file will be kept secure. It is good practice to share the transcription with the interviewee. As noted above, sharing the same background or similar identity with the interviewee may have a positive effect and facilitate a rapport between interviewer and interviewee. This may help produce a rich conversation based on empathy and mutual respect. Assumptions about sameness to interviewees, however, may be challenged during the interview.

Issues to be aware of in using interviews

Use easily understood language that is appropriate for your interviewee. Use words with commonly and uniformly accepted meanings to help avoid ambiguity.

Avoid leading questions as much as possible (that is, questions that encourage a particular response). It's important to phrase each question carefully.

All interviews will be different and have their own rhythm. Always follow the conversational flow with your interviewee. At the end of an interview, review the interview schedule to make sure all themes have been covered.

It is important to interact, and sometimes share information with interviewees, rather than solely extracting information from them. Sharing experiences and exchanging ideas and information with interviewees is widely accepted as good research practice. Be careful, however, not express views and experiences in a way that interviewees feel unable to express contrary opinions.

Suggested further reading

There are not many open-source textbooks about interviewing. One of the best chapters is Dunn, K. 2021. 'Engaging interviews', in *Qualitative*

Research Methods in Human Geography, ed. by I. Hay and M. Cope (Oxford University Press), pp. 148–185.

Another excellent publication is Valentine, G. 2013. 'Tell me about . . .: Using interviews as a research methodology', in *Methods in Human Geography*, ed. by R. Flowerdew and D. Martin (Routledge), pp. 110–127.

On both interviews and focus groups see: Longhurst, R. and L. Johnston. 2023. 'Semi-structured interviews and focus groups', in *Key Methods in Geography*, ed. by N. Clifford, M. Cope, and T. Gillespie (Sage Publications), pp. 168–183.

References cited

Chakov, A., Chang, T., Covey, H., Dickson, T., Goggins, S., Harris, N., Purna, S., Widell, S., and Druschke, C.G., Chapter 33, this volume. 'Oral history'.

Johnston, L. 2019. *Transforming Gender, Sex and Place: Gender Variant Geographies* (Routledge).

Johnston, L. and G. Waitt. 2021. 'Play, protest and pride: Un/happy queers of Proud to Play in Auckland, Aotearoa New Zealand', *Urban Studies*, 58.7, pp. 1431–1447.

Winata, F. and McLafferty, S., Chapter 43, this volume. 'Survey and questionnaire methods'.

33. Oral history

Alexandra Chakov, Tom Chang, Henry Covey, Taylor Dickson, Sydney Goggins, Nora Harris, Sujash Purna, Sydney Widell, Caroline Gottschalk Druschke

Definition

Oral history is both the process and the product of a recorded, spoken interview about someone's life or a particular event or topic. Once recorded, the interview is placed in an archive, typically a long-term, managed repository. With permission from the interviewee and interviewer, the recording can be used by others into the future. Oral history is employed across a variety of disciplines.

The basics of oral histories

At their core, oral histories value vernacular voices, insisting that the inclusion of everyday perspectives, including the emotional context they provide, is key to more fully understanding past events. Oral history can work to fill in more formal sources, like public documents, statistical data, official histories, and photographs; speak to the lived experience of historical events; offer a window onto change over time; and capture lived experience in people's own voices for future generations. Key to oral history is long term preservation and access, as well as the importance of a curated set of stories rather than a single interview with one individual.

  https://doi.org/10.11647/OBP.0418.33

Why are oral histories important?

Oral histories do important work not only through their end result, but also through their process. The collaborative work necessary between story gatherer and storyteller offers opportunities to build relationships that can last beyond the project itself. Likewise, because of oral history's emphasis on gathering multiple stories about the same historical event, the method offers opportunities to build community across contributing storytellers, as well as across communities involved in the project. The reflective work of oral history positions storytellers to look back on and process their involvement in historically significant events, while also serving a constitutive function, speaking to the future to create collective imagination. This chance to look back and imagine ahead can offer space for healing for both storytellers and story gatherers, as well as creating opportunities to speak to children, grandchildren, and future generations, while also offering homage to ancestors and place.

In addition to these process outcomes, oral histories are important for what they produce. They can supplement, complicate, and even correct the public record about historic events by providing additional content from multiple perspectives, offering emotional context and nuance, and demonstrating the complexity of lived experience and history. By providing insight into the lived experience of contributors, they can offer an avenue for underrepresented voices to influence dominant discourse. And because oral histories rely on preserving storytellers' own voices and experiences, they create opportunities for communities to build accessible archives of their own history. When considered collectively, they have the potential to inform public policy across scales, making visible community needs that might otherwise go unrecognised.

Relationship of oral histories with other methods

Oral history owes a huge debt to long oral traditions across cultures, and many community members and increasingly researchers focus on a variety of forms of storytelling for capturing community experiences and truths. A focus on storytelling and story gathering can also support a shift in perspective among researchers away from the collection of data and towards an appreciation of personal experience outside of a research transaction. Broadly, oral history is one of a suite of qualitative methodologies that focus on individual lived experience and can be adapted based on context or work well in tandem with other methodologies.

Where oral history features the retelling of a past event that the interviewee personally experienced and the curation of a collection of those stories, some researchers adapt the methodology as oral narrative to look not only at past events, but to link these explicitly with present and future. Also similar in approach, research interviews (see Johnston and Longhurst, Chapter 32) are typically somewhat more formal in nature than oral histories and narratives, focusing on a particular research question and including less personal context.

Oral history may be supplemented with other complementary research approaches, including archival research (Cope, Chapter 22), to round out various perspectives on a historical event. It can also work well with a suite of approaches to community-based research given its focus on highlighting vernacular voices and countering hegemonic sources of power. Oral history projects can support the goals of communities where they are situated, leading to organisational capacity building, knowledge sharing, social and environmental transformations, and broader community education and advocacy efforts. This emancipatory possibility exists in oral history, but oral history is not inherently a community-based practice, and community participation does not necessarily make research "participatory".

Ethical issues and oral histories

Oral history can be a challenging practice because of its inherent focus on asking individuals to reflect on personal experiences of historically significant events, which may be emotionally charged or even traumatic. Researchers need to be well-trained to deal sensitively with a range of storytellers' emotional responses, and need to remain open to responses that may go in directions they would not have anticipated.

Various countries have different approaches to ethical oversight of oral histories. At U.S. institutions, for example, while there are valid criticisms of Institutional Review Board (IRB) review, the formal group at each institution that reviews and monitors human subjects research, it is true that IRB exists to offer ethical guidance and protection for research subjects. But oral history has long sat outside these formal, structural processes, a position that places the onus on individual researchers to follow best ethical practices from guiding organisations like the Oral History Association (AHO, Chapter 5). Oral history's placement outside the IRB process was for good reason. As the Oral History Association has described, "oral historians have found the IRB process poorly suited to the consensual, shared authority interview methods that are the foundation of sound oral history practice" and "IRB policies that mandated confidentiality, and even the destruction of interviews after a period of time, directly contradict the principle of narrator ownership of copyright and best practices on archival preservation". Oral history is increasingly coming under the oversight of IRBs at various U.S. institutions with many IRBs working to update their recommendations and guidance documents to better suit oral histories.

Oral history has liberatory potential, but is not inherently liberatory or even community-driven. Oral history can lend itself to extractive, data-centric research, with huge potential for harm because of its imperative to explore sensitive topics with participants. Researchers interested in oral history methodologies should consider a number of guiding ethical questions. Is there value for storytellers in the work? How can we develop and maintain a reciprocal relationship with storytellers? Where will we place recorded stories to ensure community access and what types of protections do storytellers want in place for their stories? Have storytellers been invited into the development of the question guide for the oral history session? What intentions and biases are we bringing to this work?

Issues to be aware of in using oral histories

While oral history methodology clearly offers a suite of potential benefits, it does pose some complications. Given its focus on creating a collection of individually recorded stories, the methodology is time intensive. For the story gatherer, there's a need for time and expertise related to recording stories, but also and especially to project management, transcription, quality control, preservation, indexing, and other archival maintenance. Storytellers are also sharing their own time and expertise. Consent is crucial. It's important to decide exactly who will have access to recorded stories, how they can be shared, and what they can be used for, on a time-scale that exceeds the life of the storyteller. Those decisions need to be made in advance and be well documented. And consent is especially critical because oral history, given its focus on transparent and productive conversations about potentially sensitive subjects can be uncomfortable for all involved. Story gatherers must be prepared for this possibility, and well-trained in active listening, trauma-informed practices, and managing challenging conversations to ensure a supportive interview experience for all involved and minimise potential harm. In light of the sensitivity of this work, oral history projects work best when they are built on—and help build—some level of mutual trust. This might begin with a story gatherer who is connected to the event or community in question, or a project manager who has put in the time to get to know members of various communities related to the historical event who have asked for their communities' stories to be recorded and gathered. Beginning with this foundation, when story gatherers engage with storytellers with active listening, trauma-informed practices, genuine respect, and care, that mutual trust can grow, deepening oral history's potential to build relationships across story gatherers and storytellers, and among storytellers, that can last well behind the project.

Suggested further reading

Oral History Association. This international association, comprised of educators, policy makers, historians, and others interested in oral history, features a web site that includes comprehensive information about principles and best practices, Institutional Review Boards, funding opportunities, interviewing resources, and more.

StoryCorps:. This nonprofit organisation is committed to preserving and sharing the U.S. experience, with a collection of well over a half million stories collected in the U.S. Library of Congress. Their web site features guidance and support to contribute to, develop, and sustain oral history work.

Related reading

Gottschalk Druschke, C., M. Higgins, T. Dean, E. Booth, and R. Lave. 2022. 'Storying the floods: Experiments in feminist flood futures', *Open Rivers: Rethinking Water, Place and Community*, 22.

Gottschalk Druschke, C., T. Dean, M. Higgins, M. Beaty, M., L. Henner, M. Higgins, R. Hosemann, J. Meyer, B. Sellers, S. Widell, and T. Woser. 2022. 'Stories from the flood: Promoting healing and fostering policy change through storytelling, community literacy, and community-based learning', *Community Literacy Journal*, 16.2.

Grobman, L. et al. 2015. 'Collaborative complexities: Co-authorship, voice, and African American rhetoric in oral history community literacy projects', *Community Literacy Journal*, 9.2, pp. 1–25.

Jackson, R.C. and D. Whitehorse DeLaune. 2019. 'Decolonizing community writing with community listening: Story, transrhetorical resistance, and Indigenous cultural literacy activism', *Community Literacy Journal*, 13.1, pp. 37–54.

Riley Mukavetz, A.M. 2014. 'Towards a cultural rhetorics methodology: Making research matter with multi-generational women from the Little Traverse Bay Band', *Journal of Rhetoric, Professional Communication, and Globalization*, 5.1, p. 6.

Texas After Violence Project. *Documenting Narratives of Violence Trauma-Informed Interviewing for General Audiences*, https://vimeo.com/430151906

34. Participant observation and ethnography

Nathan F. Sayre

Definition

Participant observation is a method used in social science research in which one observes people while participating in their activities, i.e., passively, rather than actively enlisting them into a research-focused activity through an interview, a questionnaire or another participatory research method. The goal is to ascertain practical, extra-verbal, or routine aspects of peoples' lives, including particularities and contextual factors that other methods omit or exclude. Ethnography is an intensive form of participant observation, primarily distinguished by a longer duration (months to years) of sustained interaction. The goal of ethnography is to see the world through the eyes of the people being studied, so to speak, attaining effective fluency in their language and sympathetic understanding of their norms, practices, and worldviews.

The basics of participant observation and ethnography

The core principle of participant observation and ethnography (PO/E) is that many social phenomena can only be accessed for study via participation in relevant activities and development of relationships with the people involved. The reasons for this are both practical and epistemological. It would be awkward or rude to study a cultural ritual such as a marriage, funeral or religious celebration without participating in it, for example—participation here meaning (at a minimum) respectful conformity of comportment and behaviour during the event. Similarly, many things about ordinary livelihood activities can only be studied by 'going along' with the people performing them, as doing

    https://doi.org/10.11647/OBP.0418.34

otherwise would interfere with or prevent the activity itself. This is the sense in which PO/E tends to be passive. Finally, there are major dimensions of social reality, such as meaning and value, that are so semi- or subconscious, embodied, or symbolic in nature as to escape an outsider's apprehension in the absence of sustained immersion.

Participant observation and ethnography in depth

Anthropology is the discipline most closely associated with PO/E; indeed, the elevation of ethnography to the status of method by the likes of Marcel Mauss, A.R. Radcliffe-Brown, Franz Boas and Bronislaw Malinowski may be said to mark the advent of anthropology as a scholarly discipline (as distinct from the de facto ethnography practiced out of necessity by countless travellers, traders and explorers through the ages). Prolonged co-residence and acquisition of native languages were practical necessities in settings where transportation was arduous and interpreters non-existent. The scientific virtues of PO/E are not limited to such contexts, however, and the method is now used across the interpretative social sciences for understanding virtually any group of people, including political and economic elites, scientists, bureaucrats, activists, migrants, members of specific social or occupational categories, and communities impacted by climate change, conflict or other dislocations.

More so than other methods, PO/E is difficult to learn from a book or class; doing it in practice is widely seen as the only way to develop the requisite skills, dispositions and relationships. One reason for this is that all people and communities are particular and unique in at least some ways; even if a researcher wishes to study them as members of a group or tokens of a type (e.g., rural youth, tribal elders, factory workers, etc.), they are unavoidably also individuals who must be approached and treated as such. Thus, there are a handful of prescriptive steps for successful PO/E (cultivate multiple informants; take copious notes, and update them before going to sleep each night; be vigilant in securing and backing up/duplicating your notes; review your notes often, including in the field), but the key principles are social in nature. Above all, PO/E depends on trusting interpersonal relationships, which require mutual respect sustained over time; trust develops slowly but can be lost very quickly. Patience is advisable; one can expect to encounter people repeatedly, and what they share will likely deepen (and possibly change) over time. Thus, for methodological as well as ethical reasons, credibility is paramount; ethnographers must be honest about the purposes of their research, good for their word, and genuine in their self-presentation.

Why is participant observation and ethnography important?

Most simply, many social phenomena cannot be apprehended by other scientific methods. There are at least three additional reasons that PO/E is important. First, no other method is as open-ended with regard to its questions and categories. Every researcher must already have questions and categories in order to get started, but the radical inductiveness of PO/E allows for one's questions and categories to be challenged, re-worked or even discarded. It is thus more likely than other methods to uncover phenomena or hypotheses that are novel or unanticipated by the researcher. Second, all quantitative research ultimately rests on qualitative claims or assumptions (Sayer 1984), and whereas quantitative methods with large sample sizes can identify patterns in social phenomena, PO/E is more likely to reveal the qualitative mechanisms underlying those patterns. Third, PO/E is better equipped than other methods to cope with what may be the most distinctive feature of human beings as compared to other research objects: the capacity and propensity to lie or mislead, not only to the researcher but also to oneself. Intentionally or not, it is both easy and common for people to invent answers to researchers' questions in one-off encounters such as polls and surveys, especially where power asymmetries are pronounced (in either direction). It is also often true that what people say they do and what they actually do align imperfectly. By its direct observation and prolonged and iterative nature, PO/E is more likely to reveal and overcome "the chasm between discourse and practice" (Bobrow-Strain 2007: 26) that inheres in everyday life and social relations. Indeed, according to critical social theorists from Marx and DuBois to Gramsci and Bourdieu, these discrepancies are integral to the reproduction and naturalisation of social domination.

Relationship of participant observation and ethnography with other methods

PO/E is often employed in combination with archival methods (Cope, Chapter 22) to establish the historical origins or context of present-day social relations. Indeed, a similar sort of hermeneutic practice is often useful in both, working back and forth between one's own semiotic system and that of one's research subjects (living or dead). PO/E is also highly synergistic with case-study or intensive research (Lane, Chapter 24) and the various methods pursuant thereto. Fairhead and Leach (1996), for example, combined ethnography with remote sensing, archival analysis, village surveys, and ecological transects in an outstanding example of critical interdisciplinary research. Finally, semi-structured or open-ended interviews (see Johnston and Longhurst, Chapter 32) may also complement PO/E as a means of initiating longer-term relationships or for individuals whose position or circumstances do not allow for sustained co-participation. Focus groups (Longhurst and Johnston, Chapter 27) may serve a similar purpose, although in general they are almost diametrically opposed to PO/E because they intentionally remove people from their everyday lives.

Ethical issues with participant observation and ethnography

The ethical issues associated with PO/E are significant and complex. The particularity of detail made available by ethnography can make protecting the identities of research subjects difficult, even impossible, especially within their own communities. Ethnographic research can also lead to close personal friendships, introducing potential conflicts of interest or biases in interpretation. More fundamental issues arise from power asymmetries that often attend and enable PO/E. Early anthropologists routinely benefited from extremely unequal terms of engagement, from access to transport, money and market goods to membership in (and sometimes sponsorship by) colonial governments. Moreover, these power relations were ignored or omitted in the discipline's publications and self-understanding for decades. As long as academic researchers enjoy structural advantages vis-à-vis many populations, the potential for extractive relationships will remain. That said, subaltern groups may see scholars as valuable allies or spokespeople for their causes and needs, and PO/E can also be employed to expose and demystify elites by "studying up"—provided that access can be secured (Nader 1974 [1969]). Finally, if PO/E affords insight into power asymmetries, it also can give rise to political entanglements. Researchers may develop strong feelings of gratitude and obligation toward the individuals, communities, and causes they engaged while conducting PO/E, and they may come to be viewed as spokespeople and advocates rather than dispassionate scholars.

Issues to be aware of in using participant observation and ethnography

In addition to ethical issues, PO/E entails important epistemological challenges, ones reflected more widely in case study-based research (see Lane, Chapter 24). Foremost are questions of representativeness and generalisability: one cannot assume that the particular case one has chosen to study will directly reveal broader facts or substantiate claims about a larger group of apparently similar cases. Indeed, by studying one case in intimate detail, PO/E is likely to uncover any number of idiosyncrasies or particularities that problematise extrapolation. Parallel issues arise in any effort to derive causal claims, as the array of contextual or contributing factors outruns any attempt at strict reduction or control. Alternative theoretical frameworks emphasising relationality, process, and a dialectic of contingency and determination (what Anthony Giddens (2014) terms "structuration") are better suited to PO/E than positivist inference or hypothesis testing. All of these considerations must be taken into account from the very beginning to ensure that research questions are suited to the method. Finally, the effort to see the world through others' eyes often lays bare one's own preconceptions, revealing as much about oneself as about others. Such reflexivity is salutary, but it has also been known to result in navel-gazing paralysis.

Suggested further reading

There are numerous websites dedicated to PO/E, including some maintained by university Institutional Review Boards to assist researchers in complying with protocols for the protection of human subjects. For a succinct, open-source discussion of ethnography, see Sangasubana, N. 2011. 'How to conduct ethnographic research', *The Qualitative Report*, 16.2, pp. 567–73, http://www.nova.edu/ssss/QR/QR16-2/sangasubanat.pdf

References cited

Bobrow-Strain, A. 2007. *Intimate Enemies: Landowners, Power, and Violence in Chiapas* (Duke University Press). https://doi.org/10.1515/9780822389521

Cope, M., Chapter 22, this volume. 'Archival methods'.

Fairhead, J. and M. Leach. 1996. *Misreading the African landscape: Society and Ecology in a Forest-Savanna Mosaic* (Cambridge University Press). https://doi.org/10.1017/cbo9781139164023

Giddens, A. 2014. 'Structuration theory: past, present and future', in *Giddens' Theory of Structuration*, ed. by C. Bryant and D. Jary (Routledge), pp. 201–221. https://doi.org/10.4324/9781315822556

Lane, S.N., Chapter 21, this volume. 'Case studies'.

Longhurst, R. and Johnston, L., Chapter 27, this volume. 'Focus groups'.

Johnston, L. and Longhurst, R., Chapter 32, this volume. 'Interviews: Structured, semi-structured and open-ended'.

Nader, L. 1974 [1969]. 'Up the anthropologist: Perspectives gained from studying up', in *Reinventing Anthropology*, ed. by D. Hymes (Vintage Books), pp. 284–311.

Sayer, A. 1992. *Method in Social Science: A Realist Approach* (Routledge). https://doi.org/10.4324/9780203310762

35. Participatory modelling

Catharina Landström

Definition

Participatory modelling is a way of doing environmental modelling (see Krueger, Chapter 26) together with lay people. The objective is to integrate both scientific and experience-based knowledge in the computational analysis of environmental processes.

The basics of participatory modelling

Participatory modelling originates in the environmental sciences where computer simulation modelling has become a key research technique. Participatory modelling is a form of environmental participation that emphasises co-production of knowledge (Landström 2020). Motivations for scientists to involve lay people in the modelling process range from empowering local communities (Whatmore and Landström 2011) to accessing local environmental data that cannot be collected by other means (Ritzema et al. 2010) to impacting environmental management (Hare 2011). There are many different participatory modelling approaches with distinct combinations of activities, participants, and scientific models.

Participatory modelling in depth

The procedure of participatory modelling varies between projects and lay participants can be involved in different aspects of the modelling. In some projects lay participants are invited to comment on modelled scenarios, which they may, or may not, have been involved with creating.

 https://doi.org/10.11647/OBP.0418.35

Other projects ask participants to engage with the models themselves e.g., to evaluate critically the representation of the physical process investigated. This is common in projects involving experts on the real-world problem, such as water management professionals. More radical projects involve participants already when deciding which questions to address and when conceptualising, building and/or using a model (Lane et al. 2011).

The organisation of participatory modelling projects also varies depending on the background of participants, the number of participants involved, and their relation to the process modelled. Some projects invite representatives of larger populations that are geographically or demographically defined. Such projects involve large groups of lay participants and engagement with everyone is by necessity limited. For instance participation may consist of a one-day workshop in which participants only view scenarios and modelling outputs (Stave 2010).

Other projects target people with practical expertise in e.g., environmental management or environmental NGOs. These projects often aim for participant engagement with modelling as the models developed are intended to become new tools for use in relevant practices (Whitman et al. 2015). Projects of this type organise several events for the same participants, to enable dialogue about the model development process.

There are also projects that engage with smaller groups of self-selected people who are affected by the issue. Such projects may attempt to co-produce new models drawing on the knowledge of local participants; they are organised in the same sequence as scientific modelling processes, beginning with conceptualising the problem and concluding with a model that can address questions originating in the local experience of a problem rather than in scientific discourse (Lane et al. 2011).

Why is participatory modelling important?

Participatory modelling is a way to connect science and publics that can result in radically new knowledge, modelling approaches, and environmental management practices (Gray et al. 2017). Environmental modelling is poorly understood in society and often used to legitimise decisions about local resource allocation. Participatory modelling opens out this way of analysing future possibilities to everybody affected by environmental processes.

Relationship of participatory modelling with other methods?

Participatory modelling benefits from being used in combination with qualitative social science methods such as ethnography (Sayre, Chapter 34) and interviews (see Johnston and Longhurst, Chapter 32). This is because knowledge about the local context is needed at the start and the finish of a project. In the beginning of a project, it is important to know about local matters of concern to ensure the quality of the participant recruitment process. At the completion of a project in-depth understanding of the societal context can enhance knowledge dissemination.

Ethical issues and participatory modelling

Participatory modelling raises three ethical issues. First, there is a duty of care to participants (Meadow et al., Chapter 5). It is important to follow ethical protocols ensuring that participants come to no harm. This includes recognising that local environmental risks can cause vulnerability and that local environmental management can be contentious (Reed 2008). A second ethical issue concerns the expectations of participants. It is important to be clear on the expected outcome of participation from the outset (Hore et al. 2021). Third is the role of participatory modelling in democratic decision making. Involving those affected by local environmental management in the modelling process could both protect against disproportional influence of articulate, privileged minorities (Hedelin et al. 2021), but also reinforce that influence if the participatory modelling is not founded upon a carefully thought through ethical basis.

Issues to be aware of in using participatory modelling

The terms used to refer to participants vary. Some researchers use the notion of stakeholder with reference to all people not part of the scientific team and who have a relationship to the modelled environmental process (Glicken 2000). This can range from environmental management professionals in local councils, to anglers, to residents, to volunteer groups. Other projects use terms like members of the public, local community, or local people, emphasising the participants' distance from science and connection to place. It is important to be aware of this variability when reviewing the literature and planning a project.

Be clear about whether the participants are intended to represent a larger group or attending as individuals (Skarlatidou et al. 2019). Anybody can be in either category, e.g., environmental managers could participate as representatives of a company or a profession, or as interested individuals; local people can be invited because they are knowledgeable individuals or represent a local interest group or a demographic category. The definition of participants as representatives or individuals impacts on recruitment. A project looking to represent local demographics may use statistical social science methods to identify potential participants. Environmental management professionals could be appointed by their organisations. In contrast interested individuals could be recruited by local advertising in print and social media.

Participatory modelling can be done with any scientific modelling approach and projects can be successful without creating new scientific models. Scientific accounts of participatory modelling focusing on the development of a new model create the impression that the modelling approach deployed guarantees the success of the project. There is no evidence for this, and in fact everything points to the opposite—that any type of model can work well if the project is well-formulated (Voinov et al. 2018).

Although in charge of the project organisation, scientists do not have full control of the process or the outcome in participatory modelling. The content must emerge through collaborative activities and the method impacts the involved scientists as well as lay participants (Landström 2017).

Suggested further reading

Hare, M. 2011. 'Forms of participatory modelling and its potential for widespread adoption in the water sector', *Environmental Policy and*

Governance, 21.6, pp. 386–402. This article overviews participatory modelling in water research in relation to water management.

Voinov, A., K. Jenni, S. Gray, N. Kolagani, P.D. Glynn, P. Bommel, et al. 2018. 'Tools and methods in participatory modeling: Selecting the right tool for the job', *Environmental Modelling and Software*, 109, pp. 232–255. Drawing on extensive experience the authors discuss the pros and cons of different participatory environmental modelling approaches.

Whatmore, S. J. and C. Landström. 2011. 'Flood-apprentices: an exercise in making things public', *Economy and Society*, 40.4, pp. 582–610. This case study details the way in which participatory modelling can impact on local flood management politics.

References cited

Glicken, J. 2000. 'Getting stakeholder participation "right": a discussion of participatory processes and possible pitfalls', *Environmental Science & Policy*, 3.6, pp. 305–310.

Gray, S., R. Jordan, A. Crall, G. Newman, C. Hmelo-Silver, J. Huang, W. Novak, D. Mellor, T. Frensley, M. Prysby, and A. Singer. 2017. 'Combining participatory modelling and citizen science to support volunteer conservation action', *Biological Conservation*, 208, pp. 76–86.

Hedelin, B., S. Gray, S. Woehlke, T.K. BenDor, A. Singer, R. Jordan, M. Zellner, P. Giabbanelli, P. Glynn, K. Jenni, and A. Jetter. 2021. 'What's left before participatory modeling can fully support real-world environmental planning processes: a case study review', *Environmental Modelling & Software*, 143, p. 105073.

Hore, K., J. Gaillard, T. Davies, and R. Kearns. 2021. 'Participatory research in practice: Understandings of power and embodied methodologies', *New Zealand Geographer*, 77.3, pp. 221–229.

Landström, C. 2020. *Environmental Participation. Practices Engaging the Public with Science and Governance* (Palgrave MacMillan).

Landström, C. 2017. *Transdisciplinary Environmental Research: A Practical Approach* (Palgrave MacMillan).

Lane, S.N., N.A. Odoni, C. Landström, S.J. Whatmore, N. Ward, and S. Bradley. 2011 'Doing flood risk science differently: an experiment in radical scientific method', *Transactions of the Institute of British Geographers*, 36.1, pp. 15–36.

Reed, M. 2008. 'Stakeholder participation for environmental management', *Biological Conservation*, 141.10, pp. 2417–2431.

Ritzema, H., J. Froebrich, R. Raju, C. Sreenivas, and R. Kselik. 2010. 'Using participatory modelling to compensate for data scarcity in environmental planning: a case study from India', *Environmental Modelling & Software*, 25.11, pp. 1450–1458.

Skarlatidou, A. et al. 2019. 'The value of stakeholder mapping to enhance co-creation in citizen science initiatives', *Citizen Science: Theory and Practice*, 4.1, pp. 1–10.

Stave, Krystyna. 2010. 'Participatory system dynamics modeling for sustainable environmental management: Observations from four cases', *Sustainability*, 2.9, pp. 2762–2784.

Whitman, G.P., R. Pain, and D.G. Milledge. 2015. 'Going with the flow? Using participatory action research in physical geography', *Progress in Physical Geography*, 39.5, pp. 622–639.

36. Participatory methods

Jennifer Mokos

Definition of participatory methods

Participatory methods comprise a range of approaches to research in which researchers co-produce knowledge in collaboration with people and communities who are not trained as researchers and who draw primarily on their lived experiences. Researchers develop reciprocal relationships with community members, who contribute to the research as co-investigators. This enables people who are typically being researched or who will be most affected by the research to shape its process and outcomes.

The basics of participatory methods

Participatory methods are based on the idea that involving perspectives often ignored by researchers can lead to more equitable outcomes that better explain and address complex realities. This work requires a shift in power from researchers to those whom the research concerns, who are empowered to do more than just "participate" in a pre-determined research project. Instead, they shape it through meaningful engagement in the research itself. Participatory projects can involve community members throughout all stages of the research process, from identifying research priorities to collecting and interpreting data, disseminating findings, and implementing actions. This requires researchers to be humble, self-reflective, and willing to learn from others. They may also find it useful to seek out additional training in group facilitation, nonviolent communication, and cultural humility before beginning a project.

    https://doi.org/10.11647/OBP.0418.36

Participatory methods in depth

Participatory methods are primarily defined by an ethical commitment to produce knowledge through reciprocity and meaningful community engagement instead of through a specific research method. Because of this, participatory projects can draw on a wide range of methods (including qualitative, quantitative, mapping, modelling, and arts-based approaches) and can vary in their purpose, amount of community participation, and where in the process participation occurs. However, community participation should be meaningful and a formative aspect of all participatory research projects. Simply tacking on participation at the end is not truly participatory.

Participatory methods are typically carried out through an iterative and inductive process (instead of a linear series of steps) in which participants become researchers. Each cycle commonly involves shared stages of identifying a problem, negotiating participation, collecting data and observations, sharing of results, and determining follow-up actions. Notably, plans and specific details can change as the research progresses. The iterative and inductive aspects of participatory methods are analogous to the process of cooking a pot of soup where the cook can adjust the flavours and even ingredients as the entire dish develops throughout the process (except that with participatory methods, there are multiple cooks working together in the kitchen).

For participatory methods to be effective, relationships built on trust, mutual respect, and reciprocity between academically trained researchers and local community members who become researchers in the project are key. These relationships may take time to develop and can be difficult to negotiate (Armstrong et al. 2022; Cornish et al. 2023). Community members may distrust researchers and scientific experts or be wary of outsiders because of legacies of exploitation and systemic marginalisation. They may also be exhausted from repeated instances of "helicopter research" or "parachute science", in which researchers from more privileged locales travel to low-income or otherwise marginalised communities to collect samples or conduct interviews and then leave to publish results elsewhere without providing any lasting benefits to the local community or environment (Adame 2021; de Vos & Schwartz 2022; Minasny et al. 2020).

What follows are six guidelines for building relationships with communities and potential collaborators.

First, do not approach participatory methods as charity. Local people do not owe you or anyone else their participation. Even when projects are intended to be participatory, community members might be reticent to commit. Participation takes time that community members might not have available to them, or they might need to build trust before fully engaging. Moreover, communities experiencing a great deal of public attention may be approached repeatedly to participate in participatory projects, which can exacerbate research fatigue, especially if their immediate concerns continue to be unmet. Do not take it personally if you are initially turned down or find it hard to connect.

Second, go with the speed of trust. Pushing too hard may result in people hiding their true experiences, knowledge, or culture even if they do talk with you (Deptula et al. 2023). Conversely, asking permission and respecting "no" builds trust. It demonstrates respect for autonomy and boundaries, which can deepen relationships and lead to more open and honest interactions over time. Working with an interlocutor one-on-one to understand issues, discuss concerns, and suggest ways forward can be crucial for building trust and addressing concerns, especially at the beginning of a project.

Third, take time to get to know communities in their complexity before jumping in and proposing ideas or solutions. Public meetings or events, especially those organised from minoritised perspectives, are a great entry point if they are available. Listen for what community members value, what their concerns are, and for different perspectives within the community, while being aware about how your own social and cultural positions might be influencing your own interpretations, judgements, and actions.

Fourth, be a person beyond your identity as a researcher. Sharing coffee, meals, downtime, social experiences, rituals, or celebrations can build relationships. Being present (if appropriate) without a focus on collecting data demonstrates your commitment.

Fifth, seek out and respect community preferences as much as possible. Paying attention to small, seemingly mundane, details such as asking, "Where would you like me to sit?" communicates respect for autonomy and consent by subtly shifting control to community members. In addition, discussions based on sharing experiences or with brought objects, such as maps, can serve as a useful entry point to build meaningful relationships across existing power differentials.

Finally, offer something in return. Participatory research should benefit your collaborators or local community, which could take the form of co-authoring publications or co-presenting at conferences. However, these customary forms of academic currency might not hold as much value to your community partners. This is why the products of participatory scholarship can take many different forms, depending on what is useful to the community. Examples may include an interpretive exhibit, a public art project, a collection of stories, a community flood map, a database of flora and fauna, a policy brief, letter writing campaign, a grant proposal, or a plan for habitat restoration.

Why are participatory methods important?

Participatory methods seek to correct power asymmetries in how knowledge is produced. They aim to reduce extractive practices that treat local communities as sources of data or objects of research by prioritising community benefits throughout the research process and democratising its outcomes.

Moreover, participatory methods challenge norms about who can ask research questions, set research priorities, and interpret and act on data, which has the potential to change the content and values of science (Balaze and Morello-Frosch 2013). Local communities may have a more nuanced understanding of their environment or experience it through a different value system. Integrating expertise from different social and cultural standpoints can bring to light perspectives and values that would have remained invisible from a traditional researcher's perspectives alone. This can open research to new questions and framings that would have otherwise gone unaddressed by traditional, disciplinary scholarship.

Relationship of participatory methods with other methods

It can be useful to think of participatory methods as a meta-method that encompasses the entire research process, like the outside layer of a nesting doll. The various stages of conceptualising, planning, carrying out, interpreting, and acting upon the production of knowledge are all nested within participatory methods. This contrasts with non-participatory modes of knowledge production that regard methods as a specific step in the research process or the steps by which data are collected. Different research methods can be incorporated into participatory methods, depending on the project's purpose or the academic researcher's expertise.

Many different frameworks make up participatory methods, including community geography, participatory GIS (PGIS), participatory modelling, and participatory action research (PAR). The various approaches incorporate a wide variety of research methods and span multiple disciplines (see Vaughn and Jacquez 2020 for a detailed but non-comprehensive list).

Lastly, participatory methods are distinct from citizen science, which also involves people who are not formally trained as researchers. As part of citizen science, members of the public learn and carry out data collection for research that has already been designed and planned by a scientist or academic researcher. In contrast to participatory methods, citizen science affords local communities minimal influence on the research design and implementation.

Ethical issues and participatory methods

The ethics of participatory methods are messy and complex. At their heart lies a radical empathy and a deep commitment to people and relationships. This relationality requires researchers to approach projects with flexibility and an understanding that they might not always make the final call on what happens during the research.

Consent is a continually negotiated process that is not satisfied by the one-step "informed consent" mandated by Institutional Review Boards. Discussions about consent might include aspects not typically considered, such as data ownership and storage. It can also be a challenge to navigate community expectations, which are not always in line with what is feasible given researchers' timing or resource constraints. Community members can make assumptions that shape their expectations. It is important for academic researchers to listen for and to address misunderstandings as soon as possible.

Researchers also need to be careful not to romanticise the community or the idea of participation. Communities are diverse and complex. Be aware that a single community partner or organisation does not represent the entire community.

Issues to be aware of in using participatory methods

Conventional measures of academic research output and productivity do not always align with the efforts required for participatory methods. Some ways to deal with this include, first, advocating for broader definitions of what counts as scholarship and meaningful research outcomes that are more inclusive of participatory methods than traditional research indicators like academic publications; second, developing parallel research tracks (i.e., having a stream of participatory research that is separate from one's non-participatory research agenda); third, publishing on the process of conducting participatory research, which can provide a more flexible route for publication that isn't focused on research results or outcomes; and finally, participatory methods can be amenable to student involvement, especially for academics at teaching-focused institutions. However, student involvement introduces additional concerns. The community should not be the "object" of learning. It takes time and intention to prepare students to make sure they do not cause harm to the community. The uncertainty and need for flexibility inherent with participatory methods can also be a challenge for students in a course environment, especially when academic schedules and constraints don't align with community schedules or needs. It may be more challenging for students to move away from their obligations (to complete a course or earn credits) than it is for other researchers.

Suggested further reading

For a practical primer on the basics of participatory methods, see: Hacker, K. 2013. *Community-Based Participatory Research*. (Sage Publications). https://doi.org/10.4135/9781452244181

For a list of foundational readings in Community Geography, see: *Community Geographies Collaborative* website (https://cgcollaborative.org/publications/).

For open access training and resources on cultural humility, see: The University of Oregon's *Cultural Humility Toolkit* (https://inclusion.uoregon.edu/cultural-humility-toolkit).

References cited

Adame, F. 2021. 'Meaningful collaborations can end "helicopter research"', *Nature*, https://doi.org/10.1038/d41586-021-01795-1

Armstrong, A., E. Flynn, K. Salt, J. Briggs, R. Clarke, J. Vines, and A. MacDonald. 2022. 'Trust and temporality in participatory research', *Qualitative Research*, 23.4, pp. 1000–1021. https://doi.org/10.1177/14687941211065163

Balazs, C.L. and R. Morello-Frosch. 2013. 'The three Rs: How community-based participatory research strengthens the rigor, relevance, and reach of science', *Environmental Justice*, 6.1, pp. 9–16. https://doi.org/10.1089/env.2012.0017

Cornish, F., N. Breton, U. Moreno-Tabarez, J. Delgado, M. Rua, A. de-Gaft Aikins, and D. Hodgetts. 2023. 'Participatory action research', *Nature Reviews Methods Primers*, 3. 34 . https://doi.org/10.1038/s43586-023-00214-1

Deptula, S., D.L. Jefferson, M. Schneider, C.A. Baez, and A.H. Skinstad. 2023. 'Research at the speed of trust: a guide for researchers and Native communities', *National American Indian and Alaska Native Prevention Technology Transfer Center,*

de Vos, A. and M.W. Schwartz. 2022. 'Confronting parachute science in conservation', *Conservation Science and Practice*, 4.5, p. e12681. https://doi.org/10.1111/csp2.12681

Landström, C., Chapter 35, this volume. 'Participatory modelling'.

Minasny, B., D. Fiantis, B. Mulyanto, Y. Sulaeman, and W. Widyatmanti. 2020. 'Global soil science research collaboration in the 21st century: Time to end helicopter research', *Geoderma*, 373, p. 114299. https://doi.org/10.1016/j.geoderma.2020.114299

Vaughn, L. and F. Jacquez. 2020. 'Participatory research methods: Choice points in the research process', *Journal of Participatory Research Methods*, 1.1. https://doi.org/10.35844/001c.13244

37. Q method

Eric Nost

Definition of Q method

Q is a survey-based method in which researchers employ quantitative and qualitative techniques to characterise the different ways of thinking about a topic that exist.

The basics of Q method

A Q method survey typically involves a relatively small number of participants (e.g., 15–25) sorting representative statements about a topic into a series of positions spanning "most like how I think", "neutral", and "least like how I think". Each position is allowed only so many statements, meaning respondents must reflect on their beliefs and attitudes and decide, for instance, "I agree with this statement about the costs of artificial intelligence (AI) more than I agree with this statement about the ethical concerns it raises". In this way, Q method differs from conventional surveys where respondents can strongly agree or disagree with every statement they're presented with.

Q method also departs from other surveys in analysis. In conventional surveys, researchers correlate answers to statements across respondents. This allows them to make claims such as "software developers are more likely to agree that the use of AI in nature conservation is necessary." However, in Q method, researchers correlate respondents across the set of statements.

Suppose we are trying to understand what kinds of perspectives exist regarding the use of digital technologies in nature conservation. Respondents A, B, and C sort the statement, "artificial intelligence

    https://doi.org/10.11647/OBP.0418.37

presents conservation with ethical dilemmas" into position 3 (strongly agree) and X, Y, and Z sort it in position -2 (disagree). Likewise, A, B, and C sort the statement, "what conservation needs is more advanced technology" into position 0 (neutral) while X, Y, and Z all sort it into position 3 (strongly agree). In this case, A, B, and C all sorted both statements similarly, as did X, Y, and Z.

Q method identifies similar sorting patterns in order to construct statistical "factors", or shared perspectives on an issue. In our example, we might conclude that there is one perspective that worries over the ethics of data technologies in conservation and remains ambivalent about whether conservationists need more advanced tools. There is a second perspective that advocates for these technologies and does not see them raising any ethical issues. Across a fuller set of statements, we might come to define one as "optimistic" and another as "pessimistic" about digital conservation.

Q method in depth

In Q method, researchers sample discourse rather than people. Every Q survey requires collecting a representative set of statements about the research topic; this sample is known as the "concourse." Sampling must be done critically, in recognition of the researcher's own positionality vis a vis dominant discourses and the fact that "not all individuals or groups have equivalent capacity to shape discourse" (Sneegas 2019: 79). The risk is a concourse that overlooks important counter-narratives.

Survey participants themselves should be more or less representative of those with a stake in the issue, but a strict stratified sampling approach isn't necessary. Q method's rules of thumb relate the number of respondents required for a survey to the number of statements to be sorted; generally, a survey with 30 statements requires around 20 respondents (Watts and Stenner 2005). Respondents can sort the concourse in any number of ways. Live sorts, whether in-person or on video conferencing software, make it easier for researchers to ask participants about why they ranked statements the way they did.

Researchers typically use principal components analysis to distil the large number of possible factors (as many as there are unique sorts) into a smaller set of coherent but statistically differentiated ones that represent idealised sorting patterns each respondent's actual answers more or less correlate with. The number of factors a Q study should find is never fixed, though three to five is common. Choosing how many factors adequately represent the data is always an interpretative decision based on the researcher's ability to actually explain what each means. Statistical measures such as eigenvalues and variance provide rules of thumb. Another guideline is that each factor should have at least two sorts significantly correlated with it—these "define" it because they most closely represent it and no other factor. Data "rotation" techniques such as varimax can help construct factors in such a way as to avoid "confounding" sorts that define more than one factor (which would imply that those factors are not particularly distinct patterns of thought). The qmethod package for the R software language provides relatively accessible functions for completing these calculations.

Q method provides no hard and fast way of determining the meaning of each factor, even after all of these statistical operations. Generally, researchers have a few approaches they can take. For each factor, researchers should look at the position it places each statement. One factor may place the statement "wildlife cameras violate animals' rights to privacy" in position 3 (strongly agree) while another may place it in position -3 (strongly disagree), indicating that the two factors differ on this issue. Software such as qmethod can identify "distinguishing" statements that significantly differentiate a factor from others. Likewise, researchers should look at consensus statements where all the factors concur—concurrence may be in the form of shared neutrality, agreement, or disagreement with a statement.

Ultimately, the goal in Q method is to use this quantitative information to qualitatively identify different perspectives and to narrate a story about each one. It isn't simply to say factor 1 strongly agrees with statements a, b, and c and disagrees with d, e, and f while factor 2 believes x, y, and z. Instead, researchers might "axially" code the statements to reduce them into higher order themes and find that factor 1 prioritises challenges to conservation technology (whether they agree or disagree with each specific challenge), while opportunities are more important to factor 2.

Why is Q method important?

Q method helps researchers generate surprising and useful insights into discourses and peoples' relation to them (Robbins and Krueger 2001). It provides a "bottom-up" approach to classifying the social world, as the researcher must induce meaning out of participants' choices about how to rank statements, rather than make deductive assumptions.

For instance, Q method might help to characterise the field of digital conservation beyond pre-defined categories. A reasonable hypothesis is that conservation organisation staff will worry about costs, government officials will be excited about making their work more efficient, and technology developers will want to sell their products. But Q method might illustrate how the more salient distinctions aren't these different job titles, but attitudes about ethics, costs, efficiency, and innovation that are actually shared within and between such actors.

In contrast, a conventional survey would treat these job titles as variables that correlate with responses. Such a survey might support claiming, for instance, that conservation organisation staff are more likely to be concerned about the ethics of using digital technologies. Although these kinds of claims may be statistically valid, they can pit groups of people against each other. A Q method approach provides a nuanced view of dynamics and processes in (environmental) governance in ways that might better support stakeholder-oriented co-management initiatives (Robbins and Krueger 2001).

Relationship of Q method with other methods

Q method is typically paired with social science methods such as interviews (see Johnston and Longhurst, Chapter 32). For instance, researchers may use themes derived from coding interview transcripts in order to structure their concourse of statements. Statements themselves may be modified quotations from interviewees. Researchers also often conduct interviews as part of the Q sort itself, asking respondents to explain the reasoning behind their sorts.

Q method can be fruitfully paired with critical discourse analysis (see Doel, Chapter 44). For instance, a critical discourse analysis approach can be applied to creating the concourse of statements by identifying and accounting for "inconsistencies, silences, and gaps" in the source material (Sneegas 2019: 5). Q method has also occasionally been utilised alongside participatory mapping (e.g., Hawthorne et al. 2008). In these studies, statements are typically spatialised in some sense. For instance, statements may refer to land uses that can be mapped when interpreting the meaning of each factor.

Ethical issues and Q method

The ethical issues that researchers must confront in Q method are similar to those in other social science methods, especially interviews and surveys. Researchers must take care to gain and maintain informed consent from their respondents. They also have a duty to steward respondents' data. Although Q survey data is rarely big data, it is still the case that "data are people" (Zook et al. 2017). Survey respondents could become de-anonymised depending on how much of the results are shared and how large the surveyed community is. Ethical research engagements also involve some degree of reciprocity with participants and Q researchers sometimes attempt to validate their interpretations by asking respondents for feedback.

Issues to be aware of in using Q method

There are three main analytical challenges to be aware of in using Q method: whether it alone can support claims about how prevalent a shared perspective is, whether it provides "unbiased" results, and whether it is appropriate to label respondents with the identified factors.

While Q surveys can characterise different perspectives on an issue, they cannot tell us anything about how prevalent these patterns of thought are. This is because Q method does not involve sampling from a population of potential respondents.

The choices researchers make—from what statements to include to which participants to invite and what statistical measures to use— mean that Q method never provides an unbiased view from nowhere. While there are useful rules of thumb for Q researchers, these shouldn't be followed without reflection, as if they alone will ensure a study is objective (Nost et al. 2019).

Arguably, the end goal of a Q study shouldn't be to label each participant as an "optimist", "pessimist," or other label derived from the factor analysis, but to consider how these identified perspectives shed light on debates and controversies surrounding a topic (Nost et al. 2019). People only partially subscribe to any set of constructed ideas or labels. This is clear in the correlations Q method statistics produce, where factors may only explain 60 or 70% of a participant's sort.

Suggested further reading

Watts, S. and P. Stenner. 2005. 'Doing Q methodology: Theory, method and interpretation', *Qualitative Research in Psychology*, 2, pp. 67–91 https://doi. org/10.1191/1478088705qp022oa. This is a substantial primer on Q method, providing a very helpful discussion of how and why to use it. Although written for psychologists, the authors' thoughts on developing a sample of statements, recruiting participants, conducting sorts, and analysing results will resonate with social scientists.

Webler, T., S. Danielson, and S. Tuler. 2009. 'Using Q method to reveal social perspectives in environmental research', *Social and Environmental Research Institute*, https://www.betterevaluation.org/sites/default/files/Qprimer. pdf. Another substantial primer on Q method, with a specific focus on environmental social science.

The Q-Method Testing and Inquiry Platform (Q-TIP), https://qtip.geography. wisc.edu/#/ provides researchers a way of designing and sharing Q surveys online and asynchronously.

The qmethod package for the R statistical software helps researchers process and analyse Q sort data, https://cran.r-project.org/package=qmethod

References cited

Doel, M., Chapter 44, this volume. 'Textual analysis'.

Hawthorne, T., J. Krygier, and M.-P. Kwan. 2008. 'Mapping ambivalence: Exploring the geographies of community change and rails-to-trails development using photo-based Q method and PPGIS', *Geoforum*, 39.2, pp. 1058–1078. https://doi.org/10.1016/j.geoforum.2007.11.006

Johnston, L. and Longhurst, R., Chapter 32, this volume. 'Interviews: Structured, semi-structured and open-ended'.

Nost, E., M. Robertson, and R. Lave. 2019. 'Q-method and the performance of subjectivity: Reflections from a survey of US stream restoration practitioners', *Geoforum*, 105, pp. 23–31. https://doi.org/10.1016/j.geoforum.2019.06.004

Robbins, P. and R. Krueger. 2000. 'Beyond bias? The promise and limits of Q method in human geography', *The Professional Geographer*, 52.4, pp. 636–648. https://doi.org/10.1111/0033-0124.00252

Sneegas, G. 2019. 'Making the case for critical Q methodology', *The Professional Geographer*, 72.1, pp. 78–87. https://doi.org/10.1080/00330124.2019.1598271

Zook, M., S. Barocas, D. Boyd, K. Crawford, E. Keller, S.P. Gangadharan, A. Goodman, R. Hollander, B.A. Koenig, J. Metcalf, A. Narayanan, A. Nelson, and F. Pasquale. 2017. 'Ten simple rules for responsible big data research', ed. F. Lewitter. *PLOS Computational Biology*, 13.3, p. e1005399. https://doi.org/10.1371/journal.pcbi.1005399

38. Sampling

Nicolena vonHedemann

Definition

Sampling is the selection of a subset of a defined population to investigate. Sampling methods are guided by the research questions and are used because collecting data on the entire population is not feasible.

The basics of sampling

Most environmental research—both biophysical and social—requires thoughtful sampling to provide data that can answer research questions. Choosing a sampling strategy occurs after developing research questions, determining what could answer these questions, and deciding where these data can be obtained. The selection of sampling approaches is influenced by many factors, including existing literature, training programmes, structures of power, funding institution preferences, perceived data collection feasibility (funding, logistics, and time constraints), and researcher characteristics. Carefully considering how these different factors influence the selection of a study population and sampling approach may lead researchers to include new questions, approaches, people, or locations previously marginalised in research.

Once the study population has been chosen, a sampling frame should be determined that specifies entities in the population "who have a chance to be included in the sample" (McLafferty 2016: 137). Some prior knowledge about the study population's structure is necessary to determine a sampling approach because the sampling method should capture a variety of characteristics of interest but should not sample so broadly that little can be concluded from a small sample

  https://doi.org/10.11647/OBP.0418.38

size of subgroups. Additionally, the methods used to collect data can affect sampling. For example, internet-based surveys in English exclude people who do not speak English and do not have consistent internet access; additional effort must be put into including these populations if that is desired for answering the research questions (McLafferty 2016). The overall sample size is based on the research question, the level of heterogeneity and total size of the study population (the sample size will need to be larger if there is more variation), the level of precision desired, and resources available.

Types of sampling strategies

There are two major types of sampling methods: non-probability methods and probability-based methods.

Non-probability methods include case studies (Lane, Chapter 24), purposive sampling, convenience sampling, snowball sampling, and quota sampling. These methods are more common in qualitative social science approaches, but several can be used in biophysical research as well. While these approaches do not enable the researcher to utilise inferential statistics (Lane, Chapter 42) to make probabilistic statements about the broader population, they remain powerful for answering certain types of research questions. These methods can facilitate understanding causal phenomena and providing nuanced context or be used in cases where probability methods would be infeasible or culturally inappropriate.

Case studies are in-depth examinations of one or a few "cases" (i.e., spatially or socially defined units such as cities, forests, or internet chatrooms; see Lane, Chapter 24). Case studies are useful to gather scoping information, test hypotheses, explain mechanisms, provide information for local use, and understand local context in a greater depth (Rice 2010).

Purposive or judgment sampling is choosing entities for inclusion in the sample based on what the researcher believes is representative of the study population or to understand particular characteristics of interest. While purposive sampling relies on researcher discretion and knowledge (or assumptions) about the population to determine who or what should be sampled, certain research questions can only be answered through purposive sampling. For example, in seeking to understand forest management practices across land jurisdictions, the researcher might interview forest managers with expertise on different types of land tenure and major forested ecosystems based on the assumption that ecosystem type and ownership unit are major factors affecting forest management. Purposive sampling enables the researchers to home in on certain aspects of interest (interviewees knowledgeable about a topic) and can function in locations where true random selection (see probability methods below) may not be logistically feasible.

Convenience or accessibility sampling is based on collecting data from the entities that were most accessible to the researcher (i.e., interviewing everyone standing in a line). This will likely give a sample that is not representative of the whole population (unless the population is homogenous) but may be worth the minimal effort (Jensen and Shumway 2010; Rice 2010).

Snowball sampling is utilised when the target population is harder to reach. It involves asking previously sampled people for recommendations for research participants who could provide the desired information. This approach is useful when an introduction of the researcher into the social system will make research subjects more likely to trust the researcher or when other methods for contacting research participants are not clear (i.e., no published contact information). For example, if the researcher is an outsider seeking to understand the experiences of a marginalised population that has faced violence and thus fears non-community members, snowball sampling may be the best sampling method to facilitate introductions into the community. This will bias the sample towards participants who are more connected to other participants, but in some cases snowball sampling may be the only feasible and ethical approach.

Quota sampling requires establishing quotas, or subgroups that the researcher wishes to sample that are mutually exclusive and designated by characteristics that are thought to be an important influence on the data (Jensen and Shumway 2010; Rice 2010). For example, the researcher may want to sample a certain number of men, women, and other genders from different age groups based on the general structure of the population. The researcher must use their judgment to determine in advance which population characteristics are likely to matter to the data to create the quotas, and then continue sampling with one of the above other non-probability methods until each quota is full.

Probability methods permit the researcher to use inferential statistics (Lane, Chapter 42) to develop generalisations about the broader population since all entities within the sampling frame have an equal chance of being selected (Jensen and Shumway 2010; Rice 2010). This approach aims to reduce bias and produce a sample representative of the study population as a whole. Probability methods include random sampling, systematic sampling, and stratified sampling and often require larger sample sizes to generate conclusions.

In random sampling, the samples are randomly chosen through random generation of coordinates, phone numbers, addresses, or numbers assigned to an enumerated population. For a systematic sampling approach, the researcher determines a rule for sampling so the selection will not be dependent on researcher discretion but rather based on regular sampling intervals. For example, the depth of fine fuel matter (litter) can be measured every 10 cm along a transect, or every 3rd house in a neighbourhood can be surveyed. A stratified sampling approach, similar to quota sampling, recognises distinctive subgroups within a population that should be sampled (i.e., gender groups), and gathers a representative sample within each strata using random or systematic sampling (Jensen and Shumway 2010; Rice 2010). Researchers need to justify the process of defining strata when using this method.

Challenges, concerns, and ethical issues

A careful sampling approach will produce a sample that more accurately represents the population and avoid collecting a biased sample that leads to misleading or meaningless conclusions (Jensen and Shumway 2010; Johnston et al. 2000; Rice 2010). Researchers should also take care to utilise equipment and protocols used to collect data properly, follow predefined methods designed to reduce bias, and sample the heterogeneity existing in the defined study population (Jensen and Shumway 2010; Rice 2010). Additionally, the researcher should explicitly justify why they chose a particular method as well as the potential limitations it brings to the work. Every method—even well-designed random and systematic sampling procedures—has some level of bias because it is impossible for researchers to completely eliminate the integration of their perspectives in data collection. Additionally, ethical considerations mean that random selection cannot truly be achieved in some human research because, for example, random selection may not align with the values of the study population or randomly selected locations may be inaccessible (i.e., if landowners do not have the opportunity to give informed consent to grant access to their property).

Suggested further reading

For more detailed information on probabilistic methods, see Jensen and Shumway's 'Sampling our world' in *Research Methods in Geography: A Critical Introduction* or Rice's 'Sampling in geography' in *Key Methods in Geography*, 2nd edition. For a focus on survey sampling, see McLafferty in *Key Methods in Geography*, 3rd edition.

References cited

Jensen, R. and J.M. Shumway. 2010. 'Sampling our world', chapter 39 in *Research Methods in Geography: A Critical Introduction*, ed. by J.P. Jones III and B. Gomez (Wiley-Blackwell), pp. 77–90.

Johnston, R.J., D. Gregory, G. Pratt, and M. Watts. 2000. *The Dictionary of Human Geography* (Blackwell).

Lane, S.N., Chapter 24, this volume. 'Case studies'.

Lane, S.N., Chapter 42, this volume. 'Statistical inference'.

McLafferty, S.L. 2016. 'Conducting questionnaire surveys', in *Key Methods in Geography*, ed. by N. Clifford, M. Cope, and T. Gillespie (Sage Publications).

Rice, S. 2010. 'Sampling in geography', in *Key Methods in Geography*, ed. by N. Clifford, M. Cope, and T. Gillespie (Sage Publications), pp. 230–252.

39. (Critical) Satellite Remote Sensing

Andreas Ch. Braun

Definition

Remote sensing (RS) refers to the set of techniques used to obtain information about the Earth's surface by measuring and interpreting the electromagnetic waves emitted or reflected from it.

The basics of remote sensing

RS involves obtaining information about the Earth without touching it. Several distinctions are important. Active RS emits radiation to the Earth and measures its reflection. This includes laser (LiDAR) or radar (SAR, InSAR) radiation. One uses active RS mainly when the morphology of the Earth is of interest, or cloud cover is a problem (some radar waves penetrate clouds). Passive RS measures the reflected electromagnetic radiation from the sun. It is used when spectral properties (e.g., land cover) are of interest. Specific wavelength ranges (e.g., UV, visible light, thermal radiation) are grouped into bands, each band yielding a grayscale image. Passive RS data are acquired from satellites, aircraft or drones. These differences in platform, and also in sensors (e.g., camera or scanner), affect the quality, resolution and number of bands obtained. Multispectral data have few (max. 15) bands; hyperspectral data have many hundreds of bands leaving no gaps between them in the wavelength range. From the evaluation point of view, passive RS can be roughly divided into classifier-based techniques, which divide the image dataset into discrete (land use) classes. Index-based techniques use band values to estimate biophysical parameters (e.g., chlorophyll content) as continuous indices.

 https://doi.org/10.11647/OBP.0418.39

Remote sensing in depth

To use RS, a basic goal is to understand what it is you are trying to map and what kind of RS can provide that information. If the morphology of the Earth's surface is of interest or if clouds are a problem (e.g., in the tropics), one chooses active RS; if land cover is of interest, then one chooses passive RS. The data acquired using passive RS vary in the spatial extent of the imaged section of the Earth and in terms of resolution. Spatial resolution is the size of an image point (pixel) on the ground (0.5 m to kms, although developments in sensor technology are progressively reducing this lower bound). Spectral resolution is the number of bands (from one to hundreds). Temporal resolution is the revisit time of the sensor (which can be user-controlled for drones flown by an operator, through days for repeatedly-passing satellites to many years such as for national-scale LiDAR mapping). The temporal range of data availability is of additional interest (satellite data are available since the 1970s, historical aerial photographs since the middle of the 20th Century) and is where RS is integral to change detection at the Earth's surface. Radiometric resolution is the bit rate of the data set. Mostly (besides the availability and cost of the data) the first three resolution types are relevant. It is important not to choose automatically the highest resolution, because higher resolution always means higher complexity in the dataset and may lead to data management problems (for example, in 30 m spatial resolution data you might see single roofs within the urban space; in 0.5 m data every photovoltaic device, chimney and sunlounger are visible, and create more complexities when only 'roof/non-roof' is relevant). Thus, it is important to choose the resolution that is appropriate to the research question. Moreover, a tradeoff between the different resolutions has to be made. For long-term change detection, for instance, it may be better to take Landsat data, which are available with a moderate 30 m to 80 m resolution since 1972; taken in principle every two weeks; and are free of charge, than e.g., data with higher spatial resolution, which are expensive and rarely available historically. In addition, one should check whether one really needs to undertake one's own analysis, or whether data providers such as the United States Geological Survey (USGS) do not already offer ready-made products that are sufficient for the research question. Before starting a complicated analysis (possibly even programmed by oneself) one should check if online applications like the Google Earth Engine provide a sufficiently accurate analysis with a few simple clicks. If you do need your own analysis, don't forget to preprocess the data appropriately (e.g., by performing radiometric and geometric corrections). Then decide whether classes (forest, agriculture, city) are more relevant; if so, choose classifier-based techniques (multispectral data are often better). If biophysical parameters are relevant, choose index-based techniques (hyperspectral data are often better). The respective analytical algorithms are a large and very dynamic field of research. Note that if you are using RS as part of a project rather than actively working to advance RS as a methodology, you should not necessarily choose the latest methodology that the RS community is currently celebrating. This often requires programming skills and lots of computational time (hampering the number of possible repetitions of the analysis). Instead, the most advanced state-of-the-art approach within one's own applicant community (e.g., vegetation- or hydro-geography) should be chosen.

Why is remote sensing important

RS has three major advantages with respect to geography and environmental science.

The first advantage is that RS can show phenomena that are not visible in the field. These are historical conditions (e.g., a former city perimeter) in older data; electromagnetic spectra not visible to the eye (e.g., IR radiation, which is highly relevant to plant-health analyses); and large-scale spatial patterns that are not immediately visible in the field (see Figure 39.1).

Fig. 39.1 Example of a skyline logging deforestation pattern in Tierras Bajas, Bolivia This pattern could not be easily recognized in terrain, https://commons. wikimedia.org/wiki/File:Bolivia-Deforestation-EO.JPG.

The second advantage is that point observations of terrain geography can be interpolated over large areas (e.g., RS-supported erosion modeling or biodiversity estimation). The third advantage is that RS visualizes areas to which access is not possible in the field (e.g., due to impassable terrain, or political/legal prohibitions on access).

Ethical issues and remote sensing

An interesting debate has emerged in recent years about the (political) epistemology of RS, which is important for develop a critical RS (Bennett et al. 2022). The results of RS have for too long been treated uncritically as complete, linear, objective representations of the environment that are independent of and do not affect the socio-political context. None of these assumptions is true. RS results are tainted with apparent objectivity. However, Feilhauer et al. (2012) show that the discrete and categorical representation in RS results often fails to reflect gradual and continuous natural conditions. Moreover, the representation is by no means independent of context. What an RS image does and does not show depends on the maker of the image, their objectives and intentions, values and practices (Bennett et al. 2022, Braun, 2021, 2024). Thus, they are constructed to exactly the same degree that other research findings are. Equally, what is being measured in a RS study often has no "say" in whether or not it wishes to be measured. This may be of particular concern when RS measures humans and their environment. The false assumption that RS results are completely objective and apolitical thus becomes dangerous. In fact, these results are both influenced by the socio-political context and influence it. Nagaraj (2022) shows, for example, how the free availability of RS data and analytics drastically increases resource findings and hence the political struggles associated with them. If RS is to be used appropriately in environmental science, contextuality on the one hand, and the impact on what is being studied on the other, must be critically and reflexively incorporated.

Issues to be aware of in using remote sensing

The first major disadvantage is that geography uses different spatial concepts (e.g., container space, relational space, perception space, constructed space). Of these, RS can directly handle only one, container spaces, which are given a geographic/geodetic X,Y coordinate. The second major drawback is that RS can directly represent only patterns, but not processes. These have to be estimated or reconstructed from RS data, e.g., from time series. Such estimation or reconstruction may be possible, after appropriate data processing, but not always. The third and biggest disadvantage is that RS reduces the world to its spectral properties and generally at the surface. Many elements that are of interest to environmental scholars are simply not represented. In particular, social relations can be analyzed only when they materialize (e.g., through construction or destruction). What happens below a forest canopy, for e.g., may only be partially visible to a sensor, if at all (Walters et al., Chapter 13; Kelley, Chapter 19). These aforementioned advantages and disadvantages can help to critically assess the value of RS for one's own research question.

Suggested further reading

An excellent, freely available and extremely comprehensive textbook on remote sensing is the text by Tempfli et al. (2009). Although it is already 14 years old and thus does not pick up on some of the latest developments in the dynamic research field of remote sensing, it is still suitable for learning about the research field as a whole. It does not, however, include the critical epistemological discourse that has so far only been carried out in papers such as the one given here.

Tempfli, K., Huurneman, G. C., Bakker, W. H., Janssen, L. L. F., Feringa, W. F., Gieske, A. S. M., Grabmaier, K. A., Hecker, C. A., Horn, J. A., Kerle, N., van der Meer, F. D., Parodi, G. N., Pohl, C., Reeves, C. V., van Ruitenbeek, F. J. A., Schetselaar, E. M., Weir, M. J. C., Westinga, E., & Woldai, T. (2009). *Principles of Remote Sensing: An Introductory Textbook*. (ITC Educational Textbook Series; Vol. 2). International Institute for Geo-Information Science and Earth Observation. http://www.itc.nl/library/papers_2009/general/PrinciplesRemoteSensing.pdf

References Cited

Bennett, M. M., Chen, J. K., Alvarez Leon, L. F., & Gleason, C. J. (2022). 'The Politics of Pixels: A Review and Agenda for Critical Remote

Sensing.' *Progress in Human Geography, 46*(3), 729–752. https://doi. org/10.1177/03091325221074691

Braun, A. C. (2024). 'More accurate less meaningful? Why quality indicators do not unveil the socio-technical practices inscribed into land use maps.' *Progress in Physical Geography: Earth and Environment, 48*(3), 343–367. https://doi.org/10.1177/03091333241248055

Braun, A. C. (2021). 'More accurate less meaningful? A critical physical geographer's reflection on interpreting remote sensing land-use analyses.' *Progress in Physical Geography: Earth and Environment, 45*(5), 706–735. https://doi.org/10.1177/0309133321991814

Feilhauer, H., Zlinszky, A., Kania, A., Foody, G. M., Doktor, D., Lausch, A., & Schmidtlein, S. (2021). 'Let your maps be fuzzy!—Class probabilities and floristic gradients as alternatives to crisp mapping for remote sensing of vegetation.' *Remote Sensing in Ecology and Conservation, 7*(2), 292–305. https://doi.org/10.1002/rse2.188

Kelley, L.C., Chapter 19, this volume. 'Engaging remote sensing and ethnography to seed alternative landscape stories and scripts'.

Nagaraj, A. (2022). 'The private impact of public data: Landsat satellite maps increased gold discoveries and encouraged entry.' *Management Science, 68*(1), 564–582.

Walters, G., et al., Chapter 13, this volume. 'Revealing the social histories of ancient savannas and intact forests using a historical ecology approach in Central Africa'.

40. Social network analysis

Stephen M. Chignell

Definition

Social network analysis (SNA) is a form of analysis that aims to understand social relationships and the structures they produce using the mathematical principles of graph theory. It is primarily associated with social science research but many of its concepts and tools are being adopted and adapted by biophysical scientists and interdisciplinary researchers.

The basics of social network analysis

SNA researchers study social relations by conceptualising actors as points (called 'nodes' or 'vertices') and relations as lines (called 'ties', 'links', or 'edges'). Together, these nodes and ties form a social network, which can be described and analysed using a variety of metrics. Researchers often create network visualisations (or 'graphs') using layout algorithms designed to highlight different structural aspects and aid interpretation. SNA forms a major part of the broader field of network science, which uses graph theory techniques to understand a range of social and natural phenomena.

Social network analysis in depth

The relational approach of SNA differs from traditional social science approaches in that its foundational assumption is that structure matters. In this context 'structure' refers to an emphasis on explaining a node's outcomes or characteristics based on its position in its surrounding

    https://doi.org/10.11647/OBP.0418.40

environment, rather than as a function of the node's other characteristics (Borgatti et al. 2009). The structural properties of the whole network are also of interest, and SNA researchers often analyse node-level and network-level structural characteristics at the same time.

As a field, SNA is dominated by quantitative approaches. However, network models are fundamentally incompatible with some of the central assumptions of classical statistical modelling, namely that observations are randomly selected from a population and are statistically independent from each other (Tindall et al. 2022). SNA researchers have thus developed a suite of statistical techniques for characterising networks, including node-level connectivity metrics like Degree (i.e., number of ties) to more complex metrics like Betweenness Centrality (i.e., how many times a node falls on the shortest path between other nodes). There are also metrics for describing the whole network, such as clustering algorithms that identify communities of nodes based on their relations. Most of these statistics can also measure and visualise changes in network metrics and structure over time. Many can also be used to analyse more complex networks possessing more than one type of node and/or tie.

Before the 1990s, SNA largely focused on face-to-face relationships. However, the rise of the Internet and increasing use of digital social networks, alongside advances in computing has produced new datasets and lines of questions for SNA researchers (Tindall et al. 2022). In the last decade, there has been a shift in interest from description to prediction, with researchers conducting longitudinal statistical analyses of network ties to understand how and why social relations and structures come to be (Stadtfeld et al. 2018). An emerging relationalist tradition sees relationships as (dis)assembled and shaped by social context (Erikson 2013; Jaspersen and Stein 2019); these scholars often emphasise qualitative approaches focused on the embeddedness of single actors in their social environment, rather than attempting to capture and analyse the whole network (Crossley et al. 2015; Borgatti et al. 2018). Mixed qualitative-quantitative approaches are also increasingly popular, and combine the well-developed techniques of quantitative SNA with qualitative data collection and visual interpretation of network graphs (Venturini et al. 2021; Chignell et al. 2022; Chignell and Satterfield 2023).

Why is social network analysis important?

SNA is a rich and varied tradition, with nearly 100 years of history. It provides a well-established and ever-growing suite of techniques to empirically study a range of pressing sociological questions. With the rise of the Internet and information society, social networks have become ubiquitous; from the digitisation of public discourse via social media, to contact tracing in public health, to studies of scientific collaboration, networks are everywhere, framed as both threat and solution to societal concerns (Hogan 2021).

Relationship of social network analysis with other methods

Interdisciplinary and collaborative research increasingly uses the common language and techniques of network analysis. For example, SNA is used with geographic information systems (GIS) to explore network structure in a geospatial context (Ye and Andris 2021). SNA is often paired with interviews (see Longhurst and Johnson, Chapter 32), workshops, and participatory mapping; these serve as both sources of network data and qualitative information that can help to contextualise quantitative network analyses (Schiffer and Hauck 2010; Hayat and Lyons 2017). Other approaches, such as discourse network analysis and semantic network analysis, combine the quantitative metrics of SNA with qualitative content analysis, linking actors based on the concepts they discuss (Eder 2022; see Doel, Chapter 44). This can be used to understand important and marginalised themes in a discourse, levels of polarisation in a community, and how the socio-conceptual structure of a debate changes over time. SNA is also increasingly combined with historical, ethnographic (see Sayre, Chapter 34), and journalistic approaches, which use network graphs to explore and communicate narratives, frames, and their changes over time (Bounegru et al. 2017; Painter et al. 2019; Chignell and Satterfield 2023). Finally, socio-ecological network analysis combines sociological data with biophysical data to explore questions about human-non-human interactions (Sayles et al. 2019).

Ethical issues and social network analysis

Because SNA is ultimately interested in people and their relations, it raises similar ethical concerns as any research dealing with human subjects. Network projects eliciting data from interviews or surveys must undergo prior approval through institutional ethics boards and follow all related data management and privacy protocols. SNA research using secondary data scraped from the Internet or downloaded from digital databases (e.g., bibliographic data from Web of Science) can have significant gaps and biases which should be considered.

In all cases, it is important to recognise the power and performativity of SNA. Networks do not simply reflect the world, they change it, and often in unexpected ways: "Through the discretisation of social life, when we draw boundaries...we not only comment on social life, we reimagine it. And in that reimagining are threats and opportunities. We do not merely expose the powerful with networks so much as articulate the structures that can reproduce a system of power in ways both abstract and concrete" (Hogan 2021). Indeed, creating and analysing networks can have real-world impacts on the things they represent. The notion of structural holes, for example, changed the ways businesses manage their employees, and bibliometric network techniques are used and misused by university administrators and publishers in the political economy of academic rankings and impact evaluation (Chignell 2023). This power calls for "network response-ability", to "recognize, acknowledge, and respond to differences in how network analysis and visualization is put into practice" (van Geenen et al. 2023).

Issues to be aware of in using social network analysis

Many network metrics assume that the network being analysed is complete. However, this is rarely the case in social networks, and defining the boundary of the network is a key decision point in the analytical process. Most SNA scholars circumvent this issue through pragmatic boundary selection (e.g., administrative or geographic boundaries), but this remains an unsettled theoretical issue.

Network metrics and visualisations can seem objective but, like all models, have assumptions embedded in them. It is thus important to be cognizant and transparent about the decisions you make while constructing networks and to avoid atheoretical interpretation.

Because network visualisations are ubiquitous and visually engaging, it can be tempting to think of them as 1:1 models of the social networks they represent. Remember that network graphs are ultimately simplifications of the multi-faceted real-world networks we seek to understand.

What data you assign as nodes and links is non-trivial and has consequences for the patterns produced by downstream analyses. Is your basic unit of analysis an individual, an institution, a city, or something else? What is the relationship you are trying to understand? Think through this step carefully in the context of your data and research question.

Be aware that the term 'network' is used in different traditions with very different meanings and theoretical assumptions. Be careful in directly applying SNA techniques to other fields based on the homonym 'network'.

Suggested further reading

Gephi is a free and open-source network analysis and visualisation software with a large user community and tutorials, https://gephi.org/

Martin Grandjean (Université de Lausanne) has produced a great series of brief YouTube videos introducing social network analysis, https://www.youtube.com/watch?v=lnLW6ITFY3Mandlist=PL4iQXwvEG8CQSy4T 1Z3cJZunvPtQp4dRy, as well as a tutorial for how to do basic network analysis and visualisation using Gephi, https://www.youtube.com/watch?v=GXtbL8avpikandt=136s

References cited

Borgatti, S.P., M.G. Everett, and J.C. Johnson. 2018. 'Ego Networks', in *Analyzing Social Networks*, ed. by S.P. Borgatti, M.G. Everett, and J.C. Johnson (Sage Publications), pp. 305–329.

Bounegru, L., T. Venturini, J. Gray, and M. Jacomy. 2017. 'Narrating networks', *Digital Journalism*, 5.6, pp. 699–730. https://doi.org/10.1080/21670811.2016.1186497

Chignell, S.M. 2023. 'A missing link? Network analysis as an empirical approach for critical physical geography', *The Canadian Geographer / Le Géographe canadien*, 67.1, pp. 52–73. https://doi.org/10.1111/cag.12767

Chignell, S.M. and T. Satterfield. 2023. 'Seeing beyond the frames we inherit: A challenge to tenacious conservation narratives', *People and Nature*, 5.6, pp. 2107–2123. https://doi.org/10.1002/pan3.10550

Chignell, S.M., A. Howkins, P. Gullett, and A.G. Fountain. 2022. 'Patterns of interdisciplinary collaboration resemble biogeochemical relationships in the McMurdo Dry Valleys, Antarctica: a historical social network analysis of science, 1907–2016', *Polar Research*, 41. https://doi.org/10.33265/polar.v41.8037

Crossley, N., E. Bellotti, G. Edwards, M.G. Everett, J. Koskinen, and M. Tranmer. 2015. *Social Network Analysis for Ego-Nets* (Sage Publications). https://doi.org/10.4135/9781473911871

Doel, M., Chapter 44, this volume. 'Textual analysis.'

Eder, F. 2022. 'Discourse network analysis', in *Routledge Handbook of Foreign Policy Analysis Methods*, ed. by P.A. Mello and F. Ostermann (Taylor and Francis). https://doi.org/10.4324/9781003139850-39

Erikson, E. 2013. 'Formalist and relationalist theory in social network analysis', *Sociological Theory*, 31.3, pp. 219–242. https://doi.org/10.1177/0735275113501998

van Geenen, D., J.W.Y. Gray, L. Bounegru, T. Venturini, M. Jacomy, and A. Meunier. 2023. 'Staying with the trouble of networks', *Frontiers in Big Data*, 5. https://doi.org/10.3389/fdata.2022.510310

Hayat, T. and K. Lyons. 2017. 'A typology of collaborative research networks', *Online Information Review*, 41.2, pp. 155–170. https://doi.org/10.1108/oir-11-2015-0368

Hogan, B. 2021. 'Networks are a lens for power: A commentary on the recent advances in the ethics of social networks special issue', *Social Networks*, 67, pp. 9–12. https://doi.org/10.1016/j.socnet.2020.12.003

Jaspersen, L.J. and C. Stein. 2019. 'Beyond the matrix: visual methods for qualitative network research', *British Journal of Management*, 30.3, pp. 748–763. https://doi.org/10.1111/1467-8551.12339

Longhurst, R. and Johnston, L., Chapter 27, this volume. 'Focus groups.'

Painter, D.T., B.C. Daniels, and J. Jost. 2019. 'Network analysis for the digital humanities: Principles, problems, extensions', *Isis*, 110.3, pp. 538–554. https://doi.org/10.1086/705532

Sayles, J.S., M.M. Garcia, M. Hamilton, S.M. Alexander, J.A. Baggio, A.P. Fischer, K. Ingold, G.R. Meredith, and J. Pittman. 2019. 'Social-ecological network analysis for sustainability sciences: a systematic review and innovative research agenda for the future', *Environmental Research Letters*, 14.9. https://doi.org/10.1088/1748-9326/ab2619

Sayre, N.F., Chapter 34, this volume. 'Participant observation and ethnography.'

Schiffer, E. and J. Hauck. 2010. 'Net-map: Collecting social network data and facilitating network learning through participatory influence network mapping', *Field Methods*, 22.3, pp. 231–249. https://doi. org/10.1177/1525822x10374798

Stadtfeld, C., A. Vörös, T. Elmer, Z. Boda, and I.J. Raabe. 2018. 'Integration in emerging social networks explains academic failure and success', *Proceedings of the National Academy of Sciences*. https://doi.org/10.1073/pnas.1811388115

Tindall, D., J. McLevey, Y. Koop-Monteiro, and A. Graham. 2022. 'Big data, computational social science, and other recent innovations in social network analysis', *Canadian Review of Sociology/Revue canadienne de sociologie*, 59.2, pp. 271–288. https://doi.org/10.1111/cars.12377

Venturini, T., M. Jacomy, and P. Jensen. 2021. 'What do we see when we look at networks: visual network analysis, relational ambiguity, and force-directed layouts', *Big Data and Society*, 8.1. https://doi. org/10.1177/20539517211018488

Ye, X. and C. Andris. 2021. 'Spatial social networks in geographic information science', *International Journal of Geographical Information Science*, 35.12, pp. 2375–2379. https://doi.org/10.1080/13658816.2021.2001722

41. Soil toxicological analysis

Salvatore Engel-Di Mauro

Definition

Soil toxicological analysis refers to the interpretation of laboratory and/or field tests done on soil samples to determine the concentrations of organic and/or inorganic substances or elements that, depending on the form, intensity, and duration of exposure, may lead to the premature mortality or adverse health effects for humans and/or other beings or compromise ecosystem functions (see Malone, Chapter 16). Soil toxicology is associated with biophysical research, especially at the intersections of the environmental, health, soil, biological, and ecological sciences.

The basics of soil toxicological analysis

The objective of soil toxicological analysis is to determine the potential harm for given levels of different substances pre-existing or accumulated in soils. It is an assessment of toxicity arising from human impacts, which have become primary causes of soil contamination. As part of the analysis, one must control for toxicity levels originating from material out of which soils formed (geogenic sources) and/or added to soils over time from non-human sources. A main part of the analysis consists in identifying which soil and related ecological components hold what amount of which toxic substance, namely, soil mineral particle surfaces (especially clays and amorphous minerals), soil organic matter, soil water, soil air, and organisms living in and on soils (e.g., bacteria, earthworms, fungi, plants). This helps determine which organisms

 https://doi.org/10.11647/OBP.0418.41

could be most affected, the residence time and potential fate of different chemicals, the degree of availability and accessibility of the different substances for different organisms, among other issues related to toxicological assessment. Chemical and physical tests on soil samples are ideally supplemented by biological assays, developing models from dose-response tests on a variety of organisms to evaluate the degree of human health hazard or wider ecological ramifications.

Soil toxicological analysis in depth

Many kinds of substances or elements exist that can be harmful to us or other organisms. Harmful substances or elements are broadly classified as heavy metals or metalloids, solvents and vapours, radiation and radioactive materials, dioxins and furans, and pesticides. Toxic substances released by organisms are categorised according to the organism releasing such substances. There are many other environmental sources of harmful substances or elements, such as volcanic eruptions (e.g., hydrogen sulphide), large fires (e.g., polycyclic aromatic hydrocarbons from wood combustion), or sedimentary strata (e.g., lead from pozzolanic ash deposits). However, most soil toxins, degradable and persistent, originate from industrialised production systems since at least the late 1700s.

Determining whether and when a substance or element is toxic (soil-polluting) presents some challenges. It generally depends on the amount and characteristics of a substance or element, environmental conditions, the species (or type of organisms) considered, the level of accessibility to specific organisms, and how easily the toxin can be assimilated into (the dose) and expelled from an organism's body. A substance or element is considered a soil contaminant when it has the potential to undermine organisms' health or the functioning of an ecosystem. Soil contamination is when the concentration of an element or substance is higher than background levels, while soil pollution implies levels deemed harmful for humans, a given set of organisms, and/or an ecosystem.

Results of toxicological analysis are typically compared to established soil quality standards and/or soil screening levels that often vary among and even within countries (see Malone, Chapter

16 for related issues). These standards and screening levels specify concentrations for different elements and substances deemed to pose risks to human and/or other organisms' health and/or to ecosystem functioning based on assumed forms and degree of exposure and known toxicity data. Soil-borne contaminant exposure pathways include inhalation, ingestion, and skin absorption. The relative importance of each varies over time and space since, especially in areas frequently impacted by human activities, there are also indirect forms of exposure, such as consuming food produced on polluted soils or contaminated by lodged soil particles from polluted soil.

Exposure pathway processes are complicated by synergistic and/ or antagonistic interactions among pollutants and by variable soil environmental conditions (e.g., reduction-oxidation reactions, pH, microbial activities, root exudates, etc.) that can temporarily attenuate, amplify, or neutralise pollutants or, in the case of organic pollutants, degrade them to harmless or at times even more toxic by-products. Given pathway variability and contingency, additional investigations on exposure and bioaccessibility should be conducted to evaluate hazard levels and risk, which must also be related to degrees of vulnerability and capacity in an affected community.

Why is soil toxicological analysis important?

Determining soil toxicity levels is basic to general assessments of mainly terrestrial environment pollution levels. Soil toxicological analysis helps detect sources of contamination impacting other processes like air quality, food systems/webs, water supplies, and near-coastal ecosystems. It is thereby important to efforts promoting human and/or other organisms' health as well as ecosystem functioning. As terrestrial environmental pollution may be hidden intentionally, or simply not known about, soil toxicological analysis may be important in identifying instances of environmental injustice.

Relationship of soil toxicological analysis with other methods

Soil toxicological analysis necessitates, at the very least, an understanding of the characteristics and dynamics particular to the contaminated or polluted soils. Yet the interpretation of findings requires studying of wider ecological and social context and implications. This is because a substance or element found in soil is toxic in relation to the characteristics of the organism or organisms considered, to the biophysical interactions within and beyond soil that accentuate or attenuate a contaminant's diffusion and accessibility to organisms, and to the specific circumstances or environmental conditions lived by the organism or organisms affected. These factors need to include social relations that result in some people being exposed more than others to soil-borne toxicities.

A decisively influential factor, often ignored, is the social setting of the scientific investigation, from conception through interpretation of results and policy recommendations. Such activities are always shaped by social background and position, as in any human endeavour, so there ought to be explicit critical self-reflection relative to the political ramifications of findings, the way in which data are divulged, and with whom such data are shared. Hence, an evaluation of soil toxicological analysis from a critical social science perspective is important.

Further, given that biophysical interactions affecting contamination processes in most cases involve human action, studying the social forces behind the introduction and spread of contaminants in soils is necessary to address contaminant sources and effects. At a minimum, social science methodologies such as participatory approaches (see Mokos, Chapter 36), questionnaire surveys and/or interviews (see Johnston and Longhurst, Chapter 32; Winata and McLafferty, Chapter 43), and studies of local land use histories aid in evaluating contamination sources and differential health effects and vulnerabilities (as linked to social relations of power and modes of oppression). Equally important are perspectives and methodologies from other biophysical sciences. For instance, ecological research on local biological communities can reveal potentials for ecosystem cascading effects and geochemical investigations can help distinguish environmental from human sources of contaminants. Put differently, soil toxicological analysis alone cannot form the basis of toxicity assessment without perspectives and methodologies from the social sciences and other biophysical sciences.

Ethical issues and soil toxicological analysis

When studying the levels of toxicity in soils, there are at least four major issues to keep in mind, aside from considerations of other organisms' populations' survivability and overall ecosystem functioning. The first is gaining the confidence of those who might be impacted by soil toxicity and communicating with them in an accessible language (Malone, Chapter 16). Workshops focused on sharing technical knowledge are critical. A second issue is that there should be discretion about results. The form and extent of data disclosure should be discussed with the community directly affected before any soil sampling occurs, and the researcher must be careful to avoid harming the community with their findings (e.g., when findings of contamination are used to justify paving over a community garden and appropriating the land for other uses). To avoid this, the researcher should make sure they are informed about local concerns and power relations and consult frequently with the communities implicated in their research. A third concern is the potential for creating undue alarm relative to results, which can undermine the well-being of a community directly affected. Results should always be considered in relation to other pollution sources and health hazards, and to existing soil pollutant exposure levels, which may be minimal. This is linked to a fourth ethical matter of providing guidance on exposure mitigation or avoidance, which should be discussed from the beginning with the community involved.

Issues to be aware of in using soil toxicological analysis

Total element or substance data do not denote the level or form of exposure, nor the proportion of the pollutant concentration that is bioavailable (that is, in forms that can be accessed by an organism) and bioaccessible (that is, that can be metabolised and bioaccumulated). The amount of bioavailability and bioaccessibility depends on soil characteristics, varying physico-chemical conditions, and soil ecological processes, as well as on the pollutants' characteristics and how organisms are exposed and vulnerable to them. In terms of human health, social relations of power affect the degrees of vulnerability to differing exposure levels and capacity to reduce exposure (Malone, Chapter 16). Hence, total concentrations may be above soil quality standards or screening levels and nevertheless be relatively harmless. Other factors must be investigated to determine the level of hazard and risk.

Toxicity data do not encompass all organisms and cannot address the social processes involved in the kind and level of exposure. Toxicity data are incomplete, especially relative to the tens of thousands of industrially produced compounds in existence. It may sometimes be necessary to carry out additional studies to evaluate the levels of toxicity for a substance or element in a soil.

Because soils are comprised of dynamic sets of interactions and processes that vary over time and space, it is essential to obtain up to date and geographically detailed information on the main soil properties that affect the extent of a pollutant's toxicity (e.g., bioavailability and bioaccessibility). This implies carrying out analyses on soil characteristics beyond just the substances or elements being studied.

As a result of the above, pollutant toxicity levels vary over time and space according to changing soil and wider ecological and social conditions. Results of soil toxicological analyses cannot be assumed to be applicable beyond the spatio-temporal scale of the study. Studies focused on human health, sampling, and analyses over decades and in multiple areas provide much greater confidence about pollutant effects.

Suggested further reading

Some open-source overviews of soil toxicological analysis include:

Agency for Toxic Substances and Disease Registry, https://www.atsdr.cdc.gov/index.html.
   This is a public agency of the United States that provides a searchable catalogue with detailed information on the characteristics of substances and elements toxic to humans.

European Chemicals Agency , https://echa.europa.eu/search-for-chemicals.
   A searchable database for substances and elements falling under European Union regulations, which tend to be stricter than in the United States.

FAO and UNEP. 2021. *Global Assessment of Soil Pollution—Summary for Policy Makers* (FAO), https://doi.org/10.4060/cb4827en.
   The publication includes overviews of soil contaminants and their health and ecological effects in ways that are accessible to a lay readership.

Pesticides Property Database, http://sitem.herts.ac.uk/aeru/ppdb/en/.
   An online searchable pesticides database, hosted by the University of Hertfordshire (UK), providing human health and ecotoxicological data.

References cited

Alengebawy, A., S.T. Abdelkhalek, S.R. Qureshi, and M.Q. Wang. 2021. 'Heavy metals and pesticides toxicity in agricultural soil and plants: Ecological risks and human health implications', *Toxics*, 9.3.

Brevik, E.C. and L.C. Burgess. 2013. *Soils and Human Health* (CRC Press).

Davis, L.F., M.D. Ramírez-Andreotta, and S. Buxner. 2020. 'Engaging diverse citizen scientists for environmental health: Recommendations from participants and promotoras', *Citizen Science: Theory and Practice*, 5.1, pp. 1–27.

Longhurst, R. and Johnston, L., Chapter 27, this volume. 'Focus groups.'

Mokos, J., Chapter 36, this volume. 'Participatory methods.'

Ramírez-Andreotta, M.D., R. Walls, K. Youens-Clark, K. Blumberg, K.E. Isaacs, D. Kaufmann, and R.M. Maier. 2021. 'Alleviating environmental health disparities through community science and data integration', *Frontiers in Sustainable Food Systems*, 5.

Winata, F. and McLafferty, S., Chapter 43, this volume. 'Survey and questionnaire methods.'

42. Statistical inference

Stuart N. Lane

Definition

Statistical inference is a form of data collection and analysis that has the aim of sampling a population efficiently and effectively such that the sample can be used to make statements about the population. It is associated with both social science research (e.g., social surveys) and biophysical research.

The basics of statistical inference

The principle of statistical inference is that it is not necessary to measure every individual in a population in order to make statements about that population. Thus, inferential statistics are different to descriptive statistics (Lane, Chapter 25) as the latter refer simply to describing the properties of what has been observed using statistical descriptors (e.g., mean, median, etc. of what has been observed). In inferential statistics, observations about only the sample of a population are used to make statements about the entire population (e.g., mean, median, etc. of the population on the basis of what has been observed).

Statistical inference in depth

Classically, statistical inference is frequentist; it assumes that by randomly sampling an entire population, it is possible to obtain a representation of that population that is sufficiently close to the entire population that characteristics of the sample can be used to make statements about the population. However, as the sample is finite, such statements can

https://doi.org/10.11647/OBP.0418.42

only probably be correct. Statistical inference, then, also determines the probability that the statement about the population holds. This is why statistical testing using inference requires reference to sample size; broadly speaking, larger sample sizes result in a higher probability (and greater confidence) that the statement is true of the population.

Alternative approaches to inference (e.g., Bayesian inference) recognise that sometimes when we sample, we already know something about a population. Using this knowledge has advantages as by incorporating it into sampling, we can increase confidence in estimates of the population using smaller sampler sizes. For instance, in a survey designed to determine how much time is dedicated to childcare in a population, if we know that 50% of a population are women we might randomly sample the population under the proviso that 50% of respondents should be women. This is important as we know that women generally take on more childcare responsibilities and so stratification of sampling would reduce bias in the results because more or fewer women than 50% have been included in the sample.

The most common is a parametric model which assumes that the population conforms to a Gaussian or normal distribution that can be fully described by its mean and variance. Non-parametric models make fewer assumptions regarding the population.

Finally, statistical inference leads us to the notion of statistical testing: the determination of a probability that a given descriptor of a sample or samples holds true of the population. For instance, imagine that you were interested in the relationship between life expectancy and mean annual household income. You had used stratified sampling of men and women to obtain a dataset with variables life expectancy and mean annual household income. Statistical testing would then intervene in two senses. The first would be to decide if the dataset is Gaussian or not as this would determine the kind of test you could use; given the sample, can you conclude with sufficient confidence (a high enough probability) that your population conforms to a normal distribution. The second would be to see if life expectancy and income are related. Correlating the data would give you a descriptive statistic (the correlation parameter) that describes a property of the data. Inference goes one step further by looking at the size of the sample, how the statistic was calculated and the strength of the correlation to determine the probability (or confidence we can have) that the property of the sample holds also of the population.

Why is statistical inference important?

Statistical inference is crucial to making the quantitative study of large populations efficient. As both social and biophysical surveys are expensive and time-consuming, substantial savings may arise from inferring characteristics of the population from a sample. Thus, statistical inference is used routinely across the biophysical and social sciences. Equally, when we present estimates of a population derived from data we should accept that they are uncertain and so also present a measure of the probabilities that these estimates apply to the population.

Relationship of statistical inference with other methods

Statistical inference builds upon descriptive statistics (Lane, Chapter 25). Inference is, then, a step that has to be gone through in passing from the description of properties of a sample to being able to state that the description also applies to the wider population that was sampled.

Statistical inference is often at its most powerful when it is mixed with other methods. This is partly because even when statistical inference allows us to conclude that a property or relationship likely exists (quantitatively) in a population it does not mean that the property exists or is truly causal. This is often where other kinds of investigations can provide valuable meaning on whether and why this property or relationship holds (see Lane, Chapter 24). For instance, Egli et al. (2021) used statistical inference to show that there was a positive association between rising temperatures and the collapse of a set of Swiss Alpine glacier snouts. By studying one collapse on one glacier in-depth (i.e., a case study) using geophysical methods they were then able to show why rising temperatures were causing this effect. Mixing the two methods was synergistic; the geophysical study of a glacier on its own said nothing about whether or not observed processes were more general; the statistical inference revealed a pattern but said nothing about what was causing this pattern to exist.

Ethical issues and statistical inference

As statistical inference is being used to make statements about (social or biophysical) populations there are two related ethical dimensions. The first is that as data are being sampled, there are important ethical responsibilities associated with the collection and use of those data (e.g., rights to privacy). The second is that statements about a population can be taken up beyond the project that generated them, and can have surprisingly strong impacts on policy discussions because claims expressed numerically are often viewed as more precise and more accurate than qualitative claims. This means that the results of statistical analysis can have considerable societal impact and so must be considered in ethical terms.

Issues to be aware of in using statistical inference

Effective use of statistical inference is not just about analysis of a dataset that has already been collected. To be able to move from a sampled population to statements about the population also requires very careful consideration to be given to data collection (vonHedemann, Chapter 38). Thus, applying effective statistical inference should be seen as an umbrella concept in which survey design before field data collection is developed with a view to how the data will be analysed once acquired.

It is often forgotten that a statistical test only really has sense in the context of using a sample to make statements about a population. If your sample is the population (and that can happen), then statistical inference has no meaning.

A sufficient sample size is, without a doubt, a crucial element of statistical inference. Put simply, the larger the sample size, the more likely that the sample can be used to infer properties of the population. Descriptive statistics (like correlation), when applied to samples, have little meaning unless they are also interpreted using inferential methods.

It is important not to confuse confidence that a characteristic of a population can be inferred from a sample with the characteristic itself. For instance, imagine that you had a very long time series (100 years) of annual growth rates of a tree. You find that there is a correlation (i.e., normalised measure of covariance) of 0.215 between growth rates and mean annual temperature. You do a statistical test to show that you can be highly confident (perhaps to a probability of 99% or more) that this correlation in your sample holds of the population. This high probability, though, does not necessarily mean that the growth rate and climate are highly correlated—a statistically significant correlation is not the same as a high correlation. The latter is a qualitative judgement.

Suggested further reading

There are a number of open source statistical textbooks. One of the best is Lane, D., 2003. *Introduction to Statistics*, available at the Open Textbook Library, https://open.umn.edu/opentextbooks/textbooks/459. Chapter 1 covers inferential statistics and shows how these fit into wider quantitative methods.

References cited

Egli, O., B. Belotti, B. Ouvry, J. Irving, and S.N. Lane. 2021. 'Subglacial channels, climate warming, and increasing frequency of Alpine glacier snout collapse', *Geophysical Research Letters*, 48, e2021GL096031. https://doi.org/10.1029/2021gl096031

Lane, S.N., Chapter 24, this volume. 'Case studies'.

Lane, S.N., Chapter 25, this volume. 'Descriptive statistics'.

vonHedemann, N., Chapter 38, this volume. 'Sampling and surveys'.

43. Survey and questionnaire methods

Fikriyah Winata and Sara McLafferty

Definition of survey research

Survey research involves administering a questionnaire (survey) to a set of people, organisations, or entities to uncover their characteristics, attitudes, experiences, and/or behaviours on a topic of interest. Researchers use survey methods in understanding the social, urban, economic, and political dimensions of people's lives and their interactions with everyday environments.

The basics of survey research

Survey research is an approach to data collection and analysis that consists of asking a series of pre-defined questions to a set of participants (respondents); collecting their responses; and analysing the responses to shed light on a research topic. It involves designing the questionnaire (survey instrument), identifying respondents, administering the questionnaire, and analysing the responses.

Survey research in depth

Designing the survey

A clear and detailed understanding of the research topic is essential, as a questionnaire is only as useful as the questions it asks. What is the goal of the survey, and what are specific themes of interest? For example, a survey about people's experiences of green spaces might investigate themes such as: types of green spaces visited; frequency of visits; social

 https://doi.org/10.11647/OBP.0418.43

and emotional experiences during visits; and respondent characteristics. Each theme may be treated in multiple questions.

Two broad types of questions are typically used in creating questionnaires: fixed-response and open-ended (McLafferty and Winata 2023). Fixed-response questions ask survey participants to record categorical (e.g., yes/no), ordinal (e.g., high/moderate/low), or numerical responses. Fixed-response questions often are constructed using Likert scales with responses selected between two extremes, e.g, "very poor" and "excellent". A five-point Likert scale lets respondents choose among five possible outcomes with the third, middle value representing a neutral response.

Open-ended questions ask participants to respond in their own words in narrative form. Participants can write about their experiences, attitudes, feelings, and emotions. Open-ended questions can address broad topics, giving respondents some opportunity to raise issues important to them.

Both types of questions have strengths and weaknesses. If well-designed, fixed response questions generate data that is comparable across respondents and appropriate for statistical analysis and mapping (see Lane, Chapter 25). For example, the percentage of respondents rating local parks as "poor" can be quantified and how that percentage varies by neighbourhood may be determined. Fixed-response questions provide useful data for statistical inference: estimating characteristics of a population from a sample, i.e., those who respond. Most national social, economic, and demographic surveys rely on fixed-response questions.

Open-ended questions provide rich qualitative information, useful for understanding how and why, e.g., why people rate their local parks as poor. Data from open-ended questions are typically analysed using qualitative methods.

Thus, the two types of questions are complementary: fixed-response questions give limited detail about individual circumstances and the processes that shape them, open-ended questions are limited in their ability to shed light on population-level characteristics.

Validity and reliability are important when designing survey questions; validity refers to accuracy in representing the underlying construct of interest; reliability describes consistency in representing the construct. Questions can be reliable but not valid, and vice versa. To be valid and reliable, survey questions need to be clear and understandable to respondents. An extensive literature exists on designing good survey questions. Some general guidelines include keeping questions simple, avoiding jargon, and limiting each question to a single topic.

Thinking about respondents' needs, understandings, educational backgrounds, and familiarity with the topic is essential to survey design. Surveys should be created in the language respondents use and employ terms they understand. Researchers need to consider racial, ethnic, linguistic, and cultural bias in designing survey questions. Survey length and question number and ordering may affect responses. The time required to respond to a lengthy survey may be problematic. If those who have time constraints fail to respond, the result is non-response bias and the sample of respondents does not accurately represent the study population. To avoid these issues, a questionnaire should be piloted on a small number of potential respondents to assess their reactions to the questions and willingness and ability to answer them. Most questionnaires are modified after piloting.

Success of a survey also depends on accurately identifying the study population and sample of respondents (see vonHedemann, Chapter 38). With a small population, sampling is not needed. However, surveys often focus on large populations (e.g., visitors to a national park) and surveying the whole population is infeasible. Decisions about the target number of respondents (sample size), sampling procedures, and geographical extent and timing of sampling are influenced not only by the research questions and survey design, but also by practical and budgetary considerations.

Administering the survey

Generally, surveys need to address the question of sampling (vonHedemann, Chapter 38). Traditionally, questionnaires were administered in person or by mail. Online formats now eclipse traditional formats. Online surveys can be distributed via websites, email, social media, text/chat messenger groups, etc. An Internet connection is essential for online surveys. Popular online survey platforms include Qualtrics, Google Form, SurveyMonkey, SurveySparrow, and Microsoft Forms. Templates included in these platforms enable researchers to design and modify the questionnaire easily.

Online questionnaire surveys have strengths and weaknesses. First, surveys can be distributed quickly to respondents through digital platforms, reaching a large audience and increasing the likelihood of response. Second, the survey is easily completed via smartphones or tablets, allowing people without access to computers to participate. Third, respondents can complete the survey anonymously, with which respondents may feel more comfortable. Lastly, online surveys are less expensive and time-consuming to distribute than in-person surveys. However, online surveys are prone to receiving multiple entries from the same respondents. The quality of responses may be difficult to assess. Groups who do not have access to the Internet are excluded. Although Internet use has increased significantly, the digital divide remains wide across the globe.

In-person surveys are still useful for specific purposes. They can be combined with other in-person field research methods, such as participant observation (see Sayre, Chapter 34). For instance, food access researchers who conduct in-person surveys with store managers may also assess availability of healthy foods by doing a store audit during the visit. However, in-person surveys often involve significant staff time and travel costs, and safety can be a concern. In-person surveys may make reaching the desired number of responses a challenge.

Why is survey research important?

Survey research is important for understanding the social, political and geographical dimensions of people, organisations, and institutions. It can solicit both qualitative and quantitative information to assess sensitive and challenging topics. In our recent study among Indonesian female domestic workers (FDWs) in Hong Kong (Winata and McLafferty 2023), we conducted an online survey about FDWs' health, wellbeing, and access to therapeutic landscapes and networks.

Relationship of survey research with other methods

Surveys are often combined with other methods. Quantitative and qualitative data from surveys can be triangulated (see Biermann and Gibbes, Chapter 4), which reduces biases from a single method, improves data richness by mixing different datasets, and enhances the credibility and validity of data and findings. Importantly, survey data can be integrated into study design, methods, and interpretation and reporting (see Fetters et al. 2013). For instance, researchers can statistically analyse quantitative data from fixed responses, then integrate findings with those based on responses to open-ended questions. Or themes emerging from quantitative analysis of survey responses can be explored in follow-up interviews with a subset of participants. For instance, in Winata and McLafferty (2023) we also interviewed a sub-sample of survey respondents via Zoom (see Johnston and Longhurst, Chapter 32), which helped expand our understanding of FDWs' lives and clarified the quantitative modelling results from the survey. At the interpretation and reporting stage, researchers can report the results sequentially, transform variables from one type to the other, or use a joint display to combine quantitative and qualitative results (Fetters et al. 2013).

Ethical issues in using survey research

Respecting and protecting respondents' rights, including rights to privacy and confidentiality, are essential in survey research (see Meadow et al., Chapter 5). Respondents need to know the purpose of the survey, how it will be used, and whether or not their responses will be kept confidential prior to participating. Participant concerns are also crucial in designing and administering surveys. In addition to language and question wording, researchers should consider respondents' cultural backgrounds, gender and power relations, and other influences on how participants perceive, respond to, and experience the survey. Online survey methods can pose unique challenges regarding privacy, confidentiality, and respondent experiences (Madge 2007). Prior to administering a survey, researchers must obtain approval from the body in their home institution responsible for ethical approvals to ensure that the survey will meet established ethical standards.

Issues to be aware of in using survey research

Effective survey research requires detailed understanding of the research topic and population to be studied. It involves considerable time and resources, which can be wasted if the survey does not ask useful questions nor reach its intended population.

Bias is a crucial concern. It can stem from, among others, poorly worded and unclear questions, participant non-response, and failure to think about how participants will perceive and respond to questions. Piloting the questionnaire minimises such bias.

Surveys are increasingly administered online, which can reduce costs and extend the survey to a large and geographically dispersed population. Whilst convenient for participants, online surveys can also result in bias, notably due to poor access and familiarity with digital devices.

Survey research is well-suited to mixed-methods projects. Questionnaires can include both quantitative (fixed response) and qualitative (open-ended) questions. Triangulating these data can provide innovative insights.

Before conducting a survey, the researcher must secure approval from the appropriate body responsible for ethical approval. This body guarantees that research meets established ethical standards, respecting respondents, and protecting their rights and well-being.

References cited

Creswell, J.W. and M. Hirose. 2019. 'Mixed methods and survey research in family medicine and community health', *Family Medicine and Community Health*, 7.2, e000086.

Fetters, M.D., L.A. Curry, and J.W. Creswell. 2013. 'Achieving integration in mixed methods designs—principles and practices', *Health Services Research*, 48, pp. 2134–2156.

Johnston, L. and Longhurst, R., Chapter 32, this volume. 'Interviews: structured, semi-structured, and open-ended'.

Lane, S.N., Chapter 25, this volume. 'Descriptive statistics'.

Madge, C. 2007. 'Developing an online agenda for geographers' research ethics', *Progress in Human Geography*, 31.5, pp. 654–674.

McLafferty, S.L. and F. Winata. 2023. 'Questionnaire survey research', in *Key Methods in Geography*, ed. by N. Clifford, M. Cope, T. Gillespie, and G. Valentine (Sage Publications).

Meadow, A., Wilmer, H., and Ferguson, D., Chapter 5, this volume. 'Expanding research ethics for inclusive and transdisciplinary research'.

Sayre, N.F., Chapter 34, this volume. 'Participant observation and ethnography'.

Winata, F. and S.L. McLafferty. 2023. 'Therapeutic landscapes, networks, and health and wellbeing during the COVID-19 pandemic: a mixed-methods study among female domestic workers', *Social Science and Medicine*, 322.

44. Textual analysis

Marcus A. Doel

Definition of textual analysis

Textual analysis comes in three main forms. First, approaches rooted in literary criticism, literary theory, and the philosophy of language. Meaning, intention, and reference are often key concerns. Second, approaches rooted in communication studies, media studies, and cultural studies. The social life and performative work of texts are often their main concerns. Third, the analysis of machine-readable texts using computer software, which exists in its own right and is beginning to be applied to the other two forms (e.g., language models and sentiment analysis). Given the ease with which texts can be transformed into a machine-readable format, and the widespread availability of powerful computers, the third form is fast becoming the dominant one. Heterogeneous texts are now routinely processed into homogenised datasets, much like any other mass-produced commodity.

The basics of textual analysis

Analysing texts manually takes a lot of effort, and quickly becomes unfeasible as the volume increases. Consequently, computer software is increasingly taking over the arduous work of textual analysis, just as it has taken over the laborious work of reading, writing, and calculation. Sophisticated software packages now enable automatic textual analysis, often with machine learning and artificial intelligence, which extract important qualitative information and statistically significant quantitative patterns. While such automation is very efficient and highly productive, it suffers from the same problems that bedevil other computational

    https://doi.org/10.11647/OBP.0418.44

approaches, which arguably includes an inability to count, since what counts is rarely unambiguous and is often undecidable, as Derrida's (1992) attempt to count the yeses in James Joyces' *Ulysses* demonstrates so well. Such automation also arguably fails to take adequate account of the most important facets of any text—its context and its interpretation. On the one hand, as the etymology of the word makes plain, 'context' (woven together) is simply more text to be taken into account: every text refers to other texts. On the other hand, context is obviously more than merely textual, since every text refers to the world writ large. Since we are always working (with)in innumerable and illimitable (con)texts that are woven together without closure, textual analysis is arguably the most mixed up of mixed methods. It weaves together texts and contexts that always remain open to other texts and other contexts, and so demands further exegesis that may encompass every conceivable domain.

Textual analysis in depth

Rather than considering texts in the narrow sense of a piece of writing, such as a sign, a message, or a book, textual analysis considers texts in a much more expansive sense: as anything that makes sense or conveys sense; as anything that imparts meaning or solicits interpretation. Thus, when you start looking, texts appear everywhere. Many have been authored intentionally by humans, but most have not. Think, for example, of the genetic code or the signature of anthropogenic climate change. Think also of fieldwork activities that generate a wealth of textual material: from field observations and the collection of textual ready-mades, such as gathering newspaper articles and travel writing or sourcing palaeoenvironmental records (ice cores, tree-rings, and lake sediments, for example), to the production of transcripts and qualitative and quantitative datasets laden with meaning and significance (such as the proxy records of climatic change extracted from ice, trees, and sediments).

In keeping with this enlarged understanding of what constitutes a text, textual analysis tends to think expansively: every text refers to other texts and contexts with which it is interwoven. Indeed, recall that this is what the word 'context' means: woven together. Textual analysis is therefore a quintessentially mixed methodology, since its textual and analytical expansiveness encompasses the textual and the contextual, which is sometimes called the 'extra-textual' (since it is outside the text, but in the strange form of a constitutive outside: the text both refers to and depends upon its outside, and without its outside it would collapse into nonsense). Just think of the extra-textual and contextual work that goes into discerning the climatic signature archived in a polar ice core, from chemical and statistical analysis to visualisation and publication, or the extra-textual and contextual work that goes into uncovering a history of violence written in bombed-out ruins and bulldozed rubble, by way of forensic architecture, for example (Weizman 2017).

The principle underpinning textual analysis is that everything in the world makes sense, and so everything calls out for interpretation. Sometimes, textual analysis aims to pin down the meaning and the sense of a text, often with reference and deference to the intention of its author, whether human or non-human. Oftentimes, however, textual analysis aims to multiply the meanings of a text, revealing the countless, contradictory, and incommensurable ways in which it can be taken up by different interpretative communities and transformed in different contexts.

Although textual analysis comes in a great many forms, the advent of powerful computer software for textual analysis is threatening to monopolise the field, especially in the social sciences, which seem to have a newfound appetite for digesting all manner of heterogenous texts into homogenised datasets suitable for computational and algorithmic processing. Digitisation is fast becoming the universal currency through which all texts become interchangeable: they can all be rendered fungible through the binary code. The conversion of unstructured texts into structured data is a complex process, typically involving chopping them up into meaningful parts so that they can be tagged and parsed. The resulting datasets can then be subjected to processes such as sentiment analysis, topic analysis, intention analysis, and data extraction, using classification and extraction rules that are either set by the researcher or else conjured up by the software itself. When coupled with statistical packages and modelling capabilities, automated textual analysis can even go beyond the data themselves through extrapolation, generalisation, and simulation, and they can be visualised like any other product of algorithmic calculation. Word clouds, tag clouds, slope graphs, and Sankey diagrams are increasingly well-known forms of textual visualisation.

Why is textual analysis important?

Textual analysis is important because everything we study makes sense, conveys sense, and solicits sense. This is not to say that there is nothing other than bookish texts, human discourse, and linguistic expression, but rather that textual analysis investigates the embedment of texts within contexts. Whenever you are in a field, you will be embedded in countless contexts that command your attention and solicit your interpretation. So, it is imperative that you are well-versed in (con)textual analysis.

Relationship of textual analysis with other methods

The various forms of textual analysis obviously complement other methods that collect, generate, and work with texts of any description, such as the written accounts of ethnographic research (see Sayre, Chapter 34), the transcripts of interviews and focus groups (see Johnston and Longhurst, Chapter 32; Longhurst and Johnston, Chapter 27), the chatter of social media, and the records held within bureaucratic and palaeoenvironmental archives (see Cope, Chapter 22; King and Abbott, Chapter 28; Davis, Chapter 29): filing cabinets, ice cores, tree-rings, punch cards, and suchlike. Indeed, it is worth bearing in mind that textual analysis is not exclusively a qualitative research method, not only because text is so often quantified, both in real-world contexts and by researchers who make sense of them, but also because quantification and calculation are essentially textual: think, for example, of binary code, computer code, and mathematical hieroglyphics. Numbers, quantities, and statistics all make sense—in one context or another. They mean something and require interpretation.

Ethical issues and textual analysis

The first ethical issue to consider is that the selection of texts and contexts is necessarily selective and therefore partial, neglectful, and harmful. Here as elsewhere, research must take things out of context and transplant them into other contexts, transforming them in the process. The second ethical concern is that the processing and analysing of texts is also selective and therefore patrial, neglectful, and harmful. The seemingly innocuous phrase 'textual analysis' masks a process of profound transformation in which texts are stripped of many contexts, and thus of meanings that are important for some communities and contexts. A third ethical concern is the increasing delegation of textual analysis to machines, such that texts risk becoming little more than grist to the post-industrial mill of data processing, with data-protection and commercial interest considerations being perhaps the most obvious worries. For example, it is worth recalling that a great many humans and non-humans have been forced into textual analysis and both violated and transformed in the process, such as the commercial extortion and re-engineering of the so-called 'genetic code' of viruses, plants, animals, and humans. When almost everything can be transformed into a digital text and compelled to yield its senses for processing and exploitation, we should regard that transmogrification as an overarching ethical concern and remain alert to the potential horror of such 'sinister inscriptions' (Black 2001; Fleischman et al. 2013). A final ethical concern is that the delegation of textual analysis to machines has profound implications for the workforce that sustains this sector of the digital economy, much of which is precariously employed and often highly alienated and strongly exploited (Posada et al. 2023).

Issues to be aware of in using textual analysis

Perhaps the key issue to bear in mind whilst undertaking textual analysis is that making sense and working with sense is so deeply ingrained in all of us that we can take the expertise required for granted. However, the various techniques of textual analysis that are available to us as researchers are as sophisticated, complicated, and exacting as any other methodological approach. They need to be learnt and practiced. Their ontological and epistemological underpinnings need to be studied and understood. And their strengths and weaknesses, and implications and consequences, need to be appreciated and reckoned with. Texts and contexts are at work everywhere, and we underestimate their significance at our peril.

Suggested further reading

For a discussion of textual analysis in geography, and more widely, see:

Doel, M.A. 2023. 'Textual analysis', in *Key Methods in Geography*, ed. by N. Clifford, M. Cope, and T. Gillespie (Sage Publications), pp. 245–261.

McKee, A. 2003. *Textual Analysis: A Beginner's Guide* (Sage Publications).

For open-access books that give a flavour of different forms of textual analysis, see:

Helgesson, S., H. Bodin, and A. Mörte Alling. 2022. *Literature and the Making of the World: Cosmopolitan Texts, Vernacular Practices* (Bloomsbury). https://doi.org/10.5040/9781501374180

Jones, R.H., A. Chik, and C.A. Hafner. 2015. *Discourse and Digital Practices: Doing Discourse Analysis in the Digital Age* (Routledge). https://doi.org/10.4324/9781315726465

McEnery, A. and H. Baker. 2017. *Corpus Linguistics and 17th-Century Prostitution: Computational Linguistics and History* (Bloomsbury). https://doi.org/10.5040/9781474295062

Schwandt, S. 2021. *Digital Methods in the Humanities: Challenges, Ideas, Perspectives* (Bielefeld University Press). https://doi.org/10.2307/j.ctv2f9xskk

Tonkin, E.L. and G.J.L. Tourte. 2016. *Working with Text: Tools, Techniques and Approaches for Text Mining* (Chandos Elsevier).

References cited

Black, E. 2001. *IBM and the Holocaust: The Strategic Alliance between Nazi Germany and America's Most Powerful Corporation* (Little, Brown).

Cope, M., Chapter 22, this volume. 'Archival methods.'

Davis, D.K., Chapter 29, this volume. 'Historical ecology.'

Derrida, J. 1992. 'Ulysses gramophone: hear say yes in Joyce', in *Acts of Literature* (Routledge), pp. 253–309.

Doel, M.A. 2023. 'Textual analysis', pp. 245–261 in *Key Methods in Geography*, ed. by N. Clifford, M. Cope, and T. Gillespie (Sage Publications).

Fleischman, R.K., W. Funnell, and S.P. Walker. 2013. *Critical Histories of Accounting: Sinister Inscriptions in the Modern Era* (Routledge). https://doi.org/10.4324/9780203102749

Johnston, L. and Longhurst, R., Chapter 32, this volume. 'Interviews: structured, semi-structured and open-ended.'

King, G.E. and Abbott, P.M., Chapter 28, this volume. 'Geochronological methods.'

Longhurst, R. and Johnston, L., Chapter 27, this volume. 'Focus groups.'

Posada, J., G. Newlands, and M. Miceli. 2023. 'Labor, automation, and human–machine communication', in *The Sage Handbook of Human–Machine Communication*, ed. by A.L. Guzman, R. McEwen, and S. Jones (Sage Publications), pp. 384–391. https://doi.org/10.4135/9781529782783

Sayre, N.F., Chapter 34, this volume. 'Participant observation and ethnography.'

Sutherland, K. 2011. 'Marx in jargon', in *Stupefaction: A Radical Anatomy of Phantoms* (Seagull), pp. 26–90.

Weizman, E. 2017. *Forensic Architecture: Violence at the Threshold of Detectability* (Zone). https://doi.org/10.2307/j.ctv14gphth

45. Uncrewed airborne systems

Elina Kasvi

Definition

Uncrewed airborne systems (UAS; sometimes referred to as drones) involve aircraft which operate without a human pilot on board. They can be either remotely controlled or operated autonomously using pre-programmed flight plans. In environmental research they are widely used as platforms for instruments and sensors and can thus be exploited in a wide range of measurements.

The basics of UAS statistical inference

UAS have become valuable tools for environmental research. They can be equipped with a variety of sensors, such as cameras and laser scanners, allowing researchers to perform low-cost close-range remote sensing surveys (Fonstad et al. 2013; Kasvi et al. 2019; Schmucki et al. 2023), but also with physical sampling devices allowing in situ measurements from locations which are difficult to access (Shelare et al. 2021; Terada et al. 2018). UAS range in size from small handheld devices to large, high-altitude, long-endurance systems.

UAS in depth

UAS have a great potential to revolutionise environmental research by providing rapid, accurate, and relatively cost-efficient ways to study the environment. UAS are used, for example, in applications that benefit from wide coverage without losing details such as in mapping land use, land cover and water resources (Ancin-Murguzur et al. 2020;

https://doi.org/10.11647/OBP.0418.45

De Keukelaere et al. 2023; Kalantar et al. 2017) water and air quality (Afshar-Mohajer and Wu 2023), elevation, or bathymetry (Dietrich 2017; Fonstad et al. 2013). Applications that require repetition of the measurements frequently, such as change detection of vegetation or morphology (Ancin-Murguzur et al. 2020; de Almeida et al. 2020; Eltner et al. 2015; Guisado-Pintado et al. 2019) now use UAS widely. Also mapping of rapidly evolving or moving systems, which require capacity to react and to adapt the mapping relative to the target, such as natural disasters (Daud et al. 2022) or wildlife monitoring in hard-to-access environments (Geldart et al. 2022; Krishnan et al. 2023) have benefitted from the flexibility of UAS. One of its agreed benefits is that it does not disturb the research target, which has been demonstrated to apply also to wildlife: in a study of Geldart et al. (2022), the heart rates of incubating female birds did not change during a drone-based population survey, when compared to their baseline heart rate.

UAS are available in a broad spectrum of sizes. Small-scale UAS (typical weight 1 to 25 kg) are able to carry some additional load such as cameras, while medium-scale UAS (typical weight 25 to 150 kg) are often equipped with more sophisticated and heavier sensors such as laser scanners. Both are commonly used for aerial photography and surveying. Large UAS (150 to 1,000 kg) are used in carrying more substantial payloads while High-Altitude, Long-Endurance UAS (HALE UAS, typical weight > 1,000 kg) are designed to operate at high altitudes for extended periods, sometimes up to several days.

UAS provide a diverse range of sensing options and the off-the-shelf selection is growing very rapidly. Customised options combining sensors for a particular objective are also common. UAS equipped with RGB (Visible Light) or thermal cameras are used to collect high resolution aerial and orthoimages, while multi- or hyperspectral sensors can capture imagery across multiple spectral bands (Jackisch et al., 2018; Prior et al. 2021). By exploiting Structure-from-Motion photogrammetry, detailed 3D reconstructions or digital elevation models can be created from a set of aerial imagery (Dai et al. 2023; Fonstad et al. 2013). Schmucki et al. (2023) demonstrated how UAS allowed for safe surveying of debris-flow prone torrent channel properties with high resolution and reliability. They also pointed out that, in the case of dense vegetation coverage, LiDAR sensors, which are also increasingly used in UAS (Kellner et al. 2019), have clear advantages over optical imagery in DTM production.

UASs are also widely used in atmospheric, climatological, and hydrological research to measure parameters like temperature, humidity, pressure, and different concentrations. For example, they allow detection and measurement of air pollutants or other hazardous substances remotely (Afshar-Mohajer and Wu 2023; Burgués and Marco 2023). In a modelling-based study, Afshar-Mohajer and Wu (2023) pointed out that the advantages of a drone equipped with an air quality sensor as compared to a network of fixed position low-cost sensors, are its flexibility to manoeuvre over hard-to-access terrains, and its ability to pinpoint unknown emitting sources. In hydrological research, UAS are used to measure parameters directly from the water, or even to gather water samples physically for various analysis (Horricks et al. 2022; Terada et al. 2018).

Why are UAS important?

The advantages of UAS over traditional manned aircraft include faster operation and lower operating costs, increased safety for pilots and crew members, and often higher data resolution. UAS also enable environmental measurements in areas that are difficult or dangerous for humans to reach (Schmucki et al. 2023).

Relationship of UAS with other methods

In environmental research, UAS are often used in conjunction with other measurement methods. As with traditional remote sensing data (Braun, Chapter 39), data collected using UAS require reference measurements such as in situ field observations (sometimes referred to as 'ground truthing') for instrument calibration, quality assurance and to support the data interpretation. In many cases UAS are used to complement in situ measurements for example by enhancing spatial coverage, even though they may provide lower data quality compared to the in situ measurements. On the other hand, UAS-generated aerial imagery can support the interpretation of lower-resolution satellite images or supplement coarse elevation models for critical areas. In many cases, the combination of various methods is the most well-functioning approach.

Use of all UAS imagery relies upon accurate positioning and orientation of the UAS at the time of each image acquisition, as well as upon knowledge of the geometry of the imaging system itself, which aligns the analysis of UAS data with the longstanding discipline of photogrammetry. Developments in computer vision (such as Structure from Motion, Multi View Stereo photogrammetry) have rendered application of photogrammetry to this problem more straightforward, notably using oblique imagery, but also require expertise from photogrammetry and engineering surveying if this is to be done properly (cf. Dai et al. 2023).

As a summary, in addition of being able to handle the device itself, UAS users need to have the required expertise to plan and to perform the survey campaign with the attached sensor(s), and to handle and to analyse the collected data in appropriate way.

Ethical issues concerning the use of UAS

As UAS are relatively inexpensive to obtain and as they enable remote usage, including for example collection of high-resolution imagery, they raise privacy concerns and safety issues, and have the potential for misuse or abuse. Governments and regulatory bodies now strongly regulate the use of UAS to ensure their safe and responsible use. The rules, regulations, and permission systems vary considerably depending on the country and region and thus it is vital to know local conditions before UAS use.

Issues to be aware of in using UAS

The use of UAS may require a driver's license or other special permits. Its use in certain areas may be restricted or completely prohibited.

Obtaining good data requires a skilled user who understands the operation of the sensor and instrument with which the system is equipped. Data collection and post-processing require specialised expertise in many cases.

Interpreting the collected data also requires expertise, similar to interpreting traditional remote sensing data and notably in relation to long-established photogrammetric principles.

Although UAS enable data collection without physically touching the target, the data often requires in situ reference data (i.e., ground-truthing) for interpretation and quality assurance.

While UAS platforms are relatively affordable, the sensors and instruments and the software required for their handling can be expensive.

Often, the data is stored within the instrument itself. This means that the data will be lost if the device is lost.

Suggested further reading

Eltner, Hoffmeister, Kaiser, Karrasch, Klingbeil, Stöcker, Rovere (eds.) 2023. *UAVs for the Environmental Sciences, Methods and Applications* (Wissenschaftliche Buchgesellschaft, Darmstadt). ISBN: 978-3-534-40588-6, https://files.wbg-wissenverbindet.de/Files/Article/ARTK_ZOA_1028514_0003.pdf

D.R. Green, B.J. Gregory, A. Karachok (eds.) 2023. *Unmanned Aerial Remote Sensing (UAS) for Environmental Applications* (CRC Press).

References cited

Afshar-Mohajer, N. and C.-Y. Wu. 2023. 'Use of a drone-based sensor as a field-ready technique for short-term concentration mapping of air pollutants: a modeling study', *Atmospheric Environment*, 294.

Ancin-Murguzur, F.J., L. Munoz, C. Monz, and V.H. Hausner. 2020. 'Drones as a tool to monitor human impacts and vegetation changes in parks and protected areas', *Remote Sensing in Ecology and Conservation*, 6, pp. 105–113.

Braun, A., Chapter 39, this volume. '(Critical) Satellite remote sensing'.

Burgués, J. and S. Marco. 2023. 'Drone-based monitoring of environmental gases', in *Air quality networks: Data Analysis, Calibration and Data Fusion, Environmental Informatics and Modeling*, ed. by S. De Vito, K. Karatzas, A. Bartonova, and G. Fattoruso (Springer), pp. 115–137.

Dai, W., G. Zheng, G. Antoniazza, F. Zhao, K. Chen, W. Lu, and S.N. Lane. 2023. 'Improving UAV-SfM photogrammetry for modelling high-relief terrain: Image collection strategies and ground control quantity', *Earth Surface Processes and Landforms*, 48.14, pp. 2884–2899.

de Almeida, D.R.A., A.M. Almeyda Zambrano, E.N. Broadbent, A.L. Wendt, P. Foster, B.E. Wilkinson, C. Salk, D. Papa, S.C. Stark, R. Valbuena, E.B. Gorgens, C.A. Silva, P.H.S. Brancalion, M. Fagan, P. Meli, and R. Chazdon. 2020. 'Detecting successional changes in tropical forest structure using GatorEye drone-borne lidar', *Biotropica*, 52, pp. 1155–1167.

de Keukelaere, L., R. Moelans, E. Knaeps, S. Sterckx, I. Reusen, D. de Munck, S.G.H. Simis, A.M. Constantinescu, A. Scrieciu, G. Katsouras, W. Mertens, P.D. Hunter, E. Spyrakos, and A. Tyler. 2023. 'Airborne drones for water quality mapping in inland, transitional and coastal waters—MapEO water data processing and validation', *Remote Sensing*, 15, p. 1345.

Dietrich, J.T. 2017. 'Bathymetric Structure-from-Motion: Extracting shallow stream bathymetry from multi-view stereo photogrammetry', *Earth Surface Processes and Landforms*, 42, pp. 355–364.

Eltner, A., P. Baumgart, H.-G. Maas, and D. Faust. 2015. 'Multi-temporal UAV data for automatic measurement of rill and interrill erosion on loess soil', *Earth Surface Processes and Landforms*, 40, pp. 741–755.

Fonstad, M.A., J.T. Dietrich, B.C. Courville, J.L. Jensen, and P.E. Carbonneau. 2013. 'Topographic structure from motion: a new development in photogrammetric measurement', *Earth Surface Processes and Landforms*, 38, pp. 421–430.

Geldart, E.A., A.F. Barnas, C.A.D. Semeniuk, H.G. Gilchrist, C.M. Harris, and O.P. Love. 2022. 'A colonial-nesting seabird shows no heart-rate response to drone-based population surveys', *Science Reports*, 12, p. 18804.

Guisado-Pintado, E., D.W.T. Jackson, and D. Rogers. 2019. '3D mapping efficacy of a drone and terrestrial laser scanner over a temperate beach-dune zone', *Geomorphology*, 328, pp. 157–172.

Horricks, R.A., C. Bannister, L.M. Lewis-McCrea, J. Hicks, K. Watson, and G.K. Reid. 2022. 'Comparison of drone and vessel-based collection of microbiological water samples in marine environments', *Environmental Monitoring and Assessment*, 194, p. 439.

Jackisch, R., S. Lorenz, R. Zimmermann, R. Möckel, and R. Gloaguen. 2018. 'Drone-borne hyperspectral monitoring of acid mine drainage: an example from the Sokolov Lignite district', *Remote Sensing*, 10, p. 385.

Kalantar, B., S.B. Mansor, M.I. Sameen, B. Pradhan, and H.Z.M. Shafri. 2017. 'Drone-based land-cover mapping using a fuzzy unordered rule induction algorithm integrated into object-based image analysis', *International Journal of Remote Sensing*, 38, pp. 2535–2556.

Kasvi, E., J. Salmela, E. Lotsari, T. Kumpula, and S.N. Lane. 2019. 'Comparison of remote sensing based approaches for mapping bathymetry of shallow, clear water rivers', *Geomorphology*, 333, pp. 180–197.

Kellner, J.R., J. Armston, M. Birrer, K.C. Cushman, L. Duncanson, C. Eck, C. Falleger, B. Imbach, K. Král, M. Krůček, J. Trochta, T. Vrška, and C. Zgraggen. 2019. 'New opportunities for forest remote sensing through ultra-high-density drone lidar', *Surveys in Geophysics*, 40, pp. 959–977.

Krishnan, B.S., L.R. Jones, J.A. Elmore, S. Samiappan, K.O. Evans, M.B. Pfeiffer, B.F. Blackwell, and R.B. Iglay. 2023. 'Fusion of visible and thermal images improves automated detection and classification of animals for drone surveys', *Scientific Reports*, 13, p. 10385.

Mohd Daud, S.M.S., M.Y.P. Mohd Yusof, C.C. Heo, L.S. Khoo, M.K. Chainchel Singh, M.S. Mahmood, and H. Nawawi. 2022. 'Applications of drone in disaster management: A scoping review', *Science and Justice*, 62, pp. 30–42.

Prior, E.M., C.A. Aquilina, J.A. Czuba, T.J. Pingel, and W.C. Hession. 2021. 'Estimating floodplain vegetative roughness using drone-based laser scanning and structure from motion photogrammetry', *Remote Sensing*, 13, p. 2616.

Schmucki, G., P. Bartelt, Y. Bühler, A. Caviezel, C. Graf, M. Marty, A. Stoffel, and C. Huggel. 2023. 'Towards an automated acquisition and parametrization of debris-flow prone torrent channel properties based on photogrammetric-derived uncrewed aerial vehicle data', *Earth Surface Processes and Landforms*, 48, pp. 1742–1764.

Shelare, S.D., K.R. Aglawe, S.N. Waghmare, and P.N. Belkhode. 2021. 'Advances in water sample collections with a drone – A review', *Materials Today: Proceedings*, 47, pp. 4490–4494.

Terada, A., Y. Morita, T. Hashimoto, T. Mori, T. Ohba, M. Yaguchi, and W. Kanda. 2018. 'Water sampling using a drone at Yugama crater lake, Kusatsu-Shirane volcano, Japan', *Earth Planets Space*, 70, p. 64.

Contributing authors

Editors

Rebecca Lave is Professor of Geography at Indiana University and the 2022-2025 American Association of Geographers Vice-President/President/Past-President. Her research takes a Critical Physical Geography approach, combining political economy, STS, and fluvial geomorphology to analyse stream restoration, the politics of environmental expertise, and non-structural approaches to flooding. She has published in journals ranging from *Science* to *Social Studies of Science* and is the author of two monographs: *Fields and Streams: Stream Restoration, Neoliberalism, and the Future of Environmental Science* (2012, University of Georgia Press) and *Streams of Revenues: The Restoration Economy and the Ecosystems it Creates* (2021 MIT Press; co-written with Martin Doyle). She has co-edited four volumes, including the *Handbook of Critical Physical Geography* (2018, with Christine Biermann and Stuart N. Lane). Orcid id: 0000-0001-5335-9058.

Stuart N. Lane is Professor of Geomorphology at the University of Lausanne. He is a geographer and civil engineer by training who has held posts at the Universities of Cambridge, Leeds, and Durham in the U.K. and Lausanne in Switzerland. His work has sought to bring a geographical perspective to contemporary environmental concerns such as flooding and pollution. The primary focus of his current work is the environments created by disappearing glaciers in terms of ice, water, sediment, and ecosystems and the consequences of these changes for environmental management. An important thread through his most recent research criticises the current alignment of geography as a discipline with the ever more neo-liberal academy; and then argues for the rediscovery of a more scientific geographical science better able to cope with the crises the world is experiencing today. Orcid id: 0000-0002-6077-6076

Other chapter authors

- Peter M. Abbott, Advanced Postdoctoral Researcher, Climate and Environmental Physics, Physics Institute, and Oeschger Centre for Climate Change Research, University of Bern, 0000-0002-6347-9499

- Javier Arce-Nazario, Associate Professor of Geography at the University of North Carolina at Chapel Hill, 0000-0002-4051-575X.

- Dipak Basnet, Research Associate at the Social Science Baha, Kathmandu, Nepal

- Nyima Dorjee Bhotia, Research Associate at the Social Science Baha, Kathmandu, Nepal.

- Christine Biermann, Associate Professor of Geography and Environmental Studies at University of Colorado Colorado Springs, 0000-0002-0821-755X

- Ninon Blond, Maîtresse de Conférences in Geography at École Normale Supérieure de Lyon (France), 0000-0002-0149-0975.

- Eric G. Booth, Associate Scientist in Plant and Agroecosystem Sciences at the University of Wisconsin-Madison, 0000-0003-2191-6627.

- Andreas Braun, Professor of Human-Environment Interactions at University of Kassel, 0000-0002-6760-1105.

- Gary Brierley, Professor of Physical Geography at the University of Auckland, 0000-0002-1310-1105.

- Alexandra Chakov, Ph.D. Student in Composition and Rhetoric, Department of English, University of Wisconsin-Madison

- Tom Chang, Ph.D. Student in Composition and Rhetoric, Department of English, University of Wisconsin-Madison

- Stephen Chignell, Postdoctoral Researcher at the University of Bristol, 0000-0002-8277-4338.

- Meghan Cope, Professor of Geography and Geosciences, University of Vermont, 0000-0003-4000-9838.

- Henry Covey, Ph.D. Student in Composition and Rhetoric, Department of English, University of Wisconsin-Madison

- Diana K. Davis, Professor of History and Geography at the University of California, Davis.

- Taylor Dickson, Ph.D. Student in Composition and Rhetoric, Department of English, University of Wisconsin-Madison

- Marcus Doel, Professor of Human Geography at Swansea University, UK, 0000-0002-8892-2709

- Tek Bahadur Dong, Research Associate at the Social Science Baha, Kathmandu, Nepal.

- Salvatore Engel-Di Mauro, Professor of Geography and Environmental Studies, SUNY New Paltz, 0000-0001-5125-1967.

- Daniel B. Ferguson, Associate Professor of Environmental Science at the University of Arizona, 0000-0002-2197-4424.

- Andrew Fountain, Professor Emeritus of Geography and Geology at Portland State University, 0000-0001-5299-2273

- Marjolein Gevers, PhD Researcher at the University of Lausanne, 0000-0002-5866-7436

- Cerian Gibbes, Professor of Geography and Environmental Studies at University of Colorado Colorado Springs, 0000-0002-2388-0077.

- Sydney Goggins, Ph.D. Student in Composition and Rhetoric, Department of English, University of Wisconsin-Madison

- Caroline Gottschalk Druschke, Vilas Distinguished Achievement Professor of English, University of Wisconsin-Madison

- Nora Harris, Ph.D. Student in Composition and Rhetoric, Department of English, University of Wisconsin-Madison

- Adrian Howkins, Reader in Environmental History at the University of Bristol, 0000-0002-9615-2337.

- Olivier Hymas, Senior Researcher, University of Lausanne, Switzerland.

- Mrill Ingram, Participatory Action Research Scientist at Michael Fields Agricultural Institute, 0000-0002-4155-4523.

- Amy Leigh Johnson, Assistant Professor of Anthropology at Georgia College and State University, 0009-0003-8865-3523.

- Lynda Johnston, Professor of Geography, University of Waikato, Tauranga Moana, Aotearoa, 0000-0001-7686-9181.

- Elina Kasvi, Lecturer of Geoinformatics at the University of Turku, Academic Research Fellow, 0000-0002-3495-465X.

- Lisa C. Kelley, Assistant Professor of Geography and Environmental Sciences at the University of Colorado Denver, 0000-0001-6638-4017.

- Mark Kincey, Lecturer in Physical Geography at Newcastle University, UK, 0000-0002-9632-4223

- Georgina E. King, Associate Professor of Quaternary Geochronology at the University of Lausanne, 0000-0003-1059-8192

- Tobias Krueger, Professor of Hydrology and Society at Humboldt-Universität zu Berlin, 0000-0002-4559-6667.

- Catharina Landström, Associate Professor, Science, Technology and Society , Chalmers University of Technology, 0000-0001-7782-1438.

- Karen Lebek, Postdoc at the Institute of Environmental Science and Geography at the University of Potsdam, 0000-0003-4380-0422.

- Robyn Longhurst, Deputy Vice-Chancellor Academic at Te Herenga Waka-Victoria University of Wellington, 0000-0001-9192-0643.

- Melanie Malone, Associate Professor in the School of Interdisciplinary Arts and Sciences, University of Washington, Bothell, 0000-0001-7054-0924.

- Maurizio Mazzoleni, Assistant Professor at Vrije Universiteit Amsterdam 0000-0002-0913-9370.

- Sara L. McLafferty, Professor Emerita at University of Illinois Urbana-Champaign, 0000-0001-8413-7562.

- Alison M. Meadow, Associate Research Professor – Office of Societal Impact at the University of Arizona, 0000-0003-0315-5799.

- Lieke Melsen, Associate Professor Computational Hydrology at Wageningen University, 0000-0003-0062-1301.

- Floreana Miesen, Field Technician / Scientific Collaborator at the University of Lausanne, 0000-0003-2388-9853.

- Jennifer Mokos, Associate Professor of Honors and Sustainability and Coastal Resilience at Coastal Carolina University, 0000-0002-4843-6909.

- Kevin Ndong, Head of the Audiovisual Service, Agence National des Parcs Nationaux, Gabon

- Eric Nost, Associate Professor of Geography at the University of Guelph, 0000-0001-9320-072X.

- Katie Oven, Assistant Professor, Human Geography, Policy and Development at Northumbria University Newcastle, UK, 0000-0002-3363-8604.

- Sujash Purna, Ph.D. Student in Composition and Rhetoric, Department of English, University of Wisconsin-Madison

- Anuradha Puri, Research Associate at the Social Science Baha, Kathmandu, Nepal

- Nick Rosser, Professor of Geography at Durham University, UK, 0000-0002-1435-2512.

- Maria Rusca, Senior Lecturer in Global Development at the University of Manchester, 0000-0003-4513-3213

- Jennifer Salmond, Professor of Geography at the University of Auckland, 0000-0001-5670-7724.

- Nathan F. Sayre, Professor of Geography at University of California-Berkeley, 0000-0002-9243-2219.

- Sunil Tamang, PhD Candidate in Water Resource Management at the University of Canterbury, New Zealand, 0000-0002-0761-0528

- Stevens Touladjan, Community Officer, Agence National des Parcs Nationaux, Gabon

- Nicolena vonHedemann, Human Dimensions Specialist at the Ecological Restoration Institute at Northern Arizona University, 0000-0002-1376-7927

- Gretchen Walters, Associate Professor, University of Lausanne, Switzerland, 0000-0002-9772-232X

- Sydney Widell, Watershed Coordinator, Coon Creek Community Watershed Council.

- Hailey Wilmer, Research Rangeland Management Specialist, USDA-ARS Range Sheep Production Efficiency Research Unit, 0000-0003-0810-9687.

- Fikriyah Winata, Assistant Professor of Geography at Texas A & M University, 000-0002-2251-7401.

Index

stratified 536, 546, 572

systematic 546

savanna 45, 184, 207–212, 224–235

scale 1, 40, 43–46, 53, 100, 109, 190, 201, 203, 211, 226, 235, 247, 251, 261, 267, 284, 291, 294, 297–298, 315, 340–341, 361, 377–378, 391–394, 403, 406, 408–409, 413, 433, 443, 457–458, 463, 487–488, 510, 512, 550–551, 568, 578, 594

multiscalar 289, 291

scenarios 126, 177, 254, 281–282, 286, 296–298, 300, 312, 315, 317, 490, 521–522

Schmidt hammer 477

Science and Technology Studies (STS) 248–249, 251, 255, 270, 397, 485

boundary objects 72, 282, 496

knowledge politics 122, 203, 357, 379, 389

politics of science 52

science as a social practice 50

scientific protocols 420

sclerochronology 476

screening levels 312–313, 321, 564–565, 567

secondary regrowth 389

sedimentology 29, 410, 415, 417

self-determination 61, 64–65, 68, 75

semantic network analysis 557

semi-structured interviews 134, 225, 293, 334, 406, 410, 433, 472

sense 6, 15, 28, 41, 43, 87, 110, 132–133, 140, 159–160, 162–163, 174, 186, 190, 290, 301, 341, 348, 350, 356, 358, 365–366, 381, 389, 405, 428, 432–433, 449, 457, 471, 491, 496, 504, 506, 516, 539, 574, 586–589

sensitivity analysis 464, 494–495

sexism 436

sexuality 153, 159, 503

Shallow Water Equations 488

silo 26, 39, 52, 87, 120, 209–210, 233

situatedness 42, 51, 89–90, 92–94, 105–106, 111–112, 150, 249–250, 252, 268–269, 282, 309, 343, 380, 387–388, 391–394, 396–397, 420, 484–485, 511

skepticism 435–436, 460

slaves 216

slope monitoring 121, 132, 134–137

slow down 3, 31, 260

pause and reflection 140

slowing down 31, 39, 133, 184–185

slow science 371

snowball sampling 505, 544, 545. *See also* sampling

social construction 269

social learning 63, 71, 76–77, 257

social network analysis (SNA) 361, 369, 555–559

social relations 43, 123, 360, 365, 379, 392, 395, 517–518, 553, 555–556, 566–567

socio-ecology 29, 416, 557

socio-economic 102, 106, 159, 262, 264, 266–267, 412, 415, 460

socio-environmental 291, 294, 298, 403, 411

socio-hydrology 18, 248

soil 1, 17–20, 24, 92, 129, 136, 138, 248, 284–285, 310–320, 356, 360–361, 371, 411–412, 417, 462, 493–494, 496, 563–568

moisture 493–494

toxicology 563

urban soil 17, 19, 317

solvent 564

space and place 327–328, 330–331, 343, 346–348, 350

speak back 2, 18, 31, 34, 46, 381

spoken word 442

stakeholder 136, 185, 286–287, 289, 291, 294, 297–298, 332, 334, 337, 479, 484, 496, 523, 538

statistics 172, 189, 190, 191, 193, 196, 201, 203, 207, 309, 380, 428, 455, 456, 457, 458, 459, 460, 463, 540, 544, 546, 556, 571, 573, 574, 575, 588. *See also* descriptive statistics

About the Team

Alessandra Tosi was the managing editor for this book.

Lucy Barnes proof-read and indexed this manuscript.

Jeevanjot Kaur Nagpal designed the cover. The cover was produced in InDesign using the Fontin font.

Cameron Craig typeset the book in InDesign and produced the paperback and hardback editions. The main text font is Tex Gyre Pagella and the heading font is Californian FB.

Cameron also produced the PDF and HTML editions. The conversion was performed with open-source software and other tools freely available on our GitHub page at https://github.com/OpenBookPublishers.

Jeremy Bowman created the EPUB.

Raegan Allen was in charge of marketing.

This book was peer-reviewed by two anonymous referees. Experts in their field, these readers give their time freely to help ensure the academic rigour of our books. We are grateful for their generous and invaluable contributions.

This book need not end here...

Share

All our books — including the one you have just read — are free to access online so that students, researchers and members of the public who can't afford a printed edition will have access to the same ideas. This title will be accessed online by hundreds of readers each month across the globe: why not share the link so that someone you know is one of them?

This book and additional content is available at
https://doi.org/10.11647/OBP.0418

Donate

Open Book Publishers is an award-winning, scholar-led, not-for-profit press making knowledge freely available one book at a time. We don't charge authors to publish with us: instead, our work is supported by our library members and by donations from people who believe that research shouldn't be locked behind paywalls.

Join the effort to free knowledge by supporting us at
https://www.openbookpublishers.com/support-us

We invite you to connect with us on our socials!

BLUESKY
@openbookpublish
.bsky.social

MASTODON
@OpenBookPublish
@hcommons.social

LINKEDIN
open-book-publishers

Read more at the Open Book Publishers Blog
https://blogs.openbookpublishers.com

You may also be interested in:

Right Research
Modelling Sustainable Research Practices in the Anthropocene
Chelsea Miya, Oliver Rossier and Geoffrey Rockwell (eds)

https://doi.org/10.11647/obp.0213

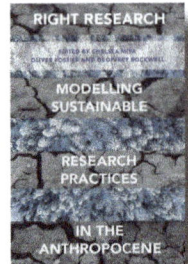

Democratising Participatory Research
Pathways to Social Justice from the South
Carmen Martinez-Vargas

https://doi.org/10.11647/obp.0273

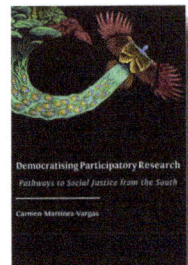

Transforming Conservation
A Practical Guide to Evidence and Decision Making
William J. Sutherland (ed.)

https://doi.org/10.11647/obp.0321

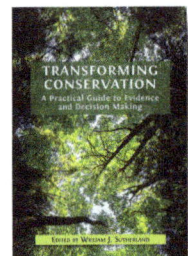

9 781805 113669